5 **A solitary predator.** Leopards hunt alone. Their prey, such as this bush pig, rarely weigh more than the hunter. Social carnivores capture prey that can weigh much more than any single predator. Alternative hypotheses on the significance of social hunting are found in Chapter 10. Photograph by Warren Garst/ Tom Stack & Associates.

6 **Simple society.** An aggregation of ladybird beetles forms a simple society. An analysis of the alternative hypotheses that have been advanced for why individuals may derive a reproductive advantage by forming groups is presented in Chapter 15. Photograph by the author.

7 **Threat display.** Chimpanzees threaten rivals with an open-mouth display that exhibits the canines. The evolutionary basis of communication signals is the subject of Chapter 8. Photograph by J. Carter.

Animal Behavior

ANIMAL BEHAVIOR

An Evolutionary Approach

FOURTH EDITION

JOHN ALCOCK
Arizona State University

 SINAUER ASSOCIATES, INC. • PUBLISHERS
Sunderland, Massachusetts

About the Cover

Most birds, including these masked booby birds from the Galapagos Islands, form a monogamous relationship in which one male pairs with one female and helps her rear their young. See Chapter 14 on the evolution of mating systems. Photograph by G. and B. Corsi/Tom Stack & Associates.

Library of Congress Cataloging-in-Publication Data

Alcock, John, 1942–
    Animal behavior : an evolutionary approach / John Alcock.—4th ed.
        p.   cm.
    Bibliography: p.
    Includes index.
    ISBN 0-87893-020-5
    1. Animal behavior.   2. Behavior evolution.   I. Title.
QL751.A58 1988
591.5′1—dc19   88-22023
                                              CIP

Printed in U.S.A.

8 7 6 5 4 3 2 1

To Sue, Nick, and Joe

CONTENTS

6 The Organization of Behavior   151

7 The Evolution of Behavior: Historical Pathways   185

10 The Ecology of Feeding Behavior 297

11 The Ecology of Antipredator Behavior 331

14 The Ecology of Mating Systems   443

15 The Ecology of Social Behavior   471

16 An Evolutionary Approach to Human Behavior   511

PREFACE

The study of animal behavior has always been good fun. But when I wrote the first edition of this book, the discipline was in a transition phase as it moved from the concerns and approach of the 1950s and 1960s to the evolutionary approach that is now widely used. This transition was dramatic, and it has been exciting to witness the transformation of the field. I have enjoyed reporting here on some of the discoveries made by my colleagues over the past 14 years. Each year has brought new insights into what animals can do, how they perform their feats, and why (in evolutionary terms) they have become endowed with these skills. The discipline of animal behavior has never been more entertaining and satisfying than it is now. I hope that this fourth edition helps my readers understand the current excitement and pleasure of studying animal behavior.

As with each preceding edition, I have taken advantage of the opportunity to revise by completely rewriting the book. In doing so, I tried to make the distinction between proximate and ultimate questions about behavior clearer and cleaner, in part by reorganizing some chapters in the preceding edition. I have moved the chapter on the historical basis of behavior so that now it leads off the second half of the book, which deals with the ultimate examination of behavior. This chapter is a logical introduction to the evolutionary aspects of behavior. Its new location allows continuity across the chapters on the ecological aspects of behavior while permitting expansion of the coverage of animal communication. I have made a number of other changes in the effort to make the material flow in a satisfying and convincing progression, one whose organizational scheme will be clear to readers.

The most important change, as far as I am concerned, is my attempt to emphasize the tentative, unfinished nature of science and the significance of using multiple working hypotheses as a method for doing science. I very much want my student readers to realize that not all the answers are in, and they never will be; I want to educate them, just as I have been educated in recent years, on the importance and beauty of simultaneously testing more than one hypothesis about how or why an animal does something.

The impetus for making revisions came from reviewers who provided me with their ideas and criticisms, and I acknowledge these individuals by name below. In addition, I have learned a great deal from my colleagues, particularly Ron Rutowski and Darryl Gwynne. Finally, I am indebted to the many researchers whose papers on animal behavior have appeared in the past four years. I have been impressed and a little awed by the number of excellent papers that have appeared in animal behavior journals during this time. More and more people are using an evolutionary approach to the study of animal behavior, and together they have made remarkable progress in discovering fascinating things about behavior.

The great number of really good new papers has created a problem for me, however, which is to limit my choices to a few to illustrate the conceptual issues that most evolutionary biologists think are important. I have no doubt that I could rewrite the book with an almost entirely different set of examples without diluting the main message.

The fact that one group of studies might be largely replaced with another without damage makes the point that the book's primary goal is not the presentation of a body of conclusions, a mass of information stacked up like cordwood that students would do well to memorize. Rather the goal is to offer a point of view, an approach to studying animal behavior. Michael Ghiselin says, correctly, that science may look like it is endeavoring to build up a body of "facts," but scientists are really engaged in asking questions. What scientists value most, therefore, are methods that enable them to ask productive, interesting questions. Evolutionary theory provides such a method for all aspects of biology, including animal behavior, and my book is first and foremost a book about the methodology of using an evolutionary viewpoint.

I am most eager to have my readers learn enough about the theory to see why it has been so useful to students of animal behavior, why evolutionary biologists ask the kind of questions they ask, and how they go about trying to answer these questions. If I have done my job adequately, my readers will come to appreciate that only a few questions about animal behavior have been answered to everyone's satisfaction. Much of the pleasure of science lies in scrutinizing existing hypotheses and the evidence on them—and not being easily satisfied, but continuing to think of different questions, alternative explanations, and better ways to test ideas. The study of animal behavior is part of the entertaining business of science, and I hope that this fourth edition contributes to a better understanding of this point.

Instructors familiar with the previous editions of this book will note that the lists of recommended films no longer follow each chapter. Now a film list by chapter is part of a new Instructor's Manual. The Manual includes questions designed for both short quizzes and longer exams as well as answers to all the discussion questions in this book. Finally, a new glossary of terms has been added to the text, recognizing the proliferation of terms and the necessity of having them carefully defined if misunderstandings are to be avoided.

Acknowledgments

To give me guidance for writing a fourth edition, the third edition was reviewed critically by Jeffrey Baylis, David Chiszar, and Donald Kramer. Revised versions of individual chapters were subjected to helpful scrutiny by Richard Alexander, Stevan Arnold, Luis Baptista, Thomas Caraco, Martin Daly, Paul Ewald, Jeffrey Hall, Ronald Hoy, Douglas Mock, Michael Moore, Lewis Oring, Michael Ryan, Paul Sherman, David Sherry, George Williams, Margo Wilson, and Ronald Ydenberg. Large sections of the book were reviewed in detail by Michael Beecher and Jack Bradbury. The book's reviewers have helped improve it a great deal. In addition, many people have taken the time to answer particular questions or to offer specific suggestions related to the book, including Jerram Brown, Luis Baptista, Ian Common, William Hamner, C. H. Johnson, Jane Packard, Floriano Papi, R. J. Putman, and Ernie Reese. All these scientists have contributed materially to my education and I am grateful to them. As the saying goes, I accept full responsibility for ignoring some of the good advice I have received, and for any errors or shortcomings that exist in the text.

Still others have helped by providing illustrations and permissions, sometimes going out of their way to get me what I asked for, and I have tried to acknowledge them at the appropriate figure. Acknowledgments to the publishers that have generously granted permission to use their copyrighted material appear on page 581.

While the fourth edition was being written, I had the support of my department chair, Ann Kammer, and the good companionship of a number of colleagues who enjoy drinking beer as much as I do. My wife, Sue, and sons, Nick and Joe, made many contributions, not the least by keeping my life interesting. My editor at Sinauer Associates, Carlton Brose, has shepherded me through the process again with the same helpfulness, interest in what I was doing, and attention to detail as in the first, second, and third editions. Thanks to all who have helped me do it again.

JOHN ALCOCK

1

An Evolutionary Approach to Animal Behavior

For hundreds of thousands of years, humans observed animals for a thoroughly practical reason: their lives depended on a knowledge of animal behavior. Today we can indulge ourselves and learn about animal behavior for the thoroughly impractical reason that this is a topic that happens to be fascinating and understandable. Of all the subdisciplines of biology, none is more accessible than the study of behavior.

Whereas it is almost impossible for the modern molecular biologist to do research without a laboratory overflowing with high-powered equipment and specially trained technicians, some behavioral biologists can still manage with a note pad, a pencil, some marking paints or bands, and maybe an insect net. And although behavioral researchers have developed their own jargon, a good many still can be understood by any reasonably educated person. The discoveries reported by persons studying animal behavior in the past 50 years—and over the past 20 years in particular—have revolutionized our view of what animals can do. We now know that the duckbill platypus can find food by detecting the extremely weak electric field surrounding its living prey. Ground squirrels can sense the difference between full siblings and half-siblings. In some species of damselflies, a male uses his penis to scrub out other males' sperm from his mate's sperm storage organ before transferring his own sperm. Learning about these things ought to be entertaining, and I hope that you will be impressed with the behavioral abilities of the animals that share our world with us. I also hope that you will learn to appreciate what an evolutionary approach is and why so many biologists use it as a guide to satisfy their curiosity about animal behavior.

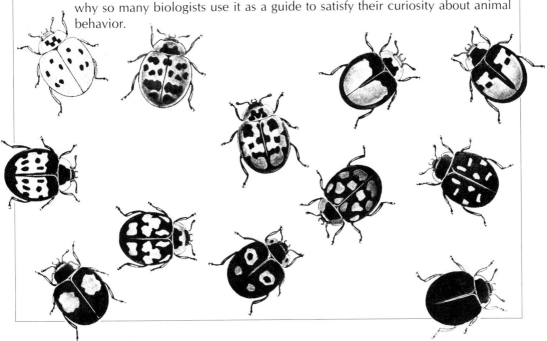

Questions about Behavior   I lived for a summer in Monteverde, a tiny community in the mountains of Costa Rica, and while I was there a friend loaned me a black light, which I hung up next to a white sheet on the back porch of our home. The ultraviolet rays produced by the black-light bulb attracted hundreds of moths on a good night. Although literally thousands of species of moths make Monteverde their home, most are so beautifully camouflaged that they are almost never seen during the day, when they sit motionless on the vegetation. But by putting a sheet behind the black light, I could each morning inspect at my convenience a delightful diversity of moths that had flown to the light at night and settled on the sheet.

Some mornings a huge bright yellow moth greeted me when I came out to check the sheet. This attractive moth, a member of the genus *Automeris*, held its forewings so that they covered its abdomen and hindwings. A sluggish animal in the chilly dawn, the moth did not struggle if I picked it up carefully. But if I jostled it suddenly or poked it sharply on its thorax, the moth abruptly lifted its forewings and held them up to expose previously concealed hindwings. The hindwings were marvelously decorated, each having a large circular patch of blue, black, and white scales set on a deep yellow background and looking like an eye. These patterns were so artfully constructed that they appeared to be three dimensional. When the forewings were quickly drawn forward, the appearance of the moth was dramatically altered, and two round eyes seemed to stare back at me (Figure 1).

A person seeing this for the first time would have to be remarkably uncurious not to wonder what's going on here. Does the resting moth cover its hindwings in order to conceal its false eyes until it is manhandled? What does the insect gain by making two "eyes" flash from under its wings when it is poked? And how does it "decide" precisely when to flip its forewings forward?

The list of questions about the simple behavior of this Costa Rican moth could be extended greatly, but no matter how long the list, each question could be assigned to one of just two categories: "how questions" about the proximate causes of the behavior, or "why questions" about its ultimate causes [533, 601].

How questions ask how an individual manages to carry out an activity; they ask how mechanisms *within* an animal operate to make its behavioral responses possible. Why questions ask why the animal has evolved the proximate mechanisms that enable it to do certain things. By distinguishing carefully between the proximate and ultimate causes of biological phenomena, we can avoid no end of confusion and wasted argument.

For example, imagine a discussion between two people on why humans eat so much candy and drink so many soft drinks. One person claims that this happens because these foods taste sweet. The sensation of sweetness is rewarding, and therefore people learn to consume foods that provide this pleasant experience. But the other person replies that this explanation misses the boat. People eat sugary foods because they provide a rich source of calories to fuel our metabolism and keep us alive.

This is the sort of dispute that comes from not recognizing the different levels of explanation in biology. The first hypothesis deals with proximate

1 **Automeris moth from Costa Rica** in its resting position, with forewings held over the hindwings (above). Photograph by the author. After being jabbed in the thorax, members of this species pull their forewings forward to expose the "eyes" on the hindwings (below). Photograph by Michael Fogden.

causes, because it focuses on psychological mechanisms within people that cause them to behave in certain ways. The second (ultimate) hypothesis concerns why humans may have evolved these internal reinforcement mechanisms in the first place. Both ideas could be correct because proximate and ultimate answers complement rather than compete with one another. This point can be reinforced by a look at the proximate and ultimate aspects of wing-flipping by *Automeris* moths.

How Questions

How does the moth decide when to pull its forewings forward, revealing its hidden hindwings? To answer this question, we would have to learn how certain kinds of tactile stimulation are detected by sensory receptors within the moth's body and how messages are then relayed to the central nervous system of the insect. In other words, we would have to study how the animal perceives that it is being poked or bumped. And then we would have to discover how this perception triggers nerve cell signals that activate certain muscles in the thorax, causing these muscles to contract and pull the forewings forward. Not a single person has examined the neurophysiological foundation of the behavior of the *Automeris* moth that I bothered in Monteverde, but these mechanisms are there waiting for future researchers to investigate.

How did the special touch receptors, brain cells, and muscle controllers used in the wing-flipping reaction come into being? The adult moth was not always an adult. It had a mother and a father, and it developed from a fertilized egg. If we are going to learn about the proximate causes of its behavior, we have to know how its physiological systems developed. The fertilized egg contained genetic instructions donated by each parent. These instructions regulated the way in which development of the *Automeris* moth larva occurred, channeling the proliferation and specialization of cells along pathways that produced a nervous system with special features in the adult insect. This is a wonderfully complex process, still poorly understood even in other, well-studied organisms.

Why Questions

Let's say that you were provided with or that you personally discovered everything there was to know about the genetic, developmental, physiological and psychological causes of the wing-flipping behavior of the *Automeris* moth. You should still not be wholly satisfied. You should still want to know why the moth has the genes it has, why its brain works one way and not another, and why this moth wing-flips but others do not.

Why questions explore the evolutionary or ultimate reasons why an animal does something. Why does the moth lift its forewings when it is molested? The British scientist David Blest suggested that the moth's action startles and frightens bird predators that find a resting moth and peck it, only to have the moth suddenly transformed into a creature with two big eyes [71]. He suggested that some birds may have a strong fear of owls, which possess large, round, staring eyes, and the moth may be exploiting this fear to scare off some birds that might otherwise eat it. In other words, the wing-flipping behavior has a possible *function*, namely, to prolong the life of the moth that uses the response in certain situations.

Blest did experiments with an "eyed" butterfly and found that the presence of large false eyes on the wings reduced the chance that one potential bird predator would attack the butterfly [71]. We can conclude that it is probable (but not yet demonstrated directly for *Automeris* moths) that their false eyes, when suddenly exposed, frighten away some of their predators.

Let's accept the hypothesis that a sudden demonstration of pseudoeyes increases the odds of survival for *Automeris* moths that have been grasped or pecked by one of their enemies. We can then ask, How did it happen that all the members of this species came to possess such beautifully complete fake eyes and the wing-flipping behavior that shows them off? To answer this why question, we must refer to the history of the species.

We can guess that in an ancestral population of the moth a particular genetic mutation appeared in an individual. This chance change influenced the development of the mutant moth so that it produced spots on its hindwings. When poked by a hungry bird, the moth might have raised its forewings to fly away, thereby exposing its spots to the bird. If the spots looked enough like eyes to cause even a brief hesitation by the predator, the insect would have enjoyed a slight survival advantage over the non-spotted members of its species. Thanks to its survival advantage, the spotted moth might have been able to leave more than the average number of surviving descendants in the next generation. Its offspring would have benefited from the trait they inherited from their parent, and they too would have left more descendants than other members of their population that lacked the genetic basis for the the beginnings of eyespots.

Because the genes of reproductively successful moths would tend to survive and replace the genes of less reproductively adept individuals, eventually the species as a whole would come to resemble the spotted ancestor. Subsequently, mutant genes that happened to promote the development of superior (more realistic and more startling) "eyes" or more effective ways to wing-flip would spread through the species in the same way, these mutations leading cumulatively to today's remarkable *Automeris* moths with huge pseudoeyes and an abrupt manner of lifting the wings when disturbed.

The *Automeris* that I manhandled in Monteverde possessed the genetic information that in the past generally conferred a reproductive advantage on its ancestors. The developmental options, and therefore the behavioral abilities, of each member of this species alive today have been defined by differential gene survival occurring throughout the history of the species.

You should now be able to discriminate between proximate and ultimate causes of behavior and so avoid the confusion that results from an inability to make this distinction. If someone were to claim that *Automeris* moths expose their hindwings when pecked because their nervous system controls the wing-flipping response, this would be a proximate explanation—as are all hypotheses based on genetic, developmental, physiological, or psychological mechanisms that are found within the body of an individual. If someone were to propose that the action conferred a survival advantage or came about because of past reproductive competition within the species, you would know that these are ultimate explanations—as are all explanations that deal with the reproductive or functional value of a trait, its historical basis, its evolutionary foundation (Figure 2).

On Solving How Questions
about Animal Behavior

Wing-flipping by an *Automeris* moth illustrates that the study of behavior involves many diverse phenomena: the action of a gene, the structure of an animal's brain, the relation of the moth to its predators, and evolutionary events occurring over millions of years. Any complete introduction to the field of animal behavior must therefore present the way in which scientists concerned with proximate questions do their research, as well as the way in which scientists that are interested in ultimate questions try to solve these very different kinds of problems. This book is divided into two major sections, with Chapters 2–6 devoted primarily to the proximate component of animal behavior and Chapters 7–16 covering primarily the ultimate aspect of animal behavior research.

Let's examine briefly the kinds of approaches that have contributed to an understanding of the proximate causes of animal behavior. Proximate questions about behavior require an exploration *into* an animal, a watchmaker's approach, a mechanic's desire to take the machine apart and understand how it runs and to analyze what factors have an immediate effect

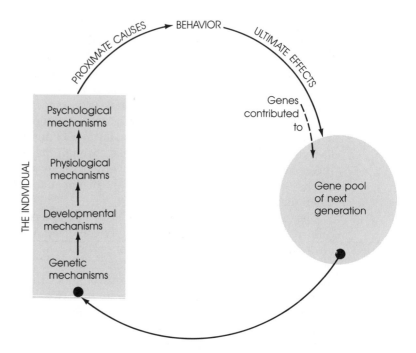

2 **Interrelationships between proximate and ultimate causes of behavior.** At the proximate level, various mechanisms internal to each individual enable it to behave. An animal's behavior determines how many copies of its genes will be passed to the next generation. Members of the next generation possess genes that have survived the process of reproductive competition in the past. Ultimately evolutionary history determines what genes are available to individuals in the present. We can therefore expect that animals will tend to develop proximate mechanisms that advance individual reproductive success.

on its operation. The complexity of biological organisms is such that there are layers upon layers of fascinating problems for the proximate biologist to investigate. At the most basic level is the relation of genes and behavior. Mendelian genetics were not rediscovered until 1900, but since then many sophisticated laboratory techniques have been developed and permit us to detect genetic differences among individuals and learn how they influence the characteristics of animals. The appropriate breeding experiments with *Automeris* could tell us many things, such as the extent to which variation in the wing-flipping response within the species is due to genetic or environmental differences among individuals, how many genes are involved in the development of the trait, and whether related species have similar hereditary factors that are linked to the behavior.

Some biologists and psychologists explore another proximate level of behavioral control, namely the development and physiology of behavior. Researchers in these two areas open an animal to chart the way in which its nervous system develops, or to determine the effect of surgical alterations of its brain, or to describe the properties of nerve cells.

Still other biologists and psychologists study the perceptual abilities of living animals, their responses to key stimuli, or the changes in behavior that are caused by experiences of individuals. One could, for example, test how much tactile stimulation is needed to trigger the wing-flipping response in an *Automeris* moth, or the extent to which the threshold of the response is affected by different light intensities, or how early experience shapes the later behavior of adult animals.

Thus, there are many levels of proximate analysis of animal behavior, and a host of methods used by biologists and psychologists as they try to find out what makes an animal behave.

On Solving Why Questions
about Animal Behavior

Why questions require an approach that is in some ways totally different from that used by researchers intrigued by proximate questions. It is entirely possible to study the immediate causes of behavior without relying on evolutionary theory. But, by definition, ultimate questions are those that cannot be answered without knowledge about the effects of history on the attributes of a species. These questions were dealt with very differently before and after 1859 [535, 536], when Charles Darwin (Figure 3) published *The Origin of Species* [170] with its explanation of how natural selection caused evolution to occur. Natural selection is recognized as one of the most important ideas of Western culture because of its simplicity, its predictive

3 **Charles Darwin** after returning from his around-the-world trip on the *Beagle* (left), and at the end of his life, well after having written *On the Origin of Species* (right). Courtesy of the Darwin Museum at Down House.

power, and its vast scope of application to biological matters. It was Darwin's genius to realize that if a few simple conditions apply, evolutionary change will occur. The Darwinian argument can be summarized as follows:

1. Variation exists in the traits of the members of most species. This is obvious in human populations and occurs wherever it has been looked for in other animals (Figure 4).

2. Some of the variation among individuals is heritable. We are all aware that human children usually resemble their parents more than other persons. The same observation applies to a host of attributes in many other species (Chapter 3).

3. The reproductive potential of animal species is staggering. Darwin calculated that a single pair of elephants would have 19 million living descendants just 750 years later, provided that every descendant along the way lived to be 100 years old and had just six surviving offspring. But elephants and other animal populations usually remain stable. Therefore, most of the young animals generated by a species die without reproducing.

4. Because of their special, inherited attributes, some individuals are likely to cope better with predators or climatic pressures or competition for food or mates. These individuals will tend to survive longer and leave more offspring than others in their species that have different and less successful inherited traits.

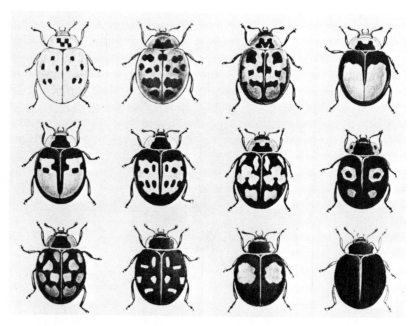

4 **A variable species.** The ladybird beetle *Harmonia axyridis* exhibits great variation in the color patterns of its wing covers. Many of the differences among these individuals are due to genetic differences.

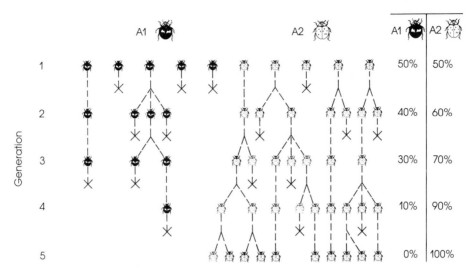

5 **Natural selection.** If individuals with the hereditary trait A1 usually leave fewer surviving offspring than carriers of the alternative trait A2, individuals with the A1 trait will eventually disappear from the population. Figures on right refer to the percentage of the population carrying the A1 and A2 traits.

The difference in the number of *surviving, reproducing offspring* produced by individuals that is caused by variation in their appearance, behavior, physiology or other traits is NATURAL SELECTION. If the variation among individuals is hereditary, evolutionary change within a reproducing group will occur as the heritable characteristics of those that reproduce best spread throughout the population (Figure 5). In contrast, the traits associated with reproductive failure will be eliminated. The logic of this process enables us to predict that animals will evolve behavioral traits that promote INDIVIDUAL REPRODUCTIVE SUCCESS as measured by the number of offspring that live to reproduce themselves (this value is an individual's DIRECT FITNESS).

The way in which a reproducing individual influences evolution is by copying its own genes and passing them to offspring, which in turn may transmit them to their progeny. Darwin developed the theory of natural selection before critical discoveries had been made on the nature of heredity. But we can now apply Darwinian logic to genes in much the same way that Darwin applied it to individuals. This will help us better understand a modern evolutionary approach to animal behavior [182].

1. Genes are present in all living things; they are nucleic acids that contain coded information about protein synthesis.

2. Many genes occur in two or more alternative forms or ALLELES. This genetic variation results in the production of slightly different forms of the same protein.

3. If one allele produces developmental effects (through its variant protein) that usually cause its bearers to replicate the allele more often than

other individuals with different alleles, then the "successful" allele will become more common in the population. Its competitors could be completely replaced if the relationship between genetic differences and reproductive success remains constant long enough. (We assume that populations cannot grow exponentially forever, so there are only a finite number of copies of a gene that can exist at any one time.)

The logical conclusion is that selective pressure on individuals will favor alleles that help build bodies that are unusually good at promoting the survival and propagation of these particular alleles. According to this view, individuals are really "survival machines" that promote the survival of their alleles [182], or as E. O. Wilson put it, a chicken is really a gene's way of making more copies of itself [853].

Darwinian Logic and the Study of Behavior   No matter how the logic of natural selection is presented, it is a blockbuster of an idea, and this was apparent to many people as soon as they read *The Origin of Species*, which quickly became a bestseller. If the concept is valid, then humans and all other living things exist to reproduce; reproduction is the biological significance of life. Furthermore, the theory of evolution by natural selection implies that all attributes of organisms have a history in which natural selection may have played a role. This notion is powerful, for it means that there is a connection between the evolutionary history of a species and its genes, the developmental patterns that have survived selection in the past, the physiological mechanisms that make behavior possible, and the visible behavioral characteristics of that species (see Figure 2).

One might assume, therefore, that biologists would have recognized the importance of examining animal behavior from an evolutionary perspective. But even though Darwin was fascinated by animal behavior and wrote several major books on the subject [171, 172], the primary focus of biology in the late nineteenth and early twentieth centuries was overwhelmingly on laboratory studies of genetics, embryology, and physiology—all centered on proximate biology. Field studies in the natural environment, vital to understanding *why* animals behave as they do, were left to dedicated amateurs, particularly those interested in bird behavior [212].

The continuing dominance of experimental, laboratory studies of proximate mechanisms of biology and the corresponding depreciation of field natural history studies may be related in part to what the great evolutionary biologist Ernst Mayr has called "physics envy." Most people, including most scientists, think that physics is an ideal science, because it uses rigorously controlled experimental procedures in the search for universal truths about the universe. Field studies of animal behavior do not fit the physics mold, and this may have contributed to the feeling that such studies were really not highly scientific. Perhaps this bias retarded the development of an evolutionary analysis of behavior. It was not until 1937 that the first journal of animal behavior was founded (in Germany), and it was not until the 1950s that the study of animal behavior was widely accepted as a serious scientific endeavor.

6 **The founders of ethology.** From left to right are scientists Niko Tinbergen, Konrad Lorenz, and Karl von Frisch.

Three men—Karl von Frisch, Niko Tinbergen, and Konrad Lorenz— played a primary role in the development of a modern approach to animal behavior (Figure 6). When these three founders of ethology received the Nobel Prize in Physiology and Medicine in 1973 for their pioneering studies, the selectors were probably influenced primarily by the proximate component of their research. But an equally important theme of ethology was an examination of animal behavior in the field to determine the adaptive (evolved) function of behavioral characteristics in free-living animals. Tinbergen described in his delightful popular book, *Curious Naturalists*, his enthusiasm for "old-fashioned" natural history, which he combined with Darwinism to explore why animals ranging from bee-wolf wasps to black-headed gulls did the peculiar things they did [774].

The early ethologists, however, seem to have forgotten a critical element of the logic of natural selection theory. In keeping with much biological work at this time, ethological research was often based on the assumption that individuals would sacrifice their own reproductive success for the general benefit of the species to which they belonged. For example, Konrad Lorenz proposed that animal aggression evolved to enable the species as a whole to "select" the superior members to reproduce and thus to pass on those traits likely to keep the species from becoming extinct.

The Problem with Group Selection

The "for-the-good-of-the-species" argument received its formal presentation in a giant book written in 1964 by V. C. Wynne-Edwards. *Animal Dispersion in Relation to Social Behaviour* [871] proposed that almost all aspects of animal behavior help species keep their populations down to avoid destroying the food base on which their survival depended. Wynne-Edwards's key point was that only those species that possessed population-regulating mechanisms survived to the present; others that lacked these mechanisms have surely gone extinct through overexploitation of the critical resources on which the species depended.

Wynne-Edwards forced biologists to examine the theoretical basis for the evolution of social behavior, and one fortunate result was G. C. Williams's reply, *Adaptation and Natural Selection*, published in 1966 [845]. Williams and a few other biologists [462] recognized the critical flaw in Wynne-Edwards's argument. Wynne-Edwards proposed that selection takes place primarily at the level of the *group*, with populations competing unconsciously to survive; those groups whose members sacrifice potential reproductive output for the benefit of the group are likely to avoid extinction. Williams explained that the theory of *group selection* is entirely different from the Darwinian theory of natural selection, and that natural (individual) selection is far more likely to cause evolutionary change than group selection.

Williams's argument goes like this. Let us say that a particular species (or group) is composed of individuals prepared to sacrifice themselves for the long-term benefit of their species. For example, imagine that lemmings do reach such high population densities that the species' survival is endangered. Imagine also, as is sometimes purported to be true, that some lemmings take it upon themselves to commit "suicide," thereby relieving pressure on critical food resources. In such a case group selection *a la Wynne-Edwards* would be said to favor the allele(s) for suicidal behavior. But in a population of lemmings with suicidal tendencies, natural selection would also be at work. A cartoon by Gary Larson, whose work generally touches in a unique way on biological themes, shows a band of lemmings heading into the water on a suicide mission—but one individual wears an inflated life ring around its waist (Figure 7). This cartoon captures the central defect of the theory of Wynne-Edwards. If there existed two types of

7 **Self-serving vs. self-sacrificing behavior.** Gary Larson's cartoon captures the essential defect of group selection, namely, the selective advantage self-serving "mutants" would have over self-sacrificing (genetically suicidal) members of their species.

lemmings—one inclined to dispose of itself at high population densities and another inclined to permit the suicidal types to carry out their sacrifice without its company—which type would constitute more of the next generation? If the two classes of lemmings were genetically distinct, what hereditary material would survive to be transmitted to following generations of lemmings? What would happen over evolutionary time to suicidal tendencies in our hypothetical population of lemmings?

The general point that Williams made is that selection among individuals will usually have a stronger effect than group selection in shaping the genetic makeup of subsequent populations. The result of his book was a much-improved awareness by biologists that, given a choice between using group selection or natural selection as a basis for a hypothesis on the adaptive value of a trait, a natural selectionist hypothesis was more likely to repay investigation.

Although research continues on forms of group selection more complex than that proposed by Wynne-Edwards, almost all behavioral biologists have accepted the arguments of Williams. As a result, by the 1970s ethology had metamorphosed into *behavioral ecology* and *sociobiology* [446]. These very similar disciplines (see Chapters 8 and 16) both employ an adaptationist approach, investigating how individuals, not groups, might derive a reproductive advantage from particular behavioral traits.

Testing Ultimate Hypotheses   There is no fundamental difference between the way in which scientists test their ideas whether they are dealing with proximate or ultimate problems in biology. Biologists of both types use a particular hypothesis to produce a prediction, which they then test in some way. If the prediction does not match reality, then the hypothesis can be rejected as false. Prediction testing provides a way to sort through alternative ideas using the process of elimination to identify the one explanation that cannot be rejected and so *might* be true. As John Hartung says, "in science you are wrong until you prove you might not be" [346].

For example, someone interested in the proximate control of wing-flipping behavior in an *Automeris* moth may locate a large nerve cell that is linked to muscles regulating the forewings. The researcher may hypothesize that this cell plays a central role in the wing-flipping response. If true, then electrical stimulation of the cell is predicted to activate the behavior. If activating the cell did not elicit wing-flipping in a restrained but living specimen, it would be logical to reject the original hypothesis and consider alternatives.

If we were interested in the reproductive value of wing-flipping, we would use fundamentally the same procedure. We might hypothesize that the eye spots deterred predators. If this hypothesis is true, we can predict that *Automeris* moths with their eye spots experimentally covered with yellow paint should survive less well than those with eye spots. Finding that this is not true would make us skeptical of the predator deterrence hypothesis.

Certainly there is no guarantee that an evolutionary hypothesis based on an individual selectionist approach will be correct. Some traits may *not* raise individual fitness at all, for a variety of reasons [184, 304].

1. The trait may have evolved through group selection, despite theoretical arguments to the contrary.

2. The trait, although once adaptive under conditions that no longer exist, is currently maladaptive. Its persistence in a present population may stem from the chance failure of a superior mutant allele to appear in the history of the species.

3. The characteristic may be a neutral or maladaptive by-product of the development of another characteristic that is selectively advantageous. As I shall describe in Chapter 3, many genes have more than one effect on the development of an individual and not all of these effects are necessarily positive.

4. The trait, an abnormal or pathological reaction, would never have occurred in the past, but it does in the present because of evolutionarily novel conditions for which the members of the species cannot be adapted.

Stephen Jay Gould of Harvard University has argued strongly [301, 302, 303] that because we can be certain that not all traits are adaptive, the approach taken by almost all persons interested in explaining the ultimate causes of behavior is mistaken. Adaptationists *assume* that wing-flipping (for example) increases individual reproductive success, and they then propose a hypothesis on how the trait might have this consequence in the environment in which the animal lives. Gould writes that adaptationists "have become overzealous about the power and range of selection by trying to attribute every significant form and behavior to its direct action." To drive his point home, Gould claims that adaptationists invent fables as absurd as the fictional "just-so stories" of Kipling, which contained inventive myths for the leopard's spots and camel's hump. These sharp criticisms have attracted numerous rebuttals [111, 184, 242, 537], the essence of which can be summarized as follows:

1. Any hypothesis rests on a set of assumptions. If we were to accept Gould's advice and propose that a particular behavior was *not* adaptive but instead, say, an incidental side effect of another trait, we would have to employ the assumption of *nonadaptation* as a foundation for the hypothesis.

2. The scientific point of a hypothesis is not to be right but to be testable. Someone who speculates that a trait may advance individual reproductive success for a particular reason does so, not because he blindly and naively believes that every trait is actually adaptive, but because this is the way to get an idea on the table for testing.

3. Many adaptationist hypotheses (and many nonadaptationist ones, too) will be wrong. Some will be silly. That is where the testing comes in; the "just-so stories" will tend to be discarded quickly when the predictions from these stories fail. Science is a self-correcting process in which prediction-testing eliminates misguided explanations.

A Weak Test of a Working Hypothesis

A WORKING HYPOTHESIS is intended to be tested, not accepted uncritically. But it is one thing to propose a hypothesis, derive a prediction, and test that prediction, and quite another thing to set up a battery of competing hypotheses and derive predictions from some that are *not* compatible with others. I shall illustrate this point by showing how infanticide in langurs has been examined from an evolutionary perspective.

The hanuman langurs of India live in bands, which often consist of one reproductively active male and his harem of adult females and their offspring (Figure 8). Various langur watchers have noted that other adult males, living apart from the troops with females, sometimes challenge and replace a harem master. One modern student of langur behavior, Sarah Hrdy-Blaffer, was especially intrigued by the sharp rise in the infant death rate following takeovers. Hrdy-Blaffer believed that the new male is responsible, because she observed the newcomer harassing mothers and biting their infants. Although females try to protect their babies, males are much larger and more powerful [393, 394].

To explain infanticide at the proximate level, we might devise hypotheses about the effects of a takeover on male testosterone levels and aggressive behavior, or we might consider the possible genetic foundation of the trait, or the early experiences of males involved in this activity. But we can propose possible ultimate causes for the behavior without knowing these things. Hrdy-Blaffer presented an adaptive argument for infanticide based on the assumption that the behavior helps maximize the reproductive success of individual killers. After the death of her infant, it is to the mother's advantage to become sexually receptive promptly, at which point she will be fertilized by the new male, who transmits *his* genes to the resulting offspring. (If the female were to nurse the baby fathered by the displaced harem owner, she would not ovulate again for two or three years, because females do not ovulate while lactating.) Thus, Hrdy-Blaffer's hypothesis is that an infanticidal male increases the number of descendants he produces by making females sexually receptive to him sooner than they would be otherwise.

A weak test of this hypothesis would be to derive a prediction from it and to test the prediction without reference to alternative hypotheses, *which might also yield exactly the same prediction*. A strong test of the hypothesis would be to compile a list of other hypotheses for infanticide by male langurs and use them to produce a set of nonoverlapping predictions that could be used to eliminate all but one of the explanations.

Let us compare the two kinds of tests. The Hrdy-Blaffer hypothesis (the sexual competition hypothesis) states that infanticidal males get to reproduce sooner and more often by disposing of infants whose presence would prevent mother langurs from coming into heat for a long time. One of many predictions taken from this hypothesis is that new resident males should stop killing infants by the time the first of their offspring are born to the females under their control. If infanticidal males more often than not killed their own offspring, their actions would reduce, not increase, their production of offspring.

The data relevant to the prediction are not completely clear-cut. Of the 60 to 70 reported cases of infanticide following a takeover, the killer of the infant was actually observed in the act in fewer than 15 cases (Figure 9)

◀ 8 **An extended family** of hanuman langur females and their offspring. Groups like this are the focus of competition among infanticidal males, who compete to acquire a harem of mates. Photograph by Sarah Hrdy-Blaffer; courtesy of Anthro-Photo.

[737, 751]. Usually infanticide was inferred from observations of a male attacking an infant and the infant's subsequent disappearance. Furthermore, the paternity of an infant must usually be guessed at by knowing who the harem master was when the mother became pregnant; there is room for error here. But even a severe critic of the sexual competition hypothesis concedes that when paternity of the deceased infant has been established with some certainty, in a large majority of the cases the killer was most unlikely to be the father [73]. In a recent study, the infanticidal male was almost certainly not the father of the victim in 10 of 11 cases, with only one case uncertain [738].

A Strong Test of a Working Hypothesis

Let us accept then that the prediction of "selective" infanticide is supported by data gathered from observations of langur bands. This does constitute a weak test of the sexual competition hypothesis in the sense that we could have thrown the hypothesis out if males routinely killed their own progeny. But the problem is that other hypotheses may exist that might not be eliminated by testing this one prediction. So it is useful to consider alter-

9 **An infant-killing male langur.** The male pursues a female who is clutching her baby, which has already been severely bitten by the male. The infant died shortly thereafter. Photograph by V. Sommer.

native hypotheses to the sexual competition argument. There are many competing explanations [737]. One is that infanticide at the time of a takeover is adaptive for the killer male because he might cannibalize the infants, gaining high-protein meals at a time of intense physiological stress and energetic demands stemming from the takeover effort. The fact that a new harem master kills infants that are not related to him could be an incidental effect, if cannibalism were only profitable during a takeover; at this time any infants killed would have been sired by another male.

A second alternative is that infanticide is not adaptive at all but is a social pathology induced by abnormal conditions for which langurs have no evolved response. According to this hypothesis, infanticide occurs because troops of langurs are fed supplementally by humans. This situation creates artificially high densities and brings into frequent contact bands that otherwise would avoid one another. Under these abnormal conditions, fighting occurs more often than in "natural" environments with the incidental consequence that infants are sometimes attacked and killed by males in their own bands. Males rarely kill their own offspring only because between-troop interactions have their greatest pathological effect on males that have recently moved into a troop. Such males may be in an especially "nervous" and stressed state at this time, which provides a proximate explanation for actions that have no adaptive significance, according to the social pathology hypothesis.

Table 1 lists some predictions that can be drawn from all three alternative hypotheses. As a result of these alternatives, we are now in a position to seek out information that can discriminate among them. The cannibalism hypothesis requires that we observe cases in which a male newcomer consumes an infant after killing it. This has never been seen (but as noted above direct observations of males in the act of infanticide are rare). The

TABLE 1

Predictions taken from three different hypotheses that account for the evolution of infanticide by male langurs

| | Hypothesis | | |
Prediction	Sexual competition	Cannibalism	Social pathology
Males will kill other male's infants	+ +	+	+
Males will consume killed infants	−	+ +	−
Infanticide in langurs will only occur under abnormal conditions	−	+	+ +
Infanticide in species other than langurs will only occur under abnormal conditions	−	+	+ +

−, Not predicted by the hypothesis.
+, Predicted by the hypothesis, given special conditions.
+ +, Critical prediction of the hypothesis, which, if not supported, eliminates it.

absence of the key evidence weakens the adaptive cannibalism argument.

The social pathology hypothesis predicts that male infanticide will not occur in low-density populations of the langurs or those that do not receive food supplements. Dispute continues on this point, but infanticide by male langurs occurs in two locations where they are not fed by humans [587, 751]. If these langurs really are not abnormally crowded, their practice of infanticide is very damaging to the social pathology hypothesis.

Furthermore, the social pathology hypothesis predicts that infanticide in animals other than langurs should be restricted to cases in which human intervention and interference has dramatically altered the social environment of a species. There is now a large body of information on infanticide in other species of primates and other animals in which the act occurs under thoroughly "natural" conditions—provided a new male has just taken potential mates from another male [350]. This evidence further reduces confidence in the social pathology hypothesis while permitting us to retain for the moment the sexual competition alternative.

Certainty and Science

Although I think that in the langur case the preponderance of current evidence supports the sexual competition hypothesis, this opinion could reflect my own prejudices. Someone may develop a competing idea that has not yet been considered and tested, and when this happens it may be that Hrdy-Blaffer's original proposal will have to be discarded. Complete certainty is *never* achieved in science. What appears to be an ironclad argument one day may prove the next to have devastating flaws. Fortunately scientists continually criticize each other's ideas in good humor or otherwise, and this exchange forces attention to alternatives—precisely what happened in the case of langur infanticide. The uncertainty about Truth that scientists accept (at least when talking about other people's ideas) often makes nonscientists nervous, in part because scientific results are usually presented to the public as if they were the Ten Commandments. But anyone who has taken even a quick look at the history of any science will learn that new ideas emerge and old ones are replaced or modified. The capacity to change, to incorporate new insights, to test an idea repeatedly and to throw out those that fail their tests is the essence of science.

For the moment, Darwinian theory offers the most comprehensive and practical way in which to approach the totality of biology. It is the premier foundation for the development of working hypotheses about the ultimate, reproductive function of biological characteristics. There is plenty of room for disagreement, exploration, and invention within its framework. If natural selection is ultimately shown not to be a major force behind the evolution of the traits of all living things, it will probably be because of discoveries made while testing working hypotheses based on the assumption that natural selection was the driving force of evolution. And until a replacement theory is available, I believe that an understanding of the operation of natural selection should be part of the education of every person, the better to enjoy the world around us.

SUMMARY

1 Basic questions about animal behavior fall into two categories. How questions require answers about how the proximate mechanisms within an individual—its genetic–developmental and physiological–psychological systems—enable the individual to behave. Why questions require answers about why these systems have evolved, why the behavior patterns of an individual help it overcome the obstacles to reproduction in its environment.

2 There are two major candidates for the selective forces that have influenced the evolution of a species. First, evolutionary change will occur by natural (individual) selection if genetically different *individuals* have different numbers of surviving offspring. Second, evolutionary change will occur by group selection if genetically different *groups*, or species, differ significantly in their long-term survival chances because of genetic differences among them.

3 A trait favored by group selection may lead an individual to pass on fewer of its genes than other members of its species with alternative genes and alternative traits. If so, the group-benefiting allele should usually be replaced over time by an alternative allele that promotes individual success in propagating genes. Therefore, the ultimate function of a behavioral trait should usually be to promote gene propagation by the individual rather than the welfare and survival of the group to which the individual belongs.

4 Because genes have been selected for their ability to survive, the working hypotheses of persons interested in the evolution of adaptive behavior assume that the traits of interest promote the survival of the genes that underlie these characteristics. The point of a working hypothesis is to lay one's cards on the table. If the predictions derived from a hypothesis are not confirmed, the hypothesis can be rejected. A hypothesis that withstands repeated tests in conjunction with results that eliminate possible alternative explanations for the same phenomenon can be accepted provisionally as true—pending new information and the development of as yet unconsidered alternative ideas.

5 The appeal of an evolutionary approach to behavior is that it is based on a logical concept (natural selection) and that it produces plausible working hypotheses about an enormous array of biological questions, and many of these hypotheses can be readily tested.

SUGGESTED READING

Books that capture the sense of curiosity and excitement that biologists feel as they study the proximate basis of animal behavior include Vincent Dethier's *To Know a Fly* [190] and Kenneth Roeder's *Nerve Cells and Insect Behavior* [669]. Niko Tinbergen's *Curious Naturalists* [772] and Konrad Lorenz's *King Solomon's Ring* [503] bridge the gap between proximate and ultimate approaches to some extent, although the emphasis is on the evolutionary basis of behavior. The same is true for books like Archie Carr's

So Excellent a Fish [139], Howard Evans's *Life on a Little Known Planet* [246], and Michael Ryan's *The Tungara Frog* [687]. For books that capture the delight of field studies of animal behavior, consider Evans's *Wasp Farm* [247], George Schaller's *The Year of the Gorilla* [699], and Tinbergen's *The Herring Gull's World* [774].

For books on the logic of Darwinian natural selection, you could start with Darwin himself, *On the Origin of Species* [170], which you will find surprisingly readable. G. C. Williams's classic *Adaptation and Natural Selection* [845] demolishes "for-the-good-of-the-species selection" in an utterly convincing way. Richard Dawkins has written lively books about the application of evolutionary theory to animal behavior in *The Selfish Gene* [182] and *The Blind Watchmaker* [185].

DISCUSSION QUESTIONS

1. How does the following quote relate to an understanding of the distinction between proximate and ultimate causation in biology: "Many researchers now question the utility of trying to find a behavioral function for dominance [in which one individual socially dominates others]. Instead of attempting to identify the reproductive value of being a dominant member of a social group, these scientists are focusing on the physiological correlates of dominance, such as seeing if dominant individuals have different hormone levels. If it could be shown that dominants did differ hormonally from the subordinate members of a group, it would be unnecessary to explain dominance in terms of any supposed reproductive effect."

2. Why does the following quote represent an example of group selectionist thinking, and what is its logical flaw: "Although many moose calves die in their first year, not many need to survive to keep the population stable, since only about 15 percent of the adult population dies each year. It is almost as if the excess infants are "programmed" to be eaten, insuring sufficient energy transfer to support the system's major predators and scavengers."

3. Imagine a species in which you identify two types of color patterns, a greenish type and a brownish type. If you were to measure the average number of young produced by the greenish types in a year, you might find that greenish individuals averaged 10.2 offspring and brownish ones only 8.7. Is this sufficient information to conclude that natural selection will cause evolutionary change in this population? If no, why not?

4. Use the method of multiple working hypotheses to begin to analyze the ultimate significance of the fact that *female* Belding's ground squirrels sometimes commit infanticide. See Chapter 15 for more information on the natural history of these small colonial burrowing mammals.

2

The Diversity of Behavior

Historically, students of animal behavior have classified all behavioral traits into two major categories, instinct and learned behavior. These traditional categories were created to acknowledge supposed differences in the proximate mechanisms underlying different kinds of behavior. Instinct was often claimed to be genetically controlled, whereas learned behavior was believed to be largely (or entirely) dependent on experience (the environment). One of the central goals of this chapter is to show that all behavior, however labeled, comes about because of an interaction between genetic and environmental factors. The influences of heredity and environment are integrated in the development of nervous and hormonal systems, and therefore one cannot legitimately separate behavior into genetically controlled versus environmentally determined categories. There are, however, important differences in the proximate mechanisms controlling the behaviors that have been called "innate" and "learned." I shall try to define these terms, and examine the diversity that exists *within* and *between* these categories. I shall show that many behaviors do not fit neatly under the label of either "instinctive" or "learned," because instincts are often modified by experience, and learning is often biased in interesting ways. The chapter concludes with an evolutionary analysis of why there are different kinds of behavior based on different kinds of proximate mechanisms.

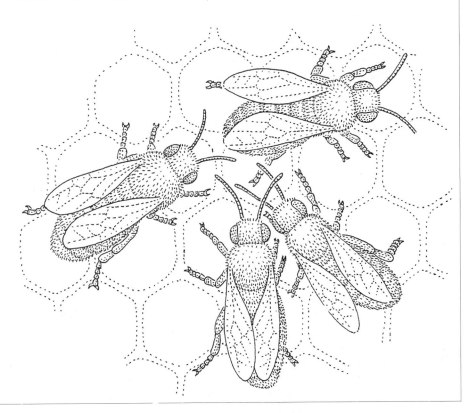

Instincts and Learned Behavior   Two large parrots, the galah *Cacatua roseicapilla*, and the pink cockatoo, *Cacatua leadbeateri*, live in the same eucalyptus woodlands in some parts of Australia. Both species use holes in trees as nest sites and very occasionally the same tree hole is taken by a pair of galahs and a pair of pink cockatoos. Because they do not start incubating their eggs until the clutch is complete, and because it takes three or four days before the female lays her last egg, the two pairs can both use the same nest for a time without interacting much. Eventually, however, conflict arises when the cockatoo and galah try to incubate their eggs. When this happens the larger pink cockatoo wins and evicts the smaller galahs, who must abandon their eggs. The galah eggs are then unwittingly incubated by the pink cockatoo.

That pink cockatoos will sometimes rear the young of another species provides a fascinating natural experiment on the effect of experience on the development of behavior. Two Australian ornithologists, Ian Rowley and Graeme Chapman, have taken advantage of the experiment to ask, How will the cross-fostered galahs behave—like galahs or like pink cockatoos [678]?

The answer is, It depends on the behavior. The nestling galah chicks give galah begging calls when they want to be fed by their parents; the pink cockatoos oblige even though these calls are decidedly different from those used by their biological offspring. Moreover, long after they have left the nest, the cross-fostered galahs give galah alarm calls when surprised by something frightening.

But when it comes to the *contact calls* that parrots use to maintain contact with other members of their flock, the foster galahs sound exactly like pink cockatoos (Figure 1). Moreover, they fly with the slow, sweeping wingbeats of pink cockatoos, not with the rapid, shallow wingbeats of their fellow galahs; Rowley and Chapman did note some lapses by the foster galahs in which they occasionally flew ahead briefly in the rapid galah pattern, only to turn back to rejoin their companions in slow cockatoo flight. And they feed on foods that pink cockatoos consume but are normally avoided by galahs.

These observations enabled the Australians to categorize the behaviors of the adopted galahs. The begging call was innate, the contact call learned. The alarm call was also innate, but the flight pattern and feeding preferences were learned. In other words, the development of the begging and alarm calls were not altered by the novel experiences provided by pink cockatoo foster parents, but the contact calls and flight pattern were modified by these experiences.

Ethology and Instincts

The early ethologists studied many examples of behavior that, like the begging and alarm calls of the galah, were not easily altered by some environmental influences that might be expected to affect their development. Niko Tinbergen, for example, found that newly hatched herring gulls somehow "know" how to get fed. They peck at the red dot on the bill of their parents, and the adults respond by regurgitating a half-digested fish or other tasty morsel [777].

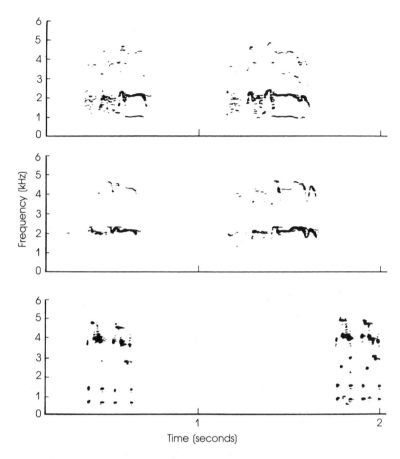

1 **Sonagrams of contact calls** of a pink cockatoo, of a galah reared by pink cockatoos, and of a galah reared by its own parents (top to bottom). *Source:* Rowley and Chapman [678].

Or if you take an egg from under an incubating greylag goose and put the egg a few feet away, the goose will retrieve the egg using a standardized behavior pattern in which it stretches its neck forward, tucks the egg under its lower bill, and rolls the egg back into its nest. This is something incubating greylag geese "know" how to do, even with their first clutch of eggs and no matter how divergent their previous life experiences [771].

We can then define an instinct as a behavior that appears in fully functional form the first time it is performed; typically such behaviors are mechanically triggered by a simple cue of some sort. Take the begging behavior of a very young herring gull. It will peck at almost any long, thin, bill-like object, provided there is a contrasting dot at the end of the "bill." Or put a portion of a stuffed female red-winged blackbird that consists only of a tail and few back feathers out in a swamp in the breeding season and the local males may rush over to "copulate" with the mount—provided the tail is raised in the position females use to signal a readiness to mate. Arrange to remove the egg a goose is rolling back into its nest when it is

2 **Instincts.** Clockwise from top left: A gull chick pecks at a long, thin stick with a striped tip—a more effective releaser of begging behavior than a realistic model of a parent gull, when the chick is very young; a red-winged blackbird "copulates" with a model that consists of a female's tail raised in the precopulatory display position; and a greylag goose rolls an egg back into the nest. The goose will complete the retrieval behavior even if the egg is taken away midway through the process.

only halfway home and the goose continues to draw its neck back just as if it was still in control of the egg (Figure 2).

For the European ethologists, FIXED ACTION PATTERNS (FAP) were instinctive responses that are played out to completion, once activated by a simple sensory cue. They called the key component of the object that triggers the FAP a SIGN STIMULUS (or RELEASER, if the sign stimulus was a signal from one individual to another).

Thus the red dot at the end of a herring gull's bill is a releaser of an FAP, the begging behavior of baby herring gulls. Figure 3 shows an especially familiar FAP, a human yawn. Yawns are very similar in appearance from person to person, last six seconds give or take a couple, are difficult to stop in midperformance, and are infectious, stimulating yawning in other humans that observe (or even hear) the yawner [646]. In other words, yawns are both FAPs and releasers. I suspect that just looking at Figure 3 and reading about yawning has caused a good many of my readers to yawn—and not, I hope, from boredom.

Code Breakers
Just as the "adoption" of galahs by pink cockatoos gives us insight into the properties of instincts, so too do cases in which one species exploits (uncon-

3 **A yawn,** a releaser of yawning in other humans. Photograph by Tim Ford.

sciously) the simple, stable relationship between certain behaviors and the cues that release these behaviors [837]. A premier code breaker is a rove beetle, *Atemeles pubicollis*, which lays its eggs in the nest of the ant *Formica polyctena* [372]. The larvae that emerge possess special glands at the tip of the abdomen. These produce an attractant scent, or PHEROMONE, that releases brood-keeping behavior in their hosts, causing the ant workers to care for the parasitic grubs in their brood chambers, where the beetles feast on ant eggs and ant larvae. Not content with raiding this larder, the parasites also mimic the food-begging behavior of ant larvae. They tap a worker ant's mandibles with their own mouthparts and so trigger regurgitation of liquid food. The larvae eventually metamorphose into adult beetles, which also perfectly mimic the food-begging behavior of adult worker ants (Figure 4).

Code-breaking species may even provide a more effective mimetic sign stimulus, or SUPERNORMAL STIMULUS than the biologically correct object. Several bird species, among them the European cuckoo and the North American cowbird, are BROOD PARASITES [336, 837]. The female of these species locates the nest of some other bird, generally smaller than herself. When the owner of the nest leaves during a pause in egg laying or incubation, the parasite that has been waiting nearby slips into the nest, quickly lays an egg, and disappears. When the owner returns, it often accepts the addition to its clutch, incubates the egg, and hatches the parasite.

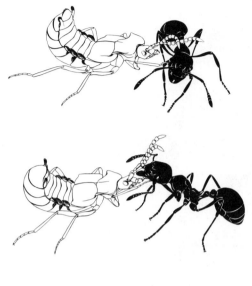

4 **Code-breaking rove beetle,** whose behavior mimics releasers of feeding behavior in its ant hosts. The beetle first taps a worker ant with its antennae, then touches the ant's mouthparts with its forelegs, and then consumes food regurgitated by the ant. Drawings by Turid Forsyth.

The young cowbird or cuckoo generally develops more rapidly than the host's offspring so the newly hatched parasite may have an opportunity to eject some of the other eggs in the nest (Figure 5). In addition, the parasite provides, by virtue of its large size and its great demands for food, a supernormal releaser of parental feeding. The relevant cues that determine which of several birds in a nest will be fed by a parent returning with a food item are how high the bird stretches out of the nest cup, how noisy its begging calls are, and how energetically it bobs its head and body. A large, voracious nestling cowbird or cuckoo is much noisier and more active than the host's own nestlings and therefore is more likely to be fed (Figure 6).

Psychology and Learning

While ethologists focused heavily on instinctive behavior, psychologists made the analysis of learned behavior their primary goal. Whereas persons interested in instincts have discovered many examples of the "blindly" mechanical nature of behavior, psychologists can point to an equally large number of cases of learning, which involves the durable modification of behavior in response to information acquired from *specific experiences*. For example, consider the kind of learned changes in behavior that occur as animals form an association between environmental events and their responses. One kind of associative learning is CLASSICAL CONDITIONING [204],

5 **Complex egg ejection.** A young cuckoo can perform a complex egg ejection response shortly after it hatches. Photograph by Eric Hosking.

a phenomenon first studied by the Russian physiologist Ivan Pavlov. Pavlov demonstrated that if you consistently made a particular sound or showed a light just as you placed meat powder in a dog's mouth, soon the dog would begin to salivate as soon as it heard the tone or saw the light even without the meat powder stimulus. Salivation is an automatic or *unconditioned response* to exposure to a food stimulus, which in this case is labeled an *unconditioned stimulus*. But dogs can learn through experience that a sound is associated with access to food, and in so doing form an association that leads to salivation, which is a *conditioned response* to sound or light, *conditioned stimuli* that initially do not trigger the response.

Classical conditioning occurs in a great variety of animals, from honey

6 **Supernormal stimulus.** The fledgling cuckoo provides exaggerated key cues that trigger parental feeding in the host more effectively than do the host's own offspring. Photograph by Eric Hosking.

bees to humans. For example, touching the front of an *Aplysia* marine mollusk always causes the animal to retract its mouthparts and cover its gills; a weak electrical shock applied to the rear of the animal does not normally have this effect. The sea hare can be classically conditioned, however, to withdraw its siphon and close its gills in response to a shock on its tail by first pairing a touch on the head with a shock to the tail over a number of trials, and then giving the animal a shock alone [132].

OPERANT CONDITIONING is another form of associative learning, but in this case the animal learns to associate a voluntary activity with the consequences that follow [727]. Imagine a laboratory rat in a Skinner box (named after the famous psychologist B. F. Skinner, who has devoted his life to the study of this form of learning). After it has been introduced into the box, the rat will wander about exploring its surroundings (Figure 7). Sooner or later, it accidentally pushes a bar on the wall, perhaps as it reaches up to investigate the side of the cage. As the bar is pressed, a

7 **Rat in a Skinner box.** The operantly conditioned rat approaches the bar (top left) and then presses it (top right). The animal awaits the arrival of a pellet of rat chow (bottom left), which it consumes (bottom right). Thus the rat is reinforced for pressing the bar, and will do so again. Photographs by Larry Stein.

mechanism releases a rat chow pellet into the hopper in the box. Some time passes before the rat (which has been deprived of food) discovers that there is something to eat in the food hopper. After this happens and the pellet is eaten, the rat may continue to explore its rather limited surroundings for some time before again pressing the bar in the course of its apparently aimless inspections. Out comes another food pellet. The rat may find it quickly this time and then turn back to the bar and press it repeatedly. It has learned to associate a particular activity with food. It is now operantly conditioned to press the bar.

Because, in their view, the control of behavior was simple (any positively reinforced behavior will be performed more frequently), Skinnerians believed that one could condition virtually any response. They pointed to the success of shaping experiments in which, by first reinforcing behavior that vaguely resembles the goal of a researcher and then rewarding better and better approximations, pigeons can be conditioned to waltz in circles or play table tennis with their beaks or "communicate" symbolically with one another [244]. Skinnerian techniques have also been used to condition humans and other animals to regulate internal activities, such as heart rate or brain electrical activity, which were at one time thought to be entirely unconsciously regulated [70].

Operant conditioning occurs outside psychology laboratories. While traveling in the company of pink cockatoos, an adopted galah is exposed to a variety of potential meals. As it samples some, it will be rewarded by the sensations provided by these foods; others will be difficult to consume or will taste bad. The galah's behavior will be shaped by these experiences. The information that it acquires through trial-and-error feeding will influence its behavior; consequently, over time it will come to concentrate on rewarding food items while ignoring those that do not offer positive reinforcement. The same is true for many other animals that feed on a diversity of foods (Figure 8).

8 **Trial and error learning.** After this experience with a foul-tasting millipede, the toad will reject this prey on sight. Photograph by Thomas Eisner.

Varieties of Learning

Classical and operant conditioning are merely two of many phenomena placed in the catch-all category of "learning." Psychologists and biologists interested in the proximate aspects of learning have discovered many ways in which animals modify their behavior in response to specific experiences. The degree to which these kinds of learning can be accommodated in the associative learning category has been a matter of debate.

For example, a digger wasp female builds a nest—typically in the ground—which she leaves to collect prey. The wasp brings her victims back to the nest to feed her offspring. Digger wasps are adept at SPATIAL LEARN-ING [772]. As they leave their nest for the first trip away they circle about the area for a few seconds and then dart off; during the brief orientation flight they store information about the landmarks about the nest, which they use to relocate the tiny, and often concealed, nest entrance. Niko Tinbergen showed long ago that if you displace objects around the nest entrance after the wasp has performed her initial orientation flight and left on a hunting trip, when she comes back she will search for her nest entrance by the displaced landmarks (Figure 9).

Similarly, those birds that store food exhibit exceptional spatial memory. A small chickadee may hide several hundred seeds or small insects in bark crevices or patches of moss on one day and be able to relocate many of these things over the next day or two. The bird stores food one item to a hiding

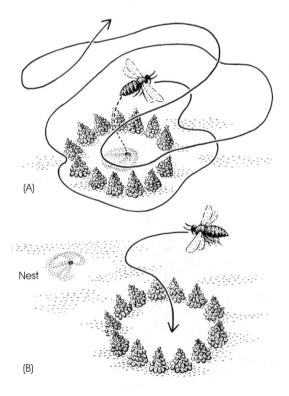

(A)

Nest

(B)

9 Spatial learning by insects. Digger wasps learn the landmarks surrounding their nest burrow's entrance (A) in a matter of seconds as shown experimentally by the response of a returning female (B) to displacement of local landmarks. *Source:* Tinbergen [772].

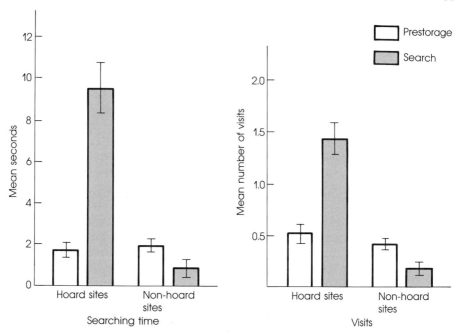

10 **Spatial learning by birds.** Chickadees spend much more time searching the parts of the cage (hoard sites) where they have stored food 24 hours previously than they did during their initial exposure to these sites. They also make many more visits to hoard sites, evidently because they remember having stored food there. In these experiments the stored food was removed from the cage during the trials that test whether the birds remember where they have hidden food. *Source:* Sherry [718].

spot, and never uses the same location twice. Experimental work by David Sherry and his colleagues demonstrates that the chickadees remember where they have stored their food and also where they have already retrieved a stored food item (which helps them avoid wasting time double-checking a depleted cache). In these experiments, chickadees were given a chance to store food in holes drilled in small trees placed in an aviary. The birds typically placed a sunflower seed in 4 or 5 of 72 possible storage sites. They were then taken from the aviary. A day later the chickadees reentered the aviary to search for the food, which had been removed in the interim. All 72 storage sites were closed with Velcro covers. The chickadees spent much more time inspecting and pulling at the covers at hoard sites than at spots where they had not stored food 24 hours previously (Figure 10). Because the storage holes were empty and covered, there were no olfactory or visual cues provided by stored food to guide the birds in their search; they could only rely on their memory of where they had hidden the food [718].

Clark's nutcrackers may have an even more impressive memory, for these birds scatter as many as 9000 caches of pine seeds (1–10 seeds per cache) over entire hillsides. The bird digs a little hole for each store of seeds and then completely covers the cache. Although a nutcracker does this work in the fall, it relies on its stores through the winter and into the spring; so months may have passed before it comes back to retrieve the seeds in a particular cache [35].

It could be that nutcrackers do not really remember where each food storage spot is but rely on some general rule, such as "caches will be made by little tufts of grass." Or they might remember the general location where food was stored but once there look about until they saw signs of caching or until they smelled buried seeds. But experiments similar to those performed with chickadees show that birds do remember precisely where they hid their food. In one such experiment, a nutcracker was given a chance to store seeds in a large outdoor aviary; after the storing activity the bird was moved to another cage. The observer, Russell Balda, mapped the location of each cache accurately, and then removed the buried seeds and swept the cage floor, removing any signs of caches. Thus there were no visual or olfactory cues available to the bird when it was permitted to go back to the cage *a week later* and hunt for food. Balda mapped the locations where the nutcracker probed with its bill, searching for the nonexistent caches. The bird's spatial memory served it well, for it dug into as many as 80 percent of the cache sites while only very rarely digging in other places [35].

Insight Learning

As a final demonstration of the variety of phenomena that fall into the category of learning, I shall describe Sue Savage-Rumbaugh's study of two captive young chimpanzees [698]. She trained Sherman and Austin to use an illuminated keyboard with a complex set of symbols to communicate with their trainers and each other (Figure 11). The behavior of the chimpanzees was shaped by conditioning techniques, for the animals were rewarded with food and praise for correct responses, that is, pressing the symbol for a wrench when they were shown a wrench and pressing the symbol for a screwdriver when they were shown a screwdriver.

What was especially interesting about the conditioning process was the way in which the chimps sometimes learned complex tasks. For example, when Sherman was being trained to name a set of three tools, his initial choices of keyboard symbols were essentially random. If the researcher held up a wrench, Sherman was as likely to press the keyboard symbol for "straw" as he was to answer "wrench." Over the first 108 trials, only 43 percent of Sherman's answers were right, and by chance alone he would have been expected to be right 33 percent of the time. However, he suddenly seemed to "catch on," and for the next 106 "questions" Sherman named the tools correctly 102 times!

If this is operant conditioning, it means that the term covers a very broad range of learning phenomena [545], perhaps including insight learning. In Wolfgang Köhler's classic experiments with chimpanzees, a favored food, such as bananas, was placed out of reach of a captive animal [439]. In the cage were boxes and sticks, which if used appropriately could enable the chimp to get food. After appearing to study the situation for a time, some chimpanzees quickly solved the problem and secured the food. This sort of

11 **Complex learning by chimpanzees.** Sherman is working at the computer keyboard, learning which symbol to press to represent a screwdriver. Once he and Austin, a companion chimp, learned the correct symbol from a human teacher, the chimpanzees communicated with each other via keyboards in two rooms. In this way, one chimp could secure a tool that only the other had in his room, a tool that enabled the requester to complete a task that yielded a food reward. After Savage-Rumbaugh [698].

problem-solving behavior seems qualitatively different from the white rat's gradual increase in bar-pressing as it comes to associate the action with access to food.

On the other hand, a group of Skinnerian psychologists have shown that pigeons will behave in a remarkably insightful manner—if they have been operantly conditioned to perform actions that will be used later in solving a problem [243]. They first reinforced pigeons for pushing a box toward a green spot on the floor of their cage and separately trained them to climb onto an object and peck at a miniature facsimile of a yellow banana hanging in their chamber. These birds were *not* trained to push the box toward the banana model, nor could they move the box that was positioned beneath the banana during this training. But when they were placed in a cage with a *moveable* box in the middle of the chamber, and a banana suspended some distance away, the birds (after some initial hesitation) began to push the box around, stopping when the box was positioned under the banana, at which point they climbed up and pecked at the banana. (There was no green spot on the floor under the suspended banana.)

Pigeons that had been rewarded for climbing onto a stationary box and banana-pecking, but had not previously been trained to push a box, failed to show the novel problem-solving ability of their fellows who had acquired the two behaviors, "climb and peck" and "push box to green spot." These findings suggest that apparently intelligent problem-solving may not be so different from operant conditioning after all.

Misconceptions about Instincts and Learning   Although both *learned behavior* and *instinct* are terms that embrace many different things, we have persisted in categorizing all behavior as either one or the other. Not everyone agrees that this is a wise thing, and in fact there has been a long and rancorous controversy surrounding the use of the word *instinct*. At its heart, this has been a controversy over definitions and, like many semantic arguments, the intensity of the debate achieved a level that seems out of all proportion to the issue being debated.

I believe that one can usefully retain the terms *instinct* and *learned behavior*, but only if one realizes that *instinct is not purely genetic, and learned behavior is not purely environmental*. A great many persons thought and some still think that the key difference between those behaviors labeled instinctive or innate and those called learned is that one class is "genetically determined" while the other is "environmentally determined." These phrases are often interpreted to mean that innate behaviors are somehow encoded in a gene or genes, while learned behaviors are independent of genetic influence and instead require only the environment—the experience—of the animal. These interpretations are simply wrong.

Consider the (innate) begging call of a foster-reared nestling galah and the (learned) contact call of the foster-reared adult galah. Both calls are produced by the vocal apparatus of the bird, which is under the control of its brain. These structures are not encoded in the genes of the bird, nor do they arise spontaneously from the environment. Both genes and environment contribute to brain development, and the galah's brain is so structured that a nestling can give a begging call when the bird is hungry and in the presence of an adult. This signal is much the same for all baby galahs; it will appear no matter what sounds are present in the prehatching environment of the birds, and it is resistant to change after the birds have hatched, even if they hear pink cockatoo communication signals rather than the sounds made by parent galahs.

But the brain of a young galah also has components that enable it to store information about the contact calls it hears from the members of the flock it joins. If it stays with pink cockatoos, it hears pink cockatoo contact calls; this acoustical experience can change the way in which its brain operates, so that it comes to use the contact calls of another species. The ability to incorporate this novel information reflects the capacity of its brain cells to function in certain ways. These cells developed because of an interaction between genetic information and environmentally supplied materials and experiences. In other words, there is a genetic foundation for the development of *all* behavior.

All right, the ability to behave instinctively or to learn something depends on nerve cells and the development of those cells depends on both genes and environment, but surely instincts must be *more* genetically determined than learned behavior, and learned responses *more* environmentally determined than instincts? Wrong again. Richard Dawkins asks whether you would be tempted to say that a chocolate cake is 60 percent recipe instructions and 40 percent flour, eggs, cocoa, salt, and sugar [184]? Or 65 percent raw materials and 35 percent recipe? He does not think you would, and neither do I. Yet to say that an instinct is more genetic than a learned behavior is to make precisely this mistake. Genetic information (the recipe) and the molecules needed to construct an animal (the raw materials) are utterly different entities. The two things play complementary roles in the development of a nervous system that permits its owner to do some things innately and to do others through learning.

Instincts Can Be Modified by Experience

In Chapters 3 and 4 I shall discuss in more detail how the genetic recipe in concert with the environment shapes the development of nervous systems. These chapters reinforce the points made here on why it is inappropriate to talk about purely genetic (or purely environmental) behaviors or even mostly genetic (or mostly environmental) behaviors. The dichotomy between instincts and learning is not as strong as it might appear at first glance, and contrary to common belief, some instincts are modified by experience, nor are learned behaviors completely flexible responses to the environment.

For an example of a modifiable instinct, consider the following case. Many species of delicate orchids grow on the sandy forest floor in the southwestern corner of Australia, and some of these attractive plants deceive certain wasps. The males of these generally nondescript wasps spend their adult lives searching for conspecific wingless females, which announce their readiness to mate by releasing a sex pheromone while perched on a stem or twig. Males fly to scent-releasing females and carry them away to copulate with them elsewhere [747].

The flowers of wasp-pollinated orchids are code breakers. They take advantage of wasp mating systems by producing a mimetic sex pheromone that smells like the scent of a particular species of wasp. Male wasps track this odor to an orchid and there they see special structures on the flower that resemble the body of a wingless female wasp. The male pounces and attempts to copulate (unsuccessfully) with the flower (Figure 12). As he struggles, the male's body comes into contact with the pollinia, or pollen-bearing sacs of the flower. The sticky pollinia adhere to the male, and when he finally gives up the futile task of attempting to carry the "female" away, the pollinia go with him. Should he be drawn to another individual of the orchid species that deceived him, he will transfer pollen to the plant, fertilizing it but not a female of his species.

If you wished to see this drama and you chose to wait by an orchid for a wasp to appear, you would have to be extraordinarily lucky or patient. Botanists who have studied the phenomenon have learned that the way to improve the odds of watching orchids fool wasps is to gather a sample of cut specimens of various orchids (which requires the appropriate permits in Australia). The flowers are then kept fresh in water-filled jars and set out in a place where other orchids are growing naturally. Even then there is no guarantee of success. When I accompanied a crew of Australian botanists in search of orchid–wasp interactions, we tried without luck for two days. But just when it appeared to me that the wasp story was a myth, many small black wasps came rushing upwind toward the flowers soon

12 **Orchid code breaker.** A male thynnine wasp is "copulating" with an Australian orchid that releases a scent resembling the sex pheromone produced by females of his species. Photograph by the author.

after they were placed out on a roadside. One after the other, the arriving males swirled around the orchids of one species and landed on the flowers. Some males attempted to copulate with the dark, wasplike structure on the labellum of the flowers, and in so doing they placed their head precisely where the orchid's pollinia could stick to the wasp's head.

Superficially the wasp's behavior appears to be that of an automaton, with its tracking and copulatory responses mechanically triggered by simple stimuli in its environment, so simple that an orchid can successfully imitate the key cues and deceive the wasp into behaving in ways that waste the male wasp's time and energy. But the fact that one almost never sees wasps visiting orchids growing naturally on the forest floor suggests that the wasps are not quite as thickheaded as we might imagine. Males searching in an area seem to become familiar with the local orchids and eventually come to avoid them. But these same wasps are highly attuned to the scent of sex pheromone coming from different places, and it is this sensitivity to *novel* sources of the scent that botanists can exploit to find out what wasp pollinates which orchid species.

The fact that the number of males coming to a sample of cut specimens declines over time is further evidence that the males learn—probably from the unrewarding consequences of grappling with orchid flowers—not to respond to odors coming from certain locations. Thus, the wasps HABITUATE to some stimuli, learning not to react to cues that originally triggered a behavior pattern.

The ability to use feedback on the consequences of using an innate response is widespread in the animal kingdom. Jack Hailman provided a classic demonstration of the modifiability of an instinct in his study of the begging response of young laughing gulls [325]. Shortly after hatching, gulls have within their brains the neural circuitry that enables them to perform a functional begging response when they perceive a relatively simple sign stimulus. But their brain tissue must also be able to store information about the effects of their begging, which they use to modify their behavior over time.

At first baby gulls will peck at models whose resemblance to an actual gull is marginal, provided the model has some key properties. Tinbergen's study, mentioned earlier in this chapter, showed that newborn herring gulls pecked at (1) long, pointed objects more than short, stubby ones, (2) red objects over all other colored objects, (3) high-contrast dots over low-contrast dots, and (4) moving objects more than stationary ones. A three-dimensional model or the stuffed head of herring gull was no more effective in stimulating begging pecks by newborn baby herring gulls than a two-dimensional pointed strip of cardboard with a red dot painted on it [777]. But indiscriminate begging eventually ceases. Hailman found that older laughing gull chicks were much more selective in their response to models than younger ones. Although one- or two-day-old gulls pecked equally at almost any pointed object presented to them, they soon began to refuse models that did not closely resemble laughing gull heads. As a result of their social interactions with their parents, young birds acquired information, both auditory and visual, that led them to identify their parents as individuals. Eventually they begged only from their mother and father [325].

Biased Learning

Male thynnine wasps and baby gulls show us that animals may have a repertoire of actions that they exercise upon encountering the appropriate sign stimuli, but the performance of an instinct may provide the animal with information that it uses to modify its behavior. The other side of the coin is that learning may be less flexible than we tend to imagine. Consider the phenomenon of IMPRINTING, in which very young animals form an association with a particular individual. Although clearly a form of learning, imprinting demonstrates the "preprogrammed" nature of learned behavior (Figure 13).

Konrad Lorenz, in his classic studies of imprinting in waterfowl, showed that baby ducks and goslings, which normally follow their mother away from the nest shortly after hatching, could be induced to follow a substitute instead [507]. If Lorenz reared the birds himself, they would waddle after him when they were old enough to leave the hatching box. By virtue of an early exposure to Lorenz, the baby birds formed an immediate attachment to Lorenz and would follow him everywhere for days thereafter instead of

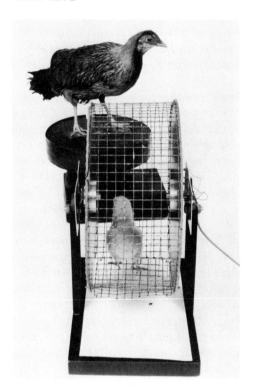

13 Experimental study of imprinting. Young domestic chickens can be induced to follow and imprint upon a stuffed jungle fowl mounted on a rotating base. The readiness of the chick to follow the moving jungle fowl is measured by the chick's activity on a running wheel. Photograph by Brian McCabe.

following an adult female of their own species or another human being. They had evidently learned to recognize him as an individual, just as they would normally have learned to identify their mother. What is even more remarkable, when they were mature many months later, the male birds would court human beings (including but not specifically limited to Lorenz) in preference to members of their own species.

Presumably a baby goose or duck can do these things because its nervous system is "primed" to be altered in a carefully defined way during the first few hours after hatching. Under natural conditions, following a moving object (the mother) during this sensitive period has two distinct developmental consequences: (1) the formation of a social attachment to a specific individual, its mother, and (2) the eventual recognition of suitable mating partners by a male.

Imprinting and imprinting-like behavior are now known to occur in many other birds. For example, the foster-reared galahs imprinted on their pink cockatoo foster parents, and as adults associated only with pink cockatoos, not with members of their own species. And imprinting also takes place in some mammals. Young shrews of a European species can be stimulated by their mother (when she wishes to lead her brood from one place to another) to hold onto the fur of another shrew (either the mother or a sibling). The mother then sets off with a conga line of babies trailing behind her (Figure 14). The role of imprinting in the attachment behavior of the young shrews has been explored experimentally [884]. When the baby

14 **Imprinting in shrews.** Young shrews form a "caravan" early in life, having learned the odor of their mother, which they will follow.

shrews are 6 or 7 days old, they will form a "caravan" by grasping a cloth and many other substitute mothers. However, in the period between 5 and 14 days, they become imprinted on the odor of the individual that is nursing them. Usually, of course, this is their mother, and she alone after this period can induce caravan formation by the young animals. However, if 5-day-old shrews are given to a substitute mother of another species, they will become imprinted upon her and when returned to their biological mother at 15 days of age will not follow her or any siblings that had been left with her. They will follow a cloth impregnated with the odor of the foster mother, a behavior that proves they had learned the identity of the female that nursed them when they were 5–15 days old.

Learning and Kin Recognition

When a baby shrew or gosling imprints on its mother, it has learned to recognize a close relative. Imprinting is then a proximate mechanism for KIN RECOGNITION, which is defined as the differential treatment of members of the same species in a manner related to their genetic relatedness [717]. The ability to identify individuals that are genetically similar is a specialized ability that occurs in a variety of animals in which parents care for their offspring (Chapter 12) or in which other relatives interact cooperatively (Chapter 15). As I shall make clear in the chapters ahead, kin recognition provides the basis for actions that can enhance the propagation of the individual's genes. Here, my focus is on a special proximate mechanism that promotes this ultimate goal, a learning mechanism, perhaps allied to imprinting, that underlies the recognition of relatives *other than* parents or offspring.

Kin recognition of siblings generally is the result of a simple "rule of thumb," which often goes something like this: "Treat individuals differently if you have been reared with them." Thus, for example, young spiny mice placed in a test arena will seek out each other's company; but if given a choice, they prefer to huddle with individuals with whom they have associated in their litter as opposed to individuals with whom they have had no prior experience. Typically this results in siblings preferring to be with siblings, but if one experimentally creates a litter composed of nonsiblings that are cared for by the same female, these unrelated littermates will treat each other as if they were sibs [641]. This "recognition error" shows that young of this species learn who their littermates are and use this information to modify their social behavior subsequently.

There are other cases, however, in which relatives that have not been reared together can still identify each other. For example, queen paper wasps that have had 3 to 5 days' exposure as emerged adults to their natal nest and sibling nestmates will subsequently show less aggression toward other unfamiliar sisters that had been reared separately than to nonrelated, unfamiliar females. Although they had no opportunity to associate directly with these unfamiliar sisters, they nevertheless treated them differently. David Pfennig and his co-workers suspected that paper wasp females absorb odor cues from their nest, and that by smelling an unfamiliar female a queen can determine whether her odor matches that which she has learned comes from her nest [632]. Under natural circumstances, siblings will share a similar nest odor, enabling females to treat sisters one way and nonsisters another.

If this hypothesis is correct, it should be possible to fool paper wasps into tolerating nonkin by transferring newly emerged queens to a foreign nest, thereby permitting them to learn the odor of this nest and nestmates. As predicted, females that participated in this experiment were especially tolerant of unfamiliar females that had been exposed to separate fragments of the nest in which they had been placed. Thus, kin recognition in paper wasps operates from a rule of thumb: "Treat as relatives those individuals that smell like the nest in which you have been reared."

A somewhat similar proximate mechanism may be responsible for kin recognition in Belding's ground squirrel. To study the phenomenon, Paul Sherman captured some pregnant females and shipped them to Warren Holmes's laboratory. When two females gave birth close in time, Holmes switched some of the pups to create four classes of juveniles: (1) siblings reared apart, (2) siblings reared together, and (3) nonsiblings reared apart and (4) nonsiblings reared together. Foster pups were readily accepted by adult females.

When the juveniles had reached the postweaning stage, pairs were placed in an arena and given a chance to interact. Animals that were reared together, whether siblings or not, generally treated each other nicely, whereas animals that had been reared apart were likely to react aggressively to one another. This study shows that the little squirrels learn something from the experience of growing up together, and they use this information in their social relations.

But perhaps the most remarkable finding of these experiments was that biological sisters reared apart engaged in a significantly lower rate of aggressive interactions than nonsiblings reared apart (Figure 15) [378]. In other words, sisters have some way of recognizing one another, perhaps an

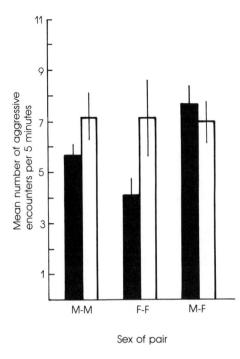

15 **Kin recognition in Belding's ground squirrels.** Sisters that have been reared apart display significantly less aggression toward each other. All other combinations of siblings reared apart (black bars) are as aggressive to one another when they meet in an experimental chamber as nonsiblings reared apart (open bars). *Source:* Holmes and Sherman [378].

odor similarity, that is not dependent on the experience of sharing a mother and a burrow. In fact, when they are permitted to interact in the test arena, even half-sisters with the same father reared apart in captivity are more nasty toward unrelated females than toward their paternal half-sisters [377].

Brothers, however, are just as aggressive toward full or half-brothers and full or half-sisters as they are to nonrelatives (Figure 15). Because in nature sisters often live in close association (Chapter 15), they have the opportunity to help one another, and they do because they can make the appropriate discriminations. Brothers live apart, so male Belding's ground squirrels rarely interact with their close genetic relatives. At the ultimate level, they do not gain by recognizing kin, and have not evolved the proximate mechanism for this ability [715].

Holmes and Sherman have given the label PHENOTYPE MATCHING to the mechanism responsible for the recognition capacity of female Belding's ground squirrels and others like them. These creatures learn something about their own phenotype, their appearance or odor or some other cue, and then discriminate among others on the basis of how similar these individuals are to them [379]. Such a mechanism can help identify relatives when relatives are similar in some way, which they often are.

Language Learning

At first glance, language learning would not seem to have anything in common with learned kin recognition or imprinting. Human languages depend on a highly flexible learning mechanism, with more than 3000 languages recorded around the world [575]. Moreover, we all know that humans can acquire new languages throughout their lives. On the other hand, all languages share certain things in common. Although the human vocal apparatus can produce a vast array of sounds, a total of only about 40 speech sounds (*phonemes*) are used to construct the thousands upon thousands of words in human languages; many of the same phonemes appear in different languages. Even more significant, all languages employ their vocabularies of words in lawful (grammatical) ways to construct meaningful sentences.

Recent discoveries have shown that the brains of human infants have properties that may facilitate the acquisition of a language. Six-week-old infants from English-speaking households are able to discriminate among many consonant sounds, including such similar pairs as "ba" and "pa." Peter Eimas discovered this by giving his infant subjects a pacifier wired to record the sucking rate of the baby. He then played tapes of artificially synthesized sounds that would be perceived by English-speaking adults as "ba," "ba," "ba,. . . ." With the first few "ba" sounds, the infant's sucking rate typically went up; but with repetition of the sound, the attention of the infant declined and the sucking rate fell. When a new consonant appeared, such as "pa" or even a novel, but very similar, consonant sound that Thai speakers use but that does not occur in English, the infants became aroused and their sucking rate promptly accelerated. This work shows that the perceptual mechanisms needed for the subtle discrimination of speech sounds have already developed in the infant's brain about the time it is born [226].

Soon after birth all young humans go through a stage in which they babble, producing a variety of unstructured sounds [474]. They seem to match the sounds they make against their stored memories of language sounds they have heard. Their ability to make fine discriminations among similar consonant and vowel sounds surely aids in this matching process. If a young child cannot hear spoken language, it has no store of information about speech sounds and it will not develop a spoken language; it will go through the babbling phase but eventually will stop producing sounds altogether.

Vocabulary development occurs remarkably rapidly in children with normal hearing. At 15 months the average child has command of 50 words; just a year and a half later the youngster will be using about 1000 words (in fact, at 30 months children appear to comprehend, if not obey, everything said to them [474]). The ability to process dozens of words in short periods of time and to understand their totally symbolic content is an astonishing human trait.

Moreover, the development of sentence structure of young children in English, Russian, Chinese, Finnish, and Zulu cultures, to name a few, follows a channeled pattern [575]. The child first uses one-word sentences ("No."; "Go."; "Mama."), then always enters a two-word stage ("Dada come."; "Bring me.") that endures for some time before the child enters the third phase: *telegraphic speech*, in which sentences consist of nouns without plurals, verbs without tense endings, speech stripped to its essentials. Yet even at the two-word stage, babies speak grammatically. As language development proceeds, the young human shows great skill in extracting (unconsciously) the rules of language use.

Without a knowledge of these rules, language communication would be impossible. (Language knowledge these a communication without impossible be of rules.) My ability to communicate with you occurs because there are only certain ways subjects, verbs, predicates, verb tenses, and prepositions can be used in English. Young children can acquire the necessary grammar skills without ever receiving formal instruction in them. In fact, correcting a young child is far less helpful than providing conversational practice.

Let me illustrate what I mean with an anecdote from an American friend who reared her children in Peru, where they became bilingual. One of her young children asked, "Mommy are my clothes planched?" The verb *planchar* means *to iron* in Spanish. The child had modified this verb using the English rule of adding the suffix *-ed* to signify the past tense. The child had extracted this rule by listening to English-speaking persons; she did not have to have the rule explained to her, nor did she have a conscious revelation about past tense. She, like all but a tiny minority of children, just did it.

The fact that the child made the error "planched," although she had never heard any adult say "planched," shows that rule extraction involves more than simple imitation of adult speech. In some way human brains acquire grammatical rules and use these rules in the construction of novel sentences, enabling even young children to communicate effectively with language, the greatest product of human evolution.

Special Features of Associative Learning

The point is that humans could not learn a language were it not for the special operating features of our brains. The same point can be made about even the most flexible of learning programs. Initially, Skinnerian psychologists claimed that it was possible with equal ease to condition almost any operant, any action that an animal could perform, but soon workers in this field began to discover contradictory cases. For example, one can readily condition a rat to do some things like running in a running wheel by sounding a warning noise and giving it an electric shock unless it is engaged in the proper operant. After a rat has had some experience with hearing the sound and receiving a shock while standing in the wheel but not while running, it will make the appropriate association and start running whenever it hears the warning cue [74]. But a rat cannot be conditioned to stand upright in the running wheel using the same procedures. Rats that happen to rear up just after the sound are not shocked; but despite this, they fail to make the association between this behavior and avoidance of electrical punishment. The frequency with which they perform the standing response when they hear the warning signal actually declines over time (Figure 16).

The psychologist who probably has done most to advance the argument that the learning process is not completely open-ended is John Garcia [273. 274]. He and his co-workers have examined the ability of white rats to

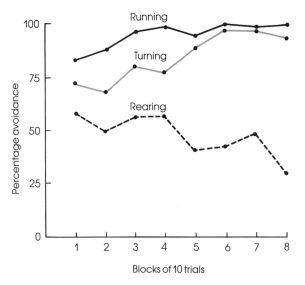

16 **Biased learning: running, turning, rearing.** Learning curves for three operants that were equally rewarded by refraining from shocking the laboratory rat when it made the appropriate response. Rats failed to learn to rear on their hindlegs to avoid a shock. *Source:* Bolles [74].

learn to avoid various sensory cues that are associated with punishing consequences. One punishment, X-ray radiation, mimics the effects of ingestion of toxic substances by causing the buildup of toxic products in irradiated tissues and body fluids. If a rat consumes a food or fluid with a distinctive taste and then is exposed to minute amounts of radiation, it will refuse to eat this material at a later time. Even tiny doses of radiation induce mild nausea, which the animal associates with any unusual tasting liquid or food it has recently consumed. The degree to which the food or fluid is avoided is proportional to (1) the intensity of resulting illness, (2) the intensity of the taste of the substance, (3) the novelty of the substance, and (4) the shortness of the interval between consumption and illness [274].

Taste-aversion learning can be seen as an example of classical conditioning, in which X-ray irradiation (the unconditioned stimulus) always elicits illness and suppresses consumption of food or liquids (the unconditioned response). The taste of the food or liquid acts as a conditioned stimulus that the animal pairs with X-ray irradiation, and the conditioned animal reduces food or fluid intake whenever it experiences the distinctive taste [204]. However, taste-aversion learning is special. For example, only a single radiation trial is needed to condition the rat to avoid food with a certain taste, unlike the more widely studied cases of classical conditioning in which repeated trials are needed before the animal forms the conditioned association.

Even more impressive, if there is long delay (up to 7 hours) between eating a distinctive food and exposure to radiation and consequent illness, the white rat is still able to link the two events and use the information to modify its behavior. Most kinds of associative learning will not take place if there is an interval of more than a few seconds between the conditioned stimulus (e.g., a tone) and the unconditioned stimulus (e.g., meat powder).

The specialized nature of taste-aversion learning is further shown by the rat's complete failure to associate a distinctive sound (a click) with internal illness. Although rats can learn to associate clicks with shock punishment (and will learn to take appropriate actions to avoid shocks upon hearing a click), when the punishment is a nausea-inducing treatment, rats fail to learn to link the sound with the treatment. In addition, rats have great difficulty in making the association between a distinctive taste as a signal that a shock is about to be delivered. If, after drinking a sweet-tasting fluid, the rat receives a shock on its feet, it often remains as fond of the fluid as it was before (as measured by amount drunk per unit time), and this is true no matter how many times it is shock-punished after drinking sweet liquids. Thus, the nature of cue and consequence determine whether a rat can learn to modify its feeding and drinking behavior (Figure 17).

The Evolutionary Basis of Instincts and Learned Behavior This chapter has been concerned with the *proximate* distinction between instincts and learning. I have argued that some behaviors develop reliably in full form, often early in an animal's life, and are played out in response to simple cues in the environment. Other behaviors do not develop fully until the animal has had the opportunity to acquire a specific piece of information as a result of interacting with its environment.

But what about the possible *ultimate* significance of the differences between the proximate categories of behavior? Why does a galah respond innately to some things and possess the capacity to learn others? Are instincts superior to learned responses in some situations and learning superior to instincts in others, in terms of contribution to individual reproductive success? Does the "directed" nature of learning actually increase an animal's chances of leaving surviving offspring?

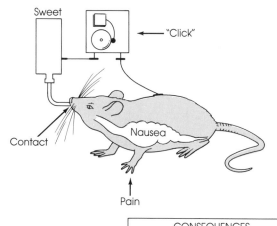

		CONSEQUENCES	
		Nausea	Pain
CUES	Sweet	Acquires aversion	Does not
	"Click"	Does not	Learns defense

17 Biased learning: tasting, hearing. Although laboratory rats can be easily conditioned to associate taste cues with nausea and acoustical cues with skin pain, they have great difficulty forming other learned associations. *Source:* Garcia et al. [274].

The Advantages of Instincts

Humans tend to think that the ability to learn is somehow inherently superior to instinctive behavior, no doubt because we are a species with an unusually well-developed capacity for learning. But this self-congratulatory attitude is probably not justified. Not all ecological conditions favor the ability to acquire, store, and use experiential inputs to modify behavior. Learning costs something, and one of the costs probably is the energy needed to manufacture and maintain those elements of the nervous system required to store information and use it to change an animal's behavior (but see [769] for another view).

Some of the most convincing evidence that learning abilities demand added neuronal tissue comes from studies of certain bird species whose males learn a complex repertoire of courtship songs. The ability of canaries to learn their song depends on a region in the forebrain called the nucleus hyperstriatum ventralis. In males this brain component is much larger than in females (which do not sing). Moreover, males that have learned more song types have a nucleus hyperstriatum ventralis that may be three times as large as that of less capable singers. Even more remarkable, canaries lose many of the nerve cells in their song system each year, investing in the costly replacement of large numbers of brain cells at the start of each breeding season (Figure 18) [593, 594, 595]. Fernando Nottebohm suggests that perhaps the loss occurs to erase old memories of learned

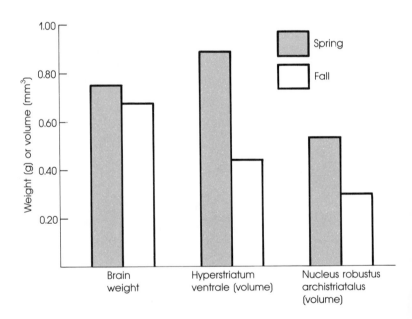

18 **Cost of a learned behavior.** The regions of the brain devoted to the song system of a male canary expand during the spring breeding season when the bird sings. The calls of canaries are learned by listening to other males, and the added neuronal tissue is thought to assist in learning. *Source:* Nottebohm [593].

songs, while the new brain cells permit the males to learn a new set of songs similar to those being produced by that year's batch of competing neighbors.

Males of the long-billed marsh wren also learn song repertoires by listening to and then mimicking other singing males. For some reason, males living in the western United States typically have repertoires that contain two or three times as many song types as those sung by their East Coast relatives. When played a tutor tape containing 200 song types, young West Coast marsh wrens learned an average of about 100 of the songs, whereas East Coast wrens incorporated only about 40 in their repertoire [450]. Here too learning ability does not come free, but requires added brain tissue. The song system in the forebrain of West Coast birds weighs 25 percent more than the equivalent region in the brains of East Coast wrens.

If the large brain of humans is related to our great ability to learn, as many people have argued, then the physiological expenses of human learning are great indeed. Although the brain makes up only 2 percent of our total body weight, it demands 15 percent of all cardiac output and 20 percent of the body's entire metabolic budget [27]. The brain is extremely sensitive to oxygen deprivation, a fact that has been responsible for the shortening of many lives. Moreover, it is also developmentally vulnerable to genetic accidents [92]; many of the known single-gene effects on human behavior involve severe brain damage resulting from a single mutant allele (see Chapter 3). Finally, our reinforcement networks, although often rewarding or punishing us in biologically appropriate ways [676], may malfunction, an event that causes various kinds of debilitating mental illnesses. A less massive brain, more simply constructed, might be less expensive to maintain and less susceptible to injury and operation failure.

Not only may simple rules of behavior be cheaper in physiological terms, they are also highly effective in situations in which there is a reliable relationship between a particular cue and a specific, productive response. Surely it is no accident that instincts are characteristic of very young animals. A baby gull or galah just out of the egg has nothing to gain from having to learn how to get its parents to feed it. A brain programmed by earlier development to react to certain cues shortly after hatching gives the young bird what it needs, an effective system of communicating with a parental adult.

Similarly, it is not surprising that when animals are confronted by lethal enemies they often evolve innate antipredator responses. Fledgling motmots (tropical birds related to kingfishers) do *not* require experience with coral snakes (Figure 19) in order to develop an aversion to them. These birds eat lizards and small snakes. Even young birds are attracted to thin snakelike objects, but as Susan Smith showed, hand-reared baby motmots readily approached and pecked at various painted wooden rods *except* those with alternating red and yellow rings, which they completely refused to touch (Table 1) [733].

The difference between Norway rat feeding behavior and that of young motmots is instructive. The dangerous foods that rats are likely to encounter are primarily poisonous plants (or baits). An interaction with one of

19 **Coral snake.** The color pattern of this snake consists of bright red, yellow, and black bands. Photograph by James D. Jenkins.

these foods is not likely to be the end of the road for a rat because rats typically eat only a very small portion of anything new. If it is poisonous, they will live and learn, avoiding thereafter any item that tastes or smells like the food that made them ill. In contrast, the life expectancy of a motmot fledgling that picks up a red-and-yellow ringed coral snake is not high. The fact that these snakes have a distinctive color pattern has enabled the birds to evolve an innate rejection of cues associated with coral snakes.

It is true that innate behaviors may be prone to exploitation by parasites and others, as in the case of ant-robbing rove beetles. But the evolutionary

TABLE **1**
Reaction of motmots to models of various sorts: Evidence for an aversion to yellow-and-red ringed pattern in young, inexperienced, hand-reared birds

Model	Model pattern	Number of pecks	Comments
⦅⦆⦆⦆⦆⦆⦆⦆⦆⦆⦆⦆⦆	Coral snake pattern: yellow and red rings	0	
⦅⦆⦆⦆⦆ ⦆	Partial snake pattern: one end yellow and red *rings*; other end plain	79	Only 15% at yellow-red end
⦅⦆⦆⦆⦆⦆⦆⦆⦆⦆⦆⦆⦆	Ring pattern control: green and blue rings	89	
⦅≣≣≣ ≣⦆	Color control I: yellow and red stripes	60	
⦅≣≣≣ ⦆	Color control II: One end yellow and red stripes; other end plain	90	47% at yellow-red end

Source: Smith [733]

maintenance of a simple, cued response requires only that the behavior be more advantageous for an individual than any alternative trait. Ants may occasionally feed a parasitic beetle, but 99 percent of the time they feed fellow colony members. Feeding the beetle is a small disadvantage of the cued behavior system that presumably is outweighed by the reliable performance of a nearly always effective behavior.

Moreover, the risk of exploitation or misapplication applies not only to instinctive behavior. It is entirely possible for an animal to modify its behavior as a result of an experience in a way that lowers, not raises, its fitness. A human being or a Norway rat may mistakenly learn to avoid an edible food if the first time he eats the food it happens to be spoiled or if he consumes it and becomes ill for some other reason [63, 143]. By associating illness with the novel item, lifetime aversions to perfectly nutritious foods can become firmly established.

The Advantages of Learning

A hypothesis to account for the specialized nature of animal learning is that these features reduce the risk that an animal will learn the wrong things or irrelevant information [413]. The ability to learn from experience does help if the individual can adjust to variable conditions that could not have been anticipated prior to its birth, such as the location of food and dens, or precisely what edible prey species are available, or exactly what its parents look like, or what language is spoken in its home. Specific learning mechanisms help focus an animal's attention on the key variables.

Why can't a white rat learn to rear up on its hindlegs when it hears a sound that will be followed by a shock unless it performs the rearing operant? An ultimate hypothesis is that running is an adaptive response to a threatening sound whereas rearing up, which is an exploratory behavior, is not [74]. Therefore, at the proximate level, rats are endowed with the neuronal circuitry that specifically facilitates the formation of learned associations between acoustical cues (like rat snarls) and escape when the reward is avoidance of painful skin sensations (caused by electrical shocks or by bites from rival rats).

Or why is it that white rats are so adept at learning to avoid novel foods that are associated with illness, even hours after ingesting the food? This skill is useful to the rat, given the food-sampling behavior of the animal. Under natural conditions, a Norway rat becomes completely familiar with the area around its burrow, foraging within that area for a wide variety of foods, plant and animal [502]. New plants and insects are constantly coming into season and then disappearing. Some of these organisms are edible and nutritious; others are toxic and potentially lethal. A rat cannot clear its digestive system of toxic foods by vomiting. Instead, the animal takes a small bite of something new, and if it gets sick later, it *should* avoid this food or liquid, because eating large amounts of the new food may kill it.

Interestingly, it is almost impossible to teach a rat to press a bar if the reinforcing food pellet is not delivered within 3 seconds of the bar press. In nature, if a rat manipulates an object with its paws and that object does not produce food quickly, the chances are high that it never will. The rat requires instant gratification if it is to learn to perform a manipulation in order to get food. On the other hand, poisonous foods rarely have an instantaneous effect, especially if eaten in tiny quantities. The rat's brain has evolved the ability to accommodate long delays between gustatory stimulation and illness because this accommodation helps individuals learn to avoid dangerous foods [274].

Evolutionary biologists argue that there should be a correspondence between the ecology of the species and the kinds of things an animal can learn easily [413]. The peculiar rules of learning of white rats become somewhat less peculiar if the feeding ecology of this omnivore is taken into account. If this argument is correct, it can be used to make any number of predictions about the occurrence of specialized learning abilities. One such prediction is that in species in which one sex travels over greater distances than the other, the sex with the larger home range will exhibit superior spatial learning ability.

Steven Gaulin and Randall Fitzgerald have examined this prediction in studies of meadow voles, *Microtus pennsylvanicus*; males of this small rodent species wander over much larger areas than the females do. As predicted, males remembered the route to a food reward through various laboratory mazes better than females of their species (Figure 20). But in two other closely related rodents—the pine vole *Microtus pinetorum* and the prairie vole, *Microtus ochrogaster*—there was no difference in the spatial learning performance of males and females. These voles are characterized by an equality among the sexes in the size of their home ranges; typically one male and one female live monogamously in the same area and so are confronted with equivalent spatial learning problems in their daily lives [275, 276]. This result is consistent with the hypothesis that learning abilities will reflect selection for skills relevant to tasks that affect individual fitness. According to this view, it is the variety of ecological problems facing individual animals that sets the costs and benefits of different degrees of behavioral flexibility and thereby generates the great diversity of animal behavior.

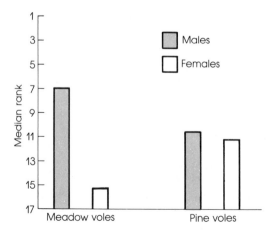

20 **Sexual differences in learning ability.** Male meadow voles have a much larger home range than the more sedentary females of their species, and males usually score higher in a maze learning test than females. In the monogamous pine vole, males and females live in the same home range and their maze learning abilities are not significantly different. *Source:* Gaulin and Fitzgerald [275].

SUMMARY

   1 The traditional classification of behavior into the proximate categories—instinct versus learned behavior—has been associated with the mistaken belief that some behavior patterns are more genetically determined or more environmentally determined than others. In multicelled animals, both instinct and learned behavior depend on neural foundations that could not develop without genetic information and environmentally supplied materials. Instincts differ from learned behaviors not in the amount of genetic or environmental influence but in the degree to which the behavior appears in complete form the first time an animal encounters a key stimulus and reacts to it.

   2 Both *instinct* and *learned behavior* are terms that cover a wide range of phenomena. Learned behavior arising from modifications traceable to a specific experience in an animal's lifetime, includes such diverse categories as classical and operant conditioning, spatial learning, imprinting, and insight learning.

   3 Innate behaviors are often modified by experience and learning is often "biased." This fact further blurs the distinction between instinct and learning because it shows that, on the one hand, animals can use feedback experience to alter certain first-time "built-in" responses to key stimuli, whereas, on the other hand, learning is not as flexible as it first appears. Indeed, in order to learn something, an animal's nervous system *must* possess "innate" properties that permit it to acquire information that is used to change its behavior in particular ways. This is most clearly seen in such things as imprinting in which the imprinted animal uses a specific bit of information from its environment acquired at a restricted period in its life to complete the development of a particular element of its behavior, such as its recognition of sexual partners later in life.

   4 An ultimate analysis of the different kinds of behavior suggests that no one category of behavior is intrinsically superior to another in terms of its contribution to individual reproductive success. Simple rules of behavior can be more adaptive than complex learning abilities when, for example, making a mistake in dealing with a key stimulus is fatal. Under other conditions, individuals gain by modifying their behavior within bounds in response to factors that could not be predicted accurately prior to the birth of an individual. Thus, for example, many animals that travel widely must learn the spatial features of their environment. An evolutionary hypothesis is that the diversity of proximate mechanisms regulating animal behavior is a response to the diversity of ecological problems confronting animal species.

SUGGESTED READING

   Readers interested in the approach and findings of traditional European ethologists should see Niko Tinbergen's *The Herring Gulls' World* [774], or the anthologies of papers by Tinbergen [776] and Konrad Lorenz [506].

   Lorenz has also written a good article on the ethological approach to learning [505]. John Garcia's review of his controversial research on white rat learning is highly recommended [274]. For two recent but somewhat different assessments of the relation between evolution and learning theory see papers by Michael Beecher [49] and Leda Cosmides and John Tooby [150].

Bert Hölldobler's *Scientific American* article on ants and their nest parasites makes excellent reading [372].

DISCUSSION QUESTIONS

1. Human brains are about *seven* times larger than expected for an animal species of our size [411]. Use the method of multiple working hypotheses to analyze the possible evolutionary significance of our extraordinarily bloated brains. You may wish to consult Richard Alexander's publications for one intriguing hypothesis [11, 12].

2. Experiments in which an animal is raised under restricted or deprived conditions and yet still exhibits certain behavioral responses to sign stimuli are sometimes used to help define whether a behavior is instinctive or not. Critics of this approach point out that you cannot remove *all* environmental influences, even in the most extreme deprivation experiment. The experiences an animal has in these experiments might still be essential for the development of the so-called instinct. These critics want to eliminate the term *instinct* from the vocabulary of behavioral biologists. To what definition of an instinct does this criticism apply? If you accept the argument that nothing can be instinctive, what categories would you create, if any, to deal with differences in the proximate mechanisms underlying behavior? Would you put gull chick begging behavior and sexual imprinting by geese in the same category?

3. What predictions would you make about the sex differences in spatial learning ability in chickadees and nutcrackers? How would you test your predictions?

3

The Genetics
of Behavior

Learning how an individual's genes influence its behavior is a
key element in understanding the proximate causes of animal
behavior. The preceding chapter made the argument that all
behavior patterns require genetic information for their development, contrary to the claim that some behaviors are "genetic"
and others are "environmental." In this chapter I want to expand on this point, using the findings of geneticists who have looked at the connection
between genes and behavior. The first goal
of behavior genetics was to show that there
is a correspondence between genes and behavior. I shall review the many techniques
used to this end, demonstrating that even a
single genetic difference between individuals
may have behavioral consequences. Having achieved their initial objective,
behavior geneticists have now begun to link genetic differences between behaviorally different individuals to variations in their nervous or hormonal systems. A few workers have even shown how ecological differences among
populations might be responsible for the spread of different genes, different
physiological mechanisms, and different behavioral traits in geographically
separated populations of the same species. The chapter concludes with a description of an innovative study of this sort that unites both the proximate and
ultimate aspects of the genetic basis of behavior.

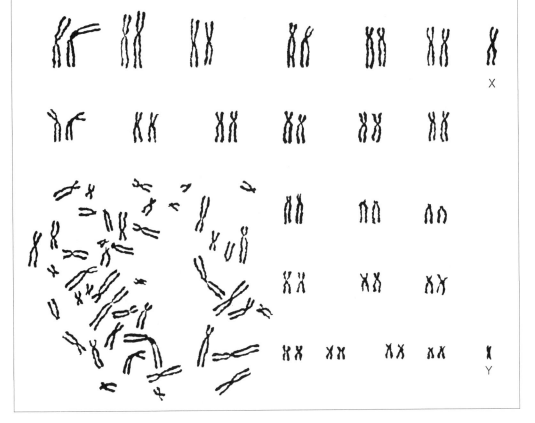

Genes and Behavior   The primary question of behavior genetics in its early years was, If individuals vary genetically, will they also show some behavioral differences? As we have already noted, no behavioral attribute can develop without both genetic information and environmental inputs for the construction of nervous systems. However, although no behavior can be purely genetic or purely environmental, the behavioral *differences* between individuals can potentially be either genetically caused or due to environmental differences between them.

This distinction may be confusing at first, so let's return to the adopted galahs whose behavior I talked about in Chapter 2. Galahs that have been reared by pink cockatoos give contact calls that are the same as the signals given by pink cockatoos. If we ask, Do genes have anything to do with the contact call given by a cross-fostered galah? we are asking a dull question, for the answer to any question of this sort will always be yes. A call given by a galah depends on the nervous system of the animal, and without genetic information present in a galah egg there would be no galah brain capable of generating the signals needed to activate a call. Without genes, there would be no galah at all.

But if we ask, Are the differences in contact calls given by a cross-fostered galah and one reared by its own galah parents genetic or environmental? we are asking a more interesting question. The evidence tells us that in this case there is a strictly environmental basis for the difference between the two galahs. Those galahs that are adopted by pink cockatoos give pink cockatoo contact calls; those that are reared by galahs give galah contact calls. There is no reason to suppose that those baby galahs that happen by chance to be adopted by pink cockatoos differ from the rest of their species with respect to the genes that influence the development of that part of the brain involved in communication. The difference between the two kinds of galahs is almost certainly due to the fact that cross-fostered and normally reared individuals are exposed to different social and acoustical environments, a difference that affects the development of contact calling. There could, however, be some other behavioral differences among galahs caused by genetic variation among them, genetic variation that affected, say, brain development, which in turn led to behavioral differences. That this is not pure speculation is shown by studies of human beings.

Heredity and Human Behavior

Oskar Stohr was raised by his grandmother as a Catholic in Nazi Germany. Jack Yufe grew up on various Caribbean islands with his Jewish father. One can hardly imagine more diverse environments for the development of two human beings, and yet these men are remarkably similar in appearance (Figure 1) and behavior. They "like sweet liqueurs, . . . store rubber bands on their wrists, read magazines from back to front, dip buttered toast in their coffee and have highly similar personalities" [370]. They are identical twins who were separated after they were born and did not meet again until they were 47 years old.

The separation of identical twins early in life provides a very rare natural

1 **Identical twins separated at birth:** Oskar Stohr (right) and Jack Yufe (left). Photograph by Bob Burroughs.

experiment with which to test a hypothesis about the genetics of human behavior. The hypothesis is that some of the differences in the behavior of humans are due to the genetic differences among them. A prediction that follows is that identical (monozygotic) twins should be more similar behaviorally than nonidentical or fraternal (dizygotic) twins. This prediction is based on the fact that identical twins are derived from the same fertilized egg (which divides and gives rise to two separate, but genetically identical, embryos), whereas fraternal twins are produced from two different fertilized eggs with different hereditary information.

In humans and other animals, each gene may be represented by one or many slightly different forms (alleles). Each gene codes for a distinctive protein such as a hormone or enzyme (e.g., amylase, which facilitates the chemical reaction that breaks down starch into smaller molecules). Each allele of the amylase gene carries information for the construction of a slightly different form of the amylase molecule, each form having its own chemical properties and abilities.

Human beings are DIPLOID organisms. Each of us has two copies of every gene. The two copies of any particular gene possessed by a person may be the same or they may be different (alleles). Imagine a population in which there were four alleles of the amylase gene (amy^1, amy^2, amy^3, and amy^4). In this population there might be a woman with one copy of amy^1 and one copy of amy^2, and she might have a husband with copies of amy^3 and amy^4. The eggs produced by the female are HAPLOID (i.e., each egg has only a single copy of each gene), so the genetic makeup, or GENOTYPE, of an egg may be either amy^1 or amy^2. These eggs will be fertilized by sperm, which are also haploid and therefore have one copy of the amylase gene, either amy^3 or amy^4 in this case. The sons and daughters produced by these individuals are equally likely to have one of the following four genotypes: amy^1, amy^3 or amy^2, amy^3 or amy^1, amy^4 or amy^2, amy^4 (Figure 2).

Pick one of these genotypes (e.g., amy^1, amy^3) and compare it with each of the four equally probable genotypes of its siblings (in the sequence listed

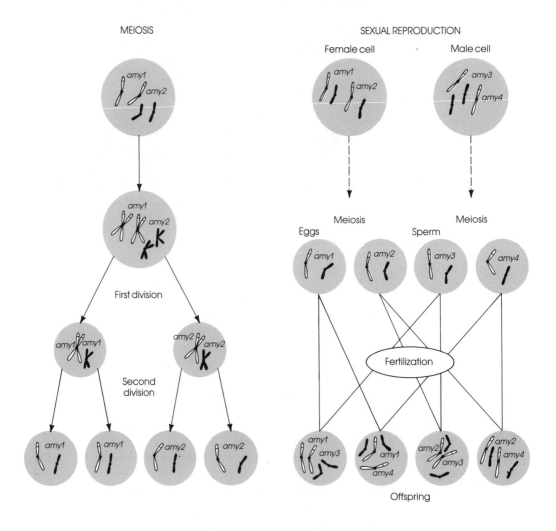

2 **Meiosis and sexual reproduction.** The cells that give rise to eggs or sperm undergo a series of meiotic divisions that produce cells with only one copy of every gene in the parental genotype. When an egg fuses with a sperm in sexual reproduction, the resulting offspring has two copies of each gene.

above). The proportion of shared alleles between them is 1.0, 0.5, 0.5, and 0.0. The average shared proportion of alleles is (1.0 + 0.5 + 0.5 + 0.0)/4 = 2/4 = 1/2. The probability that two siblings (other than identical twins) will inherit the same allele from a parent when two different alleles are represented in the parent's gametes is always 1/2. As a result, if behavioral development is influenced by the presence of a particular allele, fraternal twins should sometimes differ (because they will sometimes have different forms of a gene), whereas identical twins should be the same because they possess exactly the same genotype.

In fact, identical twins are more similar than fraternal twins with respect to a host of traits, including appearance, personality measures, and other behavioral abilities. But behavioral development is influenced not only by the genes the individual inherits, but also by his or her environment and experiences. Some people have suggested that because identical twins look alike, parents treat them more similarly than they treat fraternal twins (who are no more similar in appearance than a pair of ordinary siblings). This environmental influence, so the argument goes, could be the reason why identical twins are so similar behaviorally. The evidence that identical twins are actually subjected to more similar environmental influences than fraternal twins is not strong [638], but to be on the safe side it would be interesting to observe the behavior of identical twins who were reared apart. If they developed similar behavior, it could not be because they were encouraged to do so by their parents and a shared home environment.

This is why Jack Yufe and Oskar Stohr and others like them provide such valuable information. A recent comprehensive study of monozygotic twins reared apart (involving 44 pairs of twins separated before they were 5 years old, most at birth or shortly thereafter) demonstrates beyond reasonable doubt that genetic differences are responsible for a significant part of the differences among individuals in personality [761] and mental abilities [547]. For example, the response of identical and fraternal twins to an elaborate questionnaire designed to provide quantitative measures of personality traits (e.g. degree of aggressiveness, alienation, control, and so on) reveals that, even when separated, identical twins secure scores that are usually more similar than the scores of dizygotic twins reared together. These data show that about 50 percent of the *differences* in personality scores among all the individuals tested was due to genetic differences among them. The other 50 percent of the differences was environmental in origin.

Why IQ Is Not "Genetically Determined"

Identical twins are also more similar than fraternal twins in their scores on IQ tests [87]. When identical twins are reared together, the *correlation* between their scores is usually between 0.5 and 0.9. If the IQ scores of pairs of identical twins were always exactly the same, the correlation would be 1.0. If the correlation were 0, the score of one twin would be utterly useless as a predictor of the score of the other. As it turns out, if one twin has an IQ of 115, the odds are excellent that the other will be close to this score (say, between 110 and 120), and there would be only a very small chance that the other twin would score 95 or 135. The possibility that these similarities stem from the similar environmental experiences of twins is greatly weakened by the finding that the average IQ correlation of dizygotic twins reared together is significantly lower than that of monozygotic twins reared apart (Table 1). This is powerful evidence that genetic differences contribute to the development of IQ differences between people.

Further support for this conclusion comes from the finding that the IQs of parents and their biological children are substantially more similar than those of foster parents and their adopted children (Table 1). Each parent donates half its genotype to each child, so the proportion of shared genes between a mother and her son (for example) is 0.5. But individuals that have adopted children obviously have not endowed their adoptees with any of their genes and so have a genetic relatedness of 0.

The fact that there is a connection between the proportion of genes shared by descent and the degree of similarity in IQ scores is sometimes interpreted to mean that intelligence is "genetically determined" in the sense of not requiring environmental influences for its development. Leaving aside the serious problem of establishing the relation between IQ and intelligence (about which much ink has been spilled), this interpretation is incorrect. The results of the familial comparisons within populations outlined in Table 1 show just one thing: that genetic differences among people contribute to the differences in the scores they achieve on IQ tests. This conclusion does *not* mean that IQ is "genetic," or "genetically determined," or "inherited."

Intrafamily comparisons actually show that some of the behavioral differences between individuals are environmentally induced. If the environment plays no role in contributing to mental differences between people, then the correlations between the IQs of identical twins would be 1.0; but they are always less than this. Identical twins are not behaviorally identical because their development is shaped by somewhat different environments. Each twin experiences slightly different surroundings in the womb, each consumes somewhat different foods, and each has its own unique set of social interactions (this is especially true of twins separated early in life).

TABLE 1
Familial correlations for IQ scores: Predicted versus actual correlations

Category	Predicted correlation[a]	Number of studies	Actual median correlation
Monozygotic twins reared together	1.00	34	0.85
Monozygotic twins reared apart	1.00	3	0.67
Dizygotic twins reared together	0.50	41	0.58
Siblings reared apart	0.50	69	0.45
Nonbiological sibling pairs	0.00	5	0.29
Parent–biological child	0.50	32	0.39
Parent–adopted child	0.00	6	0.18

Source: Bouchard and McGue [87].
[a] Predicted correlations are based on the assumption that the differences between individuals are solely due to genetic differences between them.

The true importance of familial comparisons as a behavior genetics technique is simply to show that some differences in specific behavioral abilities of humans have a genetic component. For example, not only are the mental development scores of very young monozygotic twins similar, they also follow a remarkably similar, age-related pattern (Figure 3) [859]. Likewise, left-handedness is much more likely to occur in children if one or more of their biological parents is left-handed. If a child whose biological parents are right-handed is adopted into a family with a left-handed foster parent, the child's environment does not increase the probability that he or she will be left-handed [141].

The occurrence of certain mental illnesses and other afflictions has also been related to genetic differences among people. The children of a schizophrenic, a manic-depressive, or an alcoholic parent have half the parent's genotype and a much greater risk of developing the condition than the children of parents who do not have these diseases [361, 553, 706]. One might, however, argue (not unreasonably) that living with a person suffering from schizophrenia or alcoholism creates an environment that induces the trait in others. Once again, one can avoid this complication by examining the life histories of children separated early in life from their biolog-

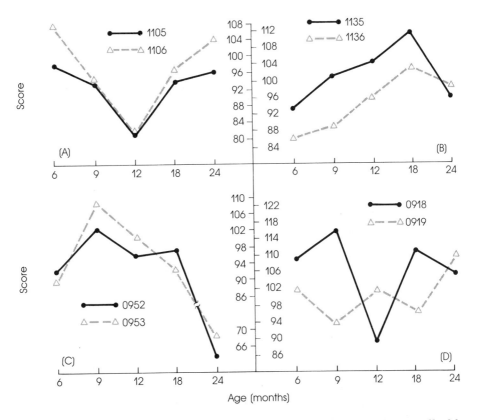

3 **Mental development scores of identical twins** usually follow a similar course (A–C) over the first two years of life, although there are exceptions (D). *Source:* Wilson [859].

ical parents (and a possibly disturbing home environment) and reared in foster homes. Adopted children with a schizophrenic biological parent show a four or five time greater incidence of the illness than adopted individuals whose parents were not schizophrenic.

I emphasize again that schizophrenia, manic-depression, and alcoholism are not "genetically determined." A child of a schizophrenic (or alcoholic or manic-depressive) may or may not have inherited the critical allele or alleles from the affected parent (see Figure 2). If the child is not a carrier, it has no genetic factors that increase its risk of developing the disease. Even if the child has received the key allele(s), he or she will not automatically be afflicted no matter what the environment. Most carriers are behaviorally normal; but, as a population, they have a somewhat greater chance of developing the condition presumably because of greater susceptibility to the environmental factors that trigger the syndrome [361].

Studies of human behavior genetics indicate that there is considerable genetic variation within modern populations, a genetic variation that contributes to the behavioral variation among humans. But studies of the behavior of relatives leaves unanswered many other intriguing questions, including, How does possession of any one allele exert its impact on the development of a behavioral trait? Little progress has been made on this point. At best there are only tantalizing suggestions.

For example, an excess of the chemical dopamine, a NEUROTRANSMITTER that relays messages between nerve cells in certain parts of our brain, has been implicated as a factor in schizophrenia [586, 745]. One study found that some schizophrenic patients have unusually low levels of an enzyme that breaks down dopamine molecules [745]. The shortage of the enzyme should logically result in the buildup of dopamine. The effectiveness of various drugs used to control schizophrenic symptoms is directly proportional to how well these drugs bind with receptor molecules on nerve cells that normally bind with dopamine [588] and that behave abnormally when subjected to excessive amounts of dopamine. By blocking dopamine receptors, antipsychotic drugs prevent dopamine from reaching them, and thus diminish schizophrenic symptoms.

A more definite relationship between the enzymatic product of a gene and psychiatric disorders has been established in the case of monoamine oxidase (MAO) [115]. The activity of this enzyme varies over a broad range among individuals. Very high and very low activity levels characterize particular families and are presumably related to the presence of specific alleles. In families with low MAO levels, the incidence of suicide and attempted suicide is eight times that of families whose members have high MAO levels.

Behavior Genetics of Nonhuman Animals   Research on the behavior genetics of human beings is exciting, but it is also marred by controversy and criticism, in part because there is no ethical way to conduct truly controlled breeding experiments with humans. Twins are rarely separated at birth and are never placed randomly in different environments. Likewise, foster children studies are complicated by the tendency to place children with foster families whose socioeconomic status is similar to that of the child's biological parents. Therefore, one cannot completely rule out the possibility that it is only environmental variation, not genetic variation, that contributes to the behavioral differences among human beings.

The great advantage in studying behavior genetics of mice and fruit flies is the freedom to carry out carefully designed experiments that control for environmental influences. For example, if one wants to test the hypothesis that the differences between two species are in part the products of genetic differences, it may be possible to create hybrids and to compare their behavior with that of the parental species. William Dilger performed such an experiment with two closely related parrots [198]. One of these, *Agapornis personata*, builds its nest by gathering bits of bark in its beak and carrying them back to a nest site. The other species, *A. roseicollis*, has the bizarre habit of transporting nest material by tucking it into the feathers on its lower back (Figure 4). By keeping females of one lovebird in cages with males of the other, Dilger produced hybrids that exhibited a remarkable intermediate pattern. These birds would pick up material in their beaks, place it in their flank feathers, remove it again, and repeat the pattern many times before finally flying back to the nest with the paper either held in their beaks or in their feathers. Because the hybrids were reared in the same laboratory setting as the parental generation, the difference in the way they transported nest material can be safely attributed to genetic differences between hybrid and parental individuals.

4 Lovebird, *Agapornis roseicollis,* the species that transports nest material in its feathers. Photograph courtesy of William Dilger.

Notice that Dilger was able to deal with two discrete behaviors, perhaps two different fixed action patterns, whereas behavior geneticists working with human behavior typically must deal with a continuous range of slight variations (such as IQ scores, most of which are clustered between 85 and 115). One of the advantages of studying behavior genetics in nonhuman animals is the possibility of examining such clearly defined elements of behavior.

The calls of frogs differ distinctively from species to species. These signals generally prevent hybridization because females go only to males that produce their species' song. You can, however, cross gray tree frogs and pinewoods tree frogs in the laboratory simply by taking males from the back of a female of their own species and switching them to females of the other species. The females will proceed to lay eggs, which the male fertilizes while gripping his substitute partner. The hybrid male offspring, when mature, sing songs intermediate to those produced by the two parental species (Figure 5). If hybrid females are given a choice between loudspeakers broadcasting the hybrid call versus either of the parental types they go over to the speaker giving the hybrid signal [201]. This is an instructive result because it demonstrates that the gene or genes that influence the development of the male calling song also play a role in the development of the acoustical system and call preferences of females.

Single-Gene Effects

Behavioral studies of hybrids give no clue as to the number of genes involved in the differences between hybrids and their parental species. Other techniques, however, have shown that single allelic differences can cause individuals to differ behaviorally. Many of these demonstrations have involved the geneticist's favorite insect, the fruit fly *Drosophila* [224]. If one

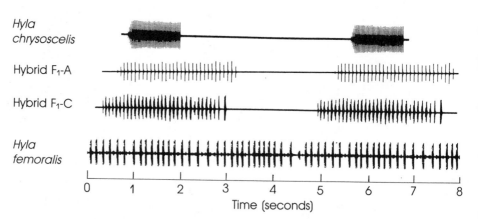

5 **Treefrog calls: parental species and hybrids.** Sonagrams of the calls of male hybrids produced by crossing the grey treefrog (*Hyla chrysocelis*) and the pinewoods treefrog (*Hyla femoralis*) show that the hybrid signals are intermediate to those of the two parental species. *Source:* Doherty and Gerhardt [201].

looks at sufficiently large numbers of fruit flies, particularly those whose parents have been exposed to substances that cause mutations, one can sometimes find behavioral mutants. By performing the appropriate breeding experiments, some behavioral abnormalities have been traced to the alteration of a single gene [448]. The mutant alleles have been named to reflect their behavioral consequences, among them *stuck* (males with the mutant gene fail to dismount after the normal 20 minutes of copulation), *coitus interruptus* (males with this allele disengage after just 10, not 20, minutes of copulating), and *bang-sensitive* (a sudden jolt causes the mutant fly to become paralyzed) [272].

Other mutant genes affect the daily activity pattern of fruit flies, which normally follows a schedule that repeats every 24 hours (Figure 6). The mutant *per°*, however, has no apparent activity rhythm at all. Flies with the *per^s* allele have a shortened activity cycle, one that repeats every 19 hours; flies with *per^L* exhibit a lengthened 29-hour cycle (Figure 6).

Using the techniques of molecular genetics, two research teams have shown that each of the mutant alleles differs from the typical or *wild-type period* gene by just one pair of nucleotides in a chain of DNA composed of more than 3500 such pairs [46, 878]. For the *per°* form of the gene, the single alteration in the DNA molecule results in the production of a protein that contains about 400 amino acids, instead of the 1200 appearing in the

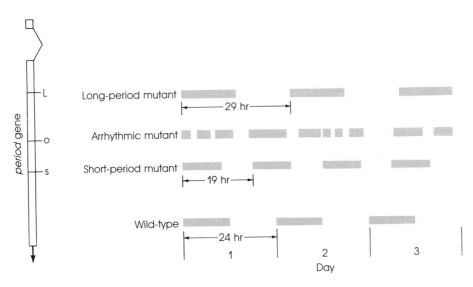

6   **Mutations and behavior.** There are many alleles of the *period* gene of *Drosophila* fruit flies. The sketch on the left shows the strand of DNA that constitutes the *period* gene. Researchers now know precisely where in the DNA molecule different mutations have occurred. Each mutant allele has a characteristic effect on the activity patterns of fruit flies, shown on the right. The normal pattern of wild-type flies appears at the bottom of the diagram.

wild-type protein. The mutation causes the cell's machinery to stop "reading" the gene's information prematurely.

The information contained in the per^s and per^L alleles is used by the fly's cells to make a protein chain with the full 1200 amino acids, 1199 of which are the same as those in the wild-type sequence. But even though each mutant chain is nearly identical to the wild-type protein, they cause dramatic differences in the activity patterns of fruit flies.

Learning Mutants

Many other mutations affect the ability of flies to learn certain things. Fruit flies are not geniuses, but normal individuals do store information about some experiences and use it to modify their behavior. For example, if males encounter fertilized (unreceptive) females and court them without success, this depresses their readiness over the next hour or two to court other females that they meet [721]. The *dunce* mutant, however, is impervious to the effects of courtship failure. Its eagerness to court females is not affected by experience with unreceptive, previously mated individuals.

The same is true for the *don giovanni* mutant, but, unlike *dunce* males, the *don giovanni* mutation does *not* affect learning ability. As it turns out, males with the *don giovanni* allele for some reason cannot induce mated females to release the special chemical that announces their lack of receptivity. Because mated females do not release their aversive scent, *don giovanni* males do not get the signal that conditions them to reduce their courtship activity. If, however, these males are paired with mated females that have just been courted unsuccessfully by normal males, the "rejection" scent is in the air and the *don giovanni* males learn, as shown by their subsequent failure to court females they encounter in the next hour or so.

Thus, the allele *don giovanni* affects male courtship behavior, but not learning ability, whereas *dunce* prevents stable associative learning because the flies almost immediately forget what they have experienced. Flies with still another allele, *amnesiac*, are capable of short-term learning. However, although they do not forget as fast as *dunce* males, they cannot retain what they have learned as long as normal flies [456, 792]. Flies carrying the *amnesiac* allele can, for example, learn to avoid places with a distinctive odor if they are given an electrical shock in these places. But within 45 minutes they are just as likely to go to the shock spot as they were before having been trained to avoid the area. Shocked normal flies stay away from the odor for hours.

The difference between *dunce* and *amnesiac* has been illustrated in an experiment in which males were paired with virgin females in a container filled with quinine, which is an aversive odor to fruit flies. This treatment reduces the responsiveness of normal males to females for a couple of hours, but has no effect at all on *dunce* males. Males with the *amnesiac* allele show a reduction in courtship enthusiasm for a short period, demonstrating that they have formed an association between the unpleasant scent and females, but they soon forget it and are responsive to females long before normal males (Figure 7) [2].

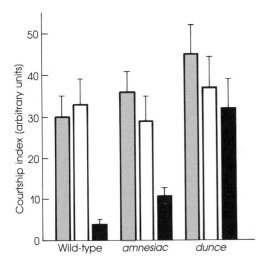

7 **Learning mutants.** Male fruit flies of three genotypes—wild-type, *amnesiac*, and *dunce*—spent 1 hour in a chamber either with a nonreceptive immature virgin female (shaded bars) or with unpleasant smelling quinine (open bars) or with both a nonreceptive female and quinine (black bars). After a short rest, the males were given a chance to court a mature virgin female. Males with the wild-type allele did little courting if they had experienced quinine and females. Males with the *amnesiac* and especially the *dunce* alleles quickly forgot the aversive experience and so did more courting than wild-type males after having interacted with a quinine-smelling female. *Source:* Ackerman and Siegel [2].

Pleiotropy

These experiments establish that olfactory learning is affected by the *amnesiac* mutation, but this is not its only consequence. Part of fruit fly courtship is a "song" that males produce by vibrating their wings rapidly and in a species-specific pattern. If one plays an electronically simulated courtship song to isolated females and then several minutes later permits males to enter the females' container, the "sensitized" females will copulate more quickly than those that have not heard song recordings before interacting with males. The fact that this procedure works for wild-type flies, but not for *amnesiac* females, is an indicator that these mutants have something wrong with their memory storage and retrieval systems. The *amnesiac* females can hear songs perfectly well, but they fail to record this information properly [456].

This example illustrates an important point. An allele may and usually does contribute to the development of more than one characteristic of an individual (Figure 8). A gene does *not* make a trait; typically it codes for an enzyme that catalyzes a particular reaction. For example, the *dunce* allele codes for a variant of the enzyme cyclic AMP phosphodiesterase, a variant that has a remarkable effect on learning.

Because a gene exerts its developmental impact via an enzyme, it may play a role in the development of more than one trait. Thus, females with the *amnesiac* allele cannot retain a learned association based on odor, nor can they be prestimulated by listening to courtship signals. Multiple effects (PLEIOTROPY) are the rule, not the exception, in the expression of a gene. Let's look at another example from fruit flies. Individuals with the allele *Hk* (*Hyperkinetic*) (1) are more active than flies without the allele, (2) exhibit rapid leg-movement when under anesthesia, (3) have a shorter life span, (4) engage in aberrant mating behavior, and (5) jump violently when an object passes overhead [328].

Polygeny

Another point that requires special emphasis is that dramatic single-gene effects are rare. This is because most characteristics, especially behavioral ones, stem from the integrated action of a large number of gene products. The technical term for the involvement of many genes in the development of a single character is POLYGENY (Figure 8). When a single-gene effect does occur, it does not mean that there is one gene for the courtship song of a fruit fly, another for the activity cycle, and still another for life span. Single gene effects on behavior occur because one gene's product plays a key role in a complex developmental process that requires the regulated interaction of dozens or thousands of genes. A mutant allele's effects may be analogous to giving a worker on a long assembly line an improper tool. Although many persons participate in building a car engine, if just one worker omits or damages even a single component of the engine, the performance of the machine may be seriously impaired. Most single-gene effects discovered by behavior genetics are, in fact, deleterious.

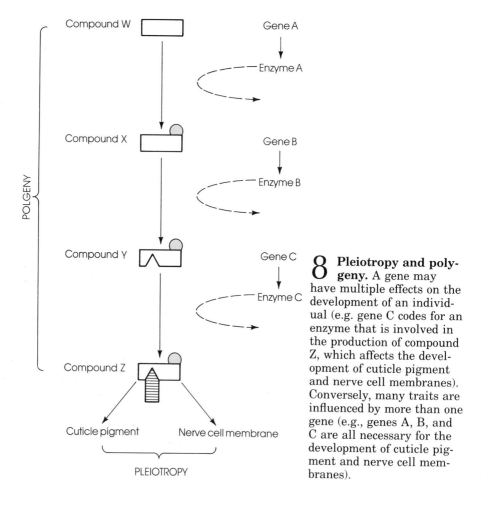

8 Pleiotropy and polygeny. A gene may have multiple effects on the development of an individual (e.g. gene C codes for an enzyme that is involved in the production of compound Z, which affects the development of cuticle pigment and nerve cell membranes). Conversely, many traits are influenced by more than one gene (e.g., genes A, B, and C are all necessary for the development of cuticle pigment and nerve cell membranes).

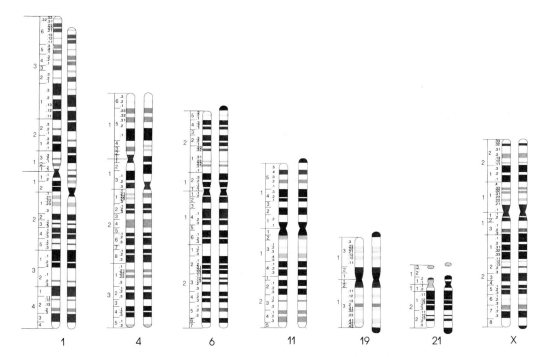

9 **Chimpanzee and human chromosomes** are extremely similar, as shown by the nearly identical banding patterns of a
selected set of chromosomes from the two species. (The human chromosome is the right-hand member of each pair.) *Source:* Yunis
[879].

Some genes exert a major role in development by regulating the activity
of batteries of other (structural) genes whose job it is to produce a set of
enzymes when so instructed. We can deduce their presence in humans
because we differ from chimpanzees even though our chromosomes (Figure
9) [879] and our structural genes are exceedingly similar. If a research
scientist from the University of Mars had only the chromosomal data, he
might conclude that chimps and humans were the same or an exceptionally
closely related pair of species. Most of us would dispute this claim, pointing
perhaps a little too indignantly to the differences in appearance and behavior that distinguish us from chimpanzees. Our differences probably
occur because we surely have different regulatory genes that control when
and where in our developing cells certain enzymes will be produced [428].
Although our cells have many of the same enzymes at their disposal, these
enzymatic tools are used in very different ways to construct the distinctive
nervous systems that underlie our separate behavioral abilities.

Artificial Selection

If a mutant gene, regulatory or structural, produces a variation in behavioral ability that confers a reproductive advantage on its possessor (rare
though this may be), the genetic basis for the change could quickly spread

throughout the population. This has been demonstrated time and again in the ARTIFICIAL SELECTION experiments of behavior geneticists [224], who permit individuals with certain attributes to breed (raising their fitness) while preventing others from breeding (reducing their fitness to 0). By rearing each generation in identical environmental conditions, one can show that any trend toward the exaggeration of a particular ability must stem from a concentration of genetic factors contributing to the trait and not because of a changing environment.

In the process of domestication, artificial selection for behavioral responses has produced homing pigeons, house dogs that are especially gentle with children, watchdogs that bark and bite readily, and so on. Artificial selection can also occur unintentionally, as the U.S. Department of Agriculture demonstrated in its attempts to develop a biological control program against the screwworm fly [120]. (The flies lay their eggs in the wounds of cattle, and the larvae feed on the flesh of the animals.) The USDA raised literally billions of sterilized flies and then released them into natural populations. In theory this would control screwworm because the eggs of a wild female inseminated by a sterile male will not hatch.

In order to rear huge numbers of flies in a small space, many sheets of paper were hung in the rearing rooms to provide more surface area on which the flies could rest. This created a maze of obstacles and evidently acted as a selective force favoring flies that walked to wherever they wanted to go. There were correlated genetic changes in the population, especially with respect to one gene whose enzymatic product affected the flight muscles of the fly. The allele of the gene present in wild flies was rapidly replaced by another form among the reared flies. Although flies with this new allele did well under laboratory conditions, they were incapable of prolonged rapid flight in the field at natural temperatures. This characteristic probably put lab males at a competitive disadvantage with wild flies in the race to find widely scattered receptive females. Thus, accidentally, the USDA selected for a strain of flies that reduced the effectiveness of their control program.

Experiments with Artificial Selection

Geneticists have performed many carefully controlled experiments in which they have deliberately attempted to select for or against a particular behavioral attribute. They usually succeed. For example, a pioneer in behavior genetics, R. C. Tryon, conducted a long series of selection experiments on laboratory rat behavior. His most famous exercise consisted of an attempt to select for maze-running ability [791]. First, he tested a diverse population of rats and divided them on the basis of their maze-running performance into three groups: (1) "maze-bright" rats, that is, those that learned quickly the characteristics of the maze during a number of initial trials and therefore ran through the maze with few detours into dead-end alleys; (2) "maze-dull" rats, that is, those that were the opposite—slow to learn the maze; and (3) an intermediate group, which was discarded.

After identifying his groups, Tryon permitted only bright males to breed with bright females and allowed dull males access only to dull females. The second generation produced by these crosses was tested in the same maze. The progeny of bright rats ran the maze with significantly fewer errors than the offspring of dull rats. (Each was given the same number of learning trials.)

Tryon repeated the procedure and by interbreeding the very best of the bright line and the very dullest of the dull line, he was able to produce brighter and brighter rats as well as duller and duller ones (Figure 10). This result is evidence that among the rats in the original population there were heritable differences related to maze-running, differences that were accentuated by artificial selection.

Tryon called his lines "maze-bright" and "maze-dull," terms that imply that he was selecting for intelligence and stupidity in laboratory rats.

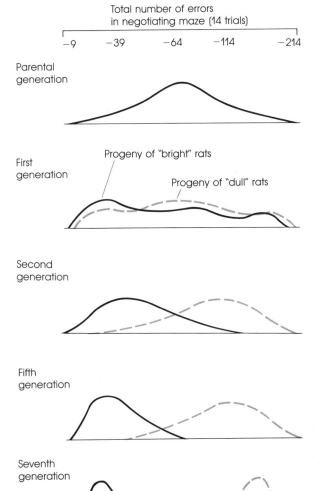

10 **R.C. Tryon's artificial selection experiment.** Through selective breeding of the bright with the bright and the dull with the dull generation after generation, Tryon produced populations of rats that either made many maze errors (entering blind alleys) or made few such errors. After Tryon [791].

However, when the same rats were tested in another kind of maze, which emphasized visual cues (psychologists have invented a wide variety of mazes), the bright rats performed no better than the dull ones [709]. This result suggests that Tryon was not necessarily selecting for degrees of all-purpose intelligence. Instead he was selecting animals with special abilities suited for the particular maze he used in his tests.

The specificity of selection has also been shown in artificial selection experiments with insects. In his field studies of crickets, William Cade discovered that some individuals chirped away for many hours each night, whereas others called for fewer hours, and still others almost never called. The differences between the male crickets might have arisen because some were well-fed and others food-deprived and unable to sing, or some other environmental influence. Or the differences might have stemmed from genetic differences among the crickets.

To find out Cade conducted an artificial selection experiment (Figure 11) that showed there is a genetic component to the behavioral differences between males that call frequently and those that rarely call [124]. By selectively breeding only the most extreme males in each generation, Cade eventually produced populations of crickets that were consistently silent or continuously calling. He could not have done this had there not been some genetic basis to the behavioral differences among males in his original population.

Genes Affect the Physiological Bases of Behavior The allele(s) "for" cricket calling behavior must somehow affect nerve cells or hormones that regulate this aspect of behavior. Unfortunately, nothing is yet known about the connection between genes and the physiological mechanisms that control

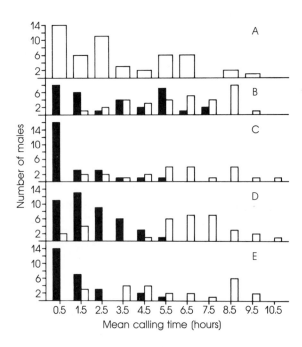

11 Artificial selection experiment with crickets. The differences among male crickets in the hours spent calling each night have a genetic component, as shown by the ability of William Cade to select for calling and noncalling behavior. *Source:* Cade [124].

cricket behavior. The same is true for most behavior genetics studies. However, some studies of *Paramecium* behavior have yielded new insights into the gene–physiology connection [221, 455].

A paramecium is a thoroughly competent swimmer, able to maneuver rapidly through its watery universe. When the anterior end of the animal collides with an obstacle, the cilia that cover its body reverse their beating stroke for a few seconds. The animal backs up during this time before the cilia resume their forward drive stroke and the paramecium starts off again. Usually its orientation has changed somewhat, and it moves past the obstacle that stopped it before. If not, it switches into reverse and repeats its avoidance response (Figure 12).

By screening large numbers of paramecia, observers have found some mutant individuals of various sorts, among them the slow-swimming *sluggish*, the extremely rapid swimming *fast*, and various avoidance mutants, including *paranoiac* (which swims backward for a much longer time than is normal) and *pawn* (which is named for the chess piece that is forbidden to back up at all). The appropriate set of genetic crosses revealed that each of these behavioral mutants was linked to its own single-gene mutation.

The normal avoidance response is mediated by an effect of a tactile stimulus on the electrical charge gradient across the membrane that surrounds the paramecium. It is possible to place minute recording wires within and without an intact paramecium and measure the charge differential across the membrane (the membrane potential). This tiny, but significant, charge difference is caused by the way charged particles (ions) are distributed inside and outside the membrane. When the paramecium touches something with its anterior end, this contact normally changes the permeability of the membrane so that calcium ions in the water enter the cell. As positively charged calcium ions enter, they change the membrane potential of the protozoan. This change causes the cilia to reverse their beating stroke, and the animal backs up. Within the paramecium, physiological systems begin expelling the calcium ions almost at once; and when the original membrane potential is achieved (after a second or two), the cilia start beating the forward stroke and the animal moves forward.

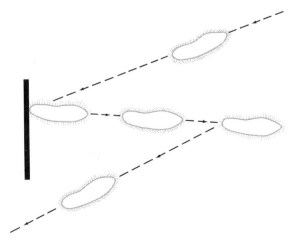

12 **Avoidance response of *Paramecia*.** When a paramecium runs into an obstacle, its cilia reverse their beating stroke for a short time and then resume the forward stroke pattern.

The behavioral effect of a *pawn* mutation arises because the membrane does not respond to a tactile stimulus in the normal fashion. There is no influx of calcium ions and thus no membrane signal to the cilia. The cilia do not reverse their beating stroke, and the creature does not back up.

Genes and Nervous Systems

Comparative physiological studies of mutant and normal individuals are also being pursued with animals that have nervous systems. For example, anesthetized *Drosophila* fruit flies with the *Shaker* allele continue to move their legs rhythmically, unlike normal flies [328]. This allele causes nerve cells to develop in ways that affect their permeability to potassium ions. As a result the cells are highly excitable, and fire more often than normal cells, affecting the mutant fly's locomotion as well as its ability to learn [151].

Another *Drosophila* mutation whose physiological consequences are partly understood is one that affects a particular neuron in a sensory organ found in fruit fly feet. This organ contains five nerve cells that respond to different substances that the fly may touch with its feet. In typical flies one of the five cells reacts exclusively to sugar in liquids, and when this cell fires rapidly, feeding is activated with the result that typical flies consume sugary materials. But flies possessing the mutant allele (s) feed avidly on salty solutions that would repel normal individuals. The mutant flies have an abnormal sugar-sensitive neuron, one that is excited by salts and so provides signals that cause them to consume salt solutions that other flies avoid [32].

One special technique for pinpointing the physiological substrate of a particular behavior comes from the experimental production of GENETIC MOSAIC flies [58, 780], which are composed of two types of genetically different cells (Figure 13). One can create mosaics that consist partly of cells with two X chromosomes and partly of cells with just one X chromosome. Cells with two X chromosomes develop into female tissue; cells with a single X chromosome develop into male tissue. Using sophisticated breeding experiments that produce flies with male and female tissues of different colors, persons can tell which parts of a fly have one or two X chromosomes. When different kinds of mosaic flies interact with one another their behavior can reveal the effects of the genetically different male and female tissues on the reproductive behavior of fruit flies.

A fly whose upper brain lies within a zone of male tissue will pursue any other fly whose posterior abdomen is chromosomally female [326, 385, 721]. The pursuer will also initiate wing-waving courtship even if the rest of the courting fly is composed of two X tissue, so that it has female antennae, eyes, wings, and genitalia. In fact, only half the dorsal posterior brain need be composed of male cells for the mosaic fly to court. A fly may even exhibit male courtship behavior sometimes and be courted by and copulate with other males (if this fly has key male tissues in its brain and female tissues in its abdomen).

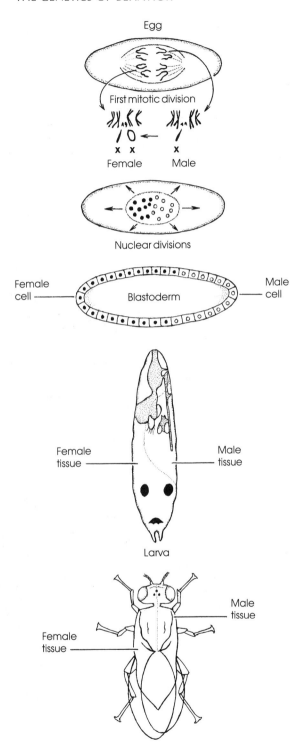

Egg

First mitotic division

x x   x
Female   Male

Nuclear divisions

Female
cell   Blastoderm   Male
cell

Female
tissue   Male
                                                      tissue

Larva

Male
tissue

Female
tissue

Adult

13 **Genetic mosaic fruit fly.** To create a fly whose body is composed of male *and* female cells, experimenters use eggs that have a special XX (female) genotype that during an early cell division gives rise to a cell with only one X chromosome. All the descendants of this cell will be male, while those coming from XX cells will be genetically female. The resulting fly is a mix of male and female cells and tissues. After Hall et al. [328].

These results show that there are no overriding hormonal influences from the female's reproductive tract that regulate behavioral decisions made by the brain. Moreover, female antennae, like male antennae, must be capable of detecting key olfactory cues that receptive females provide to attract males. But the female's brain must operate differently from that of the male, because only brains whose nerve cells carry a single X chromosome will order a fly to move after a female and wave its wings in courtship. Thus, through mosaic fly studies one can locate exactly where one or two X chromosomes have their specific effects on the development of nervous systems.

Genetics and the Evolution of Behavior   We now turn to a beautiful investigation by Stevan Arnold of the physiological consequences of naturally occurring genetic variation, a study that makes evolutionary sense of the genetic, physiological, and behavioral differences between two populations of a garter snake [29]. *Thamnophis elegans* occurs over much of western North America in a wide variety of habitats including both foggy, wet, coastal California and the drier, elevated, inland areas of that state. There are marked differences in the diets of snakes living in the two areas (referred to hereafter as "coastal" and "inland"). Coastal snakes search about in humid areas where they find their major prey, slugs. Their ability to consume these creatures (Figure 14) arouses my bewildered admiration. I made the mistake once of picking up a banana slug, and then had to spend 10 minutes at a kitchen sink scrubbing off the repulsively sticky mucus that the slug applied liberally to my hand. Slugs do not live in inland northern

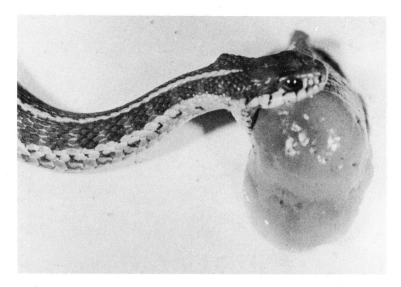

14 **Coastal Californian garter snake** consuming a chunk of banana slug, favored food of these snakes. Photograph by Stevan Arnold.

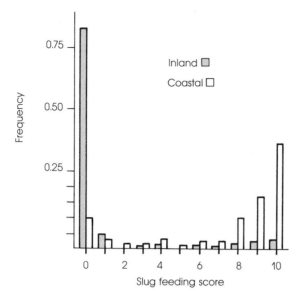

15 **Response of newborn, naive snakes to slug chunks.** Garter snakes from coastal populations tend to have high feeding scores (e.g., a score of 10 indicates that the snake ate a slug cube on each of the 10 days of the experiment). Inland garter snakes rarely eat even one slug chunk (which would yield a score of 1). *Source:* Arnold [29].

California, and not surprisingly the inland snakes find something else to eat—primarily fish and frogs, which they capture in lakes and streams.

But will snakes from inland locations eat slugs if they are given the opportunity? If one presented slugs to adult snakes from the inland population, their failure to eat the prey might mean only that their long feeding experience with fish and frogs had biased them against a strange food. The more interesting experiment would be to give newborn baby snakes from inland and coastal areas a chance to eat slugs, because this test would eliminate the possibility that prior feeding experience had shaped dietary preferences.

Arnold conducted the more interesting experiment. He captured pregnant females from each location and took them to a laboratory. After they gave birth, he placed each young animal in a separate cage away from its littermates and mother to remove this possible environmental influence on its behavior. Some days later he offered each baby snake a chance to eat a small chunk of freshly thawed banana slug by placing it on the floor of the young snake's cage. Naive young coastal snakes usually ate all the slug hors d'oeuvres they received; inland snakes usually did not (Figure 15). In both populations, slug-refusing snakes did not even make contact with the slug food but ignored or avoided it completely.

Arnold took another group of isolated newborn snakes that had never fed on anything and offered them a chance to respond to the *odors* of different prey items. He took advantage of the readiness of newborn snakes to flick their tongues and even attack cotton swabs that have been dipped in fluids from some species of prey (Figure 16). By counting the number of tongue flicks that hit the swab during a 1-minute trial, he measured the relative stimulation provided by different odors to an inexperienced baby snake. (Snakes have an organ in the roof of their mouth that analyzes odors

16 **A tongue-flicking newborn garter snake** senses odors from a cotton swab that has been dipped in slug extract. Photograph by Stevan Arnold.

carried to it by the tongue. When the animal's tongue touches a source of chemicals, it carries some molecules back to the organ, which plays a role in prey recognition.)

Populations of inland and coastal snakes responded about the same to swabs dipped in toad tadpole solution (a prey of both groups) but reacted very differently to swabs daubed with slug scent (Figure 17). Within each group, there was some variation in response; but most inland snakes were unresponsive to slug odor, whereas most coastal snakes responded strongly

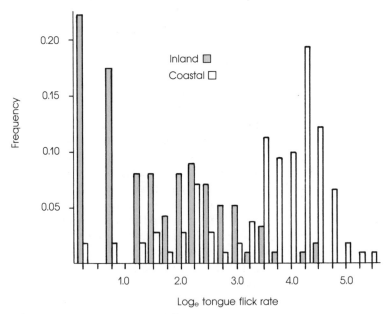

17 **Odor preferences of inland and coastal garter snakes** as measured by the frequency of tongue-flicking in response to cotton swabs dipped in slug extract. Coastal snakes tongue-flicked much more than inland snakes as a rule. *Source:* Arnold [29].

to it. By comparing the tongue flick scores of siblings within each population, Arnold determined that in both populations only about 17 percent of the differences in chemoreceptive responsiveness to slug odor was due to genetic differences among individuals. This result shows that almost all the genetic variation related to reactivity to slug odor has been eliminated within each population. Most coastal snakes have the allele or alleles that enable them to sense slugs and attack them. In contrast, snakes of the inland population have alternative alleles that result in a low rate of tongue flicking and a low attack probability when presented with a slug.

If one crosses snakes from the two populations (Arnold did that, too), there is much more variation in the resulting group of offspring than in either parental population, although the majority of the snakes are slug-refusing. These results confirm that the differences between populations have a strong genetic component, and they also indicate that the allele or alleles that affect development of slug-refusing are dominant to the alternative(s) that promote acceptance of slugs. A reasonable scenario for the evolution of the feeding differences between inland and coastal populations can be invented. Suppose that among the original colonizers of the coastal habitat were a very few individuals that carried the then rare allele(s) for slug-accepting. (There are reasons for thinking that coastal California was inhabited by *T. elegans* more recently than inland western North America.) These slug-eating individuals were able to take advantage of an abundant food resource in the new habitat. If, as a result, their reproductive success was as little as 1 percent higher than that of their slug-rejecting fellows, the coastal population could have reached its present state of divergence from the inland population in less than 10,000 years.

It is easy to imagine why slug-accepting alleles might enjoy an advantage in coastal populations. But why have they been actively selected against (and nearly eliminated) from inland populations? Arnold showed that the alleles that promote slug-eating also enhance the acceptance of aquatic leeches. These blood-sucking animals are absent in coastal California but plentiful in inland lakes. It is possible, but unproved, that snakes that try to eat leeches (which slug-acceptors will do) may be damaged by their "victims." Leeches can live even after they have been swallowed by a garter snake, and if they attach themselves to the wall of the snake's digestive tract, they might injure their consumer seriously. Even if this happens only rarely, it would exert a negative selection pressure on snakes with leech-accepting alleles. Because these alleles also lead to the development of the ability to detect and attack slugs, the ability to eat slugs would be disadvantageous in inland populations.

Thus, the geographical differences in the feeding behavior of the snakes can be explained in terms of their proximate genetic basis (different alleles predominate in the two populations), their proximate physiological basis (the chemoreceptors that detect certain molecules associated with both slugs and leeches are more prevalent in coastal populations), and ultimate ecological basis (inland snakes must contend with potentially dangerous leeches whereas coastal snakes are exposed only to edible slugs). This case study shows how behavior can evolve both at the genetic and physiological level in response to ecological differences that affect the fitness of individuals with different behavioral abilities [29]. Future research in behavior genetics that integrates genetics, physiology, and ecology will lead to a much improved understanding of the interplay between the proximate and ultimate causes of behavior.

SUMMARY

1 The behavioral differences among individuals may be the result of genetic and/or environmental differences among them. Behavior genetics research suggests that numerous behavioral differences among humans have a genetic component. The same is true for many other animal species.

2 To say that a particular allele contributes to the development of a behavioral characteristic is not to say that the trait is "genetically determined." The statement, "There is an allele for IQ score or banana slug eating or time until acceptance of a second mate," is shorthand for the following: "The presence of a particular allele in an individual's genotype provides information for the production of a distinctive protein whose contribution to the chemical reactions within cells may influence the development of the physiological foundation for a behavioral ability."

3 Studies of animals known to differ by just one allele show that behavioral differences can be caused by even a single genetic difference between individuals. Artificial selection experiments confirm that there is a genetic component to most behavioral differences, because almost all such experiments succeed in altering the behavior of the experimental population.

4 Genes have their effects on behavior by altering the physiological foundation of behavior, especially by affecting the development of nerve cells and nervous systems.

5 A study of wild garter snakes in two locations shows that behavior can rapidly evolve under natural selection. The differences in feeding behavior in the two populations arise in part from their genetic differences. The distinctive genetic features of the two populations affect the development of the chemoreceptive system of the snakes, which in turn affects their perception of prey. Individual selection has favored different perceptual and behavioral attributes in the two areas because the ecological pressures in the two areas differ. Ecological differences affect the relative reproductive success of snakes with and without the ability to detect and attack slugs.

SUGGESTED READING

Stevan Arnold's comprehensive study on the genetics, physiology, and ecology of garter snake feeding behavior [29] should be read by everyone interested in behavior genetics. Jeffrey Hall has written a review of the behavior genetics of insects, which covers the large body of work on *Drosophila* fruit flies [327]. The entire field of behavior genetics is reviewed in books by Lee Ehrman and Peter Parsons [224], and by Jeffrey Hall, Ralph Greenspan, and William Harris [328].

DISCUSSION QUESTIONS

1. Two inbred lines of white rats are reared in the same environment, but when tested on maze-running ability, the average scores of the two lines are very different. Why doesn't this hypothetical experiment prove that maze-running ability is *genetically determined*? If we were to run the experiment again, rearing the two groups of rats in an environment different from that used initially—by altering the food given the animals and adding many more objects to their cages—we might find that this time there was no difference between the two groups in their maze-running scores. If there were no maze-running differences between the two lines, would this show that there are no genetic differences between individuals in the two lines? What would the two experiments really demonstrate?

2. Describe how you would conduct an artificial selection experiment on slug-eating behavior using garter snakes taken from a single population only—the inland California population. What evidence exists that indicates that you would be able to produce via artificial selection a line of slug-accepting snakes from this population?

3. Let's say that harmless techniques existed for the exploration of most aspects of the physiology of human behavior. Let's say that a sample of monozygotic and dizygotic twins agreed to permit you to study the physiological basis of their personalities. Where would you begin? What questions would you ask? What would you look for if you were permitted to devise a research program on the relation between genes, physiology, and human personality?

The Development
of Behavior

The preceding chapter established that genetic differences among individuals can contribute to differences in their physiological and behavioral attributes. But the question of precisely how the behavioral characteristics of an individual develop remains to be explored. This is a problem that requires an analysis of the complex interactions between a growing organism, with its genes and developing structures, and the environment in which it is located. Unfortunately, few topics in biology are less well understood than the developmental process, which is staggeringly complicated. The song of an adult male white-crowned sparrow, for example, is the product of the integrated action of hundreds of millions of nerve and muscle cells. Tracing the pathway followed by even one of these cells as it developed from a single fertilized egg cell would be a monumental achievement, and yet even this accomplishment would be the smallest of first steps to an understanding of how bird singing behavior develops. Despite the difficulties of this research, some useful things have been learned about the interactions between genetic systems and the environment during behavioral development. This chapter focuses on how the differences between the sexes, which are often dramatic and usually biologically significant, come about in some species. Key environmental influences on behavioral development will be discusssed and at the same time the resiliency of the process in overcoming some environmental obstacles to normal development will be demonstrated. This apparent paradox—the sensitivity of behavioral development to some experiences *and* its capacity to achieve an adaptive end point in a broad range of environments—will be the focus of the chapter's conclusion.

The Development of Sexual Differences in Behavior

Because white rats (the domesticated laboratory variant of wild Norway rats) breed prolifically in the laboratory, can be kept in small spaces, and have relatively short generation times, they have been used extensively in experimental research in psychology and physiology. Thanks to these characteristics, much is known about all aspects of their biology, including the development of their behavior.

Adult male and female rats behave differently. Females, for example, have a specific copulatory position (Figure 1), which they assume at the proper time in the ovulatory cycle (when mature eggs are ready to be fertilized) if grasped on the flanks by a sexually eager male. Males never assume the female copulatory position; females relatively rarely mount other rats and do not exhibit the full male copulatory performance [47].

Females also differ from males in their parental behavior. Immediately after giving birth, the mother rat remains in very close association with her young for several days, keeping them warm and permitting them to suckle. If her pups are moved, the female will quickly retrieve them. In contrast, male rats are rarely solicitous toward their progeny [673].

These differences between the sexes have a proximate basis in the hormonal makeup of male and female rats [477]. If one removes the ovaries from an adult female, she will no longer copulate unless she receives injections of estrogen and progesterone (two hormones produced by the ovaries). Likewise, removal of the male testes eventually eliminates male copulatory behavior, unless one gives the castrated individual injections of testosterone (the major hormone produced by the testes).

Males and females that have had their hormone-producing reproductive organs removed when they are adult generally cannot be induced to behave like members of the opposite sex. Injecting testosterone into a female whose

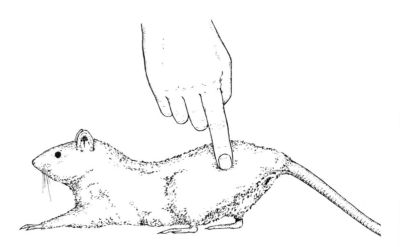

1 **A sexual difference in behavior.** Female rats, but not males, adopt a specific precopulatory position in response to pressure on the flank.

ovaries were removed past puberty does not cause her to copulate like a male. This experimental result suggests that the nervous systems of adult males and females, which are structurally different, differentiate into mature brains that are not capable of responding to hormonal signals of the opposite sex.

The developmental effects of testosterone on the central nervous system take place in a brief CRITICAL PERIOD [514]. If one removes the gonads from a newborn rat and then waits for a week before injecting testosterone into the animal, the hormone will not trigger the events that lead to development of a "male" brain. In the short interval between 15 and 27 days after conception, the brain of the developing rat is sensitive to testosterone in its bloodstream. (A rat is born after a gestation period of 22–23 days.) Certain cells within the young brain have the capacity to bind with testosterone molecules. At this early stage, both male and female brains possess these cells. If testosterone is present, either naturally because the animal is an intact male whose immature testes have begun to produce the substance or unnaturally through experimental injection, the hormone will be drawn into its target cells. Once there, testosterone affects the genetic activity of the target neurons, triggering the production of new enzymes that change biochemical activity within the cells and influence their further development. Thus, testosterone molecules in the bloodstream act as a signal for a distinctive pulse of chemical activity within certain components of the immature brain [544].

The development of testes and the production of testosterone in 16-day-old male fetuses are also correlated with the inhibition of the development of the ovaries. In an embryonic mammal, cells that have the potential to become ovarian or testicular tissue are both present at a very early stage of development. But in a male embryo, the cells that give rise to the female reproductive organs fail to multiply and differentiate in the presence of testosterone. Conversely, in fetal female rats testosterone is not present to initiate the male developmental pattern in the brain and gonads. The progenitor cells of ovaries are free to give rise to millions upon millions of additional cells, which form the mature ovaries. These structures produce estrogen and progesterone in specific patterns related to ovulation, thereby supplying the hormonal basis for adult female reproductive behavior.

Even if the key to the development of male or female behavior can be traced back to the presence or absence of a few testosterone-producing cells in a early-stage embryo, we have not uncovered the reason for their existence in male embryos but not in female ones. The bottom line is that males and females differ genetically from the moment of conception. In mammals, rats included, there are identifiable chromosomes whose presence or absence is critical for the sexual differences between males and females [348]. A male mammal typically has one X and one Y chromosome, whereas a female has two X chromosomes. Maleness arises from the presence of a gene usually carried on the Y chromosome. (Very rarely human beings are formed with only a single X sex chromosome. These individuals develop into females. There are also some rare XXY genotypes, which produce males with testes and other secondary masculine characters.)

The Y chromosome of some mammals carries very little genetic information [348]. Recently, however, a research team using sophisticated molecular genetics techniques located a gene on this chromosome in humans and other mammals that appears critical for the development of male testes [615]. Perhaps in mammals generally a gene of this sort provides information for the production of a key substance in the early embryonic cells of males only. The testis-determining factor (TDF) is then transported to other cells whose receptor molecules bond with the chemical. The receptor–TDF complex may migrate to the nucleus of target cells, where its presence activates previously quiescent genes whose enzyme products initiate changes that result in male gonads, male brains, and male sexual behavior (Figure 2).

The subtlety of the interplay between hormones and embryonic environment has been highlighted by some recent discoveries on how fetuses of laboratory mice influence each other's sexual development [802]. In a litter

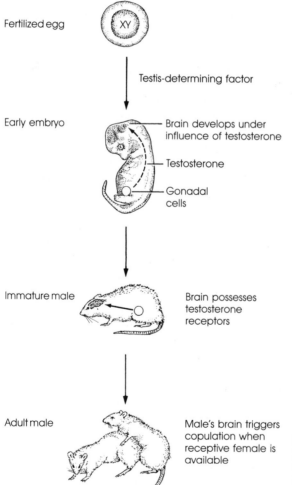

Fertilized egg

XY

Testis-determining factor

Early embryo

Brain develops under influence of testosterone

Testosterone

Gonadal cells

Immature male

Brain possesses testosterone receptors

Adult male

Male's brain triggers copulation when receptive female is available

2 **Development of male copulatory behavior** in the laboratory rat is dependent upon genetic information carried by the Y chromosome, hormones, and environmental factors.

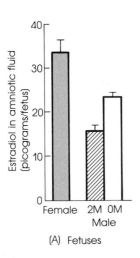

(A) Fetuses

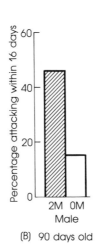

(B) 90 days old

3 **Effect of embryonic influences** on the development of male behavior in the laboratory rat. (A) The levels of estradiol, a female hormone, are higher in the amniotic fluid of fetal 0M males (sandwiched between two sisters) than in 2M males (surrounded by brothers). (B) Ninety-day-old 2M males are more aggressive toward a male intruder than 0M males are. (Mice were tested by placing them with a strange individual for a 10-minute trial on alternate days.) All males had been castrated when newborn and months later given implants of testosterone. *Source:* vom Saal [802].

of mouse embryos some males will by chance be sandwiched between two sisters while others rub shoulders with one male and one female sibling; still others will happen to be between two brothers. As fetuses develop, they release any number of biochemical products that their cells are manufacturing, among them sex hormones. These chemicals make their way into fetal siblings via the amniotic fluid that surrounds the embryos and circulates in their guts. Consequently, a 2M embryo (one between two other males) and a 0M embryo (one between two females) are exposed to different concentrations of male and female sex hormones.

To determine whether these slight differences might have a later impact on the behavior of males and females, Frederick vom Saal and his colleagues delivered mouse pups by means of a Caesarean operation (to document the position of the embryos in the uterus). Males were then castrated and later given hormonal implants, either the male hormone testosterone or the female sex hormones estrogen and progesterone. Mouse 2M males, those whose brains had developed under the influence of a little extra testosterone, behaved more aggressively than 0M males when given replacement testosterone (Figure 3). Apparently the brains of 2M males were *more* masculinized during the embryonic phase than the brains of 0M males, which had experienced the feminizing influence of embryonic estradiol received from their sisters.

Hormones and the Development of Maternal Behavior

The mouse study shows how factors as subtle as the position of a fetus in the uterus can affect behavioral development. And, obviously, the interaction between environment and genotype does not end with the birth of an animal. Consider the development of maternal behavior in rats [673]. As noted already, a female rat treats her newborn pups in a distinctive manner, keeping them close to her body and retrieving them if they are displaced.

These attributes normally first appear when mature females give birth to a litter. But one can induce guarding and retrieval behavior in nonpregnant females, indeed even in sexually immature female rats, by repeatedly exposing the female to 1- to 2-day-old pups taken from other females. If the "sensitization" process is maintained for a week, many females will stop attacking or avoiding strange pups and begin "adopting" them as their own. The sounds and tactile cues provided by neonatal rats stimulate the development of maternal behavior.

The role that hormones play in this process has been studied by removing the ovaries from immature rats [673]. If the treated rats are later given "sensitization" training, they are only one-third as likely to retrieve foster pups as intact females of the same age that have gone through a similar training procedure. If, however, one injects the ovarectomized females with estrogen, they respond to their foster pups in much the same way as sensitized nonpregnant females with ovaries. This experiment establishes that ovarian hormones circulating in the blood of nonpregnant females promote the development of maternal behavior patterns.

Figure 4 summarizes the developmental sequence that leads to normal maternal behavior in the rat. The rat example illustrates the inescapable interdependence of genotype and environment in the development of a trait. In a fertilized egg, a Y chromosome with its relatively few genes initiates a cascading series of events underlying the development of many profound behavioral differences between male and female mammals. Note that the Y chromosome does not code for a complete set of masculine characters. Instead, it carries information that shifts development from one track to another, committing gonadal and neural cells to male properties instead of guiding these cells onto a pathway that produces corresponding tissues with female attributes.

The Development of Singing Behavior in Birds

In most songbirds only the males sing a complex song. A number of developmental psychologists and biologists have explored the development of this behavior in male birds, and their extraordinary discoveries have put new life into the cliché that all behavioral development depends on a complex interaction between genetic and environmental components.

Zebra finches are representative songbirds whose males court females with species-specific vocalizations that differ from the songs of all other finches [397]. Adult females never produce the courtship song in nature. Even if one implants testosterone in an adult female, she will not sing. This result suggests that by the time a zebra finch has become adult its brain has become sexually differentiated, with different abilities for males and females.

The brains of male and female zebra finches exhibit structural differences related to song production [321]. The *song system* of these birds consists of a chain of distinctive neural elements that run from the front of the zebra finch brain to its union with the spinal cord, where it connects to neural pathways to the syrinx (the organ that produces vocalizations). The song system's components are all much larger in males than in females. The developmental basis of these brain differences can be traced back to chro-

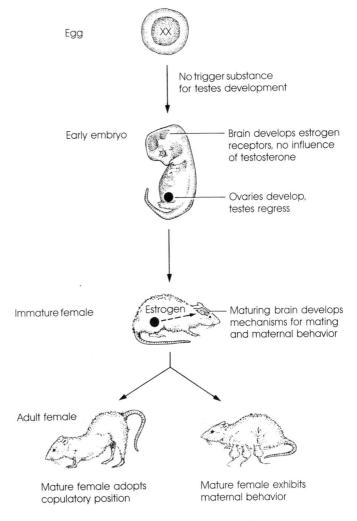

Egg

No trigger substance
for testes development

Early embryo — Brain develops estrogen
receptors, no influence
of testosterone

— Ovaries develop,
testes regress

Immature female   Estrogen — Maturing brain develops
mechanisms for mating
and maternal behavior

Adult female

Mature female adopts   Mature female exhibits
copulatory position   maternal behavior

4 **Development of maternal behavior** in the laboratory rat. In the absence of a chemical signal that activates testes development, the rat develops ovaries and a brain that promotes behavior appropriate for females.

mosomal differences between the sexes. (In birds, it is the female that has the Y chromosome, whereas males are XX.) In the absence of the Y chromosome, an embryonic male's gonads apparently produce estrogen, which acts as the critical signal for the masculinization of the brain. If one implants a newly hatched male with pellets containing estrogen, the development of the male's song system and singing behavior is not affected. But estrogen applied to a very young nestling female enlarges her song system. Estrogen appears to be the ORGANIZING SUBSTANCE that activates the development of a male-type song system by directly or indirectly stimulating growth and differentiation in critical portions of the brain [321].

A female that has had an estrogen implant as a nestling will not sing when she reaches adulthood *unless* she also receives testosterone treatment as an adult. Although her brain has been masculinized, the song system requires male hormones to prime it for song production. Cells within the song system of males (and masculinized females) have receptors that bond with testosterone, when it is present in the blood circulating through the brain.

It is also instructive that the timing of estrogen implants is important in shaping the development of singing ability in females (and thus, by implication, in males as well). Nestling females that receive their experimental hormone implants 4 days after hatching are far less capable songsters later in life [639]. Thus, brain development follows a schedule. In recently hatched birds, brain cells are sensitive to their hormonal environment with large effects coming from estrogen in the bloodstream. With the passage of time, however, cell differentiation takes place, closing off some avenues of behavioral ability.

Learning and Song Development in White-Crowned Sparrows

Zebra finch singing involves genetic information, hormone production, brain development, and the behavioral specialization of the sexes. And the story gets more complicated and interesting still, because certain kinds of experiences are essential if the finch is to sing its species' courtship song. A socially isolated bird that is unable to hear other zebra finches singing will never develop a normal courtship song. So learning must play a key role in the acquisition of this element of the zebra finch's behavioral repertoire, and the same is true for many other songbirds.

A bird species that has contributed greatly to our understanding of how learning influences song development is the white-crowned sparrow. Under natural conditions, young males will spend the first few months of their lives in an environment in which adult males are singing their distinctive territorial song. A young male, however, is usually several months old before he begins to sing himself, and his first efforts are not very promising, a twittering *subsong* that has only a vague similarity to a mature male's *full song*. But over the next 2 months, the song of the juvenile male becomes more and more complete until it closely resembles a normal, full song.

White-crowned sparrows have geographically stable dialects [523] because males learn the details of their songs by listening to other males in the place where they were born [441]. Sometimes males in two populations separated by only a few miles have their own easily recognizable song patterns (Figure 5). These dialects retain their distinctive properties from year to year as the young recruits to the population sing the song of their area.

A series of experiments, many done by Peter Marler, established the role that learning plays in song development by white-crowned sparrows. The first step was to find the nests of wild sparrows from which the eggs were removed to a laboratory incubator. Once the infant birds hatched, they were hand-reared and maintained in soundproof chambers in which it was pos-

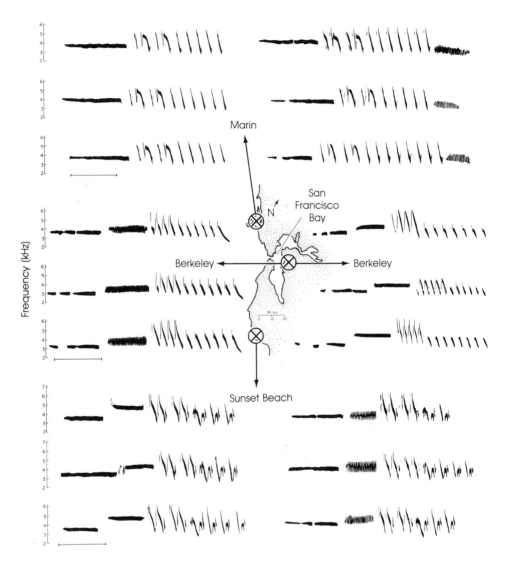

5 **Song dialects in white-crowned sparrows** from Marin, Berkeley, and Sunset Beach, California. Males in each location have their own distinctive song as revealed in these sonagrams of six singers from each location. Courtesy of Peter Marler.

sible to control exactly what sounds reached a young male. One group of isolated sparrows were not permitted to hear the songs of white-crowned sparrows as they matured. Another group of young birds listened to tapes of their species but then were surgically deafened at 5 months of age. Birds that were not permitted to listen to white-crowned sparrows' songs or were not permitted to hear themselves during the subsong phase failed to produce a normal full song (Figure 6).

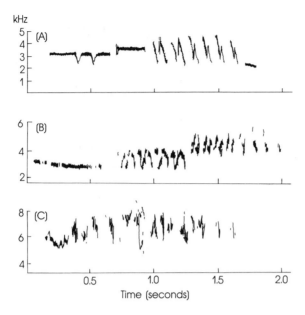

6 **Experience and song development.** Sonagram of songs produced by white-crowned sparrows that (A) were reared under normal conditions, (B) were socially isolated from members of their own species, or (C) were deafened at an early age.

These results and other laboratory experiments yield the following conclusions about what acoustical experiences are necessary for normal song development (Figure 7):

1. A male white-crowned sparrow cannot acquire the song of any bird species from tapes except its own. Isolated birds subjected to tapes of song sparrows or Lincoln's sparrows develop aberrant songs that generally resemble those of birds that have had no songs of any sort played to them.

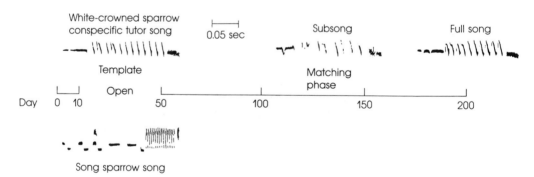

7 **Learning template hypothesis for song learning** in white-crowned sparrows in which young birds are thought to have a critical period from day 10 to day 50 for acquisition of information from a *conspecific* song tutor. Later in life, the bird matches its own subsong with its memory of the tutor's song, and eventually imitates it perfectly, unless deafened. Based on a diagram by Peter Marler.

But if the bird hears tapes with both song sparrow and white-crowned sparrow songs on it, it will develop normal white-crowned sparrow song, incorporating white-crown elements from the tape while ignoring song sparrow sounds [442].

2. The male must hear white-crown song in the period from 10 to 50 days posthatching if it is ever to sing a normal full song. This is a critical period for song learning, for the socially isolated, tape-tutored male.

3. During the interval from 150 to 200 days after hatching, when the young bird begins subsong, it must be able to hear itself sing if it is ever to produce normal full song. If it is deafened before subsong begins, it cannot match its vocal output with the memory of its species song acquired when it was 10–50 days old. As a result, development of the song is unguided, and the end product remains a twittering variable vocalization without species-specific properties.

Marler interpreted the results of his experiments to mean that portions of the white-crown's brain are primed for the acquisition of highly specific kinds of acoustical information during a limited critical or sensitive period in the juvenile male's life [520]. But is song-learning as restricted as it seems to be from the laboratory experiments with isolated, tape-tutored birds? Marler's famous experiments were all done with birds deprived of social interactions with adults. What would happen if acoustical experience were supplemented with social information? Something interesting and revealing happens. If a fledgling white-crown is placed in a cage next to a strawberry finch, a member of a totally different genus of birds, in an aviary where it can hear members of its own species but only interact with the strawberry finch, it will learn the song of its social tutor, the finch (Figure 8)! Furthermore, if you start the experiment when the bird is already 50 days old, past the apparent critical period for song learning, it still learns strawberry finch song. This outcome shows us that the rules of song learning derived from birds exposed only to acoustical stimulation apply only to those conditions [39].

Researchers now believe that in nature young male white-crowned sparrows probably pay special attention to the *social* and acoustical information provided by adult males of their species, with the result that they usually acquire the proper species-specific song (not that of another species). In a few species the most effective social tutor may be the young bird's father, as has been shown for several finches living on the Galapagos islands [559]. For white-crowned sparrows, however, the father's role is overshadowed by that of other adults on neighboring territories.

Moreover, the rules of song learning deduced from laboratory experiments with zebra finches and white-crowned sparrows do not apply to all birds. Marsh wren males that hatch late in the breeding season generally do not learn vocalizations by listening to tapes in their first year. Instead these birds have a sensitive period for learning that extends into the next spring, when they acquire information from males singing at this time. Moreover, they can acquire information (in captivity) from the taped songs

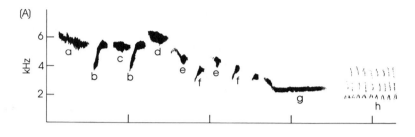

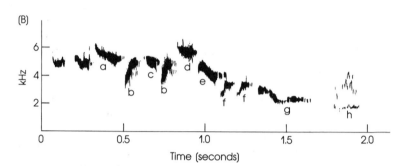

8 **Effect of a social tutor on song development.** A white-crowned sparrow that has been caged next to a strawberry finch will learn the song of its social tutor. (A) The song of the tutor strawberry finch; (B) the song of the sparrow. Letters refer to specific song syllables. Sonagrams courtesy of Luis Baptista.

of other species, elements of which they ultimately incorporate in their songs (although they do show a strong learning bias in favor of song syllables of their own species [451]).

The discovery that many songbirds need to learn things from social and acoustical tutors if they are to complete their songs should not obscure the point that in most species males appear to be predisposed to sing in certain ways. The incomplete songs of isolated swamp sparrows and song sparrows are different and the differences reflect the distinctive features of the full songs of wild, naturally reared birds. For example, wild song sparrow males have a much greater repertoire of song types than swamp sparrows [522]. The song repertoire of acoustically deprived song sparrows is about three times as great as that of isolated swamp sparrows (Figure 9).

Peter Marler and Virginia Sherman suggest that these differences between the species reflect basic developmental differences, with song sparrows primed to acquire a large variety of song types and swamp sparrows innately predisposed to sing a much more limited set of songs. Learning refines the final outcome, providing the information that establishes the precise nature of the repertoire, the duration of key notes, and the organizational features of songs. This information is in a sense superimposed on an underlying neural framework, whose properties may make some kinds of learning far more easy than others for young male sparrows.

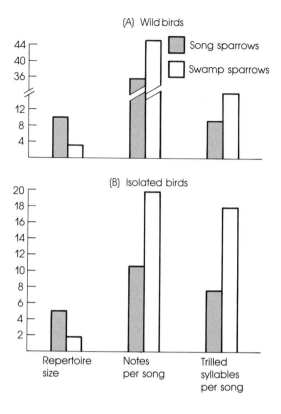

9 **Song sparrow and swamp sparrow song compared.** (A) Wild birds of these two species differ in the average number of songs in their repertoires, the notes per song, and the number of trilled syllables in their songs. (B) The pattern of differences is the same for captive song and swamp sparrows reared in acoustical isolation from others of their species. *Source:* Marler and Sherman [522].

Learning and Song Development in Cowbirds

The full song of a naturally reared song sparrow represents a complex amalgam of instinctive signaling altered adaptively by biased learning. But what about cowbirds? Because this is a parasitic species whose females deposit their eggs in the nests of other species, the young birds are reared by adult red-eyed vireos, yellow warblers, bluebirds, and over 100 other host species but never by members of their own species. If fledgling male cowbirds had brains capable of being influenced by the social and acoustical stimuli offered by their immediate surroundings, one would expect that they would develop vireo or warbler-influenced songs. But they do not. No matter what the foster species, male cowbirds will sing cowbird song, and not vireo or warbler song. Moreover, a young male cowbird growing up in a laboratory cage with no social or acoustical companions of any sort will produce a song that is nearly identical to that uttered by naturally reared cowbirds. The development of their song appears to be resistant to acoustical influences either from other birds or even from potential cowbird tutors.

But two developmental psychologists, Meredith West and Andrew King, have found that even here certain kinds of experience subtly shape the songs of cowbird males. They discovered that adult females were *more* likely to mate when they heard the songs of acoustically and socially isolated males than when they heard songs sung by various naturally reared males.

Female cowbirds signal receptivity upon hearing a song in the standard bird fashion, by crouching and raising the tail, a behavior that is called the precopulatory display. The fact that females find the songs of "experience-deprived" males more arousing than those that have developed under natural conditions suggested to West and King that the full range of experience included some things that slightly modified the song of wild birds (Figure 10).

One thing that naturally reared males learn from experience is their place in a DOMINANCE HIERARCHY. Cowbird males flock together in nature; under these conditions, or when placed together in captivity in an aviary, one male in each group physically dominates the others and will attack or displace subordinates from perches. The other males rank second, third, fourth and so on in their ability to dominate others in their group. The dominant male in a captive flock sings high-potency songs that effectively stimulate females to adopt the precopulatory display. Thus the dominant male monopolizes the sexually receptive females in the aviary. The other males learn from the attacks and threats of more dominant males not to sing the most stimulating song but to modify it, so as to avoid being set upon or chased away. A socially isolated male does not have the chance to acquire the experience that teaches it to restrain its potent song until it becomes dominant, when it is safe to sing the most sexually stimulating song [828]. If naive, socially deprived males are introduced into aviaries containing a flock with an established dominance hierarchy, the isolates will sing their high-potency songs, triggering assaults by the dominant male, assaults so violent that the introduced bird may be killed.

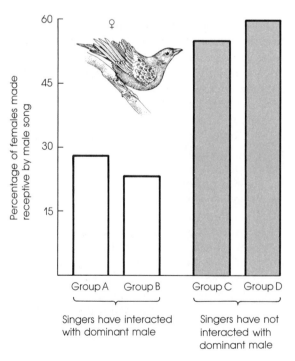

10 Sexually stimulating song. Female cowbirds find the songs of males that have not been reared with other males more sexually stimulating than the songs of males reared under more typical conditions. Males in Groups A and B had at least visual contact with a dominant male in another cage; males in Groups C and D lived in partial or complete social isolation from other cowbirds. *Source:* West et al. [828].

Social learning, male to male, affects the song development of young male cowbirds. Males also modify their songs in response to the reaction they get from females. West and King took advantage of the fact that there are two subspecies of cowbirds, geographically separated from each other and with slightly but consistently different song types. We shall call them subspecies A and B. When an A male sings his song to a B female, she is far less likely to respond with the precopulatory crouch than when she hears the song of her own subspecies. When A males are given opportunities to interact only with B females they gradually change their song toward a pattern characteristic of B males. They do this even though they *never* hear the B subspecies song. Females do not sing, but they do react more positively to some variant songs than to others, and this kind of social feedback is enough to shape the development of cowbird singing behavior (Table 1) [427, 827].

Developmental Resilience and Behavior   One conclusion from studies of song development in birds is that many subtle experiential factors have an effect on the singing behavior of a male. White-crowned sparrows learn a dialect by listening to nearby adult males; cowbirds alter their song in response to the reactions of acoustically silent females. But it is equally impressive that in nature there are so few male cowbirds or white-crowned sparrows that fail to develop a fully functional species-specific song, despite the complexity of song development.

The acquisition of "normal" behavior in the proper sex becomes all the more puzzling when one considers that each white-crowned sparrow and each wild Norway rat has an unique genotype drawn from the sample of genes contained in its parents' bodies. Moreover, no two sparrows (and no two rats) eat exactly the same foods, experience identical climatic conditions, hear the same sounds, or encounter identical social situations. The development of each individual is therefore the result of an interaction between a unique genotype and a unique environment. The capacity of the developmental process to buffer itself against potentially disruptive genetic and experiential influences is called DEVELOPMENTAL HOMEOSTASIS.

TABLE 1

Effects of social experience with females of two subspecies on the song development and potency of male cowbirds of subspecies A

| Female companion | Song development of males | | Percentage of trials in which females of subspecies A give copulatory display to a recording of the male's song |
	Number of notes in phrase 1 of song	Percentage of subspecies B song	
Subspecies A	4.8	0	54
Subspecies B	2.6	64	23

Sources: King and West [427]; West and King [827].

Development Is Reliable

Although there has been debate between ethologists and psychologists [470, 471, 504] on the flexibility of behavioral development, there is little argument that developmental homeostasis promotes the appearance of adaptive traits. Embryologists can predict with some confidence *when* certain structures will first appear in a rat or chick embryo and can chart with precision the sequence of changes in these structures over time. There is developmental variation among individuals, but it is usually modest. In a few cases it has proved possible to follow the fate of identifiable single cells, and this work, too, has shown that development is predictable and repeatable [789].

In the grasshopper *Schistocerca nitens*, for example, certain distinctive cells can be located in essentially every 5-day-old embryo when one removes the limb buds from the thorax of the embryo and examines the exposed tissue with a microscope. These cells occur between two tissue aggregates that give rise eventually to components of the thoracic ganglia (a part of the central nervous system of the insect). By studying a large number of embryonic grasshoppers of different ages, Corey Goodman traced what happens over the course of development to one of these central cells [290]. At a particular stage, the cell begins to fission, a process that produces new cells. These daughter units gradually change their shape, growing into specific neurons with their own predictable schedule of growth, structural change, and alterations of electrical activity. Eventually one particular descendant of the early-stage "mother cell" becomes a fully differentiated motor neuron running between certain cells in the thoracic ganglia to muscle cells in the limb bud destined to become a leg. Figure 11 shows the typical pattern of development of these cells, a pattern that occurs despite considerable genetic diversity in the species and the great environmental differences affecting its members.

It is almost as if the nerve cells of the grasshopper "know where they are supposed to go." This property is not unique to insect neurons; researchers studying marine mollusks [538], amphibians, birds, and mammals [197] have all documented the impressive capacity of neurons to migrate to and make appropriate connections with the proper target tissues. How this happens is still largely mysterious, although it may depend in some cases on the distinctive chemistry of the growing point of the nerve cell, which enables it to identify appropriate guidepost cells along the way to an ultimate target cell [57].

Despite the dynamic interaction between genetic information and environmental influences in cell development, the regularity with which functionally effective cells, tissues, and whole bodies develop shows that some developmental outcomes are far more probable than others. Perhaps regulatory genes activate different elements within the genome to cope with different problems confronting a developing cell [44]. Should a mutant gene's product interfere with the manufacture or degradation of a material, a regulatory mechanism might trigger other genes to provide substitute enzymes for the task. Should the animal find itself in an environment deficient in a key substance, a regulatory gene might activate a combination of genes whose products would lead to the synthesis of the missing com-

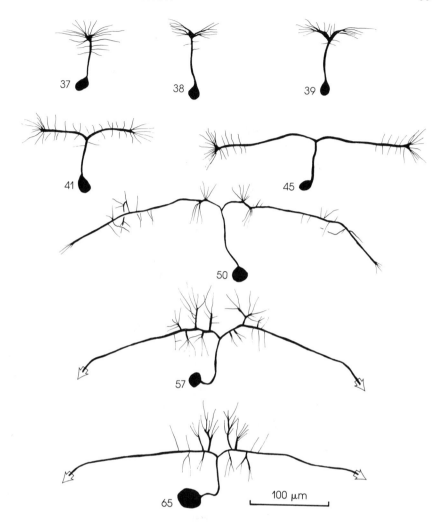

11 **Developmental pattern of a grasshopper nerve cell.**
One identifiable cell is shown as it grows in size and complexity from its early embryonic stages (low numbers) to its more mature stages (high numbers). *Source:* Goodman [290].

pound. By adjusting the biochemical responses of its cells to the particular conditions encountered, the organism would have the flexibility to direct its future development toward the same conclusion in a range of different environments.

Backup systems that enable the developmental process to contend with varied conditions have been discovered in some animals. Alain Ghysen performed a surgical maneuver on mature larvae of the fruit fly *Drosophila* [280]. Ghysen misdirected a portion of an incompletely developed nerve cell into a region of the fly's thoracic ganglion where it normally would not occur. The misrouted nerve cell grew into the ganglion at the "wrong" spot, but it then grew a much longer projection than usual, a projection that made its way through the "foreign" region of the central nervous system until it reached its usual target point. Then it stopped growing and made the normal connections with the cells in this area.

Behavioral Development in Abnormal Conditions

The ability of a nerve cell of a fruit fly to compensate for its physical displacement is a prime example of developmental homeostasis, in which a structure needed for normal behavior develops in an abnormal (to say the least) environment. This speaks to the resilience of the developmental process, a phenomenon well documented in behavioral studies. For example, social deprivation does not always lead to behavioral abnormalities. Male crickets that have spent all their lives isolated from their fellows will sing a normal species-specific song despite their severely restricted social and acoustical experiences [56], and the same is true for ring doves that have been deafened at 5 days of age [596]. European squirrels reared in isolation in a cage with a bare hard floor will nevertheless perform acorn-burying movements, when first given a supply of acorns, using the same actions that wild animals use [225]. Baby rats that have been separated from their mother and sibs when less than 2 days old and fed artificially for 3 weeks still approach scents from the anal excreta of female rats, just as naturally reared rat pups will when temporarily separated from their mother [270]. Captive, hand-reared female cowbirds that have never heard a male cowbird sing nevertheless will adopt the appropriate precopulatory pose when they hear this song for the first time if they have mature eggs to be fertilized [426]. These examples illustrate that behavioral development sometimes is uninfluenced by experiences that might be thought essential for the full expression of a trait.

Two case studies will reinforce this point by showing that even a bare minimum of critical experience can suffice to trigger normal or near normal visual development in cats and social behavior in rhesus monkeys. There is a critical period early in a kitten's life when its brain cells must receive patterned visual stimulation if they are to develop normally. By stitching shut the eyelids on one eye of a newborn cat, researchers permit only diffuse faint light to enter the shut eye; this prevents receptor cells in this eye from sending normal messages to the parts of the brain that process information from the eye. If after the first 3 months the sutures are removed and the kitten is able to use the previously blocked eye, it will never be able to see normally with this eye. The brain cells that analyze signals from this eye have passed the period when stimulation from discrete images in the environment can affect their development [576].

But one can delay the critical period for brain cell development by rearing kittens in complete darkness. In one experiment, kittens were kept in darkness with both eyes open but unstimulated for the first 4 months of their life. The animals then had one eye sewed shut before they were transferred to a room with light. When the sutures were removed 3 months later, the months of monocular vision in light had caused the same kind of abnormal development that occurs when the suturing is performed on newborn kittens (Figure 12). In the absence of any visual stimulation for the first 4 months, the brain cells of light-deprived kittens had put themselves

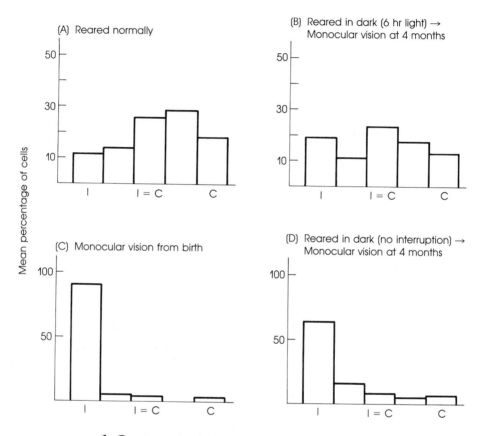

12 **Developmental homeostasis in kitten vision.** (A) The distribution of cell types in the visual cortex of a kitten that has been reared with normal visual experience. (B) Animals that are exposed to light for only a few hours during 4 months of rearing in the dark and then subjected to closing of one eye for an additional period nevertheless develop a battery of cell types the same as that of kittens reared normally. (C) Kittens with one eye sealed for a period of months after birth develop a much different set of cells, as do (D) kittens that are reared in the dark for 4 months, then subjected to experimental deprivation of the vision from one eye. Cell types: I, Cell only responds to stimulation from the eye on the opposite side of the head (e.g., a cell in the left visual cortex that reacts only to stimulation of the right eye); I = C, Cell responds equally to stimulation of both eyes. C, Cell responds primarily to stimulation of the eye on the same side of the head (e.g., a cell in the left visual cortex that reacts only to stimulation of the left eye). *Source:* Mower [576].

on hold, so to speak. They remained capable of reacting to light stimulation when the animals were 4 months old, when light was finally made available and the developmental sequence could begin.

The interesting question then becomes, How much experience with light is needed to set in motion the normal developmental sequence? Suppose that instead of keeping young kittens in unbroken darkness for 4 months, the experimenter interrupts the treatment midway with one block of "lights on" for 6 hours, at which time the kittens are able to look about their laboratory home *with both eyes open*. Then the lights go off again until the kittens are 4 months old, and at that time one eye is sewed shut before the animals are exposed to light for 3 months of monocular vision. This treatment has no effect on the development of the cortical brain cells that monitor inputs from the eye that has been sealed (Figure 12) [576]. Just 6 hours of light was enough to trigger the start of normal postbirth development of the cortical cells associated with vision, and once underway, the process proceeded to completion without any additional patterned visual stimulation. This result is a powerful example of developmental homeostasis acting on a key physiological foundation for behavior.

Developmental Homeostasis
and Social Behavior

The classic experiments of Margaret and Harry Harlow on the development of social behavior [343, 344] involved separation of the infant rhesus from its mother shortly after birth. The baby was placed in a cage with an

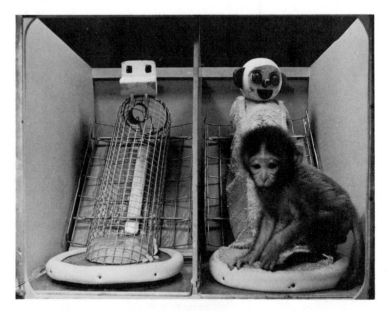

13 **Social deprivation experiment.** The isolated baby rhesus monkey prefers the terry cloth as its surrogate mother to the wire cylinder surrogate, although only the wire surrogate can provide milk. Photograph courtesy of Harry Harlow.

artificial "surrogate" mother (Figure 13), which might be a wire cylinder or terry cloth figure with a bottle from which the baby was able to nurse. The young rhesus gained weight normally and developed physically in the same way that nonisolated rhesus infants do. However, it soon began to spend its days crouched in a corner, rocking back and forth, biting itself. If confronted with a strange object or another monkey, the isolated baby withdrew in apparent terror.

The maternal deprivation experiments demonstrated that the experience of interacting with a mother provided critical experience for the full social development of rhesus monkeys. But even when young rhesus monkeys were reared alone with their mothers, they did not develop truly normal sexual, play, and aggressive behavior. When they encountered animals their own age later in life, these exclusively mother-reared monkeys were likely to react with excessive fear or inappropriate violence. Early contact and interactions with peers are essential ingredients for normal behavioral development in this species.

But how much social experience is needed to promote development of normal social behavior? The surprising answer is that otherwise completely isolated infants developed essentially typical social behavior if they were introduced into a cage with three other infants for just *15 minutes* each day [343]. At first, the young rhesus monkeys simply clung to one another (Figure 14), but later they began to play. In their natural habitat, rhesus babies start to play when they are about 1 month old, and by 6 months they spend practically every waking moment in the company of their peers

14 **Socially isolated rhesus monkey infants** that are permitted to interact with other social isolates for short periods each day at first cling to each other during the contact period. Photograph courtesy of Harry Harlow.

[739]. Yet the experimentally deprived monkeys required a mere fraction of what would be normal social experience in order to develop critical social traits.

The compensatory resilience of the developmental system of this primate is further illustrated by more recent studies on the effects of early experience. William Mason devised an experiment in which socially isolated monkey infants were reared with a *mobile* mother surrogate (a wheeled plastic baby horse with a rug saddle) that was moved several times a day from one spot to another in the infant's cage [528]. When the mobile-mother infants were introduced to other monkeys for the first time at age 14 months and for a second time at 4–5 years of age, their social behavior was fairly normal, so much so that several individuals even copulated successfully with their new social companions.

Mason also reared some rhesus monkeys with mongrel dogs as surrogate mothers (Figure 15). This mother substitute is far more active and interactive than a plastic horse. Infant rhesus monkeys quickly adopt a dog as an attachment figure and through their social interactions with this being come to develop close-to-normal social behavior, despite having an extremely atypical "mother" [529].

Mason tested the intellectual development of the monkeys reared with

15 **Rhesus infant with a dog as a surrogate mother.** Rhesus monkeys reared with these companions develop nearly normal social behavior. Photograph by W. A. Mason.

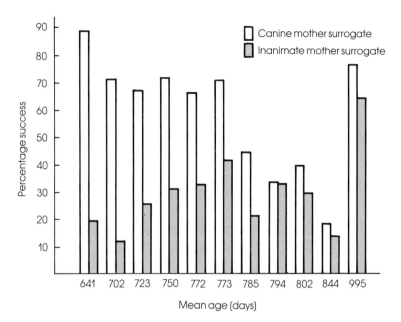

16 **Effect of rearing experience on learning ability.** At first, young rhesus monkeys that have been reared by an inanimate mother surrogate do poorly in learning tests, but the deprived group quickly catches up with rhesus monkeys reared with dogs as mother substitutes. *Source:* Mason [529].

and without dog mothers by presenting the monkeys with an opportunity to receive a food reward through performance of a simple task, such as pulling or pushing at a novel object. The group with canine companions initially did much better at this learning task (Figure 16), but monkeys with inanimate mother substitutes caught up with the other group over a period of about a year. The recovery of the problem-solving ability of the monkeys with inanimate mother surrogates is further evidence for developmental homeostasis.

Developmental Homeostasis
and Human Behavior

Naturally, one wonders about the relevance of these studies for human beings. Many people believe that the experiences very young children have greatly affect the development of their intelligence, personality, and other attributes. But there is growing evidence that the development of human behavior is also buffered to some extent against some environmental shortfalls, even if these occur early in life [211]. Consider, for example, the results of a study of the mental performance of a group of Dutch teenagers who were born or conceived at a time when their mothers were being starved as a result of the Nazi transport embargo during the winter of 1944–1945 [742]. The embargo prevented food from reaching the large

Dutch cities during this time. Deaths from starvation were common, and many persons lost one-fourth of their body weight. For most of the famine period, the average caloric intake was about 750 calories per day, although at times it fell lower still (1500 is the absolute minimum needed to sustain an individual on a long-term basis [877]). Thus, pregnant women living in

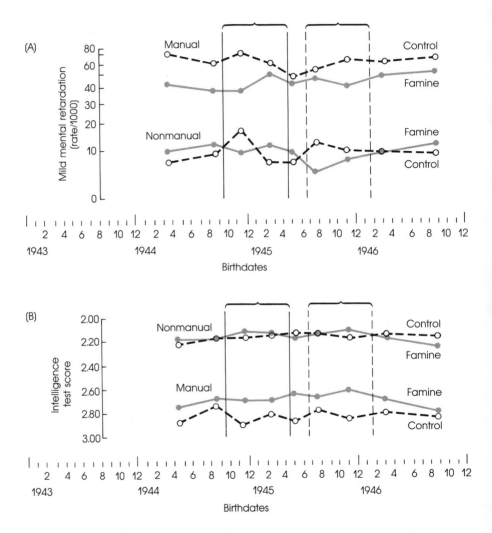

17 **Developmental homeostasis in humans.** Maternal starvation has slight effects on intellectual development in humans judging from the evidence on (A) rates of mild mental retardation and (B) the intelligence test scores of 19-year-old Dutch men. Individuals who were born or conceived under famine conditions had the same rates of retardation and test scores as other men who were born or conceived at the same time to rural mothers who were not starving. All the subjects of the study were placed into one of two groups, depending on the occupation (manual or nonmanual) of their father. *Source:* Stein [742].

cities were subjected to famine conditions for months. In contrast, rural women were less dependent on food transported to them and their babies weighed much more than infants born or conceived at the same time to women living in the cities.

One commonly reads that poor nutrition during pregnancy will inevitably result in intellectual damage to the offspring because this is a critical time for brain development. However, the famine babies did not exhibit a higher incidence of mental retardation at age 19 than those born or conceived at the same time in nonfamine areas (Figure 17). Nor did these deprived babies score more poorly than relatively well nourished infants on the Dutch intelligence test administered to men of draft age. Despite having had a lower weight at birth, the children suffered no permanent intellectual damage [742]. These data indicate that our developmental programs have the capacity to overcome some deficient prenatal environments.

I am not saying that humans are impervious to all environmental influences or that it is a good idea for pregnant women to starve. However, our developmental systems appear remarkably resilient. Jerome Kagan, a child psychologist, went to Guatemala, expecting (according to his own account) to find that the child-rearing techniques practiced in rural Guatemala would permanently stunt the intellectual growth of children. He knew that villagers there provided little social stimulation for their babies during the first year of life. Despite this (and Kagan says that if he had seen North American children treated similarly he would have called the police), by the age of 11 the Guatemalan children scored as well on his intellectual development tests as North Americans of the same age [417].

The Adaptive Value
of Developmental Homeostasis

We have seen that the developmental process is a dynamic interaction between genetic information and the environment in which the individual exists. But not all developmental outcomes are equally likely. We can visualize the course of development in a manner suggested by C. H. Waddington with a fertilized egg or a developing tissue represented as a ball at the top of an inclined plane (Figure 18) [807]. As time passes, the ball rolls down the landscape, which is not a perfectly level slope but is highly contoured. The position and shape of the valleys constrains the path followed by the ball and represents the regulatory forces that restrict development. Thus, the presence of testosterone in the environment of brain cells in the very young rat acts in concert with a genetic switch mechanism to channel the development of these cells down a valley leading toward an end point that consists of male gonads and male brain tissue. Further influences along the way will determine which of various likely pathways (valleys) the growing tissue will travel along. But only very extreme environmental or genetic forces can cause the descending ball (growing tissue) to depart from a valley once it has become committed to a particular path.

We can extend Waddington's metaphor further to describe the ability of individuals to compensate for deficits in their background by suggesting that many of the valleys on the developmental landscape must come together at various points (i.e., there are many routes to the same end point).

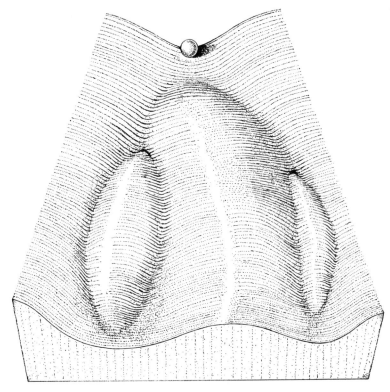

18 **Developmental "landscape."** The diagram illustrates the guiding constraints on the course of development of a cell, tissue, or individual—constraints that steer development toward an adaptive end point. *Source:* Waddington [807].

Without getting too carried away with this metaphor, it helps remind us that a developing organism is not a passive billiard ball capable of responding equally to every possible environmental influence or genotypic peculiarity. Such a system would often lead to reproductive disaster. Individuals might fail to develop critically useful traits as the result of a transitory environmental deficit, such as a temporary shortage of food or lack of social stimulation. The genes of these developmentally "sensitive" individuals would surely be less likely to survive than those that had the capacity to overcome obstacles to normal, fitness-elevating development. Although much remains to be learned about the proximate basis of developmental homeostasis, there can be little doubt that it is a widespread phenomenon and that in an ultimate, evolutionary sense individuals benefit by being able to withstand disruptive influences on their behavioral development.

SUMMARY

1 The development of any trait is an interactive process involving the genotype in a fertilized egg and the environment of the developing organism.

For example, the sexual differences between males and females of some birds and mammals have their roots in relatively small chromosomal (genetic) differences between males and females. These lead to differences in the hormones produced by embryonic gonadal tissues. The presence of a key hormone acts as an environmental trigger for other cells, altering their genetic and biochemical activity with cascading effects throughout the body. The result is a spectrum of physiological and behavioral differences between males and females.

2 The environment of a developing organism consists not only of the metabolic products of its cells and the food materials it receives but also of its sensory and social experiences. All of these factors can potentially act as cues (sometimes only if they occur during a restricted critical period) that have long-term developmental consequences (as in the acquisition of bird song in the white-crowned sparrow).

3 Despite the great variation in the genetic and environmental influences operating on the individuals of a species, most animals develop into functionally competent, reproductively capable creatures. Regulatory mechanisms must make certain developmental outcomes more likely than others and confer upon the developing organism a certain protection against environmental or genetic perturbations that might prevent the development of adaptive abilities.

4 The channeled or buffered aspect of development (developmental homeostasis) is seen in (1) the constancy with which individuals of a species pass through certain species-specific stages and (2) the development of normal physiological and behavioral characters in animals placed experimentally in highly abnormal environments. The resilience of the developmental process helps individuals develop the key characteristics that promote reproductive success.

SUGGESTED READING

The March 20, 1981 issue of *Science* magazine has a collection of superb review articles [581] on the relation between genes, hormones, and sexual differentiation in animals. The work by Peter Marler [520, 522], Luis Baptista and Lewis Petrinovich [39, 40], and Meredith West and Andrew King [427, 827] on song learning by sparrows and cowbirds has greatly improved our understanding of the role of environment and heredity in the development of behavior. Mazukazu Konishi's review of song learning research is comprehensive and helpful [442].

DISCUSSION QUESTIONS

1. Many mammals communicate with species-specific odors. For example, tigers mark plants in their territories with special secretions from various glands [89]. Imagine that you wished to explore the possible experiential effects on the development of this kind of social signaling. Knowing

what you know about bird song acquisition, develop at least two hypotheses on the proximate role experience might play in the development of scent marking and then design experiments on young tigers that would test your hypotheses.

2. White-crowned sparrow song learning is influenced by both social and acoustical experiences. How would you test the hypothesis that at certain stages of the developmental process (particularly the 10-to-50-day period after hatching) white-crowned males were more susceptible to specific kinds of *social* experience than at other stages?

3. One can talk about developmental homeostasis as resulting from "constraints" on the developmental process, limitations that steer development toward a particular outcome. Speaking of such a constraint may imply that the organism would have a wider range of abilities, more behavioral options, if developmental homeostasis did not exist. Explain why the removal of the "constraints" inherent in developmental homeostasis might *not* have the effect of widening the options of an individual.

4. What prediction would you make about the development and size of song control systems in those bird species that perform complex duets, an activity that requires as much singing ability in females as males? See [95] after constructing your prediction.

5

Nerve Cells and Behavior

The three previous chapters on the proximate causes of behavior presented the proposition that genes influence behavior *indirectly* by affecting the development of neural and endocrine systems within individuals. We have already noted the profound behavioral effects of hormones, which direct the sexual differentiation of the brain and prime mature animals to respond to certain environmental cues. This chapter will focus on how nerve cells enable animals to respond to what they perceive in the world around them. Nerve cells are highly specialized for the detection of certain events in an animal's internal and external environment, and they order rapid responses to this information. Although all nerve cells share some fundamental properties, what they do varies enormously both within individuals and across species, thanks to great differences in their structure and function. This chapter contains some examples of this diversity, showing how nerve cells permit their owners to perform biologically useful activities. These examples may persuade the reader that our own perception of and response to our environment are not necessarily identical to that of animals generally, but that each species has a unique nervous system with adaptive specializations, biases, distortions, and abilities. The chapter concludes with a section on how animals navigate from place to place, a fascinating ability whose neural basis is still largely mysterious, therefore demonstrating how much we still have to learn about nerve cells and behavior.

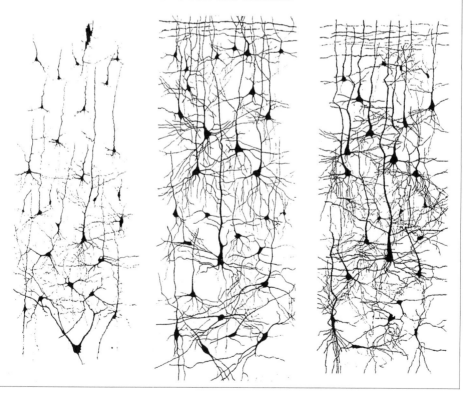

How Do Moths   Many of the principles of this chapter will be illustrated with
Evade Bats?   an analysis of how the nervous system of certain noctuid moths
helps them avoid a deadly enemy—nocturnal insectivorous bats—and
thereby improves the adults' chances of surviving long enough to reproduce.
On a summer evening, one can sometimes watch bats hunting moths over
open grassy areas. As Kenneth Roeder notes, all that is required is "a
minimum amount of illumination, perhaps a 100-watt bulb with a reflector,
and a fair amount of patience and mosquito repellent" [669]. A patient
observer will sometimes see a bat catch a moth in its tail membrane and
fly off with his catch. But if he or she were acute enough, the bat watcher
would also see some flying moths turn abruptly even before a bat came
rushing into view. Moreover, the observer might see moths dive or cartwheel out of the grasp of an approaching bat. These observations indicate
that some moths have the ability to detect a bat at a distance and can make
themselves difficult to capture.

How do they do this? It is hard to believe that moths can see bats far off
at night, and there are for human observers no other cues that could alert
the moth to approaching danger. But this is because we cannot hear high-
frequency vocalizations, which bats produce in abundance as they fly along.
These high pitched sounds are, however, precisely what moths can hear.

Why do bats vocalize as they cruise the night sky? In the 1950s, Donald
Griffin proposed that bats employ a form of sonar, that is, they emit pulses
of high-frequency sound and then listen for the weak echoes reflected back
from objects in their flight path [310]. He tested this hypothesis by placing
the little brown bat, a common New England species, in a room with wire
obstructions strung from the ceiling to the floor [311]. He released fruit
flies in the room, and some bats cooperated by flying about, uttering their
cries and gobbling flies. Griffin then proved that a bat's calls were critical
for its navigational abilities by turning on a machine that generated high-
frequency sounds above 20,000 hertz, in the range of those produced by the
predators themselves. As soon as these extraneous sounds began to bombard
them, flying bats began to collide with obstacles and crash to the floor,
where they remained until the jamming device was turned off. In contrast,
noisy low-frequency sounds of 1000 to 15,000 hertz had no effect, because
these sounds did not mask the high-frequency echoes that the bats require
if they are to fly safely and find food in the dark.

The bat's reliance on a sonar system has favored individual moths that
can detect the pulses of sound produced by bats navigating at night [670].
The detection device consists of a pair of ears, one on each side of the thorax
(Figure 1). A moth ear has a tympanic membrane on the outside of its body;
to the membrane are attached two sensory receptor cells, the A1 and A2
fibers. When high-frequency airborne vibrations strike the moth, they may
cause the tympanic membrane to vibrate. The mechanical energy in these
vibrations reaches the receptors and may induce them to respond. The
response is manifested in a change in the permeability of the membrane of
the receptor cell to sodium ions. These positively charged ions enter the
cell at a point near the tympanum, altering the charge differential across
the neighboring portions of the receptor membrane. Sodium ions enter at

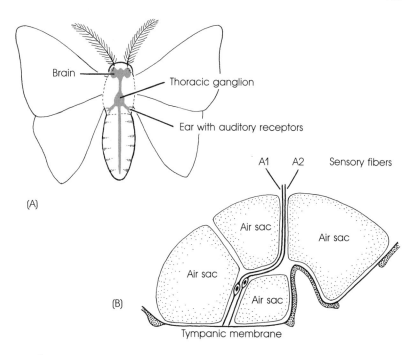

1 **Ears of a noctuid moth.** (A) The location of the ear and (B) its design, which features two sensory fibers (A1, A2) linked to a tympanic membrane. After Roeder [669].

these sites and repeat the effect, causing changes in membrane permeability to sweep around the cell body. Depending upon the intensity of the activating stimulus, the change in cell permeability to sodium ions may then affect the axonal membrane sufficiently to induce the cell to fire (Figure 2). A neural message, or ACTION POTENTIAL, is a brief, standardized change in membrane permeability to sodium ions that travels the length of the axon to the point of near contact (the SYNAPSE) with the next cell in the network. Nerve cells (or NEURONS) communicate with one another in a number of ways. A common method is for the arrival of an action potential at a synapse to cause the release by one cell of a neurotransmitter, which diffuses across the synapse to the neighboring cell(s). Neurotransmitters may affect the membrane permeability of the next link in the network in ways that increase or decrease the probability that the cell will produce its own action potential(s). (See Camhi [127] for a fuller description of the elements of neural action and interaction.)

The cells that relay receptor information to the brain (or its equivalent) are called SENSORY INTERNEURONS. Their messages can change the activity of other cells in the central nervous system, which in the moth consists of the aggregates of neurons in the thoracic ganglia and other ganglia in the head. Ganglion cells can be thought of as decoders, analyzing sensory inputs and making "decisions" about which reactions to order.

Certain patterns of activity in the thoracic ganglia affect MOTOR INTER-

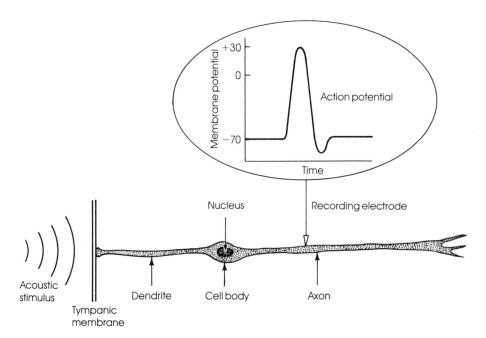

2 **Structure of a nerve cell.** The electrical activity of this acoustical receptor depends first on the effect of acoustic stimuli on the dendrite. Changes in the electrical charge differential of the dendrite's membrane can, if sufficiently great, trigger an action potential that begins near the cell body and travels along the axon of the receptor toward the next cell in the network that processes information about acoustic stimuli.

NEURONS, whose action potentials in turn reach nerve cells (motor neurons) that are connected with the wing muscles of the moth. When a motor neuron fires, the neurotransmitter it releases at the synapse with a muscle fiber induces complementary changes in membrane permeability in muscle cells. These changes regulate the contraction or relaxation of the muscle, with consequent effects on the movements of the wings and the moth's behavior.

How a Moth Uses Its Nerve Cells

The nerve cells of noctuid moths operate in basically the same way as neurons of other animals. For all species with nervous systems, an action is the product of an integrated series of changes in cell chemistry, initiated by receptor cells and carried on by sensory interneurons, brain cells, motor interneurons and motor cells, and muscles (Figure 3). Because these changes occur with remarkable rapidity, an individual can react to changing stimuli in its environment in fractions of a second.

Although there is nothing extraordinary about the design of noctuid moth neurons, they perform very special tasks for their owners. For example, A1 and A2 receptors gather critical information about bats. We know this thanks to the simple but elegant experiments of Kenneth Roeder

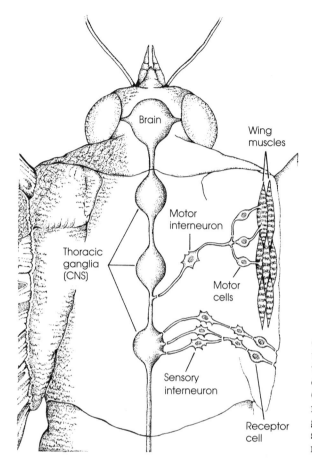

Brain

Wing muscles

Motor interneuron

Thoracic ganglia (CNS)

Motor cells

Sensory interneuron

Receptor cell

3 **Neural network of a moth.** This schematic diagram focuses on receptors in the ear that relay their information to the central nervous system (CNS) via sensory interneurons. Cells in the thoracic ganglia communicate decisions to wing muscles via motor interneurons.

[669, 670]. He attached recording electrodes to each sensory fiber in a living, but restrained, moth and projected a variety of sounds at the ear. The electrical activity that resulted was relayed to an oscilloscope, which can convert action potentials into a visual pattern on a screen and on a paper reel. The record of activity of the two receptors in response to different kinds of acoustical stimulation revealed the following points (Figure 4):

1. The A1 cell is sensitive to low-intensity sounds. The other receptor is not and only begins to produce action potentials when a sound is loud.
2. As sounds increase in intensity, the A1 neuron fires more often and with a shorter delay between arrival of the stimulus at the tympanum and onset of the first action potential.
3. The A1 fiber fires much more frequently to pulses of sound than to steady uninterrupted sounds.
4. Neither neuron responds differently to sounds of different frequency over a broad ultrasonic range. A burst of sound of 20,000 hertz elicits much the same pattern of firing as an equally intense sound at 40,000 hertz.

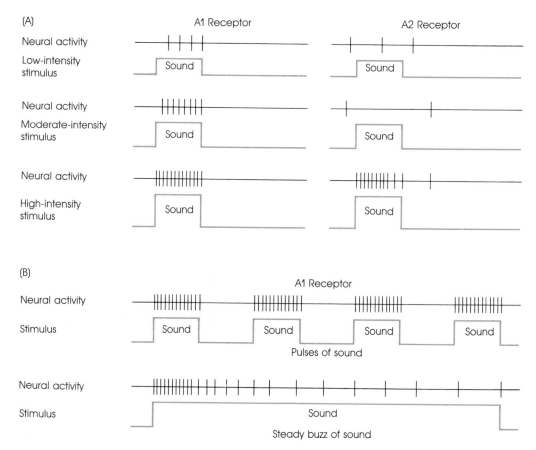

4 **Properties of the auditory receptors of a noctuid moth.**
(A) Sounds of moderate or low intensity do not generate action potentials in the A2 receptor. The A1 fiber fires sooner and more often as sound intensity increases. (B) The A1 receptor reacts strongly to pulses of high-frequency sound but ceases to fire after a short time if the stimulus is a steady hum of sound of the same frequency and intensity.

5. The receptor cells do not respond at all to low-frequency sounds. The moths are deaf to stimuli that we can easily hear.

Although each ear has just two receptors, the amount of information they can provide the central nervous system about sonar-using bats is impressive. The key property of the A1 fiber is its great sensitivity to sound frequencies higher than 20,000 hertz, particularly to *pulses* of ultrasound. Bat orientation cries consist of pulsed, ultrasonic sound. The highly sensitive A1 fiber begins firing in response to cries from a little brown bat that is 100 feet away, long before the bat can detect the moth. Because the rate of firing in this cell is proportional to the loudness of the sound, the insect has a system for determining whether the bat is coming toward it.

In addition, the moth's ears gather information that could be used to locate the bat in space (Figure 5). For example, if a hunting bat is on the right, the A1 receptor on the right side will be stimulated sooner and more strongly than the A1 receptor in the left ear, which is shielded from the sound by the moth's body. As a result, the right receptor will fire sooner and more often than the left receptor. Thus, according to one hypothesis, the brain's decoder neurons, by comparing sensory inputs from both ears, could place the bat in the horizontal plane.

The moth's nervous system can also potentially determine whether the bat is above it or below it. If the predator is higher than the moth, then with every up and down movement of the insect's wings there will be a corresponding fluctuation in the rate of firing by the A1 receptors when exposed to and then shielded from bat cries by the wings. If the bat is lower than the moth, there will be no such fluctuation.

As waves of neural activity initiated by the receptors sweep through the moth's nervous system, they may ultimately generate a battery of motor messages that causes the moth to turn and fly directly away from ultrasound [672]. When a moth is moving away from a bat, it exposes less echo-reflecting area than would be exposed if it were flying at right angles to the predator and presenting the full surface of its wings to the bat's vocalizations. If a bat receives no echoes from its calls, it cannot detect a prey. Bats rarely fly in a straight line for long, and therefore the odds are good that a moth will remain undetected if it can stay out of range for a few seconds. By then the bat will have found something else within its 8-foot moth detection range and will have veered off to pursue it.

In order to implement its antidetection response, a moth need only orient so as to synchronize the activity of the two A1 fibers. Differences in the rate of action potential production by the receptors in the two ears are probably monitored by the brain, which relays neural messages to the wing muscles via the thoracic ganglia and allied motor neurons. The resulting changes in muscular action steer the moth away from the side of its body with the ear that is more strongly stimulated. As the moth turns, it will reach a point where both A1 cells are equally active. If it then maintains this condition, the insect would be flying in the same direction as, and away from, the bat.

Although this reaction is effective if the moth has not been detected, it is useless if a speedy bat has come within the 8-foot detection range. Moths that are close to bats do not try futilely to outrace them, but employ evasive responses, including wild loops and power dives, that make it relatively difficult for bats to intercept them. A moth that executes a successful power dive and reaches a bush or grassy spot is safe from further attack because echoes from the resting place mask those coming from the moth itself [672].

Roeder has speculated that the physiological basis for the erratic flight of the moth lies in circuitry leading from the A2 fiber to the brain and back to the thoracic ganglion [671]. Whenever a bat is about to collide with a moth, the intensity of sound waves reaching the insect's auditory receptors is high. It is under these conditions that the A2 cells are stimulated sufficiently to fire. Their messages are relayed to the brain, which in turn may

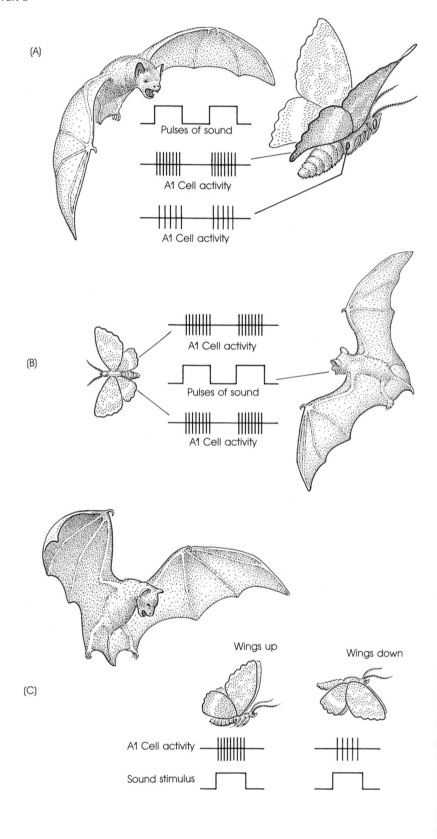

(A)

Pulses of sound

A1 Cell activity

A1 Cell activity

(B)

A1 Cell activity

Pulses of sound

A1 Cell activity

(C)

Wings up

Wings down

A1 Cell activity

Sound stimulus

◀**5** **How moths locate bats in space.** (A) A bat is to one side of
the moth; the receptor on the side closer to the predator fires
sooner and more often than the shielded A1 receptor in the other
ear. (B) A bat is directly behind the moth; both A1 fibers fire at the
same rate and time. (C) A bat is above the moth; activity in the A1
receptors fluctuates in synchrony with the wing beat of the moth.
Figures not drawn to scale.

shut down central steering mechanisms that regulate the activity of motor
neurons (Figure 3). When the steering mechanism is inhibited, the moth's
wings begin beating out of synchrony or irregularly or not at all. As a
result, the insect does not know where it is going, but neither does the
pursuing bat, whose inability to plot the path of its prey may permit the
insect to escape.

Stimulus Filtering and Behavior

The noctuid moth's auditory system and antipredator behavior
provide a classic example of the relation between STIMULUS
FILTERING and adaptive programmed responses to simple sensory cues. The
moth's ear does not relay information about a host of acoustical stimuli in
its environment. Low-frequency sounds audible to us do not trigger membrane changes and action potentials in the A1 and A2 fibers. Ultrasonic
sounds of different frequency do not elicit different patterns of activity by
the receptors and so cannot be discriminated perceptually by the moth
(whereas humans can easily tell the difference between C and C sharp).
Prolonged steady sounds are quickly ignored by the receptors. The ear
appears to have one task of paramount importance—the detection of cues
associated with its nocturnal archenemies. To this end its auditory capabilities are distorted and limited, sensitive to pulsed ultrasonic sound at
the expense of most other sounds. Likewise, its behavioral repertoire of
responses is simple. It turns away from low-intensity ultrasound and dives,
flips, or spirals erratically when it hears high-intensity ultrasound.

Bat Detection by Crickets

Neuroethologists have hypothesized that stimulus filtering and selective
responses eliminate the need to process biologically irrelevant material and
therefore increase the probability of detecting critical events. If true, species
unrelated to moths but faced with bat predation should have independently
evolved similar perceptual systems for dealing with these deadly hunters.
Most of us are unaware that many crickets fly at night when bats are
searching for food. Ron Hoy and his associates have discovered that a
specific nerve cell —interneuron-1, or int-1—in one species of cricket,
Teleogryllus oceanicus, apparently plays a critical role in helping crickets
foil bat predators.

A pair of the key interneurons is located in the central nervous system,
one on each side of the cricket's body. The central interneurons are linked
to peripheral auditory receptors that, unlike moth ears, reside in each
foreleg, not in the thorax. The cricket ear responds to a considerable range

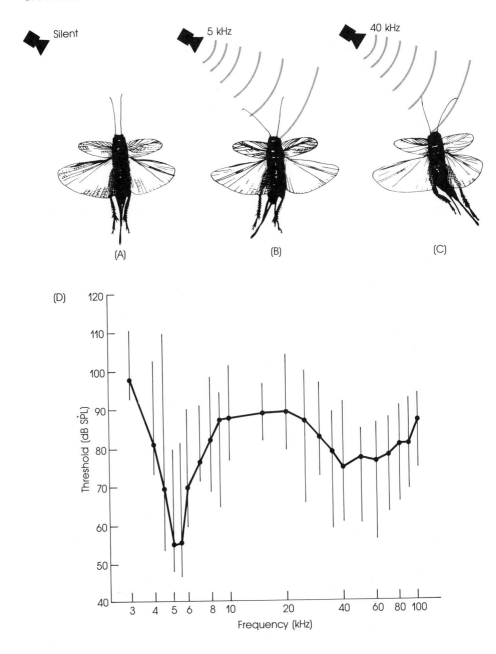

(A)   (B)   (C)

6 **Flying tethered crickets** hold their abdomens straight (A) unless they hear low frequency sounds, in which case they turn toward the sound (B), or unless they hear high-frequency sounds, in which case they turn away (C). Males of this species produce sounds in the 5-kilohertz range; bats produce high-frequency calls. (D) The tuning curve of the cricket: sounds in a low-frequency band and a high-frequency band need be less intense, as measured in decibels (dB SPL), to elicit a neural response. After Moiseff et al. [564].

of acoustic stimulation, with sensory cells relaying their messages to the central nervous system. However, cricket ears are not equally sensitive to all frequencies of sound, but have lower thresholds for sounds in the two frequency bands (Figure 6). The int-1 cells contribute to the selective perception of sounds because they become highly excited only when the cricket's ears are bathed in high-frequency sounds (over 15,000 hertz). The more intense a sound in the 40 to 50 kilohertz range, the more action potentials produced, and the shorter the latency between stimulus and response—two properties that exactly match those of the A1 fiber in noctuid moths.

If an int-1 cell fires only sporadically, there is no behavioral reaction by the cricket. But if one of the two neurons fires steadily in response to relatively intense high-frequency sound, then the dorsal longitudinal muscles on one side of the abdomen will contract. This contraction bends the cricket's abdomen, which acts like a rudder, steering the animal away from a source of intense high-frequency sound coming to it from the side (Figure 6).

The int-1 cells are sensory interneurons. These neurons do not innervate the dorsal longitudinal muscles directly; instead, they send their signals to the cricket's brain where other cells integrate inputs from several cells, including int-1. The decision-making capacity of central neurons is most clearly evident when the cricket is walking at which time even extreme int-1 activity has no behavioral effect. Only when the cricket is flying does information from int-1 serve to steer the insect, a discovery that was made by tethering a cricket and suspending it above the ground with a stream of air blowing over it. Under these conditions, flying crickets bend their bodies in response to certain kinds of acoustical stimulation.

The functional significance of this system seems clear. Bats cannot detect walking crickets, but flying crickets are anything but safe if a bat is headed their way. By having a network of cells that turns them away from high-frequency sound, flying crickets do what noctuid moths do—they move away from an approaching but still distant bat, keeping out of detection range longer and so improving the odds of surviving to fly another night.

Bat Detection by Lacewings

Lacewings are delicate little insects that also fly at night. Although much smaller than most noctuid moths and crickets, they are entirely edible to bats, and they, too, have evolved a bat-detecting ear. Their ears are not in the thorax or forelegs, however; instead the lacewing's ears lie within an enlarged vein in each forewing. Here there is a tympanum linked with 25 receptors that respond to airborne high-frequency sound [558]. The electrical messages from stimulated receptors are rapidly relayed through the lacewing's body, eventually triggering motor neurons attached to muscles that fold the wings over the insect's back.

Intense orientation cries of bats (which are about 50 centimeters from the insect) are sufficient to stimulate the receptors. As a consequence, the flying lacewing closes its wings and plunges downward at 2 meters per second (Figure 7). The lacewing's response is analogous to the antiinterception behavior of noctuid moths.

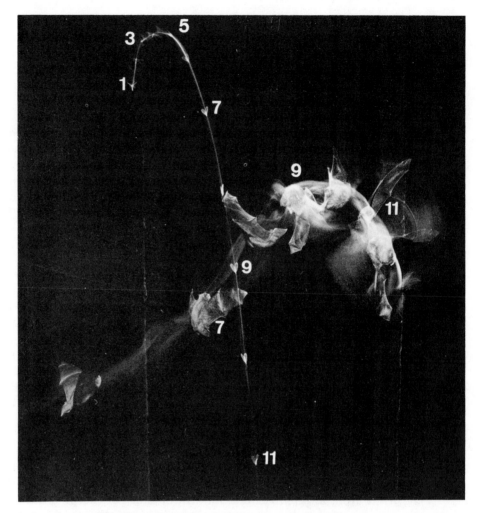

7 **Antiinterception response of lacewing.** A multiple exposure photograph whose numbers show the relative positions of the bat and lacewing over time. The bat missed the power-diving lacewing despite performing an aerial somersault. Photograph by Lee Miller.

The power-dive response unmodified might save some lacewings, but bats easily detect and track falling objects 50 centimeters from them. However, Lee Miller found that only 30 percent of power-diving lacewings were caught by captive hunting bats. This finding led him to take a closer look at the interaction between predator and falling prey. By examining sequences of strobe-flash photographs, he found that 50–100 milliseconds before a bat attempted to sweep up a lacewing, the insect often performed a "wing flip," sharply beating its wings once.

The wing flip interrupted and altered the lacewing's descending path. The bat, therefore, often missed the insect because it attacked where the lacewing appeared to be going before the wing flip. Miller believes that the wing flip is triggered by the intense terminal buzz of cries made by a bat that has tracked a prey and is on the verge of capturing it. Although the receptors responsible for detecting the buzz have not been precisely identified, the point is that the lacewing, like the noctuid moth and cricket, has a selective auditory system that is tuned to high-frequency sound and linked to interneurons that command anti-bat responses [557, 558].

The Ecological Significance of Audition

Noctuid moths, lacewings, and crickets need not have the vaguest mental image of a bat nor any idea of the danger they represent. It is sufficient that they have inherited the neural "wiring" that automatically helps them avoid predators better than individuals with different nervous systems. The neural units that have survived to the present provide moths, lacewings, and crickets with an incomplete picture of their world and with a limited behavioral repertoire—but one that works in the competition to survive and reproduce in a nocturnal world filled with bats.

Animals do not usually perceive stimuli unless they are associated with events that affect individual reproductive success [521, 585]. If there were giant bats that swept humans off their feet, I suspect that we would have evolved the capacity to detect ultrasound with our ears and that our response to these stimuli would be dramatic. But, luckily for us, man-eating bats do not exist, and we lack the ability to hear ultrasonic sounds.

Interestingly enough, some noctuid moths cannot hear bats either—during the larval stage, when the caterpillars are under no special risk of attack by aerial hunting, nocturnal bat predators. The cabbage moth larva, bane of backyard gardeners, is a small green caterpillar that spends its days feeding voraciously on cabbage leaves and the like. It has an "ear" of sorts, which is capable of detecting airborne sounds [516]. The ear consists of eight thin hairs located in little sockets on its anterior segments (Figure 8). If one projects noises at the caterpillar, some frequencies will cause the grub to stop feeding; it may then begin to squirm on its leaf, regurgitate a greenish fluid, and eventually drop to the ground. The frequencies that have this effect are all under 1000 hertz, well within our hearing range and well outside the 20,000 to 100,000 hertz produced by aerial hunting bats.

The larva's low-frequency hearing ability helps it avoid its enemies, which are not bats but paper wasps. These wasps hunt for moth larvae, which they sting, rip apart, and feed to their brood within a nest (see Figure 17 in Chapter 15). A flying paper wasp's wings buzz, producing a low-frequency sound of about 150 hertz. Sounds in the 100 to 700 hertz range readily cause the moth larva's filiform hairs to vibrate (Figure 9). Information from the stimulated hairs in some way activates defensive responses described earlier. The reactions reduce the risk of predation. If one removes the hairs from a larva, it cannot detect wasps at a distance and so does not do anything until a wasp touches it, and then it may be too late. In one experiment, "deaf" caterpillars were 30 percent more likely than intact individuals to be captured by hunting paper wasps [760].

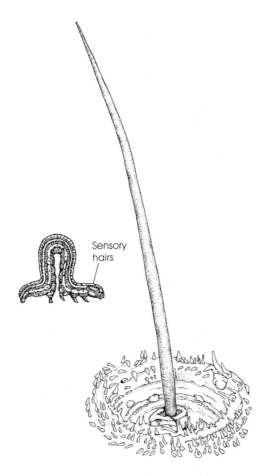

Sensory
hairs

8 **Acoustical hair of a noctuid moth larva.** The larva has eight hairs on each side of its body. Each hair is poised in a socket. Airborne vibrations can cause the hair to move, thereby triggering receptors associated with the hairs.

Stimulus Filtering by Vertebrates   A skeptic might say that it is all well and good to show that insects, with their simple nervous systems, exhibit extreme stimulus filtering, but the perceptual abilities of vertebrates, which have much larger and more complex nervous systems, should be essentially complete. This argument is not without merit. After all, the human ear has millions of receptor cells, not just two, and we can discriminate among many sounds far more skillfully than a noctuid moth. But even vertebrate systems have their limitations and specializations. Our acoustic system does not respond to high-frequency sounds that are routinely detected by a moth or bat. Therefore, our auditory system performs stimulus filtering. In addition, if we were to analyze any one of our auditory receptors, we would find that it could detect certain stimuli much more readily than others. These two general properties of nervous systems—stimulus filtering by the system as a whole and specialized feature detection by certain of its component cells—are also evident in the way bats perceive sounds.

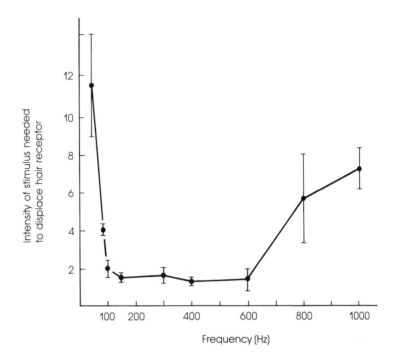

9 **Tuning curve** of the acoustical receptors of a noctuid larva, showing that this species is most sensitive to sound in the 100–600 hertz range. *Source:* Markl and Tautz [516].

Superficially, a bat's ear is similar to our own. Sound waves reaching a bat enter an impressively large external ear and pass along the middle ear canal to the inner ear. At the end of the ear channel, they strike the membranous eardrum, causing it to vibrate. These vibrations are transmitted via middle ear bones to the auditory sensory unit, the cochlea, which contains the primary receptor cells. The cochlea is a fluid-filled organ with a thin basilar membrane running through the middle of the entire structure. Vibrations reaching the cochlea cause the fluid to oscillate. This motion in turn vibrates specific portions of the basilar membrane; the portion of the membrane that vibrates depends on the frequency of the sound waves entering the ear. The mechanical energy present in these movements deforms receptor cells attached to the membrane; this energy is transformed into a receptor signal that is relayed to sensory neurons associated with each receptor. When these neurons fire, their messages are carried away from the cochlea along the fibers that constitute the auditory nerve. This cable runs to the lateral lemniscus and inferior colliculus in the bat's brain (Figure 10), where cells analyze input from the auditory sensory system and make decisions that control the bat's movements [360].

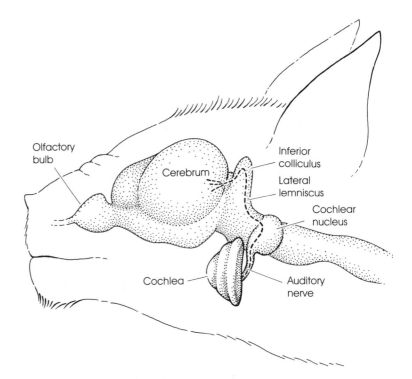

10 **Components of a bat's auditory system.** Sound waves entering the ear are relayed to the cochlea, where they may stimulate receptors within this organ. Auditory signals are transmitted via the auditory nerve to the cochlear nucleus, lateral lemniscus, inferior colliculus, and then to regions of the cerebrum for analysis.

The Detection of Ultrasonic Echoes

If a bat is to catch a flying insect, its nervous system must be capable of detecting very weak ultrasound reflected from prey items. The trouble is that these faint echoes return to the bat within milliseconds after it has produced an orientation vocalization about 2000 times as intense as the echo is likely to be [310]. An auditory cell that has just been exposed to a loud sound becomes temporarily insensitive to soft noise. But bats preserve the sensitivity of their auditory receptors in the following way. Just before cells in the bat's brain order muscles controlling the larynx to produce an orientation cry, they also send messages that cause a muscle in the middle ear to contract. The muscle damps vibrations of the middle ear bones that transmit the energy in sound waves to the cochlea. The muscle relaxes 2 to 8 milliseconds after an orientation pulse; consequently, the bat's auditory receptors are not recovering from a recent vocalization blast but can detect weak reflected sound waves coming from objects in the bat's acoustical field [407].

The middle ear muscle is not the only mechanism for protecting the bat's auditory system against self-generated sounds. In the lateral lemniscus, there are neurons that block transmission of auditory messages to higher regions of the brain during a vocalization [750]. The combined effect of the middle ear muscle contraction and the inhibition of relay neurons in the lateral lemniscus is to prevent signals arising from self-produced sounds from ever reaching the inferior colliculus.

In addition to these two mechanisms designed to protect the sensitivity of the auditory system to echoes, bat brains have echo-detector cells [640]. These neurons respond more intensely to the second of two separate pulses of sound, one following the other in close succession, just as an echo would follow the vocalization pulse. Certain nerve cells in the inferior colliculus of the Mexican freetail bat react to sounds at the very threshold of bat hearing if, and only if, the low-amplitude sound has been preceded by a loud sound, preferably of the intensity and frequency of a typical Mexican freetail orientation pulse (Figure 11).

The Identification and Interception of Prey

Echo-detector cells have been discovered in the brains of mustache bats as well. These cells do more than merely detect reflected ultrasound. Some units are "tracking neurons" that remain active only if there is a continuing

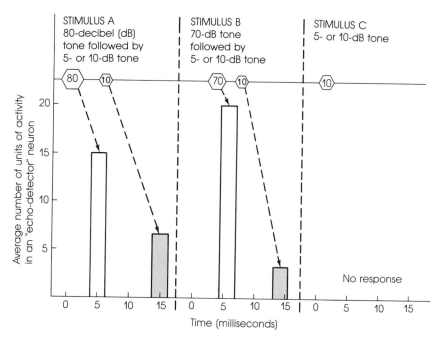

11 **Echo-detector cells** in a Mexican freetail bat respond differently to three stimuli. The cell does not respond to isolated low-intensity sounds (stimulus C) but will produce neural impulses when these same tones are preceded by a loud burst of noise (stimuli A and B). *Source:* Pollack [640].

decrease in the interval between the bat's emitted orientation sound and the echo. Continuing activity in these cells would inform the bat that it was getting closer and closer to an echo-reflecting source. Needless to say, this is what the bat "wants" to do if the object is a moth or other edible flying insect [600].

Still other cells in the mustache bat's brain are "range-tuned." These cells respond most actively to one constant interval between a vocalization and an echo. For example, one of these cells might fire at its highest rate only when the delay between orientation pulse and returning echo is 3 milliseconds; another might be tuned to a 6-millisecond delay; a third might fire most rapidly when the interval is 9 milliseconds. These nerve cells are arranged in linear sequence, running from short- to long-delay neurons in the bat's cerebral cortex. In effect, sensory inputs are received by neurons that are nonrandomly distributed on the surface of the cerebral cortex; this distribution creates a brain map that may facilitate decoding of spatial information contained in the sensory messages [749]. The brain map of a mustache bat may help it identify the source of the signals in space, ordering adjustments in a bat's flight path needed to keep the predator moving toward its targeted meal. Computational "maps" of this sort are common in the brains of animals that perform elaborate analyses of spatially distributed stimuli [434].

There is no doubt that the auditory system of the bat is more complex and provides much more information than the ear of the moth. In part, this difference exists because the bat relies much more on auditory inputs to control its behavior and because the tasks that it undertakes are exceptionally demanding. But despite the differences, there is a critical similarity between bat and moth audition. In both animals the biochemical and functional properties of the auditory system are exceedingly specialized. Moths construct an acoustical world from bat cries, while bats live in a world of echoes that contain the information for a richly detailed image of the environment. A noctuid moth's nervous system helps it locate bats in the night sky. A bat's nerve cells help it focus on and interpret biologically vital information, especially echoes from pulses of sound it emitted, while ignoring less relevant sounds (such as its own vocalizations).

The Perception of Electric Fields

Our perceptual appreciation of echoes is limited, thanks to the many proximate differences in the nervous systems of humans and bats—differences that are based on our separate evolutionary histories. To reinforce the point that the perceptual world of each species has its own unique features, let us consider electrical perception and its behavioral uses in certain fish.

Just as we find it difficult to imagine what it would be like to navigate through the air in complete darkness by listening to the echoes from our vocalizations, so too it is humbling to realize that there are animals who can detect and make sophisticated use of electric fields so weak that you and I are totally unable to sense them. The platypus is such a creature. This supposedly primitive mammal hunts underwater at night in murky streams for small aquatic animals like crayfish and freshwater shrimp—

and judging from a captive specimen's daily intake of a pound of earth-worms, large numbers of mealworms, two frogs and "two coddled eggs" [312], they have a substantial appetite. One reason they are able to find sufficient prey has to do with their capacity to detect the faint electric fields inadvertently produced by their prey. Free-swimming animals will investigate batteries placed in their pools and will turn over bricks concealing electrodes that create a weak electric field of the sort that shrimp produce [701]. The electroreceptors of the platypus dot the edge of its soft, cartilaginous bill. Messages from the receptors reach regions of the brain cortex that analyze the signals and direct the creature to its hidden prey.

Two groups of electric fishes occurring in South America and Africa take electroreception one step further by creating a weak electric field about themselves with their own electric organs [384, 488]. Objects in the field distort the self-generated lines of electrical force, either concentrating the current flow (if the object is a better conductor of electricity than water) or dispersing it (if the object is a poorer conductor) (Figure 12). These fishes have rows of electroreceptors in pores along the length of the body. The receptors respond to changes in the electric field by changing their rate of firing. Cells in the central nervous system of the fish monitor receptor activity in the zone affected by the "shadow" cast by an intruding object. An increase or decrease in the firing rate of a group of neighboring receptors results in new patterns of activity in certain brain neurons. Electric fishes

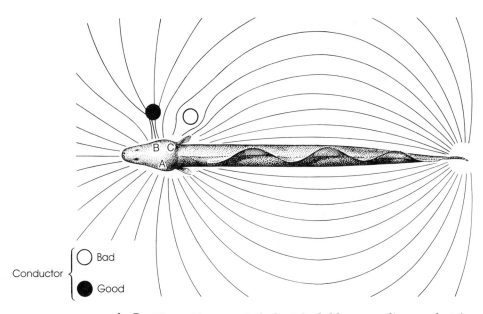

12 **The self-generated electric field** surrounding an electric fish with an electric organ in its tail. Objects near the animal distort the field, and these changes are detected by electroreceptors in the fish's skin. Receptor activity is normal at A, elevated at B, and depressed at C.

are remarkably good at detecting changes in their electric fields. One species can sense a change in electric current as small as 0.03 microvolts and can be trained with food rewards to discriminate between glass rods that are 0.5 or 2.0 millimeters in diameter [488].

In many respects the electroreception systems of these fishes are similar to the sonar apparatus of the bat at both the proximate and ultimate levels. Functionally, electroperception is the kind of ability that helps an animal navigate and search for prey in the dark. Both groups of electric fishes live in turbid, muddy waters or hunt primarily at night for prey that do not advertise their position, as it is not to their advantage to be eaten. A bat's sonar and a fish's electric field "forces" prey items to give up information about their location.

How to Avoid Electrical "Jamming"

At the proximate level, the extraordinary abilities of the fishes reside in the specialized nature of their receptors and decoders. Just as insectivorous bats devote a large portion of their brains to analysis of echoes, so too the brain of an electric fish has an unusually well developed region that analyzes inputs from electroreceptors. Bat and electric fish brains must also cope with a similar problem—what to do about the potentially interfering signals ("jamming") generated by other individuals in their environments. Some bats have brain cells that respond selectively to echoes from their own cries while ignoring ultrasound produced by other vocalizing bats. Electric fish of the genus *Eigenmannia* have evolved a similar solution to the problem [530]. In this fish, two isolated individuals often generate electric pulses at about the same rate (say, 370–375 discharges per second). But if they are placed together so that their electric fields overlap and interfere with one another, the fish have a *jamming avoidance reflex* (JAR) in which they automatically adjust the rate of electric organ discharge up or down away from their neighbor's frequency. When each fish has its own private channel, brain cells can selectively monitor distortions in the distinctive baseline activity of its receptors without risk of interference from a neighbor.

There is one weakly electric fish (*Sternopygus*) that lacks a JAR and yet is able to locate objects in its environment, even in the presence of a discharging neighbor. In its brain are cells that are insensitive to the large-scale distortions in its electric field caused by a neighbor's electric activity [530]. When such cells receive inputs from many receptors whose firing rate has been altered, they do not respond. But if just a few receptors fire among the many that feed their signals to a type 3 brain cell, it then becomes active (Figure 13). Type 3 brain cells are designed to detect small local distortions and to ignore irrelevant widespread changes in electrical inputs; the small-scale changes are likely to be caused by biologically significant objects.

Here again we see an example of a specialized decoder cell that receives messages from many receptors, integrates their information, and responds maximally to a few patterns while ignoring many others. Selective reaction helps the fish perceive critical cues in an electrically complex environment.

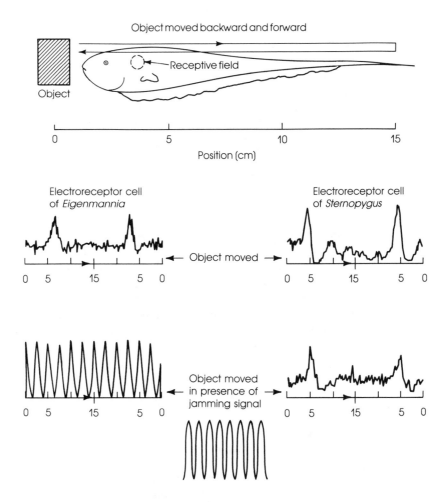

13 **Nonjammable electroreceptors** of the electric fish *Sternopygus*. Recordings were made from a cell whose receptive field is 5 cm from the fish's head. As the object passes this point, the cell's activity changes. If a jamming electrical signal is turned on as the object is moved, some cells of the fish are unaffected, because they ignore widespread background electric activity while continuing to respond to small localized changes in their electric field. *Source:* Matsubara [530].

Selective Visual Perception   We never take for granted perceptual systems like echolocation and electroreception, which are very different from our own. We immediately recognize them for what they are: highly specialized ways of gathering and processing information from the environment. In contrast, vision is so natural to us that it is easy to assume that what we see must be exactly what other animals see. We can quickly disabuse ourselves of this notion by examining visual perception in the common toad of Europe. This study teaches us that what an animal sees and how it responds are every bit as much a reflection of specialized discriminating cells as the sonar system of a bat or the electric field detection system of a platypus.

The toad possesses a visual network similar to our own with two large eyes and their retinal surfaces exposed to the environment. The retina contains a layer of receptor cells that detect the energy in the light reflected from objects. Changes in membrane permeability of these receptors generate messages that are relayed to the next link in the network, bipolar cells, which in turn feed their output to ganglion cells. The long axonal projections from ganglion cells run together to form the optic nerve, which carries action potentials to regions within the optic tectum and thalamus of the toad's brain (Figure 14). Information that is received provides the basis for decisions that the animal makes in responding to prey or predators.

Jorg-Peter Ewert has devised a sophisticated apparatus that can record the activity of single cells in the optic tectum of a toad [252, 253]. The apparatus can be mounted on a living toad that is free to move about and respond to stimuli in its (laboratory) environment. This device informs us that, when a toad is sitting still (something toads are good at) and there is nothing moving in its surroundings, the animal apparently does not see a thing! Photons of light are striking its retinal receptors, but the optic nerve does not relay receptor messages to the optic tectum.

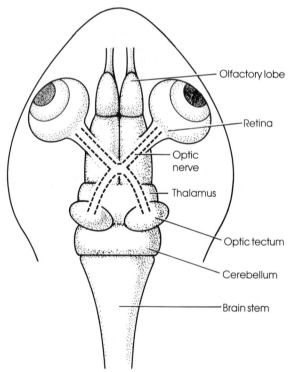

Olfactory lobe

Retina

Optic nerve

Thalamus

Optic tectum

Cerebellum

Brain stem

14 **Visual system of a toad.** This view from above shows the toad's brain with the optic nerve carrying visual input to the optic tectum and thalamus.

The ganglion cells in the visual system perform stimulus filtering for the toad. These cells fire only when a moving stimulus passes through their receptive field. The RECEPTIVE FIELD of a single ganglion cell is that area of the retina whose receptors feed messages back to the cell via bipolar neurons. In the toad and numerous other animals, each ganglion cell monitors a small elliptical portion of the retina; its receptive field typically is organized into a central excitatory region surrounded by an inhibitory ring (Figure 15). Imagine a brown beetle moving over dark soil in front of a toad. The beetle will cast a very small image on the surface of the retina as the light waves reflected from the beetle enter the lens and are focused on the back of the eye. The image moves over clusters of receptors. Some of these clusters will constitute the excitatory central area of the receptive field of one (or more) ganglion cells. These cells will fire and information will be sent to the brain.

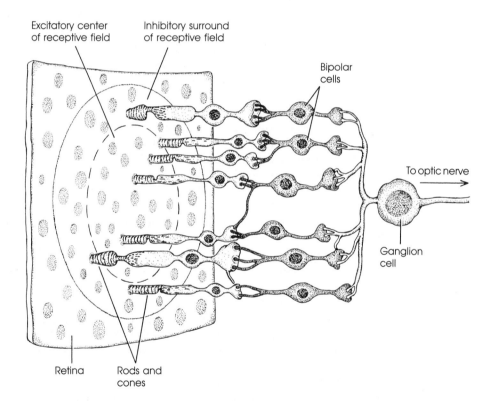

15 **Receptive field of a ganglion cell.** The ganglion cell receives signals from many bipolar cells, which in turn are linked to many receptor cells. As a result, the ganglion cell integrates information from a portion of the surface of the retina. Objects that cover the entire receptive field elicit little response in the ganglion cell. Small objects that move through the center of the field may trigger many action potentials in the ganglion cell.

Imagine a large object such as a human hand passed close to a toad's eye. The hand will cast a relatively large image on the retina, an image that will cover the entire receptive field of most ganglion cells. For these neurons, the inputs received from the inhibitory surrounding ring will largely or entirely cancel any messages from the excitatory central zone. These cells are not likely to fire.

Thus, ganglion cells are movement detectors that respond to different aspects of changes in light intensity that reach the retina. From a toad's perspective, small moving images on its retina are much more likely to be caused by its prey (nearby beetles and worms) or its enemies (distant herons or hedgehogs) than by objects that cast large stationary images on its receptor surface.

The selective analysis of visual stimulation does not end at the level of the ganglion cells. Neurons within the optic tectum receive messages from many neighboring ganglion cells. Each brain neuron has, therefore, a receptive field of its own, a field consisting of that area of the retina monitored by the ganglion cells linked to it. The properties of the receptive field of a tectal cell can be studied by applying a recording electrode to the cell and then moving various objects in front of the eye of a live, but immobilized, toad. Some stimuli will elicit a considerable response; others will generate a few action potentials; and still others will have no effect on the activity of the neuron.

Some cells in the European toad's tectum respond most to long, thin objects that move horizontally across the toad's visual field (Figure 16). These cells have a roughly circular receptive field that consists of an excitatory central strip lying horizontally in an inhibitory surrounding region. Objects that happen to move primarily through the excitatory area of the receptive field will cause the cell to fire. Objects that move through both the excitatory and inhibitory areas produce few or no neural impulses because the excitatory effect of the stimulus is canceled by its inhibitory or blocking effect.

The consequences of the design of this class of tectal neurons become clear if one imagines how they would respond to a nearby moving worm. The worm creates a small, 5- to 10-millimeter image on the toad retina. This horizontally oriented image passes through the excitatory central strip of the receptive field of many tectal cells, causing them to fire rapidly. These messages travel to certain brain cells that also receive inputs from another part of the brain, the thalamus. Some thalamic cells respond strongly to moving objects, like herons and storks, that cast a *perpendicular* image on the retina. If both the worm- and stork-detector cells are active, the decision-making cells in the toad's brain tell the toad's muscles either to hold their present position or to contract, causing the toad to crouch inconspicuously. If, however, the input is largely from the worm-detectors, the toad begins to turn toward the object. In the optic tectum there are three other kinds of cells that sequentially provide the motor commands that cause the toad (1) to stop turning when both eyes are fixed on the potential victim, (2) to lean closer within tongue range, and (3) to open the mouth, flip out the tongue, and snap up the worm [252].

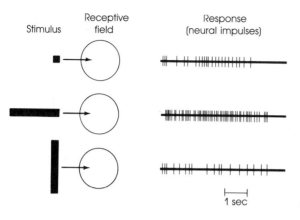

Stimulus   Receptive field   Response (neural impulses)

1 sec

16 **Worm-detector of the European toad.** A cell in the optic tectum of the toad responds differently to different kinds of stimuli that are moved through the cell's receptive field on the retina. Note that long, thin objects moved horizontally elicit the maximum response from this class of cell. *Source:* Ewert [253].

Specialized Visual Perception in Humans

The optical illusions illustrated in Figure 17 tell us that our visual mechanisms, like those of toads, have special features that "encourage" us to detect specific stimuli. Certain of our visual cells react exceptionally strongly to edges, and actually enhance the contrast between neighboring dark and light areas. There is no more light reaching our eyes from the white paper right next to the black squares than from the central grayish regions that are not bordered directly by black. But our brain tells us that there is, creating a useful illusion of the sort that contributes to our ability to see the outlines of objects.

The other illusion in Figure 17 also arises because of the interpretative capacity of human visual mechanisms. We see objects where none actually exists in the drawing, but where they logically might be. The brain is far more than a passive reporter of environmental stimulation. It assists us in the perception of the partly hidden body of an animal in vegetation, a sharp twig projecting into a path we are traveling, the subtle change of expression in a companion.

One of the most fascinating cases of specialized visual perception involves the analysis of faces. A few humans who have suffered damage to a particular part of the brain's temporal lobe cannot recognize faces. They can see things, they can recognize other objects, they can identify particular individuals by the sound of their voice or by familiar clothing, but when shown an image of their friends, their spouses, even themselves, they are at a complete loss.

The region of the temporal lobe that appears to be critical in the perception and recognition of faces receives input from other parts of the brain that receive sensory information from the optic nerve; it appears to be equipped to perform special analyses. Given the significance of recognizing faces and interpreting facial expressions for humans, perhaps it is not surprising that we devote a certain amount of brain tissue specifically to this task.

There is still controversy about whether a face-detecting "center" exists in the human brain. To test whether such a mechanism might be present

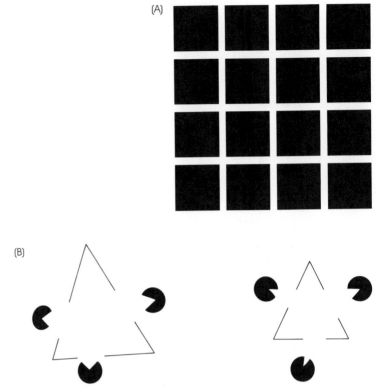

17 **Optical illusions** that illustrate selective perception in humans. (A) Illusory gray spots appear at the intersections of the white bars. Our brains exaggerate the contrast wherever there is an edge (black bordered by white) and therefore white areas without edges appear darker. (B) We perceive white forms overlying the triangles even though the lines that we see are not there. *Source:* Marr [524].

in other animals, David Perrett and Edmund Rolls conducted experiments with a primate relative of humans, the rhesus macaque monkey [628]. They recorded the activity of single neurons with microelectrodes placed in the region of the macaque's temporal lobe that corresponds to the area implicated in facial analysis in humans. The monkeys were conscious during the experiment, in which they were presented with a variety of visual stimuli.

Perrett and Rolls discovered a category of individual cells that responded two to ten times more frequently to images of human or monkey faces than to all other objects. Within this category, different cells reacted somewhat differently. One cell responded equally strongly to faces whether close up or distant, upside down or covered with red filters, but it barely increased its firing rate at all to a face in profile (Figure 18). Another cell fired just as much to an image of two eyes as to a complete face, but ignored a face in which the eyes had been covered. Macaques apparently do have a battery of cells whose integrated activity provides an analysis of faces, an analysis pertinent to the lives of these highly social primates.

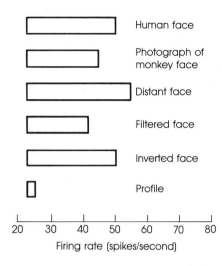

18 **"Face detector" cell** in the cortex of a rhesus monkey. The cell fires at a high rate whenever the monkey sees a human's face or the photograph of a monkey's face, whether close up or distantly, whether seen through colored filters or inverted. A face in profile, however, does not stimulate the cell. *Source:* Perrett and Rolls [628].

Mechanisms of Locomotion and Navigation The examples of neurophysiological mechanisms that I have discussed illustrate the general point that perceptual systems are *not* designed to provide a neutral, complete representation of the "real world." This does not mean that different animals have utterly different perceptions of their environment, nor does it mean that the specialized features of perception create a sensory world that has little relation to the external world. No doubt many species use their different sensory systems to detect accurately some of the same elements of their environment. But it is revealing that biases of perception are linked to the special problems of survival and reproduction that members of a particular species must solve.

We have focused heavily on sensory systems, showing that these are specialized for the detection of critical stimuli. It is also true that motor systems consist of nerve cells with special properties that enable animals to respond adaptively to the environment. An aquatic leech may be a humble and not altogether appealing creature, but it is a perfectly competent swimmer, undulating its body and moving forward in response to a variety of sensory cues. A number of researchers have asked, How does a leech sustain the rhythmic movements that make swimming possible? By probing the relatively simple nervous system of the medicinal leech (used in great frequency by nineteenth century doctors to remove blood from patients), neurobiologists have discovered that this animal, like any number of others, possesses a CENTRAL PATTERN GENERATOR [127, 313].

In each of the segments of a leech, there is a set of identifiable cells in the central nerve cord that can generate and sustain a specific pattern of activity once they begin firing in response to a triggering stimulus. Even if one severs all incoming and outgoing nerves linked to the central nerve cord, thereby eliminating the possibility of sensory feedback arising from the activity of central cells, the systematic pattern of firing persists. If one electrically stimulates certain interneurons in the central nervous system,

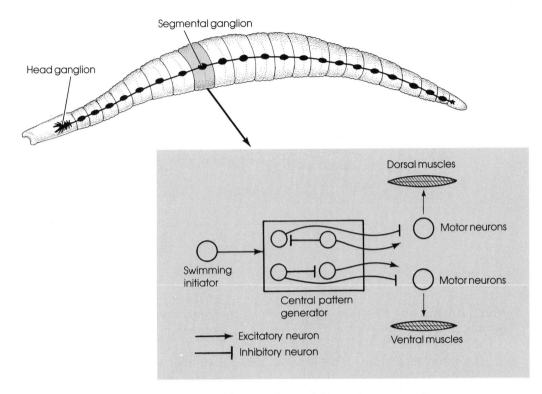

19 **Swimming movements and central pattern generators**
in the medicinal leech. A leech swims by undulating its
segmented, wormlike body. Swimming movements are coordinated
by a neuronal complex found in each segmental ganglion. When
activated by a swimming initiator cell in the nervous system, the
four neurons in the central pattern generator produce an integrated
series of signals that travel to key motor neurons attached to the
dorsal and ventral muscles of the leech. These signals cause the
muscles to contract in ways that contribute to swimming movements.

these cells "turn on" the central pattern generator, and subsequently a
prolonged and precisely organized volley of signals comes from the cells
involved.

In nature a leech may receive sensory signals that activate specific cells
in its brain. These *trigger interneurons* send out messages that reach the
swim-initiator cells in each of the segments; these in turn excite the central
pattern generator neurons, which begin to play out their "tape" of rhythmically organized signals that travel to motor cells controlling muscles on
the back and venter of the leech (Figure 19). The motor output that results
has precise timing: first the muscles on the back of the leech contract, while
the ventral muscles are inhibited from contracting. This bends the segment
one way. Then the ventral muscles are excited, while the excitatory dorsal
motor cell is inhibited, and so cannot interfere with the contraction of the
ventral muscles, which bend the segment in the opposite direction. After

the appropriate (very brief) period, the dorsal cell is reactivated, and the sequence repeated over and over. Because the firing pattern in all segments is coordinated, the leech's entire body moves as a wave of bending travels from its head to tail. As a consequence of the patterned undulation of its flattened body, the leech swims.

We shall have more to say about the organization of behavioral output in Chapter 6, but for the moment the important point is that certain of the cells regulating locomotion in leeches have an inherent timing capacity and their own preprogrammed firing instructions. In the case of the leech and other animals with central pattern generators, the rhythm of neuronal output translates into the rhythmic pattern of muscular contraction that makes coordinated movement possible [127]. Just as sensory cells are not neutral reporters of environmental events, so too central neurons regulating movement have their own unique properties that "bias" behavioral outcomes in particular ways.

Navigation

Exploring the "simple" movements of a leech has revealed an intimidating complexity of neuronal control [598], but there are many other complex locomotory abilities whose neural control is a thousand times more mysterious. A premier example is the ability of some animals to navigate through unfamiliar terrain to reach a goal, as when a migratory animal or a displaced bird returns home after a trip of hundreds of miles over a route that the individual has never followed before.

If you or I were dropped off in a strange spot even a few miles from home, we would probably have no idea which way to go. Nor would we be any better off if we were given a compass by the persons studying our homing abilities. We would have to be told which direction to go; only then would the compass be helpful. Navigation to a destination requires both a compass sense and a map sense (Figure 20). You have to know where you are relative

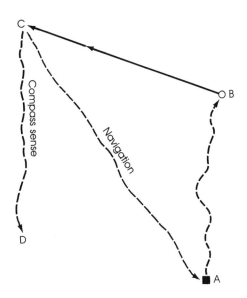

20 A compass sense is necessary but not sufficient for navigation. Here an animal that has moved from its homesite A to B is captured and transported to point C. If the animal is able to return directly to its home from C it demonstrates an ability to navigate. If the animal moves to point D it possesses a compass sense, but has not navigated to its home site.

to your goal and you have to be able to orient your movements accordingly. How do animals do this?

Although we know something about how a few species use visual landmarks to construct mental images that assist them in their travels [216, 374, 818], in general our understanding of navigational maps is rudimentary. However, the compass sense of a few species, including honeybees and homing pigeons, has been partly analyzed. Both bee and pigeon are skilled navigators, as demonstrated by a honeybee's ability to make a beeline back to its hive after a meandering outward journey in search of food and a homing pigeon's ability to make a pigeonline back to its loft after having been released in a distant and strange location. Both animals have participated in ingenious experiments that have revealed a good deal about the basis of their homing ability.

Honeybees and homing pigeons are active during the daytime, and, as one might suspect, both are able to use the sun's position in the sky as a directional guide [420, 804]. Even we can do this to some extent, knowing that the sun rises in the east and sets in the west—provided we know approximately what time of day it is. Every hour the sun moves 15° on its circular arc through the sky. One has to adjust for the sun's movement if one is to use its position as a compass. A bee leaving its hive notes the position of the sun in the sky relative to the hive and flies off on a foraging trip. It might spend 15–30 minutes on its trip and move into unfamiliar terrain in a search for food. If it were to try then to return home, orienting as if the sun were where it was at the start of its travels, the bee would get lost because the sun's position had changed with the passage of time.

Honeybees rarely get lost, however, in part because they learn visual landmarks in foraging areas but also because even in unfamiliar places they use their internal clock mechanism (see Chapter 6) to compensate for the sun's movement [485]. This mechanism can be demonstrated by training some marked bees to fly to a sugar-water feeder some distance from the nest (say, 300 meters due east of the hive). One then can trap the workers inside the hive and move everything—lock, stock, and barrel—to a new location. At this site, a new feeder is set up at a different orientation from the nest (say, 300 meters southeast). One observer stands at this feeder and another observer watches an empty feeder 300 meters due east of the hive. After 2 or 3 hours have passed, the hive is unplugged and the workers are free to go in search of food. They do not have familiar visual landmarks to guide them and yet the marked individuals remember that food is found 300 meters due east. They fly to the spot where the food source "should have been." They do not go to the new sugar feeder, a finding that shows that they are not tracking feeders by olfactory or other cues. Instead, they have compensated for the 30° or 45° shift in the position of the sun that has taken place during the hours of their confinement.

Pigeons, too, can be tricked into demonstrating how important a clock sense is if they are to orient accurately by the sun. The birds' compass orientation can be disturbed if they are induced to reset their biological clock [808]. This can be done by placing a pigeon in a closed room with artificial lighting and then shifting the light and dark periods in the room out of phase with sunrise and sunset in the real world. For example, if sunrise is at 6 a.m. and sunset at 6 p.m., one might set the lights to go on at midnight and off at noon. A pigeon exposed to this routine for a number of days would experience a *clock shift* of 6 hours out of phase with the natural day. If taken from the room and released at 6 a.m. at a spot some distance from the loft, the bird will behave as if the time were 6 hours later (noon) and orient improperly. For example, let us say that the pigeon is released at a place 50 miles due west from its loft. Its map sense somehow tells it this and it attempts to orient itself to fly east. As Charles Walcott points out: "To fly east at 6:00 a.m., you fly roughly toward the sun, but because your clock tells you it is really noon, you know that the sun is in the south and that to fly east, you must fly 90 degrees to the left of the sun. And this is exactly what the birds do, although they presumably do not go through the reasoning process I have described." Figure 21 illustrates this story.

Backup Orientation Mechanisms

If a sun compass were the only mechanism available to bees and pigeons, they would not be able to home on cloudy days. But both species navigate successfully under total overcast [215, 860]. In fact, some pigeons can home accurately at night or when they have had frosted contact lenses placed over the eyes, devices that reduce their vision to a murky blur! Thus, these species have more than one compass mechanism, one of which may be a sensitivity to the weak lines of magnetic force created by the earth's mag-

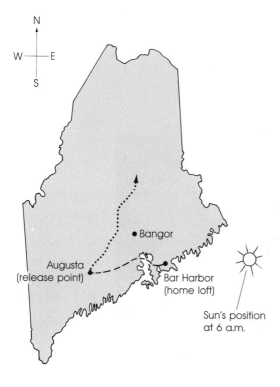

21 **Clock-shifting and altered navigation** in homing pigeons: results of a hypothetical experiment. The pigeons were released at 6 a.m. near Augusta, Maine. The lower track was followed by a bird previously kept in natural light; the upper dotted line was the path taken by a bird whose clock had been shifted by 6 hours (see text).

netic field. These lines run roughly north–south and, if detected, might help orient a homing animal. Both pigeons and honeybees have magnetic compounds concentrated in certain tissues of their bodies; these compounds may be part of a magnetism detector [293].

Magnetite particles have also been found in a space within a skull bone of yellowfin tuna, a migratory fish. The particles are organized in chains that may change position in response to minute changes in the earth's magnetic field. Detection of the movement of the particles could give the fish a magnetic sense useful in its long-distance journeys [810].

The value of a magnetic sense as a backup compass has been unequivocally demonstrated with homing pigeons and some other birds [241]. Charles Walcott altered the magnetic field about some pigeons, either by strapping a magnet to the birds or by outfitting his subjects with helmets consisting of a battery-powered Helmholtz coil (Figure 22). The altered magnetic field about the pigeons disoriented them when they were released far from home—but only on overcast days. When the sun was shining, the birds attended to the cues provided by sun position and ignored the signals detected by their magnetism sensors. But when reliable information from the sun was missing, the birds used the next best thing, information about lines of magnetic force, as an aid to getting home [808].

Nocturnal Navigation

A magnetic compass should also be useful for animals that travel long distances at night, something many migratory songbirds do. The indigo bunting, a representative nocturnal navigator, does sense and use the earth's magnetic field as a compass guide during migratory seasons. This has been demonstrated by the bird's ability to orient at night in the correct direction in completely enclosed cages, where they have no access to visual cues of any sort. Experimental deflection of the magnetic field about them alters the orientation of these birds, which has been measured in an ingenious fashion (Figure 23) [241, 322].

22 **Homing pigeon with a Helmholtz coil** fitted on its head to disrupt its detection of the earth's magnetic field. Photograph by Charles Walcott.

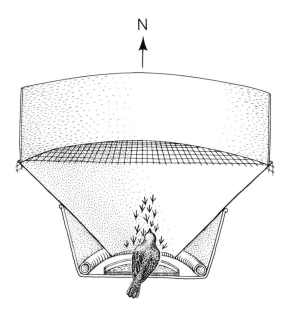

23 **Measuring the orientation preference** of captive indigo buntings. As the bird jumps up from an inked pad base in its cage, it lands on a paper funnel and leaves marks showing the orientation of its leaps. After Emlen [234].

The magnetic sense of indigo buntings and other nocturnal migrants is probably a backup system used on overcast nights when the pattern of stars in the sky is obscured [234]. A number of experiments have shown that birds held in cages in a planetarium will orient themselves during migratory periods according to the celestial cues provided by stars. One can project on a planetarium ceiling a star pattern rotated 90° from the actual pattern, and the birds will shift their nighttime hoppings accordingly. This experiment shows that when celestial information is available birds will attend to it even if it is not in harmony with magnetic field information.

Stephen Emlen's study of indigo buntings is an especially instructive case history of how a songbird uses star position as a compass [234]. He demonstrated that when birds in a planetarium can see several major stars that are located in the northern sector of the sky—among them the North Star and the stars of the Big Dipper—they orient appropriately for the season even if the other stars of the sky are artificially rearranged. The spatial relationship between the North Star and several major constellations clustered about it does not change during the night as the earth rotates (Figure 24), so this cluster provides a reliable fixed compass cue.

Olfactory Navigation?

The ability of nocturnal migratory birds to use star position as a compass guide and diurnal travelers to use sun position for the same purpose illustrates again the connection between the ecology of species and their special neural capacities, even though exactly how nerve cells give birds and bees a compass sense is barely understood. Furthermore, as mentioned previously, even less is known about how animals acquire a map used in getting from one place to a goal.

One line of investigation into the mysterious map sense has been devel-

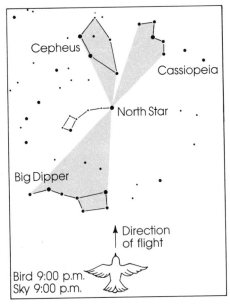

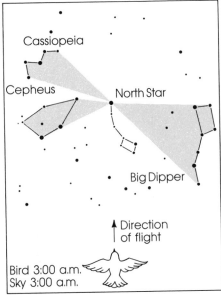

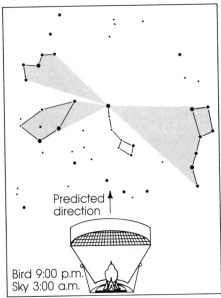

24 **Orientation by star position.**
The positions of most stars in the sky change over the course of a night, but not the North Star. Indigo buntings held in a cage in a planetarium in the spring were still able to orient in a northerly direction even when star positions were altered, provided the relationship between the North Star and the major constellations about it were not changed. *Source:* Emlen [234].

oped especially by Floriano Papi and H. G. Wallraff. They hypothesize that some birds may use olfactory cues to construct a map of their environment [621]. This hypothesis has been the subject of mildly acrimonious debate, as can be seen in the reviews of research on olfactory navigation by Klaus Schmidt-Koenig [703] and Floriano Papi [620].

The hypothesis that homing pigeons form a mosaic map of odors (Figure 25) by learning which scents always come on winds from the north, which from the west, and so on generates testable predictions. Obviously, if one

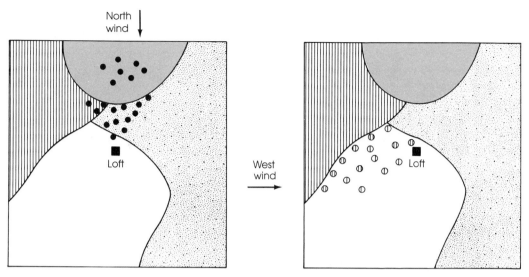

25 **A possible olfactory map.** If different locations have different odors, a north wind and a west wind would each have a distinctive odor. Pigeons at a home loft might be able to store this information to construct an olfactory map used in local navigation to the loft. After Papi [619].

interferes with the olfactory system, either by cutting key olfactory nerves or by applying a temporary local anesthetic to olfactory receptors, the smell-blind pigeon should not be able to navigate home. Experimental manipulations of the olfactory network do often reduce the accuracy of the initial orientation of released homing pigeons (Figure 26), and homing is poor, especially for inexperienced smell-blind birds released more than 50 kilometers from their loft [811].

Despite altered initial orientations, some released pigeons return home eventually, indicating that the birds can perhaps compensate for an absence of olfactory information—a result not predicted by the olfactory hypothesis. And debate continues on whether alterations of the olfactory system might not have side effects, such as changing the birds' motivation and attention, which might independently affect the initial orientation and homing ability of the experimental pigeons.

The olfactory hypothesis also predicts that if one were to deflect the route taken by winds arriving at a loft, the homing pigeons would construct a "deflected" olfactory map that would shift the route they take to return home. In other words, if winds from the west were deflected by wooden baffles to enter the loft from the south, the affected pigeons would be deceived into thinking that odors from places in the west were in the south, and this would cause them to shift their orientation when they were released away from the loft and smelled the local air. Experiments involving birds reared in deflector lofts (Figure 27) yield the expected result. The initial orientation of birds at the release site is shifted by the angle predicted if they have been fooled into constructing a deflected odor map.

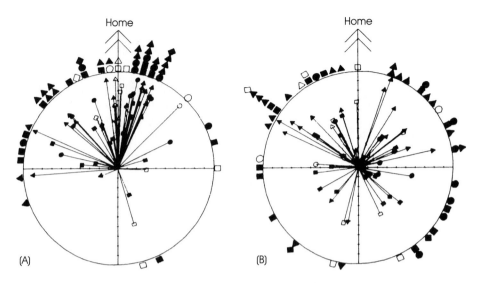

26 **Summary of olfactory experiments** with homing pigeons. (A) Control pigeons with unaltered olfaction tend to orient toward their home loft when released in a strange location. (B) Pigeons with blocked olfaction show much greater scatter in their initial orientation upon release. Each symbol presents the mean orientation for the pigeons tested in a given study. Different locations for the experiments are represented by different symbols (e.g., solid triangles stand for those studies done in Ithaca, New York). *Source:* Papi [620].

Once again, strong differences of opinion exist on the validity of these experiments as a complete test of the olfactory hypothesis. Perhaps the deflector panels also deflect light entering the loft and this visual effect alters the orientation mechanism of homing pigeons. If true, birds whose olfactory nerve has been severed before rearing in a deflector loft should also exhibit shifted orientation when released in an unfamiliar spot, even though they cannot have acquired a shifted olfactory map because they cannot smell anything. Schmidt-Koenig claims that smell-deprived birds reared in deflector lofts show the same shifted orientation as birds whose olfactory system is intact [703].

Although the olfactory hypothesis has been challenged, even severe critics accept the results of experiments by J. Kiepenheuer [425], which seem to demonstrate a role for olfaction in orientation. Kiepenheuer took air samples from four future release sites (A, B, C, and D). Pigeons were then introduced into airtight containers with one of the four samples and kept there for a time, after which they were removed and given a local anesthetic that knocked out their sense of smell. The researcher then packed up the birds and took them either to site A, B, C, or D for release. Only the birds that had smelled air from A before being released at A tended to fly off in the direction of home when let go. Birds with experience with B, C, or D

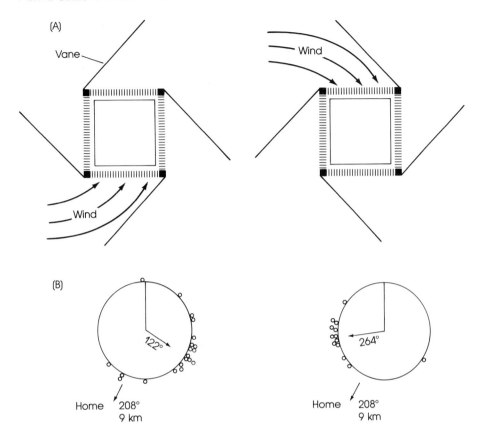

27 **Deflector lofts.** (A) Baffles deflect the wind 90 degrees, thereby altering the pigeon's perception of the direction from which certain odors come. (B) Pigeons that have been kept in different deflector lofts orient differently when released to home back to the loft. Open circles at the periphery of the circle diagrams show the angle of orientation taken by each bird upon release; the arrow within the circle shows the mean orientation direction of all the birds used in the experiment. *Source:* Baldaccini et al. [36].

air oriented randomly upon release at A, but pigeons that had last smelled B air oriented accurately *when released at site B* and so on (Figure 28).

Birds whose last olfactory experience was with air from site A apparently interpreted their point of release as being at site A wherever they were released. It is as if birds actually released at site B (or C or D) that had been kept in a container with site A air before being made anosmic said to themselves upon release, "I last smelled A air; site A is located west of the home loft; therefore in order to return to the loft, I must go east." Again, they presumably do this "unconsciously" or automatically; but however they do it, the results are consistent with the hypothesis that they are using an olfactory map.

The fact that in this experiment all the birds *eventually* returned home

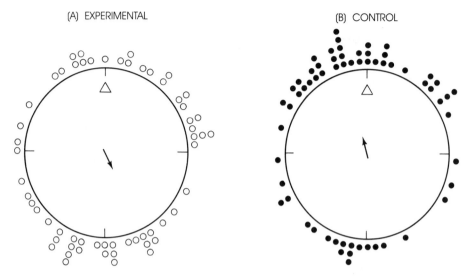

28 **Olfactory cues and orientation in homing pigeons.** (A) Experimental pigeons released at sites where they had not been exposed to the air at the release point before their olfactory receptors were anesthetized (see text) rarely flew in the direction of the home loft (triangle). (B) In contrast, the mean direction (arrow) of the initial orientation of control pigeons that had experience with the air at the release site was toward the home site. *Source:* Kiepenheuer [425].

has been interpreted on one hand as evidence that the olfactory effect is transitory and the pigeons are able to use some other cues to correct their initial mistaken orientation. On the other hand, Kiepenheuer's pigeons were released no more than 15 kilometers from the home loft, and the anesthetic blocked their sense of smell only temporarily. Therefore it may not be surprising that they all made it back to the loft sooner or later. What does seem clear is that more studies of the hypothetical olfactory map sense are justified, particularly since it now appears that the Italian pigeons studied by Papi and the German pigeons studied by Schmidt-Koenig may differ in their navigational reliance on olfactory cues [861]. Learning what parts of the nervous system control navigation and how they work together poses an exciting challenge for the future.

SUMMARY

1 The sensory and behavioral abilities of different animal species are often highly distinctive, as demonstrated by bats that can catch prey by echolocation, fishes that can perceive weak electric fields, and birds that can sense magnetic fields and generate olfactory maps.

2 Neurophysiologists have defined the special properties of nerve cells that contribute to the unique behavioral capacities of some species. An auditory

receptor in a noctuid moth, a cell in the optic tectum of a toad, a neuron in the brain of a bat each serves as a feature detector, with special sensitivity to a few patterns of stimulation and insensitivity to most others.

3 Stimulus filtering is a universal attribute of perceptual systems. The design of an animal's receptors limits what information they can collect and relay for further processing. Furthermore, neural systems typically consolidate complex patterns of sensory information into simpler patterns (for example, when a visual ganglion cell integrates inputs from many receptors into a few action potentials of its own). The result is that animals screen out many biologically insignificant stimuli.

4 Behavioral responses, as well as sensory perceptions, are specialized and restricted because of the way an animal's nervous system operates. Animal species often have discrete reactions to key sensory cues, reactions that may be as simple as the power dive of a noctuid moth or as complex as the navigational skill of an indigo bunting migrating over long distances at night.

5 There is a correspondence between selective neural units (e.g., worm-detectors in the toad), specialized behavioral responses (e.g., the actions that are involved in prey capture by toads), and the ecology of a species (e.g., European toads live in an environment in which worms are an abundant, edible, and vulnerable prey). This correspondence supports the conclusion that species evolve proximate neural mechanisms that enable individuals to detect and respond to those events in the environment most likely to affect reproductive success.

SUGGESTED READING

Kenneth Roeder's *Nerve Cells and Insect Behavior* [669] is a wonderful book, an understandable, entertaining, and exciting account of how to conduct research on the physiology of behavior of moths, bats and other animals. Donald Griffin's *Listening in the Dark* [310] is the classic book on bat sonar.

Jorg-Peter Ewert [253] and Jeffrey Camhi [127] have each written a helpful book called *Neuroethology*, books that summarize how nerve cells promote adaptive behavior. *Mechanisms of Animal Behavior* by Peter Marler and W. J. Hamilton [521] provides many examples of the relation between physiological mechanisms and animal ecology. Good general reviews of animal orientation include those by Stephen Emlen [233], James Gould [294], William Keeton [419], and Rudiger Wehner [819]. Competing views of the significance of olfaction on bird navigation are presented by Klaus Schmidt-Koenig [703] and Floriano Papi [620].

DISCUSSION QUESTIONS

1. When cockroaches are attacked by a toad they turn and run away. A roach has wind sensors for detecting a puff of air created by the abrupt movement of an attacking predator. These sensors are concentrated on its cerci, two thin projecting appendages at the end of its abdomen. One cercus points slightly to the right, the other to the left. Use what you know about moth orientation to bat cries to suggest how this simple system might provide the information needed to orient the roach so that it turns away from the toad, rather than toward it? How might you test your hypothesis experimentally? See [127] after developing your answer.

2. How might we test the hypothesis that stimulus filtering is an adaptive property of nerve cells and nervous systems because it promotes efficient response to key stimuli? A suggestion: How might you use detection times or reaction times to different stimuli by individuals of two closely related insects, each known to have different predators and specialized detection units for their special enemy?

3. Draw what you imagine to be the receptive field of the stork-detector cells in the thalamus of European toads. How would you test your hypothesis?

Animals are complex machines endowed with neuronal computers that identify key stimuli and order complex adaptive responses to these "important" things and events. A flying moth or cricket has sensors that relay messages about acoustic stimuli produced by bats; these messages elicit a turning response that (sometimes) guides the insect out of range of the enemy. When a toad's worm-detectors fire, the animal reacts by orienting to and then flipping its tongue at the stimulus that activated the detectors. The capacity to filter out irrelevant stimuli, to perceive some things very readily, and to order effective re-

The Organization of Behavior

sponses to perceived events is a thoroughly admirable and useful property of the physiological mechanisms that control animal behavior. But this is not the end of it. Neuronal computers are even more sophisticated, enabling an animal to do more than merely detect various key stimuli and activate response X for stimulus Y. Imagine that you are watching a flying male moth hot on the scent trail of a female perched in a distant tree. If the male's nervous system operated simply by turning on flight behavior when female scent was present in the air, and only turning it off when the male reached the female, then the moth would be unable to dive out of the sky away from an attacking bat. But the moth's nervous system does not operate in such a simple-minded fashion. Instead it is able to integrate inputs from many sources, making decisions about which of several possible behavioral options to command. The ability of nervous and endocrine systems to deal with potential conflict and to help the animal to use its behavioral repertoire in a biologically sound fashion is the central topic of this chapter. The fundamental problem to be dealt with is how an individual's behavior can be organized—from moment to moment, over the course of a day, over weeks or a breeding season or a whole year.

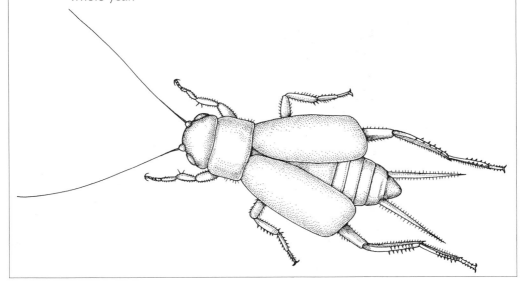

Avoiding Conflicts: Setting Priorities When a European toad is simultaneously presented with two different objects—one that stimulates a set of worm-detector cells and another that stimulates a set of predator-detector cells—thalamic neurons inhibit those tectal cells responsible for ordering the first step in prey catching (Figure 1). The key tectal cells become less likely to fire, and the toad becomes less likely to orient toward the potential prey item and more likely to carry out the crouch avoidance-response.

The unequally weighted inhibitory arrangements between cell clusters in the thalamus and optic tectum set behavioral priorities for the toad [253]. All other things being equal, escape takes precedence over feeding. On the other hand, if the stimulation from worm-detector cells vastly exceeds the signals provided by enemy-detectors, the toad will be able to carry out the complete and productive sequence of prey catching actions without interruption. A toad that oriented to a crawling worm, and then crouched —or one that tried to combine the two activities—would be a boon to the worm population, but not for long.

The Organization of Mantis Behavior
The European toad is a representative animal in that its nervous system is endowed with properties, particularly the capacity of "command centers"

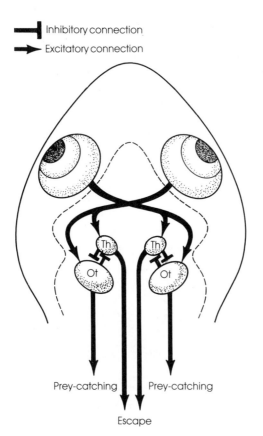

Inhibitory connection
Excitatory connection

Prey-catching   Prey-catching

Escape

1 **Inhibitory connections** between the thalamus (Th) and the optic tectum (Ot) of the European toad. Simultaneous strong activation of both brain structures by different visual stimuli will result in the inhibition of optic tectal cells, thereby freeing the thalamus to order escape behavior.

to inhibit each other, that keep toads from incapacitating conflicts in making decisions. The utility of inhibitory connections between nerve cells in the central nervous system applies to invertebrates, too, as Kenneth Roeder's work on the praying mantis demonstrates [669, 671]. A mantis perched on the leaf is bathed in a rich assortment of stimuli, many of which the insect can detect with its receptor systems. Most of the time, however, the mantis does absolutely nothing, remaining motionless until an unsuspecting prey wanders within striking distance. Once this occurs, however, the insect makes very rapid, accurate, and powerful grasping movements with its front pair of legs. If it does not encounter a meal for a long time, the mantis is likely to move to another waiting site. Moreover, reproductively mature males also search for females. The mantis is capable of sorting out these options because of a simple but elegant set of controls imposed by the brain and subesophageal ganglion on the thoracic ganglia that control the walking and striking legs (Figure 2).

Roeder explored this system of controls by surgically cutting the connections between the various components of the mantis's nervous system and

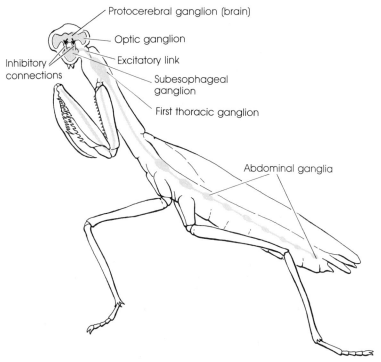

Protocerebral ganglion (brain)

Optic ganglion

Inhibitory connections

Excitatory link

Subesophageal ganglion

First thoracic ganglion

Abdominal ganglia

2 **Nervous system of a praying mantis.** The upper brain of the mantis generally inhibits the subesophageal ganglion, thus preventing the animal from attempting more than one activity at a time. If the connections between the upper and lower portions of the brain are cut, the subesophageal ganglion sends a stream of excitatory messages to the thoracic and abdominal ganglia; the mantis then attempts to do several competing activities simultaneously.

observing the behavioral effects of different operations. If he separated a single segmental ganglion from all others, the movements in that segment ceased. However, if the ganglion was stimulated electrically, the muscles and any limbs in that segment made vigorous, complete movements. Therefore, each segmental ganglion is responsible for the motor output that drives a limb forward or contributes to the movement of the abdomen.

If the segmental ganglia are responsible for telling individual muscles how to contract or relax, what is the brain doing? Removal of the protocerebral lobes produces a mantis that walks and grasps simultaneously. This behavior is disastrous because eventually the mantis's forelimbs find something to grasp firmly while its walking legs pull the animal ahead. This dilemma graphically illustrates the importance of avoiding the simultaneous activation of competing responses. The protocerebral lobes make certain that a mantis either walks or grasps, or the insect does nothing.

What are the commands originating in the brain? Roeder answered this question by removing the entire head of the mantis, a procedure that eliminates the subesophageal ganglion as well as the protocerebral lobes. Under these circumstances the animal is immobile. Single, irrelevant movements can be induced by poking the creature sharply, but that is the extent of its behavior.

These experiments provide a picture of the total control system of the praying mantis (Figure 2). Most of the time, environmental information received by the protocerebral lobes does not affect the normal stream of inhibitory messages that this region sends to the subesophageal ganglion. Particular patterns of stimulation, however, block the relay of inhibitory messages from protocerebral units to other areas in the subesophageal ganglion; these regions, freed from suppression, send messages to the thoracic ganglia where new signals are generated that order muscles to take specific actions. Depending on what sections of the subesophageal ganglion are no longer inhibited, the mantis walks forward or strikes out with its forelegs.

There is one exception to this scheme. If a mature male's head is removed, the animal, instead of losing its ability to behave, performs a series of rotary movements that swing its body sideways in a circle. While this is happening, the mantis's abdomen is twisting around and down. These actions enable a mate-seeking male that has literally lost his head to a cannibalistic female (Figure 3) to effect copulation even if he has been spotted by a female, captured, and partly eaten *before* he mounts her. Headless, his legs carry him around in a circular path until his body touches the female, at which point he climbs onto the back of his would-be mate. The male's abdomen probes about professionally, and copulation occurs.

Males can copulate without losing their head [487], and under natural conditions it may be that only a small minority of males end life mating with a cannibalistic partner. But the fact that headless males of some species are able to perform a completely functional copulation suggests that the risk of being cannibalized by their mates has favored mature males with a control system that differs from that of other mantises, one in which the thoracic and abdominal ganglia can independently order the headless animal to copulate (although see [301, 487]).

3 **Sexual cannibalism by a female mantis.** The female of this pair photographed in the field in Africa has reached back to consume the head of her partner while they copulate. Photograph by E. S. Ross.

In any event, the nervous system of the mantis, like that of many animals, appears to be functionally organized as a cluster of "centers," each with specific responsibilities for certain activities and decisions. Some clusters of neurons manufacture their own output: a stream of inhibitory messages that control the activities of other groups of cells. The inhibition of competing centers is the basis for the mantis's ability to do one thing at a time. However, after receiving selected patterns of sensory information, inhibitory neurons in the brain lobes are themselves inhibited, thereby freeing special circuits in the subesophageal ganglion that relay triggering signals to the segmental ganglia for the performance of a particular behavior.

The Timing of Behavior: Short-Term Cycles The ability of animal nervous systems to block all but one selected series of motor commands is obviously adaptive. This is not to say that hybrid or CONFLICT BEHAVIORS do not sometimes result from the interaction between competing systems (see Chapter 7). In general, however, animals perform behaviors in sequence rather than simultaneously.

One aspect of the sequential organization of behavior is the capacity of animal nervous systems to produce repeating cycles of behavior, something

4 The sea slug, *Pleuro-branchaea californ-ica.* Photograph courtesy of W. J. Davis.

the sea slug, *Pleurobranchaea californica* (Figure 4), can do. This is not the loveliest of animals but thanks to its beautiful, simple nervous system, W. J. Davis has been able to learn a great deal about the organizational scheme that enables the slug to organize its life. These slugs are voracious carnivores, and feeding normally is very high on their list of things to do —so much so that a slug given a choice between mating and eating usually chooses to feed. The slug also has several simple protective responses, one of which involves drawing back its sensitive oral veil, an anterior part of its body, when touched there. Food-deprived slugs, however, ignore touches to their oral veil. When a hungry sea slug is in the midst of grappling with a squid, its oral veil will most likely be struck in the process. Great sensitivity to this stimulus might well lead it to lose contact with its prey.

But a slug that has been fed to repletion on strips of squid meat withdraws its oral veil when this part of its body is tapped, even in the presence of squid juice [443]. Perhaps as the gut becomes distended with eaten squid, mechanoreceptors in the digestive tract are stimulated. Their signals directly or indirectly lead to the production of messages that travel to and inhibit the feeding network. This scheme has two consequences. Because the feeding neurons are blocked, (1) they are less likely to order a feeding reaction to chemosensory cues signaling "squid present," and (2) they are less likely to inhibit the competing network controlling the withdrawal response. The slug will become more "cautious," readily withdrawing its sensitive oral veil from mild tactile stimulation. As the mechanical bulk in its gut declines during food processing, the feeding response gradually regains its usual high-priority position.

A similar mechanism for the cyclical control of feeding behavior exists in blowflies [191, 192], which feed not on squid, but on various exudates from plants, juices of liquefying animal corpses, and other savory fluids rich in sugars and proteins. During the night, nutrients collected in earlier meals are metabolized to provide energy for the insect. By morning, an internal monitoring system detects low blood sugar levels and orders the fly to fly. When it smells certain odors, it heads upwind until olfactory stimulation from key stimuli becomes so intense that the muscles controlling the wings are shut down and the fly alights.

Once on a leaf, the insect walks until it steps into a solution that contains carbohydrates, at which time tarsal sugar receptors in its feet fire. Action potentials from these cells reach the brain, where decision-making neurons

5 **A blowfly extends its proboscis** in response to stepping into sugary liquid. The labellum at the base of the proboscis has many sensory hairs whose signals influence the feeding behavior of the insect. Photograph by George Gamboa.

command muscles to extend the proboscis. The lower surface of this apparatus comes into contact with the liquid, the labellum is spread (Figure 5), and liquid flows into the oral groove, stimulating still more sugar receptors there. Input from the oral receptors is routed directly to the brain; central motor neurons order the fly to begin drinking. The speed with which fluids are imbibed and the duration of sucking are proportional to the concentration of sugar in the solution. If the liquid is not very sugary, the oral receptors ADAPT (cease firing) quickly and sucking ceases. If sugar concentrations are high, adaptation of the oral receptors may not occur for 90 seconds or thereabouts. After a while, the oral sensory cells recover and will respond again to stimulation, leading to a new bout of sucking.

Eventually, however, feeding will cease entirely, even when the fly's tarsal taste receptors have regained their sensitivity to sugary stimuli and are firing enthusiastically in response to an appropriate fluid. Drinking stops when enough food is consumed to fill the crop (a storage area) to overflowing, at which point fluid is forced back into the foregut, distorting stretch receptors attached to this part of the digestive tract. The receptors fire and relay their messages to the recurrent nerve, which runs between the foregut and the brain (Figure 6). While the recurrent nerve fires, extension of the proboscis is permanently inhibited, no matter how active the sugar receptors are in the feet of the fly. Cutting the recurrent nerve produces a fly that literally cannot stop drinking, until the orgy of consumption causes its body to burst.

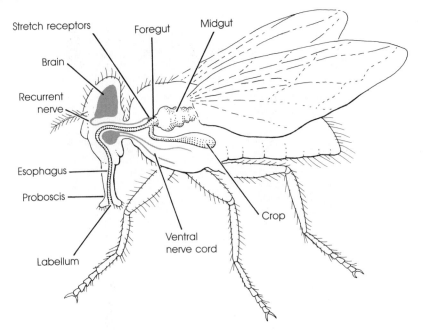

6   **Nervous system and digestive system of a blowfly.** Severing the recurrent nerve eliminates feedback to the brain on the degree to which the crop is filled with fluid; this eliminates the signals that would eventually have blocked feeding in a full blowfly.

The fly's feeding control system prevents its life from coming to a spectacularly explosive conclusion. The fly feeds in sensible bouts when it has a nutritional deficit. This coordinated behavior requires that it be able to integrate information on its internal state with diverse external stimulation encountered on its travels. Its nervous system contains the appropriate receptors and integrators that will enable it to feed cyclically in response to its fluctuating needs [191].

Circadian Rhythms

The fly's feeding behavior is organized in part by its ability to make rapid changes in response to external cues. But there are also internal mechanisms that act independently of environmental changes to provide the fly with a 24-hour cycle of physiological, biochemical, and behavioral activities. These daily cycles enable the fly to *anticipate* the predictable environmental changes that follow a daily rhythm, preparing it to behave in ways that are appropriate for the changes that inevitably occur as the earth rotates about its axis. Blowflies are inactive at night; only during the day when light and temperature conditions make searching easier do they hunt for food, mates, egg-laying sites, and so on.

Because night and day, and the seasons of the year, are so predictable, many animals besides blowflies have evolved internal anticipatory systems

7 **A male cricket calling at its burrow.** Photograph by E. S. Ross.

[616]. The most widespread and intensely studied of these are the CIRCADIAN RHYTHMS of many animals, cycles that persist even under constant conditions. For example, male *Teleogryllus* crickets have a daily bout of singing (to attract females), starting at about the same time in the evening and continuing for more or less the same number of hours before stopping (Figure 7) [498, 499, 501]. The cycle of singing behavior could arise because male crickets use declining light intensity as a cue to begin calling. If one takes a population of males into a laboratory and leaves the light on permanently (or places the crickets in continuous darkness), however, the males will continue to sing only for a limited block of hours each day. Under conditions of constant light, singing starts about 25.5 hours later than the onset of singing the previous day (Figure 8). Individual variation in this *free-running cycle* means that some males begin each bout at 26 hour intervals, whereas others held in the same location begin every 25 hours. This finding is evidence that the cycle is truly internal and not triggered by some subtle environmental cue overlooked by the experimenter.

Thus, without the cues provided by nightfall and sunrise, a male cricket will continue to sing in a repeating cycle, but he gradually drifts out of phase with the actual nighttime hours because his circadian cycle is not quite 24 hours. The slight divergence of a free-running cycle from 24 hours is typical of animal species generally [617, 618].

Now suppose our crickets have been changed to a regimen of 12 hours of light, 12 hours of dark. The switch from light to dark offers a cue that males will use to ENTRAIN the circadian pattern of singing. In a few days, they will all start to sing about 2 hours before the lights go off, accurately

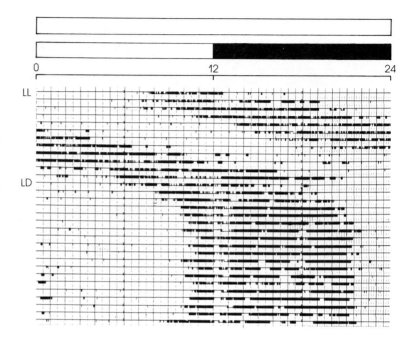

8 **Circadian rhythm.** Male crickets held under constant light (LL) still exhibit a daily cycle of singing and nonsinging, although the start of singing shifts later each day. After 12 days of continuous light in this experiment, the day was divided into 12 hours of light and 12 of dark (LD). These cues entrain the calling rhythm, which stops shifting and begins an hour or two before the lights are turned off each day. *Source:* Loher [498].

anticipating nightfall, and they will continue until about 2.5 hours before the lights go on again in the "morning" (Figure 8). This cycle of singing matches the natural one, which is synchronized with dusk; it does not drift out of phase with the 24-hour day but will continue to be reset each day so that it begins at the same time in relation to lights-out [498]. Therefore the mechanisms regulating the singing behavior of male crickets consist of a timer or biological clock and a device for synchronizing the clock to local time.

Female crickets do not sing, but they do travel to singing males at night. An unmated female that has recently metamorphosed into an adult will begin moving just after the lights go out in a laboratory that is set on a 12L:12D schedule. Her activity can be monitored if she is placed in a special cage with a running wheel, an apparatus rather like a tiny treadmill, whose revolutions activate an electronic circuit and generate a pen mark on a slowly moving paper strip. This apparatus creates a permanent record of when the female is walking (Figure 9). The female need not hear a male calling in order to begin or continue her search, which lasts for most of the night.

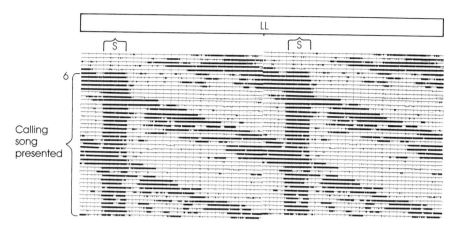

9 **Circadian rhythm.** Cricket female activity patterns follow a daily cycle even in continuous light. After day 6, male song was played to the females for 3 hours twice daily (S). The calls stimulated female locomotion but did not abolish their clock-triggered locomotory rhythm. *Source:* Loher [498].

The female's activity is also influenced by an internal timer, as can be shown by holding females in constant light. They will enter a free-running cycle with a period of about 25.5 hours, the same as the male's free-running calling cycle [499]. As in the case of the male, the switch from light to dark, if it is available, will entrain the female's circadian rhythm so that she begins searching the moment the lights go out each day.

To study the circadian mechanisms of crickets and other animals investigators have employed two main tactics. One is to infer something about the properties of the system by examining how it reacts to various environmental manipulations, usually involving changes in light and dark regimes, as just illustrated with *Teleogryllus* crickets. The other tactic is similar to that used by Roeder in his studies of mantis nervous systems, namely to disconnect various parts of the nervous system surgically. If one cuts the connections that carry sensory information from the eyes of a male cricket to the optic lobes of his brain (depriving him of his vision), he enters the free-running pattern. Visual signals are needed to entrain the daily rhythm, but a rhythm persists in the absence of this information. If, however, one separates the optic lobes from the rest of the brain, there is a complete breakdown of the singing cycle and a male will sing with equal probability at any time of the day (Figure 10).

James Truman and Lynn Riddiford's study of adult emergence in two species of silk moths offers a particularly ingenious example of a surgical study of biological clocks [790]. Adults of one species usually emerge from their pupal cocoons right about dawn; the other species enters the world as adults in the middle of the night. The removal of the brain from silk moths of either species destroys these patterns, but if the brain is transplanted from the head of the pupa to its abdomen, the animal is likely to emerge at the customary time for its species.

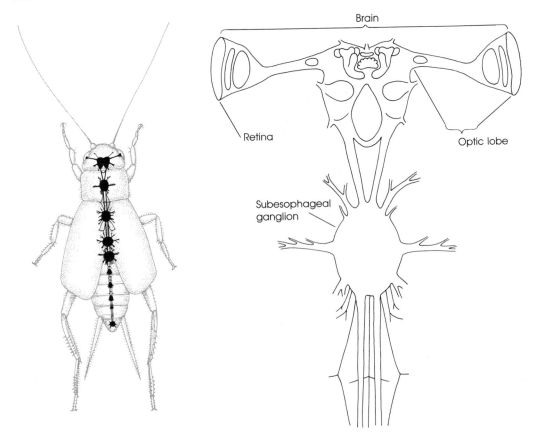

10 **Cricket nervous system.** Visual information from the eyes is relayed to the optic lobes of the brain in the head of the cricket. If the optic lobes are surgically disconnected from the rest of the brain, a cricket loses its capacity to maintain a circadian rhythm. Based on diagrams by F. Huber and W. F. Schürmann.

Truman and Riddiford then performed their neatest trick by transplanting brain tissue from species A to the abdomen of brainless species B, and vice versa. The subjects in this experiment adopted the emergence pattern of the other species, a result that clinched the argument that the brain is the site of a biological clock that controls emergence in the two silk moths.

The general picture is that insect brains contain a major clock mechanism. The clock is usually entrained by signals received from the insect's eyes or other photoreceptors, and in turn the clock sends messages to one or more systems located in other regions of the brain (and perhaps elsewhere) that drive the various circadian rhythms exhibited by the species (Figure 11) [412, 616].

The search for the location and operating rules of circadian systems in vertebrates has revealed a broadly similar pattern of cyclical control. For example, if one damages the suprachiasmatic nuclei (SCN) in the brains of hamsters and Norway rats, these animals lose circadian rhythms in a host

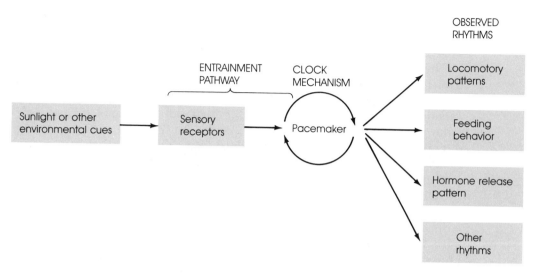

11 **Circadian rhythm pacemaker.** A master clock may, in some species, act as a pacemaker to regulate the many other clocks that control the circadian rhythms of an individual. After Johnson [412].

of attributes such as heart rate, hormone secretion patterns, locomotory cycles, and feeding behavior [885]. In addition, electrical stimulation of cells in the SCN can disrupt the circadian rhythm of some rodents [679]. Although destruction of the SCN does not eliminate all physiological cycles, nevertheless it appears to play a major timing role, perhaps by influencing the activities of a set of what have been called "slave oscillators," timing mechanisms responsible for rhythm of one specific activity, such as loco-motion or heart rate or feeding.

In some cases it is possible to remove a structure thought to contain a clock mechanism and test whether it continues to produce a circadian rhythm even in isolation from the rest of the body from which it was taken. One example is the eye of some marine snails, which exhibits a beautiful circadian rhythm of neural activity in vitro, even when isolated from the brain. Likewise, the pineal gland of lizards and birds (the so-called third eye) can be removed from anoles and kept alive in a container for as much as a week. In constant darkness an isolated pineal gland secretes the hormone melatonin on a daily cycle. This rhythm can be entrained to light–dark cycles, so the pineal gland includes a photoreceptor in addition to the clock. Moreover, the interval between peaks of melatonin release from in vitro pineal glands is little affected by different temperatures. This prop-erty, termed temperature compensation, is characteristic of all circadian rhythms, even those exhibited by isolated neural tissues [552].

Lunar Cycles
The circadian changes in responsiveness to certain stimuli that occur on a 24-hour repeating cycle are not the only behavioral rhythms in the animal

kingdom. There are some species whose neural mechanisms enable them to match their behavior to repeating lunar cycles, such as the 29.5-day cycle of moonlight intensity or the 14.8-day cycle of tides.

The occurrence of high "spring tides" every 14.8 days has great reproductive significance for the grunion, a small silvery fish that mates and lays its eggs on the beaches of California and Mexico. Grunion only come ashore in large numbers shortly after high tide on nights immediately following nights with a full moon or a new moon (Figure 12). These are the days when the difference between high and low tides are at their maximum. Females let a wave carry them up the beach, where they bury the lower part of their body in the moist sand; males release their sperm as the eggs are laid. Waves then carry the fish back to the ocean, leaving behind buried eggs that are covered with sand deposited by the receding tide. The eggs will not be exposed for about 10 days until, in the course of a new tide cycle with high tides of increasing magnitude, the surf uncovers the eggs (a process that stimulates the mature embryos to hatch) and carries the fry out to sea. The precise means by which grunion determine when to come

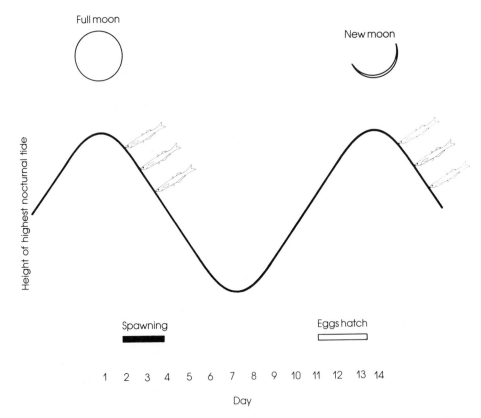

1 2 **Lunar cycle of grunion.** The grunion spawn at 2-week intervals, when the nocturnal high tide has just passed its peak. When their eggs hatch some days later, the fry are uncovered at a high tide, and swept out to sea.

ashore and breed is not known, but it would not be surprising to find that the fish possess a clock that is set to a fortnightly cycle and primes them to become sexually active at the optimal time for successful reproduction [617]. Such an internally generated cycle has been very well studied in a marine midge, a little fly that also breeds at a particular phase of the tidal cycle [616].

A very different creature whose behavior is also temporally regulated by the phase of the moon is the banner-tailed kangaroo rat [496, 497]. Here laboratory evidence suggests that the behavioral cycle followed by the rodents is not set by an internal clock based on lunar phases, but results from direct monitoring of moonlight conditions. Robert Lockard and Donald Owings recorded the activity of individual rats with an ingenious food dispenser–timer invented by Lockard. The device released very small quantities of millet seed at hourly intervals. To retrieve the seed, the animal had to walk through the dispenser. When it entered, the animal depressed a treadle, which moved a pen, which made a mark on a paper disk, which was turned slowly through the night by a clock mechanism. When the paper disk was collected in the morning, it carried a temporal record of all nocturnal visits to the dispenser.

Data collection was sometimes frustrated by ants that perversely drank all the ink or by Arizonan steers that trampled the recorders. Nevertheless, Lockard's records showed that in the fall after the animals had accumulated a large cache of seeds, the rats were selective about foraging, usually coming out of their underground bunkers only at night when the moon was not shining (Figure 13). Because the predators of kangaroo rats (coyotes and owls) locate their prey more easily when visual cues are available, banner-tails probably are safer when foraging in complete darkness.

Reproductive Cycles and Changing Behavioral Priorities   We have seen that some animals possess internal clock mechanisms that cycle on a daily or lunar period, enabling them to anticipate the appropriate (safe or productive) time to do certain things. Animals also monitor their internal environment in order to match their behavior to their physiological state. When a *Teleogryllus* cricket female that has been searching each night for a male finally finds one and mates, she instantly ceases to wander about and instead devotes her time to egg laying (Figure 14). The underlying mechanism regulating this obviously functional shift in priorities is activated by the transfer of an enzyme in the materials donated by a male to his partner, an enzyme that turns on prostaglandin (PG) production in the female. PG in turn stimulates egg laying and suppresses locomotion [500].

The ugly sea slug, *Pleurobranchaea californica*, offers another fine example of how animals with a reproductive cycle change their priorities in ways that mesh with that cycle. The slug produces eggs in batches, and it takes time to capture enough squid to manufacture and mature each clutch. As noted before, feeding behavior has very high priority for *P. californica*. At the ultimate level, this helps slugs get enough food to produce their clutches quickly. At the proximate level, specific "feeding" neurons suppress other components of the slug's nervous system in ways that put feeding high in the behavioral hierarchy of *P. californica*. All this changes, however,

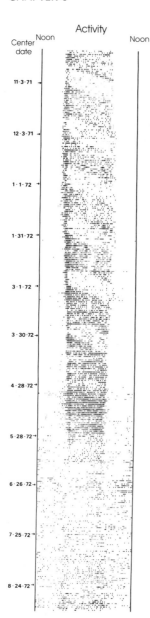

13 **Lunar cycle of banner-tailed kangaroo rats.** Each black mark represents one or more records made by rats feeding at a timer device. In the period from November to March, the rats were active at night when it was dark (the white diagonal bands correspond to the hours when the moon was shining). A shortage of seeds later caused the animals to feed throughout the night, even when the moon was up, and later still to forage during all hours of the day. Courtesy of Robert B. Lockard.

at the time the eggs are laid. A slug offered a chance to strike at a food stimulus (homogenized squid) that normally sends it into a feeding frenzy will usually ignore the stimulus for a few hours after egg laying. As eggs develop, a hormone is produced and circulates in the bloodstream. At the time of egg laying, concentrations of this hormone peak and suppress the activity of nerve cells associated with the feeding response [179].

W. J. Davis and his co-workers tested the hypothesis that hormones were involved in feeding inhibition by injecting into other individuals blood samples taken from slugs that had just laid their eggs [656]. The transfu-

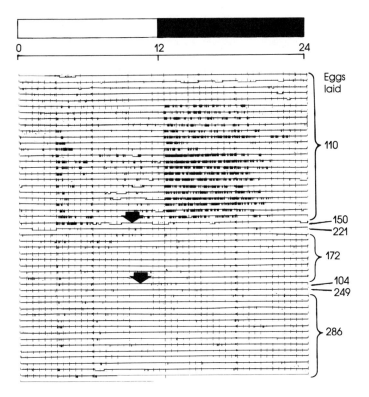

14 **Activity patterns of a female cricket** before and after copulation. After mating (arrows), the female ceases nocturnal locomotion, an activity useful in finding a mate but irrelevant to egg-laying. *Source:* Loher [500].

sions induced the recipients to lay their eggs (whether the eggs were mature or not) *and* reduced their response to squid homogenate. Normally hormonal suppression of feeding has the ultimate value of reducing the probability that a slug will cannibalize its own progeny; by the time hormonal levels fall and feeding regains its dominance, the slug will usually have left its clutch.

Hormones, Reproductive Cycles, and Anoles

The green anole of the southern United States is a little lizard capable of doing a great many things: stalk a fly, dart for cover when a sparrow hawk swoops, go dormant for months, fight with rivals, lay eggs, and copulate. All of these actions are reproductively useful to the animal—if the lizard sets its priorities correctly. For example, during the summer in South Carolina, sexually mature females regularly encounter sexually mature male anoles. A male usually responds with a courtship display in which he extends his dewlap (Figure 15) and bobs his head up and down, flagging the female with his dewlap. If the female does not flee, the male bites her behind the head, crawls onto her back, and inserts one of his two penises into her cloacal opening.

15 **Dewlap display of a male Jamaican anole.** Photograph by Thomas A. Jenssen.

But females often ignore a male's invitation to mate. Each copulation consumes 5–20 minutes, during which time the immobilized female cannot feed or accomplish anything else. Female anoles avoid superfluous matings; they only copulate when they have a mature egg ready to be fertilized. Green anoles, like sea slugs, have proximate mechanisms that keep track of egg maturation and use this information to regulate their behavioral priorities.

In green anoles only one egg develops fully at a time (over a period of 10 to 14 days). When it is mature and the female is about to ovulate, she becomes receptive; only then will she exercise her precopulatory neck-arching behavior in response to a territorial male's courtship signals. Immediately after mating, the female becomes unreceptive again and her rejection of males persists for the next week or two while a new egg develops. She eventually regains her willingness to copulate, and the cycle is repeated (Figure 16).

What regulates the female anole's reproductive cycle? How does she know when to respond and when not to accept a male's advances? The regulatory mechanisms that control receptivity are complex; but as in the sea slug, hormone–nervous system interactions play critical roles in determining what stimuli she responds to. As an egg develops, cells in the ovary communicate hormonally with the brain of the anole by releasing two sex hormones—estradiol and progesterone—into the bloodstream.

When ovarian hormones reach target cells in the brain, they activate the release of certain other hormones from the pituitary gland on the underside of the lizard's brain. The pituitary hormones travel back to the ovaries, where they help maintain the schedule of production of estrogen and progesterone. This feedback loop eventually alters those brain cells that react to courtship stimuli. Precisely how this happens in anoles is not known, but in white rats estrogen enters specific nerve cells in the brain, particularly hypothalamic cells, where it binds with receptor molecules in cell nuclei. These events alter the production of proteins in the target cells,

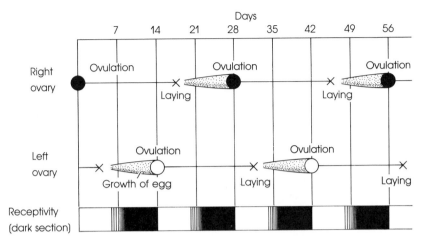

16 **Repeating cycle of sexual receptivity** in female American anoles during the breeding season. One egg matures at a time. As an egg develops, the female gradually becomes receptive and will mate before ovulation, which causes an abrupt cessation of receptivity. *Source:* Crews [155].

with the result that the cell's electrical properties are changed, ultimately changing their responsiveness to key stimuli [631]. If something similar happens in anoles about the time when ovulation occurs, hormones have altered part of the female's brain, causing her to arch her neck when she sees the bobbing dewlap of a male.

One can experimentally demonstrate the importance of estrogen and progesterone in the control of female receptivity by removing an anole's ovaries [549, 778]. The ovariectomized female will never respond sexually to male courtship again unless she receives compensatory hormone injections, such as a shot of estradiol 24 hours before she receives an injection of progesterone. Given the proper concentrations of hormone doses, almost all females will be ready to mate in about 24 hours, despite their eggless condition.

The Control of Receptivity

Once copulation is completed, one can arrange (experimentally) to have a new male presented to an intact female. If this is done within a minute or two after her first mate dismounts, a female may copulate once more. But if 5–7 minutes have passed, she will religiously refuse the male and will remain unwilling for 10–14 days, until she is about to ovulate again. The transition from receptivity to nonreceptivity is therefore extremely abrupt [155].

The onset of nonreceptivity is closely related to intromission by the male. If a female is courted and then mounted by a male anole whose double penis has been surgically removed (the life of a laboratory lizard has its unpleasant moments), she will retain her willingness to copulate after the

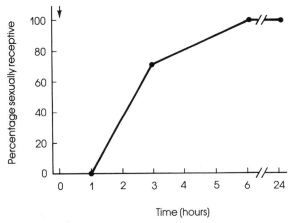

17 **Effect of injection of prostaglandin** on the receptivity of female American anoles. When injected (arrow) into seven females, prostaglandin temporarily abolished their readiness to copulate. *Source:* Tokarz and Crews [779].

male dismounts. As in the case of crickets, normal matings result in transfer from the male to the female of materials that alter PG concentrations in females, and these chemicals trigger a change in female behavior [779]. Richard Tokarz and David Crews injected minute amounts of PG in females whose receptivity had been established by their willingness to permit a mature male lizard to grip them by the neck, after which the male was removed. (The females already had been ovariectomized and had received estrogen–progesterone treatment to prime them to mate.) Shortly after PG injection, the females were no longer receptive (Figure 17). When they saw a bobbing dewlap, they did not wait to be mounted but either departed in a flash or bit the male, treating him as an aggressor rather than as a sexual partner.

Hormones and the Annual Cycle of Behavior   In addition to short-term behavioral cycles like that underlying the sexual activity of female anoles in the summer, many species exhibit a pattern of changing behavioral priorities that spans more than a day or a few weeks. The anole, for example, has an annual cycle of behavior (Figure 18). Just as we can ask what regulates the 10- to 14-day cycle of receptivity and nonreceptivity in females, we can also ask what mechanisms are responsible for the fact that anoles have a spring/summer breeding season, followed by an autumnal nonbreeding period, then inactivity in the winter with emergence in the spring and eventual reentry into the breeding cycle.

Perhaps anoles and other species that exhibit seasonal shifts in behavior possess an internal annual clock mechanism, analogous to the clock that drives circadian rhythms. A CIRCANNUAL RHYTHM would produce an annual cycle of behavioral events even in the absence of environmental cues correlated with seasonal changes. The only definitive demonstration of an annual clock, however, involves the golden-mantled ground squirrel [627], which in nature spends the late fall and winter hibernating in an underground chamber. Five members of this species were born in captivity, and blinded, and held throughout their lives in constant darkness and constant temperature while being provided with an abundance of food. Year after year they entered hibernation at about the same time as their fellows living in the wild (Figure 19).

The appropriate experiments to test whether anoles have a circannual rhythm independent of external cues have not been done. But ample evidence now exists to show that a host of factors, many of them mediated by hormonal changes, influence the timing of the main events in an anole's annual cycle of behavior (Figure 18). When male anoles emerge from winter dormancy in February they soon begin to exhibit aggressive behavior toward other males, claiming foraging territories, from which they repel other males with bobbing display threats and occasional physical combat. These activities are correlated with a sharp rise in testosterone levels in the bloodstream and brain tissue, evidence leading to the hypothesis that male territoriality and subsequent sexual behavior are testosterone-dependent.

Females emerge in March and reproduce for the first time in late April. The control of the start of the breeding season in females is also linked to changes in hormonal activity, which in turn reflect a variety of environmental influences that either accelerate or inhibit ovarian development and reproductive activity. Temperature is of paramount importance in this process [479]. Anoles possess temperature receptors that relay information about increasing temperatures in the early spring to the brain. The pitui-

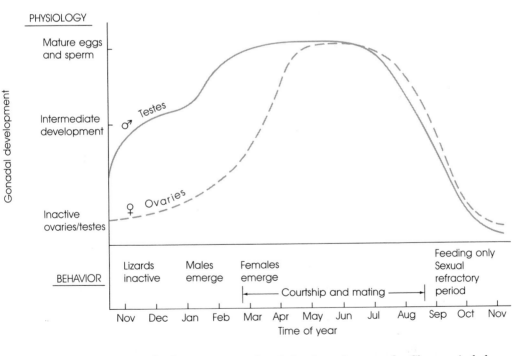

18 **Annual cycle of the American anole.** Changes in behavior of males and females are correlated with changes in the development of hormone-producing ovaries and testes. *Source:* Crews [155].

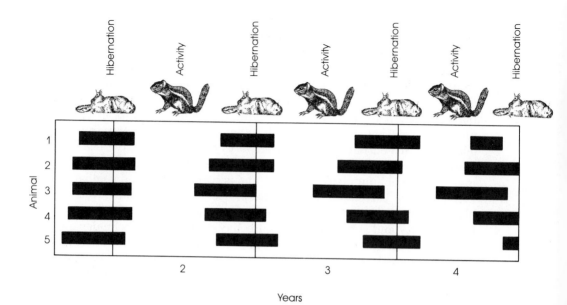

19 **Circannual cycle of the golden-mantled ground squirrel.** Animals held in constant darkness and temperature nevertheless enter hibernation (black bars) at a certain time year after year. *Source:* Pengelley and Asmundson [627].

tary gland then releases gonadotropic hormones that stimulate the growth of the ovaries in females and testes in males. The precise pattern of temperatures in any one spring determines exactly when an individual emerges from its hibernaculum and begins to undergo gonadal growth. Colder weather delays emergence and the start of reproduction; warmer days end dormancy sooner.

Reproduction is regulated not only by temperature but also by an anole's social environment. Recently emerged females encounter male territory owners and perhaps other males as well. The behavior of males can have a profound influence on the onset of reproduction in female anoles, as the following experiment illustrates.

Crews removed dormant females from their winter hiding places, moved them to the laboratory, and placed them in cages in groups (1) with no males, or (2) with castrated males only (which will not court), or (3) with males whose dewlaps had been surgically removed, or (4) with intact males, one of whom had established territorial dominance over the other males present [155]. Females in the first three groups did not have the opportunity to observe the courtship display of a territorial male, whereas females in the fourth group regularly saw spread dewlaps of the territorial males. The ovaries of the fourth group developed much more rapidly than those of the "dewlap-deprived" groups. Thus, visual stimulation from a displaying male not only triggers the neck-arching response in a receptive female, it actually prepares recently emerged females to become sexually receptive. Females that are courted frequently secrete greater quantities of pituitary gonadotropins than uncourted anoles. The increase in pituitary hormones stimulates ovarian development and the production of estrogen, which in turn causes females to become receptive, synchronizing their reproductive state with that of courting males.

In summary, the activity of the hormonal system of an anole is regulated by neuronal mechanisms that receive inputs from various elements of the physical environment (i.e. temperature), social environment (i.e. the behavior of potential mates), and internal environment (i.e., stage of maturation of ovaries and eggs). The resulting pattern of hormone release integrates the information received and sets the animal on a behavioral course that is appropriate for the season of the year, the social circumstances that it happens to experience, and the developmental stage of its gonads.

Hormones and the Annual Cycle of the White-Crowned Sparrow

White-crowned sparrows, like anoles, are faced with profound seasonal changes in conditions that make changes in behavioral priorities highly adaptive [257, 863] (Figure 20). These sparrows employ hormonal mechanisms that are similar in many ways to those that control the annual cycle of behavior in green anoles (and many other animals). Imagine a white-crown from a population that breeds in central Alaska. Food is abundant there in the late spring and summer but not in the winter. Sensibly enough, white-crowned sparrows abandon Alaska in the fall and travel as much as 2400 miles to wintering spots in southern United States and Mexico.

But in order to reproduce successfully in Alaska, a white-crown has to time its return trip just right, so that it arrives when it is not so cold that it will starve, but not so late in the season that all choice territories are occupied. And when a bird gets to Alaska it would be helpful if its ovaries or testes were ready to become functional again. During the nonreproductive seasons, the gonads of songbirds atrophy, often to less than 1 percent of their maximum weight. Full-sized ovaries and testes are, however, indispensable for reproduction and must be redeveloped at the appropriate time. Because white-crowns wintering in Mexico or Arizona have no direct information about conditions in Alaska, they need some sort of calendar to tell them when it is time to leave. Their calendar lies in the ability to sense annual changes in the length of the PHOTOPERIOD, the number of hours of daylight in a 24-hour period. Because daylight increases and decreases on a predictable seasonal pattern, white-crowns can use this cue to regulate their physiological and behavioral priorities [257].

Information about the length of the photoperiod is collected by light-sensitive receptors present somewhere in the brain itself. If the white-crown's visual system is destroyed or if its eyes are covered by light-proof goggles, the bird's circadian rhythm is not eliminated [876]. It will continue to wake up in the morning, become active for some time, less active during the middle of the day, and more active again in the early evening before going to sleep once more.

The bird's circadian clock may work in the following fashion. Built into its time-measuring system is a daily cyclical change in sensitivity to light,

BEHAVIORAL CYCLE

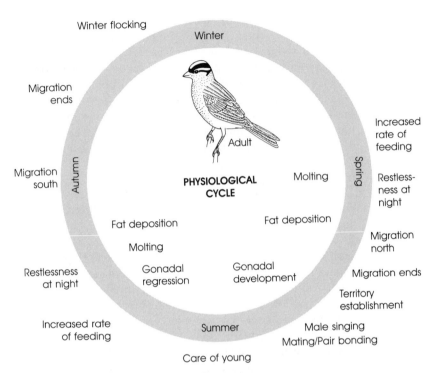

20 **Annual cycle of the white-crowned sparrow.** The seasonal sequence of behavioral changes is correlated with hormonal changes in the physiological cycle of the bird.

a cycle that is set each morning at dawn. During the initial 13 hours or so after the clock is set, the animal's control system is highly insensitive to light; this insensitivity then steadily gives way to increasing sensitivity, reaching a peak in the 16- to 20-hour period after the starting point in the cycle. Sensitivity then fades very rapidly to a low point 24 hours later, at the start of a new day and a new cycle. If the days are 12 or 13 hours long and nights 12 or 11 hours long, the photosensitive part of the system simply is never activated because there is no light during the light-sensitive phase of the cycle. However, if the days are 14 to 15 hours long, light reaches the bird's brain during the photosensitive phase. This event causes the release from the hypothalamus of hormones that stimulate the anterior pituitary. Cells there release the hormone prolactin and assorted gonadotropins, which travel by the bloodstream to the gonads of the bird, where they initiate development of the reproductive equipment of the sparrow.

If this model of the clock system is correct, it should be possible to deceive it experimentally. William Hamner, working with house finches [338], and Donald Farner, in similar studies of white-crowned sparrows [286], were able to stimulate testes growth by giving captive birds light during the hypothesized photosensitive phase of their circadian rhythms. In Farner's experiment birds that had been on a regular schedule of 8 hours of light and 16 hours of darkness (8L:16D) were shifted to a 8L:28D schedule. Because the light periods were now out of phase with a 24-hour cycle, these birds received light during the time when their brains were thought to be highly photosensitive. The male birds' testes grew under these conditions, even though there was a lower ratio of light to dark hours than during the 8L:16D cycle, which did not stimulate testicular growth (Figure 21) [286]. The same effect can be achieved by switching a bird from an 8L:16D cycle to an 8L:16D cycle that is interrupted by a 2-hour light period 17 to 19 hours after the start of the day. This coincides with the time of peak photosensitivity during the circadian cycle. The result is marked testicular development.

By the time a white-crown arrives on the Alaskan breeding ground, it has been exposed to steadily longer photoperiods and is primed for rapid gonadal growth [862]. In both males and females, testosterone levels become much higher than they were during migration. Males aggressively attack territorial intruders, sing constantly, and attract a mate. The female also aggressively defends the territory after she joins a partner. As she feeds and acquires sufficient supplies for egg production, the concentration of estrogen increases in her blood. Females then begin to solicit copulations

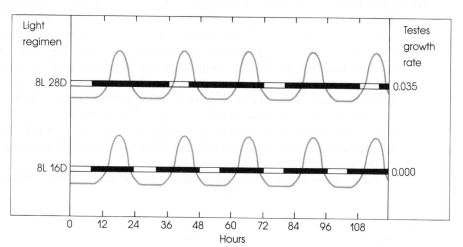

21 **Test of the hypothesis** that white-crowned sparrows possess a daily cycle of photosensitivity. The open and black sections of the two bars show the light and dark periods of two different L:D regimens. The curves represent hypothetical cycles of "photosensitivity," with a peak period of photosensitivity occurring every 24 hours. Sparrows in a 8L:28D experimental schedule are exposed to light occasionally during the photosensitive phase of the cycle, and they respond with testicular growth. *Source:* Farner [286].

from their partners, who are quick to respond positively to these displays.

The biological priorities of a pair change once eggs are laid. It is thought that in both sexes prolactin secretion becomes dominant as gonadotropin and testosterone levels wane. The gonads decline in weight, the birds become far less aggressive, and the males stop singing [862]. As the eggs of a white-crowned female hatch, the nestlings begin to beg for food and are fed by both parents; they grow up, leave the nest, and soon become independent. The adult's tendency to provide parental care declines correspondingly, perhaps linked with a decline in prolactin secretion. The birds, like green anoles, then enter a SEXUAL REFRACTORY PERIOD in the fall, during which time captive birds experimentally exposed to long photoperiods simply do not respond with gonadal growth. The hypothalamus, which in the spring reacts to increasing day lengths by stimulating the pituitary to release certain hormones, fails to do so in the fall under the same external stimulation.

Sexually refractory birds feed voraciously, put on weight, molt, and migrate (south) in time to avoid being caught in an early fall snowstorm. If they are able to survive the journey and the winter, their sensitivity to photoperiodic change and their hormonal mechanisms will act in concert to trigger a new series of behavioral changes in the spring, one cascading after the other in an adaptive sequence [863].

Diversity in the Organizing Mechanisms of Behavior   In both anoles and white-crowned sparrows, changes in external environmental conditions (temperature and photoperiod) and in social conditions (the social behavior of potential sexual partners) change the hormonal state of individuals. Hormones control gonadal development and reproductive behavior, so that reproduction occurs when physical and social conditions are most favorable. The same is true for many other reptiles, birds, and mammals, a fact that has led most biologists to accept the view that hormones are the ultimate arbiters of sexual behavior (Figure 22). Although it seems clearly adaptive in most cases for animals to have hormones that can communicate between the gonads and the brain, permitting the brain to tell when the gonads are functional, the model shown in Figure 22 ignores some interesting diversity in the organizing mechanisms of behavior.

First, the particular environmental and social conditions that affect hor-

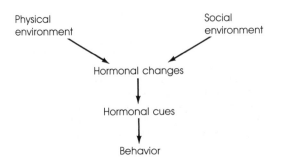

22 The traditional view that environmental cues trigger internal hormonal changes, and that these changes are necessary to initiate behavioral responses.

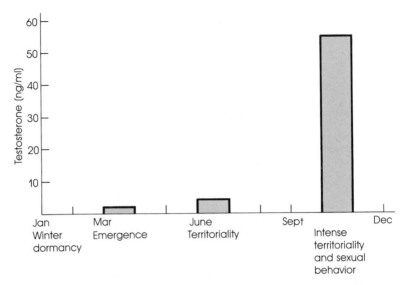

23 **Annual cycle of male mountain spiny lizards** and its relation to testosterone levels. In this species, aggressive territorial behavior begins when testosterone levels are low. *Source:* Moore [569].

monal systems often differ from species to species. We have already noted that increasing temperature plays a key regulatory role in starting the reproductive season for green anoles, whereas photoperiod is more critical for the year-round active sparrow. Moreover, not all lizards use temperature information in the same way green anoles use it. Yarrow's spiny lizard of the mountains of southeastern Arizona emerges from winter dormancy in March, about the same time as green anoles emerge in South Carolina, but then the spiny lizard waits for 5 months, not several weeks, before breeding, which takes place in the fall—just before they go dormant again (Figure 23). For much of the 5-month prereproductive phase male lizards are territorial, and the onset of this behavior in late April is marked by a slight rise in blood testosterone levels. This increase fails to activate sexual behavior, however, which does not occur until September and October, at which time male testosterone levels skyrocket [569]. Thus, the same basic seasonal changes in temperature and photoperiod induce distinctively different patterns in two lizards.

In addition, not all birds respond to photoperiodic cues in the same way white-crowned sparrows do. For example, Eleanora's falcon does not lay its eggs until late July or early August, long after white-crowned sparrows have begun to breed in most portions of their range [812]. The falcon breeds on islands and sea cliffs, primarily in the southern Mediterranean, where it captures migrant songbirds to feed to its nestlings. Migration does not begin until August, and the hawk times nesting so that its young will hatch precisely when food is most available.

Unlike Eleanora's falcon, some species live in environments in which there is no such consistent relationship between a particular season of the year and plentiful food. These animals often require access to food itself before coming into reproductive condition rather than anticipating the presence of food by using photoperiodic, temperature, or other seasonal cues. For example, the pinyon jay is an extreme specialist on the seeds of pinyon and ponderosa pine. Some years there is a bumper crop of cones; in others, the trees may produce practically no seeds. The jays, therefore, have evolved physiological mechanisms that tie reproduction directly to food supply. In good years the birds begin to court and to build nests much earlier than in years of reduced seed production [481]. Simply seeing numerous green (unripe) cones of pinyon pines in the summer reverses the gonadal decline typical of fall, a reaction that prepares the birds to become sexually active in the late winter and early spring, when they will be feeding largely on stored seeds gathered from the then-mature cones.

Diversity in the Social Regulation of Reproduction

Just as species vary in what elements of the physical environment have an impact on hormones and reproductive cycles, so, too, social stimuli can have variable effects on reproductive timing. For example, Barbara Brockway found that gonadal development in budgerigars was influenced by the kind of songs the birds heard from their companions. Although many Americans believe that budgerigars live exclusively in department stores, this small parrot does occur in the wild as well, notably in the arid center of Australia. The birds travel in large flocks, searching for the rare places where it has rained recently. Upon finding a suitable location, the birds come into reproductive condition promptly. But social factors influence the rate of reproductive development. Brockway found that in the laboratory the testicular growth of males was promoted if they heard tapes of the "loud warble," the male territorial call of this species. Unpaired females, however, that heard only this aggressive call experienced *reduced* ovarian growth. Ovarian development was stimulated only by the male's courtship signals, particularly the "soft warble," which is given when a male perches beak-to-beak with a female [97, 98].

In white-crowned sparrows, hearing male territorial song slightly speeds up ovarian development of females, showing again that the same kind of cue can have different effects on different species [574]. And whereas the key social cues may be acoustical in birds, and visual in lizards, olfactory stimuli are the critical carriers of social information for many mammals. In the house mouse, for example, the scent of a male's urine has a remarkable variety of effects on the physiological status of the animals around him. Thus, a familiar dominant male's odor promotes reproductive cycling in the mature females he lives with; the absence of a male, or the scent of a strange male, blocks female sexual activity. Furthermore the odor of a dominant male speeds sexual maturity in young females—unless the dominant male is the father of the young females (Figure 24) [472].

I interpret the ultimate significance of these effects on females in the

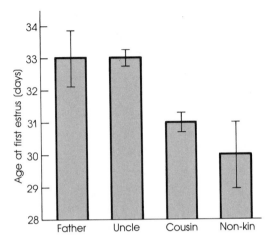

24 **Effects of social experience** on the sexual maturation of female mice. The first estrus of females reared with uncles and fathers is delayed compared to that of females reared with cousins or genetic strangers. *Source:* Lendrem [472].

following way. A dominant male mouse makes a superior mate from a female's perspective because he will protect her offspring against other males who may kill them. If a dominant male is present, there is no reason to postpone reproduction, and females do not, unless mating with the male means mating with one's father, with the attendant likelihood of producing defective inbred offspring (Chapter 9).

Variation in Hormonal Control of Sexual Behavior

Having dispelled the notion that all animals regulate hormone production and sexual behavior by relying on the same cues from the physical and social environment, it is important also to stress that there is variation in the degree to which sexual behavior is dependent on elevated levels of sex hormones. Contrary to the widespread impression that testosterone is always essential for the expression of male sexual behavior, there are species in which some elements of male reproductive behavior are independent of testosterone. For example, male white-crowned sparrows that have been castrated as young individuals before they engaged in reproductive behavior will nevertheless mount females that solicit copulations, provided the males have been exposed to long photoperiods [570]. These males have no source of testosterone but they are still able to copulate.

The uncoupling of testosterone from sexual activity in white-crowned sparrows is further illustrated in Figure 25, which shows the levels of testosterone in males from single- and multiple-brood populations over the course of a year. In some places, pairs raise two or three broods of young per breeding season, unlike Alaskan white-crowns that have time only for a single brood. Males that copulate with a partner to produce a second, or third, clutch of fertile eggs do so at times when they have relatively low testosterone concentrations in their blood.

John Wingfield and Michael Moore hypothesized that the primary function of testosterone for white-crown males is to enhance their aggressive

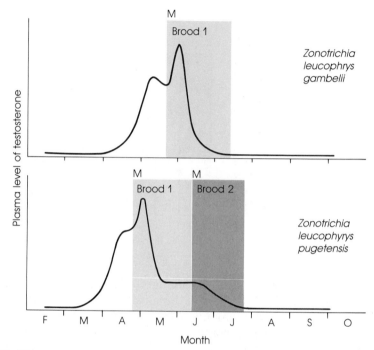

25 **Hormonal and behavioral cycles** in single- and multiple-brooded populations of white-crowned sparrows. Increased testosterone levels occur shortly before the time when males mate (M) with females in their first breeding cycle, but in populations with two breeding cycles, copulation occurs again even though testosterone levels are declining in males. *Source:* Wingfield and Moore [864].

behavior, which is useful when males defend their territory and guard their mates from other males [864]. Later in the breeding cycle, high testosterone concentrations and increased aggressiveness could interfere with a male's paternal behavior toward his first brood of fledgling sparrows, which he cares for, while in those populations where several broods are possible, his mate begins a second clutch. In experiments with the pied flycatcher, Bengt Silverin found that males that received testosterone injections when they normally would have been incubating eggs devoted themselves to singing and territorial defense but neglected their offspring. The result was that the reproductive success of their mates fell [723].

Hormones and Garter Snakes

The red-sided garter snake offers an even more dramatic case of male sexual behavior that does not depend on direct hormonal activation. This snake lives as far north as southern Canada; as a cold-blooded reptile, it spends much of the year dormant in sheltered underground hibernacula. Many individuals gather together in good over-wintering sites. In the late spring when temperatures have warmed, the snakes begin to stir, and soon they

26 **Male red-sided garter snakes** in a spring mating aggregation. This snake's sexual behavior is completely independent of circulating testosterone levels. Photograph by David Crews.

emerge en masse (Figure 26). Before going on their separate ways they engage in an orgy of sexual activity, with males slithering after females and attempting to copulate with them. Examination of the sex hormone concentrations in their blood reveals that males have almost no circulating testosterone or equivalent substance. Yet they have no trouble mating [157].

A wide variety of hormonal manipulations have been performed with these snakes without the slightest effect on their sexual behavior, which is entirely dependent on temperature-sensitive neurons. In the fall, testosterone plays its role in the formation of sperm, which are stored within the snake over the winter in anticipation of the spring mating frenzy.

David Crews argues that the extremely short season of above-ground activity available for this northern snake has favored females that mated as soon as possible after emergence in the spring. This in turn favored males that had sperm ready as soon as possible to meet the needs of their partners. Mutant male snakes that produced their gametes in the late summer and were able to engage in sexual behavior immediately after emergence independent of testosterone levels enjoyed a reproductive advantage, according to this scenario.

SUMMARY

1 Physiological mechanisms not only provide animals with the ability to perform special behavior patterns in response to key stimuli, they also

structure the use of the animal's entire repertoire of behavioral abilities. One aspect of the organization of behavior is the ability of an individual to carry out an action in complete form without interference from other competing behaviors. Typically, for example, escape behavior takes precedence over all other possible actions, with neural activity in the escape system suppressing activity in other centers.

2 The nature of the inhibitory relations between neural elements may change in response to changing internal states. These changes may involve short-term behavioral cycles, as occurs in a blowfly with a full digestive system in which feeding temporarily takes a back seat to resting or grooming behavior. Daily cycles of behavior often occur and are regulated in many cases by internal circadian clocks whose manner of operation still remains largely unknown.

3 Patterns of behavioral organization regularly follow a seasonal or annual cycle, as in female anoles, which alternate between receptivity and nonreceptivity in relation to the state of egg maturation. Hormones are generally important in the long-term organization of behavior, especially in timing reproductive activities so that young are produced when they are most likely to survive. Often seasonally related events, such as increasing day length or warmer temperature, are detected by neural mechanisms and translated into a sequential release of hormone messengers that activate one previously blocked neural system after another in the biologically correct order.

4 The sequence of behavioral events that occurs over a season or year and is typically induced by photoperiodic or food changes in the environment is in many species subject to fine adjustment in response to a host of other cues. In particular, social interactions provide information that animals use to regulate their hormonal state, and thus their behavioral priorities and tendencies.

5 There is great diversity among species in the organizational weight given to various kinds of environmental cues. For example, regulatory mechanisms in "cold-blooded" species are often tuned to temperature changes whereas photoperiodic changes may have greater effect on behavioral timing in "warm-blooded" animals. Moreover, precisely the same cues may have very different hormonal effects and behavioral consequences. Finally, the degree to which reproductive behavior is dependent on key sex hormones also varies among species. The evolutionary basis for differences among species in the proximate mechanisms that control behavior may be related in part to differences in the ecological pressures they face.

SUGGESTED READING

Kenneth Roeder's *Nerve Cells and Insect Behavior* [669] and Vincent Dethier's *The Hungry Fly* [191] discuss how some animals avoid conflict and structure their behavior over the short haul. Terry Page has written a comprehensive review of biological clocks and circadian rhythms in insects [616]. David Crews and his colleagues are responsible for a beautifully detailed picture of the organization of anole behavior [155, 156]. David Crews and Michael Moore discuss the evolutionary significance of diversity in the hormonal mechanisms controlling behavior [157]. *Psychobiology of Reproductive Behavior* (edited by David Crews) contains a series of chapters on the proximate and ultimate basis of behavioral organization [156].

DISCUSSION QUESTIONS

1. California quail live in environments that experience great fluctuations in rainfall and seed production. Some years California quail do not even attempt to breed. What are some proximate factors that might cause the birds to suppress their reproductive cycle? How would you go about testing which of the potential factors was actually responsible for regulating reproduction in this species? See [475] after planning your tests.

2. Can you think of an ultimate hypothesis for why kangaroo rats would use a circadian rhythm to time their daily activity, rather than simply checking from time to time on whether night had come?

3. In studying the hormonal control of behavior, it is common to remove an animal's ovaries or testes and then inject the creature with assorted hormones to see what behavioral effect they have. What advantage does this technique have over another approach, which is simply to measure the levels of specific hormones in the blood of animal subjects at different times? The direct measurement approach would show, for example, whether mating usually occurred when circulating testosterone or estrogen levels were elevated.

7

The Evolution
of Behavior:
Historical Pathways

Our attention until now has been centered largely, but not exclusively, on the proximate mechanisms of behavior. Ultimate or why questions have often surfaced during our examination of these mechanisms; for example, the discovery that white-crowned sparrows do not need high testosterone levels to copulate raises the ultimate question of why they differ from other animals that do. Knowing something about the history of white-crowned sparrows, the behavior and physiology of the species they evolved from, would be helpful here. Is this a novel feature of white-crowned sparrows, or did they inherit their distinctive behavioral physiology from the ancestor of all sparrows? What were the stages in the evolution of this trait? This chapter looks at how researchers try to trace the history of selected behaviors. If one could go back in time to follow the evolution of a species, it would be relatively easy to establish how the behavior of ancestral populations changed step by step to give rise to the behavior of a current species. In the absence of a time machine, however, one way to determine what extinct animals were doing thousands or millions of years ago is to make inferences about behavioral abilities from fragments of fossil bone. This process can be surprisingly revealing. But another way to go back in time is to compare behaviors within a group of living species in order to reconstruct a plausible evolutionary sequence leading to a modern behavior of particular interest. This chapter describes how comparative procedures help investigators trace the history of an unusual feeding pattern and some remarkable communication signals. An examination of these examples indicates that even the most bizarre behaviors probably evolved gradually, with ancestral traits acting as a foundation for evolutionary changes, not revolutionary ones.

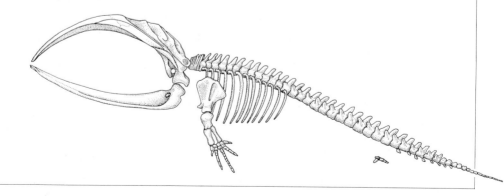

Tracking the History of Behavior in Fossils   A direct way to untangle the history of behavior is to try to learn from fossilized remains how extinct animals behaved and then to use this information to piece together a sequence leading to a modern species' behavior. To use this approach, however, you have to figure out how an extinct species behaved from fossil bones or other signs, like footprints. This might seem a tall order, but one can learn a surprising amount from a bone or two, as the case of the Irish elk makes clear. The fossil evidence indicates that males of this immense deer weighed as much as 1600 pounds, 90 to 100 pounds of which were antlers that spanned 11 feet (Figure 1) [277]. This animal deserved its Latin name, *Megaceros giganteus*. Its dramatic dimensions have fascinated biologists and the public alike for centuries and have led to the popular misperception that the antlers grew so large that they dragged the species to extinction, an event that occurred only about 10,000 years ago.

If large antlers were a liability, then we might expect to find large antlered individuals overrepresented in the fossil record. But this is not true. In his analysis of over 100 fossil skulls retrieved from clay deposits in the Ballybetagh bog near Dublin, Ireland, Anthony Barnosky found that almost all the skulls were those of males with antlers smaller than average for their body size, a finding suggesting (by analogy with living deer) that these were undernourished or sick animals [42]. From his fossil collection, Barnosky deduced that Irish elk stags herded together, probably in winter, and moved to boggy lakes, leaving females and their young elsewhere. Males of big-antlered living deer often gather on frozen lakes in the winter, where they can easily see approaching wolves and run from them on the even-surfaced ice [277]. In the Irish elk, weak young males apparently were at greater risk during the winter than the mature stags, and they died not from the tribulations of carrying huge antlers but from starvation and predators.

Although the Irish elk's antlers were not responsible for their extinction (the deer probably succumbed as a result of climatic changes that reduced the rich steppe grassland on which they and the now equally extinct mammoths fed), the question of how the deer used their great racks has generated considerable debate. In modern deer species whose males and females form separate herds during winter and spring, males compete during the fall rut to monopolize herds of females, with whom the winners mate. Males of all modern deer use their antlers to fight with other males in this competition, but did Irish elk do likewise? Some researchers believe that the broad, palmate antlers could not have been effective weapons because the antler's sharp tines pointed harmlessly backward when the deer's head was upright (Figure 1). Instead, these workers argue that the antlers were displayed purely to impress females.

However, males of the closest living relative of the Irish elk, the fallow deer, also have palmate antlers that they employ in combat with their rivals for mates [649]. They, and other fighting deer, do not fight with the head held upright, but instead lower the head and tuck the chin in to their chest. In this contorted position, the sharp and dangerous tines of the Irish elk's immense antlers would have been pointed directly at an opponent (Figure 1). If the other male had also lowered his head, the antlers would

1 **A male of the giant "Irish elk."** These extinct animals had the largest antlers of any known deer. They almost certainly used them to attract mates and fight with rival males, as do modern deer.

have interlocked, as is typical for living species as well. Andrew Kitchener claims, therefore, that male Irish elk used their great antlers to push, shove, and stab rival males during the breeding season [431].

Antlers are extremely expensive (in metabolic terms) to produce. Had they only been used for display purposes, they would have been constructed so as to withstand the modest forces of gravity but nothing more. However, when Kitchener examined the biomechanics of Irish elk antlers, he found that they could withstand more than 65 times the amount of stress they would have experienced had they merely been passive display ornaments [431]. In other words, not only were the antlers shaped as weapons, they were also constructed to withstand such use, as shown by their internal structure. Otherwise the deer would not have invested in costly materials that permitted the antlers to cope with the fierce impacts and twisting forces associated with battles between rivals weighing three-quarters of a ton.

The Evolution of Flight in Birds

The case of the Irish elk illustrates how detective work with fossils coupled with information on living animals can reveal a good deal about the behavior of an extinct species. Thus, one can hope to reconstruct an evolutionary pathway using information on the behavior of animals that no longer exist. One pathway that would be especially interesting to know about is the sequence of events that led to flight by birds.

The extinct *Archeopteryx* is the most ancient feathered, winged animal. Its fossilized remains occur in rock strata 150 million years old. There has been considerable discussion about whether this bird ever flew, in part because its breast bone (sternum) lacks a prominent keel for the attachment of the large breast muscles that power flight in modern birds. Some researchers have argued that it was really a small dinosaur that used its feathers for insulation and perhaps employed its wings as insect traps, but not for flying (Figure 2) [605].

The argument that *Archeopteryx* ran along the ground chasing down insect prey was the basis for an ingenious hypothesis on the evolution of flight developed by Gerald Caple, Russell Balda, and William Willis [129]. Their pro-avis was a runner who jumped into the air after its insect snacks, which it captured with its mouth, not with its feathered forelimbs. Wings, according to their view, developed to give the jumper better control over its movements when airborne, thereby enabling it to adjust to escape maneuvers of the flying insects it was attempting to retrieve (Figure 3). Subsequent selection on wing surface would favor wings that could generate lift to further increase the jumper's control of its body's pitch and roll. Even-

2 **Evolution of flight: one hypothesis.** Did *Archeopteryx* run along the ground, using its wings to trap prey?

tually wings that had evolved to provide body control during jumping could be flapped to produce and sustain flight. Only when lift and power are combined with control of body orientation can an animal fly effectively, and the "ground-up" hypothesis provides a scenario for the gradual evolution of the requisite skills for controlled flight.

The ground-up hypothesis for the evolution of flight convinced me completely—until I read a more recent article by D. W. Yalden [872] on the structure of the claws on the forelimb and feet of *Archeopteryx*. The not highly curved claws of *Archeopteryx* feet suggested to him that the bird could have scampered over the ground better than it could have gripped arboreal perches in the manner of modern birds. Unlike almost all living birds, however, *Archeopteryx* also had claws on its wings. John Ostrom, the inventor of the hypothesis that *Archeopteryx* used its wings as insect catchers, proposed that these highly curved claws served a predatory function. Yalden countered with the observation that *Archeopteryx*'s wing claws were delicate and needle sharp, like the claws on the feet of tree-climbing birds and unlike the robust, moderately sharp claws that predators use to stab prey forcefully. Moreover, the claws are positioned on long forelimbs in exactly the way required for use by an animal pulling itself up the trunk with its wings (Figure 4)— an activity reminiscent of the climbing behavior of the only living bird with clawed wings, the South American hoatzin.

If *Archeopteryx* relied heavily on its forelimbs for climbing, then its feet could have evolved the generalized shape that Ostrom and others interpreted as designed for a cursorial lifestyle. Ulla Norberg argues that *Archeopteryx* and other early birds were arboreal climbers that glided cheaply (and safely) from one tree to the next. The "trees-down" approach to the evolution of flight is given a lift by Norberg's analysis of the aerodynamics of gliding, which shows that an expansion of wing area greatly enhances the distance that can be covered by a glider. In addition, even very slight

3 **Evolution of flight: an alternate hypothesis.** Did *Archeopteryx* use its wings to assist it in jumping into the air to catch flying insect prey?

4 **Evolution of flight: a third alternative.** Did *Archeopteryx* use its wings to glide from tree to tree?

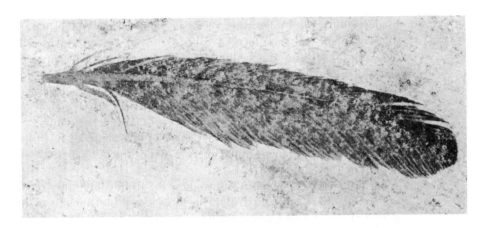

5 **Feather of *Archeopteryx*.** The curved, asymmetrically placed vein, or rachis, suggests that the wings of this bird served an aerodynamic function. Photograph courtesy of John Ostrom.

flapping generates sufficient lift and thrust to further extend the range of a gliding animal [591]. If in the early stages of gliding flight, the landing animal did not land with its feet on a limb but used its clawed wings to grip the trunk as it swooped up at the end of the glide (Figure 4), the glider could have landed without possessing the sophisticated body control mechanisms required for a branch-to-branch trip.

There is no doubt that *Archeopteryx* could fly, because the central support (or rachis) of the feathers of *Archeopteryx* (which have been beautifully preserved in some cases) is not located in the center of the feather but is set off to one side (Figure 5). Modern flightless birds have symmetric feathers, but all those that can fly have asymmetric wing feathers, a design feature related to their aerodynamic function. Whether *Archeopteryx* flew by gliding or flapping remains uncertain [259]. If the "trees-down" hypothesis is correct, then the history of bird flight originated with climbing, which set the stage for passive gliding, then powered gliding, and finally full, flapping flight. Truly understanding the behavior of extinct *Archeopteryx* will help us accurately reconstruct the historical sequence of events that ultimately resulted in the superbly professional flying behavior of so many modern birds.

The Evolution of Bipedalism in Man

Another unusual mode of locomotion is bipedal striding by human beings. Although a few other primates occasionally walk upright for short distances, we are the only living primate to habitually walk in this manner [795]. Our skeleton is greatly modified, particularly with respect to our feet and pelvis, in ways that facilitate our distinctive locomotory behavior. But how did our hominid ancestors walk? The general feeling has been that members of the genus *Australopithecus*, from which we are believed to have descended [809], probably were bipedal but were less efficient striders than modern man (Figure 6). More recently, however, fossil footprints have been discovered in the rock formed from a rain of volcanic ash dated between 3.5 and 4.0 million years old. The makers of these footprints were almost certainly australopithecines, and their footprints are almost identical to those of their modern living descendants (Figure 7). The tracks have a prominent heel mark, modern placement of the big toe in relation to the rest of the foot, and an alignment of prints that is the same as those made by living people [351, 831]. Thus, these creatures, whose skulls were very different from our own, apparently walked bipedally nearly as efficiently as you or I.

Or did they? Here again the issue is far from completely resolved. There is no doubt that some australopithecines as well as ancient members of the genus *Homo* walked in an upright position. But students of the toe bones of these hominids note that they are strongly curved, much more so than the corresponding bones in modern humans. Curved toes (and fingers) are the mark of arboreal primates, not terrestrial ones. Their occurrence in our fossil ancestors may indicate that full bipedalism did not evolve until about 1.6 million years ago and that earlier hominids did a great deal of climbing in trees. Curved digits are useful in this endeavor [478, 754]. Therefore, the fossil evidence suggests that the evolutionary sequence leading to bipedal locomotion began with walking on all fours, as is true for all our living relatives and almost all fossil primates. In our lineage, quadrupedal locomotion presumably went through a stage of occasional bipedalism, followed by increasing bipedalism with tree climbing, and eventually complete and highly specialized bipedalism.

6 **Evolution of bipedalism.** Did our australopithecine ancestors walk upright clumsily, as suggested in this reconstruction by Zdenêk Burian, or were they efficient striders?

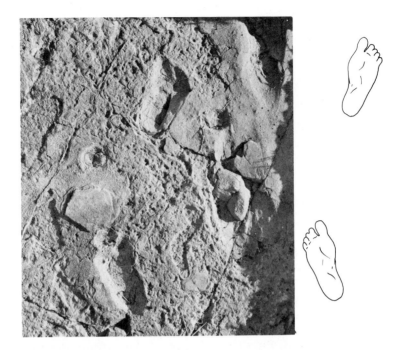

7 **Fossil footprints of hominid ancestors.** These footprints suggest efficient bipedal locomotion similar to that of modern man. Photograph courtesy of Mary Leakey and R. I. M. Campbell.

Tracking History by Comparing Living Species It is possible to gain insight into the historical pathway of a behavior by finding the right fossils and figuring out how the owners of the bones might have behaved. But the examples just discussed show that fossil evidence does not necessarily reveal the whole historical sequence leading to an existing behavior. For example, some intermediate stages of birds have not yet been discovered, and may never be. The incompleteness of the fossil record and the possibility that some transitions may have occurred very rapidly means that we cannot always rely on geologic evidence to trace the evolutionary steps leading to a behavior of interest.

But this does not mean that it is impossible to make educated guesses on the history of a trait. Consider the example of blood-sucking by a moth, about which the fossil record has nothing whatsoever to say. One night, while in the course of his superbly exotic occupation as an observer of moths on Malayan water buffaloes, Hans Bänziger captured a specimen of *Calpe eustrigata* (Figure 8). Suspecting that it might be a species that consumes droplets of mammalian blood excreted by mosquitoes gorging themselves on a host, he made a slight cut on his finger and offered it to the caged moth: "The moth climbed onto my finger and did in fact plunge its proboscis into the blood, but it appeared to imbibe none. Instead it stuck its straight, lancelike proboscis into the wound and, without any regard for the donor, penetrated the flesh. The pain I felt caused me to utter a cry of—joy! I had discovered a moth which pierces to obtain blood" [38].

8 **A moth sucking blood from a Malayan tapir.** Photograph by Hans Bänziger.

Only four of some 200,000 moths and butterflies feed in this manner [37]. The vast majority of Lepidoptera use a long and delicate proboscis to suck the nectar of flowers. Some species, however, exploit another supplier of sugary juices—overripe fruits. Still others consume the exudate from surfaces of fruits partly eaten by birds, bats, or rodents. In certain populations of these species in the past, it is probable that individuals that happened to have a heavier than usual concentration of small rasping hairs at the tip of the proboscis enjoyed a selective advantage. They could scrape at damaged surfaces and extract the fruit juices that flowed from the scraped site. This, in turn, is an adaptation that could become further modified in some populations for piercing the skin of soft fruits.

A few moths are capable of stabbing into blackberries and raspberries, not relying on other animals to wound the fruits. The kind of proboscis that can do this may, through selection for variants with more powerful components and cutting rasps, become capable of probing through thick-skinned fruits to the juices below.

9 **Male swallowtails "mud-puddling."** By taking up salty moisture butterflies may gain useful chemicals. Photograph by the author.

But many moths and butterflies, particularly the males (for reasons that are unknown), also seek out sources of sodium salts. In the course of sweaty field work, I have occasionally had butterflies land on my hands or arms for a salty drink, an enjoyable experience that always makes me feel a bit like a latter-day St. Francis of Assisi, the patron saint of animals. Those of you who have not been visited by thirsty butterflies may nevertheless have seen a cluster of butterflies by the margin of a mudpuddle, probing the soil with their proboscises (Figure 9). For at least one species, mudpuddling males are attracted to sites with high concentrations of sodium [26]. It is not surprising, therefore, that many Lepidoptera also feed on the exquisite ambrosia that is found in liquefying carcasses, in moist piles of animal dung, or in pools of urine (all fine sources of sodium ions). Still others visit living animals to drink oozing salty blood, eye secretions (Figure 10), or even crocodile tears [793].

Calpe eustrigata probably evolved from a fruit-piercer, inasmuch as other members of the genus are fruit specialists, and *C. eustrigata* itself feeds in this manner at times. If individuals in a fruit-piercing population were also attracted to animals to consume eye secretions or blood excreted by mosquitoes, they might initially have supplemented their diet with some additional fluids extracted by rasping the skin. Later mutants might have employed their powerful proboscises to cut straight through the skin of tapirs and other animals to extract blood directly from their victims [37].

Scenarios of this sort are built on certain assumptions and rules of logic [38, 207]. The first assumption is that the behaviors whose evolutionary history is being analyzed have a genetic component. In other words, there

10 **Moths drinking the salty eye secretions** of a banteng (a wild cow of northern Thailand). Photograph by Hans Bänziger.

were genetic differences among individuals of a species that contributed to behavioral differences among them (Chapter 3). This is a necessary condition for evolutionary change within a species (Chapter 1).

The second assumption is that those species that share common ancestry (and thus elements of a genetic program) are likely to behave in a similar manner. Behavior patterns may often be retained in new species as they evolve from an ancestral population in which the traits originated. It is an empirical observation that species whose close relationship has been established by traditional taxonomic techniques (the examination of anatomical and external characteristics) are likely to behave similarly. Most moths do, in fact, feed upon the nectar of flowers.

It follows that behaviors shared by many closely related species are probably all derived from a behavior pattern employed by a distant ancestor. If several related species share a behavioral characteristic, it is unlikely (but not impossible) that each has evolved the trait independently. The more parsimonious explanation is that each has retained the adaptation (a conservative pattern) from the time when their ancestors belonged to a common species. Thus, the first mothlike insect was probably a nectar feeder. Species that evolved from this ancestral population retained the trait because there were many sources of nectar that could be exploited by different species.

If some members of a cluster of related species possess distinctive behavior patterns, it is probable that these traits arose more recently in

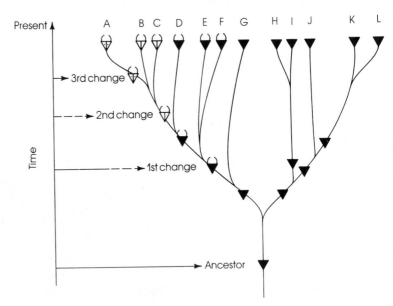

11 The history of a trait can be constructed from comparisons among living species. A hypothetical phylogenetic tree shows how a trait that arose in the distant past may appear in many currently existing species. More recently evolved traits should appear in fewer modern species.

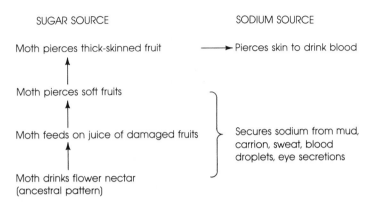

SUGAR SOURCE   SODIUM SOURCE

Moth pierces thick-skinned fruit   ⟶ Pierces skin to drink blood

Moth pierces soft fruits

Moth feeds on juice of damaged fruits   Secures sodium from mud,
                                               carrion, sweat, blood
                                               droplets, eye secretions

Moth drinks flower nectar
(ancestral pattern)

12 **Historical pathway** that may have led to blood-sucking by a modern moth.

evolutionary time. The more uncommon a character (as measured against the presumptive ancestral pattern), the more likely it is to have been a recent modification of a less radically altered version of the original behavior (Figure 11). The basic sucking proboscis and digestive system of Lepidoptera need be little modified to enable feeding on fruit juices in place of flower nectar. A modest mutation of these adaptations could result in fruit-scraping and then fruit-piercing. The kind of proboscis capable of stabbing through an orange could also readily be employed to penetrate the skin of mammals to drink blood.

Each step of this behavioral sequence could easily have been adaptive in its own right, with each modification acting as the foundation for still another modest change. The accumulation of many such changes can produce a trait, like blood-sucking in moths, that is dramatically different from the original characteristic in its lineage (Figure 12).

The History of a Fly's Courtship Signal

The remainder of the chapter will be devoted exclusively to the evolution of communication signals so strange and elaborate it is difficult to imagine at first what their antecedents might have been. The comparative method, however, offers solutions to these problems, just as it did for blood-sucking moths. Consider the balloon display of a small empidid fly, *Hilara sartor*, the males of which construct a delicate, but large, silk balloon. Balloon-carrying males gather in a swarm, there to circle about until a female arrives (Figure 13). She selects a mate from the group, accepts his balloon, and the two drift away from the group to copulate [423].

When these flies were first discovered in the nineteenth century, all that was known was that a group of males gathered together carrying silk balloons. Naturally this bewildered everybody. The finding that the female accepts the balloon as a condition for mating hardly dispels the mystery. Why should copulation depend upon receipt of an empty ball of silk? Why does the male make it in the first place? Without comparative data these puzzles would remain unsolved. There are, however, literally thousands of

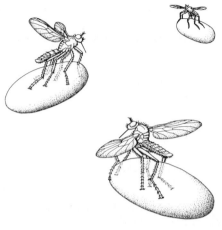

13 **Balloon flies.** Male empidids carrying empty silken balloons for their mates.

species of empidid flies (the family to which *H. sartor* belongs). E. L. Kessel made good use of this diversity to solve the riddle of the balloon fly [423].

Kessel categorized empidids into eight behavioral groups, which he then arranged into the following phylogenetic sequence:

1. The first group is composed of carnivorous species that hunt small flies, like mosquitoes and midges. Males of these species exhibit ordinary, solitary courtship of females that they come across.

2. Males of species in this group take a captured prey to a female as a courtship gift; the female eats the prey during copulation.

3. Males with captured prey form aerial swarms. A female attracted to the swarm selects a male, receives the prey, and mates (Figure 14).

4. In some species, males swarm with prey gifts to which they have applied some restraining strands of silk.

5. The same as in group 4, except that the male wraps the prey entirely in a heavy silk bandage before offering it to a female.

6. The same as in group 5, except that the male removes the juices from his offering prior to wrapping it. As a result, the female receives a non-nutritious husk.

7. These species, unlike all the preceding ones, feed only on nectar, not on flies. However, prior to courtship the male finds a dried insect fragment and uses it as a foundation for the construction of a large silk balloon, which he then presents to a female before copulating.

8. *Hilara sartor* and a few other nectar-feeding species omit an insect fragment as the starting point for balloon construction.

As the entomologist Harold Oldroyd points out, it is as if one first presented a fiancee with elegant diamonds, then plied her with diamonds in an exquisite case, and finally offered her an elaborate, but totally empty, gift-wrapped box [599]. One can speculate about the impetus for each evolutionary change, but the point is that the sequence outlined by Kessel is an eminently logical one. It is reasonable that the most ancient pattern is the one shared by the majority of insects, with the male searching out the

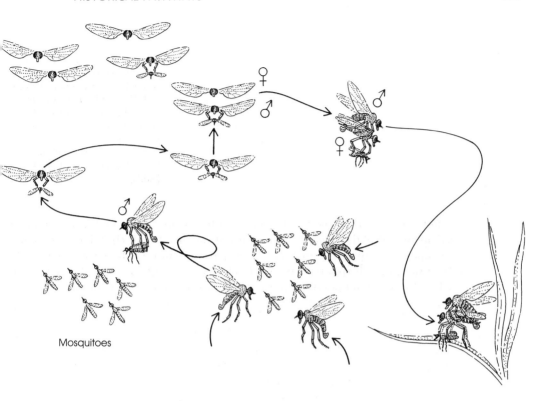

14 **Male empidid flies** offer their mates gifts of mosquitoes captured from nearby swarms. After Downes [206].

Mosquitoes

female, courtship occurring without the unusual addition of a prey offering, and so on. For logical reasons, type 5 courtships could hardly have preceded type 4 in evolutionary time; and type 8 courtships involve only the loss of a small component of type 7. Just as in the scenario for blood-sucking moths, each step of the sequence requires only a modest change based on what was present in the preceding stage.

One could test this behavioral phylogeny by examining an evolutionary arrangement of species based on characters unrelated to courtship and feeding behavior. Kessel's historical scenario would be strengthened if a second independent phylogeny matched the pattern required by his behavioral sequence—if, for example, the species placed in the type 6 category have other attributes (wing venation, leg shape) that indicate they are close relatives and that they had an ancestor older than the species that gave rise to type 7 empidids. This test remains to be done for the balloon fly story.

The History of Honeybee Dances

A honeybee colony is a remarkable society of 10,000 to 40,000 individuals, most of them worker females who will spend much of their life ranging for pollen and nectar over 100 square kilometers or more. The members of the

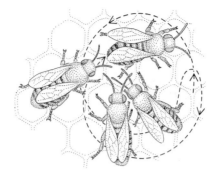

15 **Round dance of honeybees.** The dancer (the uppermost bee) is followed by three workers, who may acquire the information that a profitable food source is located within 80 meters of the hive. *Source:* von Frisch [803].

colony will in the course of a year collect something on the order of 25 kilograms of pollen and 35 kilograms of nectar, a task that Thomas Seeley estimates requires 4.3 million round trips of about 4 kilometers each [711]. Bees accomplish this monumental achievement with great efficiency, daily adjusting the food patches that they exploit en masse and concentrating on the richest sites currently available to them within a radius of 6–10 kilometers from the hive [801]. Much of their efficiency arises because they possess an astonishingly sophisticated communication system, whose meaning was decoded after 20 years of intensive study by the Austrian biologist Karl von Frisch. He concluded that the special dances that bees perform on the vertical surface of combs in their hive contain symbolic information about the distance and direction to food sources, information that other bees could use to locate good supplies of pollen and nectar [803, 804].

When a forager bee has found a rich new food source (a patch of flowers, a watch glass filled with honey), it will return to the hive and perform a dance. If the worker has collected high-quality nectar within roughly 50 meters of the hive, it will execute a *round dance* (Figure 15). Because it is normally dark inside the hive, other bees do not watch a dancer from a distance but instead follow it about on the comb as well as sensing vibrations produced by it. The followers may fly out of the hive to search more or less randomly for the nectar or pollen, keeping relatively close to the hive.

The search will not be entirely random because they will have tasted the nectar regurgitated to them by the recruiter and they will also have smelled the floral scent of the dancer. These tastes and odors identify the flowers that the dancer had been visiting. Experienced bees learn the locations of specific patches of flowers whose nectar supply fluctuates. A worker following a recruiter that bears the odor and nectar associated with a familiar food source will make a beeline to this location. However, an inexperienced worker just joining the forager force will "know" not to wander too far from the hive if recruited by a round-dancer.

If the bee has found a rich food source more than 50 meters from the hive, it will perform a *waggle dance* (Figure 16). This action conveys information about the relative distance of the food source from the hive in the range of roughly 50 to 600 meters. The information is coded (1) in the number of times the bee runs through a complete dance circuit in a unit of

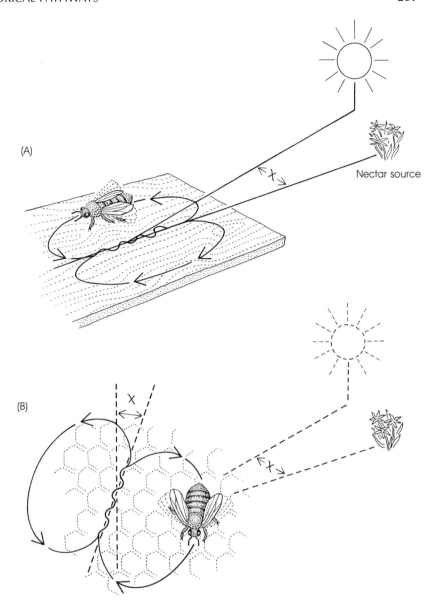

16 **Waggle dance of honeybees.** As the insect performs the straight run, her abdomen waggles; the number of waggles and the orientation of the straight run contain information about the distance and direction to a food source (see text). (A) The directional component is most obvious when the display is performed outside the hive on a horizontal surface, in which case the bee runs right at the food source. (B) On the comb, inside the dark hive, the same dance is oriented with respect to gravity so that the displacement of the straight run from vertical equals the displacement of the location of the food source from a line between the hive and sun. In this illustration, workers attending to the dancer learn that food may be found by flying 20° to the right of the sun when they leave the hive.

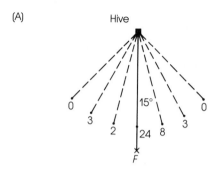

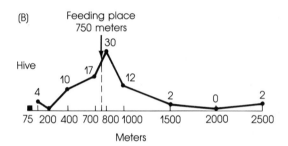

1 7 **Experimental demonstration of communication** by honeybees about distance and direction to a food source. (A) A "fan" test for directional communication. After training recruiters to come to a food dish at F, all newcomers are then collected at seven tables with equally attractive sugar water. Most new bees arrive at the site on line with F. (B) A test for distance communication. Recruiters are trained to come to a feeding place 750 meters from the hive. Thereafter, all newcomers are collected at feeding tables placed at various distances from the hive. In this experiment, 30 newcomers were collected at the site nearest 750 meters, far more than came to any other location in the same time. *Source:* von Frisch [803].

time, (2) in the number of abdomen waggles given during the straight-run portion of the dance, and (3) in the frequency with which sound bursts are produced while dancing. Figure 17 shows the results of an experimental test of the ability of a forager to recruit others to a specific food source. One bee can transfer information to others about the approximate distance to a resource [804].

The energy expended in going to a food source on the outward flight determines the nature of the dance. If a bee flies into a headwind or up a hill, it will perform fewer waggles and a slower dance than would a bee flying the same distance with a tailwind or along level ground. The less energy expended to get to a particular location, the more animated the dance, the more waggles, more revolutions per unit of time, and a higher frequency of sound bursts.

Communication about Direction
to Food

Honeybee dances indicate not only how far food is from the hive but also in what direction it lies. A forager on the way to a discovered food source notes the angle between the food and the sun (Figure 16). During a waggle dance (not during round dances), it transposes this information onto the vertical surface of the comb when it performs the straight-run portion of the dance. If the bee walks waggling straight up the comb, the nectar or pollen will be found by flying directly toward the sun. If the bee waggles straight down the comb, the food is located directly away from the sun. A patch of nectar-producing flowers positioned 90° to the right of a line drawn between the hive and the sun triggers waggle runs oriented at 90° to the right of vertical on the comb, and so on. In other words, sun-compass information is converted into a code based on gravity. A recruit following a dancer determines the angle of its movement with respect to vertical with special sensory hairs on the back of its head. As gravity pulls the bee's head down or to one side, different hairs are pressed, enabling the bee to assess (unconsciously) its position on the vertical comb [485]. The bee can use the information gained by following a dancer to narrow its field of search.

Von Frisch and others [485] have tested the directional component of dance information by training some workers to come to one feeding station. Bees that find the station are marked with paint dots on their thorax and permitted to return to the hive to dance. The researchers then place a series of feeders equidistant from the hive but in various directions from it. By collecting each unmarked honeybee as it arrives at one of these feeders, one can determine the effectiveness of the marked recruiters in transmitting directional information to other workers. Feeding stations initially visited by recruiter bees have many more visitors than those without them (Figure 17). But von Frisch's work did not categorically rule out an alternative hypothesis for the ability of recruited bees to find food sources advertised by other foragers. Adrian Wenner and his colleagues argued that odor cues from the source adhering to the body of the recruiter were sufficient to identify a feeding site [824]. Other bees could find the site simply by tracking familiar scents in the neighborhood of the hive.

It remained for J. L. Gould to trick some bees into confirming that the dance language really did have meaning for the bees that attended a dancer [292]. Gould took advantage of the ability of honeybees to orient their dances directly to the sun if they can see it. Only when foragers cannot see the sun (or a substitute light) will they transpose directional information into a gravity-based code. If one paints over the three simple light-receiving organs (the ocelli) on the head of a bee, the insect becomes much less sensitive to light even though its two much larger compound eyes are not blocked. By placing a weak light as an artificial sun inside the hive near a comb, Gould provided a cue that recruits could see but that the ocelli-blocked dancers could not. Thus, when the foragers returned from a rich food source set up by the experimenters, they oriented their waggle runs with respect to gravity. But the bees that followed them could see the "sun,"

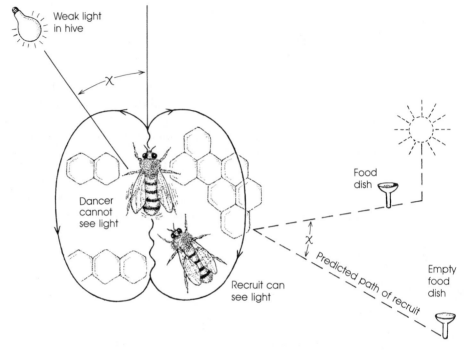

18 **Symbolic communication by honeybees** was confirmed in this experiment that used ocelli-blocked dancers. The dancer could not see the weak light above the comb; it therefore danced using gravity to indicate that food could be located by flying directly toward the sun. The recruits could see the light in the hive, and they interpreted the dance to mean that they should fly *x* degrees to the right of the sun, a path leading them to the empty food dish, where the experimenter was waiting for them.

so they interpreted the dance as if the waggles were oriented with respect to light, not gravity (Figure 18). Off they went in a direction that did not lead them to the feeder found by the dancing foragers. Because Gould could predict where the (marked) recruited bees would go, he was at this spot to record their arrival. He proved that workers can be directed purely by the symbolic information contained in the dances of their fellow bees.

The Evolution of the Dance Displays

Martin Lindauer has used the comparative method creatively to analyze the evolution of this highly specialized behavior [485]. Because all populations of the honeybee perform round and waggle dances [245], the logical group to examine first is the genus *Apis*, to which the honeybee (*Apis mellifera*) belongs. The three other *Apis* species are tropical bees, and all except *A. florea* perform dance displays on vertical surfaces. *Apis florea* dances on the horizontal surface of a comb built in the open around a limb (Figure 19). To indicate the direction to a food source, this bee simply orients

19 Nest of an Asian honeybee, *Apis florea*, has a relatively flat upper surface. Because workers use this surface when dancing, they can point in the direction of a food source when performing the waggle component of the dance. Photograph by Martin Lindauer.

her waggle run directly at it. Because this is a less sophisticated maneuver than the transposed pointing done on vertical surfaces, it is probably a relatively ancient form of dance communication.

All *Apis* species have both round and waggle dance displays, and therefore we must look to other bees for patterns than may be similar to antecedents of *Apis* dancing. Stingless bees of the tropics have proved to be a rich source of information on diverse communication systems. Martin Lindauer did for these bees what E. L. Kessel did with empidid flies; he surveyed the diversity of behavior in the group and organized categories in a plausible evolutionary sequence.

Possible ancestral pattern. Workers of some species of *Trigona* stingless bees perform unstructured excited movements coupled with a high-pitched "humming" when they return to the nest with high-quality nectar. A "dancer's" behavior arouses its hivemates, who request food samples from the dancer and detect the odor of the source on its body. With this information, they leave the nest and search for similar odors. The actions of the recruiter do not offer any specific cues about direction of and distance to the desirable food.

Modified pattern. Workers of other species of *Trigona* convey information about the location of a food source but in a manner very different from the dances of the honeybee. A worker that makes a substantial find marks the area with a pheromone produced by its mandibular glands. As the bee

20 **Communication in a stingless bee.** In this species, workers that found food on the opposite side of the pond could not recruit new foragers to the site until Martin Lindauer strung a rope across the pond. Then the workers placed scent marks on the vegetation hanging from the rope and quickly led others to their find. Photograph by Martin Lindauer.

returns to the hive, it continues to chew on grasses and rocks every few yards. At the hive entrance, there may be a group of bees waiting to be recruited. The forager cooperates by leading these individuals back along the trail it has marked (Figure 20).

A still more modified pattern. A number of stingless bees in the genus *Melipona*, unlike *Trigona* bees, separate distance and direction information. A dancing forager communicates that food is near by producing pulses of sound of short duration. The longer the pulse of sound, the farther away the food is. In order to transmit directional information, the recruiter leaves the nest with a number of followers and performs a short zigzag flight that is oriented toward the source of nectar. The lead bee returns and repeats the flight a number of times before flying straight off to the nectar site with the novice bees in close pursuit.

Trends in Bee Communication

The comparative evidence suggests a possible evolutionary sequence of events leading to the honeybee dances (Table 1) [485, 852]. Communication about the distance to a food source by an ancestor of the honeybee probably first involved relatively unorganized, agitated movements by a food-laden worker that remained highly active back at its nest. Other workers that happened to be easily socially stimulated would leave the hive to forage.

TABLE 1
Summary of the possible evolution of honeybee displays

Stage	Information about direction to food	Information about distance to food
1	Generalized activity upon returning to nestmates with high-quality food; no direction or distance information in "dance"	
2	Forager personally leads recruits to site	
3	Forager leads recruits part way to site	The closer the food, the more active and noisy the forager
4	Forager dances on horizontal surface, pointing in the direction of food	Distance information coded in speed of dance, number of waggles, and sound production
5	Forager dances on vertical comb, with transposed pointing in the direction of food	Same as 4

In some species, selection may subsequently have favored standardization of the sounds and movements made by an "excited" worker, as in *Melipona*. This trend culminates in the round and waggle dances of *Apis* bees, which convey precise information about how far food is from the hive.

Communication about the direction to a food source appears to have originated with personal leading, a worker guiding a group of recruits directly to a nectar-rich area. Here the evolutionary sequence has involved less and less complete performance of the guiding movement—in effect, selecting queens that produced workers with a tendency to perform incomplete leading. At first this may have taken the form of partial leading (as in *Melipona*) and then just pointing in the proper direction, with a very restricted movement toward the feeding place (as in *A. florea*). The final step has been the transposed pointing of *A. mellifera*, in which directional cues based on a sun compass are converted into cues based on gravity.

The History of a Hyena's Display

Insects are a rich source of puzzling communication signals, but for sheer perverseness of display, the spotted hyena is hard to top. When spotted hyenas get together, they erect their penises, which they then present to their companions for nuzzles and sniffs (Figure 21). You may be surprised to learn that female spotted hyenas participate fully in this display, for they have an essentially perfect mimetic penis (and a scrotum to boot)! In fact, it is difficult for a human to determine the sex of a hyena without being able to inspect its genitalia at very close range, not the easiest inspection to perform on your average hyena. For years, the similarity in the external secondary sexual characteristics of hyenas baffled observers.

The pseudopenis of female spotted hyenas is really an enlarged clitoris, and so the historical question becomes, How did it happen that the clitoris became penis-like and was used in the social displays of this species? When we look at other species of hyenas, we find that they are all species that live in pairs or small groups and scavenge alone for carrion or hunt for

21 **Greeting ceremony of female spotted hyenas.** Here one female with an enlarged clitoris (left arrow) has pushed her head under the leg of the other hyena to inspect the clitoris of her companion (right arrow). Photograph by M. G. L. Mills.

small game. Females of these hyenas have a perfectly ordinary-size clitoris. So here a comparative approach appears to fail to offer evidence on the phylogeny of the trait.

However, in 1973 R. F. Ewer [251] hypothesized that female spotted hyenas must have high androgen levels in their blood, a point confirmed by P. A. Racey and J. D. Skinner in 1979 [654]. In almost all other mammals, males have high concentrations of testosterone and other androgens, whereas females have trace amounts of these hormones. Ewer was aware that testosterone has key affects on the masculinization of mammalian embryos (Chapter 4). The tissues of a very early stage mammalian fetus that are destined to become reproductive organs develop into testes and penis if the embryo is a male and can manufacture testosterone. If the embryo is female, the hormone testosterone is absent and ovaries and the secondary sexual characteristics of females will develop. The same tissues that develop into a penis under the influence of testosterone become a clitoris in the absence of the hormone.

Now when a female hyena is pregnant, her embryonic offspring, male and female alike, are exposed to high levels of androgens, because the mother's androgen-laden blood reaches her embryos. This hormone activates the basic mammalian system for developing male traits and masculinizes females, producing an enlarged and malelike clitoris [299].

But why did the initial hormonal mutant enjoy greater reproductive success than the typical females in her species? K. J. Stewart notes that, unlike the other hyenas, spotted hyenas live in large bands, hunt big game, and compete fiercely for meat from the animals they kill [746]. Dominant females gain more food than subordinate ones and produce male offspring that are dominant to all others in their bands. When fully mature, these offspring emigrate to another clan of hyenas. If the son of a dominant female becomes the top male in his new clan, he will enjoy exceptional reproductive success, because only the dominant male mates.

Thus, the social system of spotted hyenas rewards dominant females exceptionally. Dominance is associated with aggressiveness, and aggressiveness is linked with hormonal state. In mammals, testosterone plays a key role in the development of aggressive behavior. A mutant female with higher than average testosterone levels would probably be more aggressive, given the basic mammalian way of regulating this behavior. To the extent that her aggressiveness led to better success in competing for food and producing dominant sons, selection would favor the androgenized female. She would tend to produce daughters with masculine secondary sexual attributes as a side-effect of her hormonal state during pregnancy [251].

The pseudopenis of masculinized females was then available for use in communication with other members of the band. The fact that inspection of the anogenital region occurs in hyenas other than the spotted hyena [454, 607] indicates that the penis of the ancestral hyena probably was used by males as a communication device, among other things [453]. The first masculinized female spotted hyenas that happened to employ their enlarged clitoris in penile displays could have gained a reproductive advantage from participation in these displays, although their current function for females and males alike is not thoroughly understood. Nor is it clear why spotted hyena females have evolved enlarged clitorises when females of other aggressive social carnivores—for example, lionesses—have not. Leaving these troubling issues to one side, we can still claim that the origins of the pseudopenis and penile display in the hyena probably lie in the spotted hyena's novel use of the basic mammalian rules of genitalic development [299].

The Lessons of History The examples given here of the behavioral reconstruction of fossils and the comparative approach to tracking historical sequences illustrate a fundamental point about the history of behavior, namely, that it is not necessary to invoke dramatic shifts or revolutionary changes in the evolutionary development of a trait. Complex, strange, puzzling behaviors have a step-by-step history; one can often plausibly show how such behaviors might have arisen gradually from less unusual, less strange behaviors. There is no need to propose that bird flight arose in perfect form in a single step from a nonflying ancestral state. It might have arisen from a jumping stage, or more probably still from a weak flapping glide that in turn was preceded by a nonflapping gliding behavior. The amazing dances of honeybees have a series of possible antecedents, and it is unlikely that they arose de novo in a single mutational stroke.

When we look at the origins of elaborate communication signals in particular, we see that they arise from ordinary, everyday activities. When a water strider moves across the water's surface, it creates little waves that carry information about its presence. Stimson Wilcox has shown that water striders currently make use of wave-generation and wave-detection in complex communication among individuals [843]. Even urinating or defecating produces materials that contain visual or olfactory information that may be useful to a signaler and to a receiver that detects this information. Spotted hyenas mark the boundaries of their territories with dung and urine [797]. These materials clearly announce the presence of hyenas and warn intruders to stay away or risk dismemberment at the jaws of the clan. Just by passing the tissue of a tree through her digestive tract, a female bark beetle generates some volatile metabolic by-products that carry information about her existence and location. Originally, males that happened to detect these chemicals and track them to their source would have derived a benefit and so would the receptive females that secured a helpful partner in this manner [690].

Once a communication system has been established, subsequent selection can favor mutant signalers that happen to produce a signal that is more distinctive, more informative, more easily dispersed, or harder to exploit. The sex pheromone currently used by bark beetles is often a highly complex combination of chemicals [66]. The boundary markers of hyenas (and other carnivores) include a mix of materials from special secretory glands whose products supplement the chemicals in waste materials, elaborating on the communication function served by feces or urine [89, 453].

Another rich source of raw material for incipient communication signals are the actions of animals that are placed in situations that elicit two opposing tendencies [225]. We noted in Chapter 6 that nervous systems have properties that reduce the frequency of "conflict behavior," but such activities may still occur (Figure 22). Thus, two rival herring gulls that meet at the border of territories may attempt to attack and to escape from their opponent. The resulting conflict may cause a male to raise his head back, as if he were about to peck at his opponent, without actually carrying out the action. Movements of this sort, which are made in preparation for a response, are called intention movements. We have just discussed such a behavior in those social bees that move toward a food source and then back toward the hive and the workers they are attempting to recruit.

Another category of conflict behavior is redirected behavior. A male gull facing a rival may attack a clump of grass rather than his neighbor (redirecting his aggression onto a safe object rather than the stimulus that aroused aggression in the first place). Still another frequently observed response is an apparently irrelevant behavior unrelated to either attack or escape motivation, such as preening of the feathers (a so-called displacement activity). Various ambivalent movements may convey subtly different messages to a rival. If certain actions deter one male from entering another's territory and both individuals gain from this (the deterred male avoids a beating, the resident saves his energy), then communication advantageous to both has taken place.

Ethologists have long argued that conflict behavior provides the simple raw material for many communication signals that, once variable, gradually evolved into ritualized and stereotyped displays [773]. If the value of a communication signal lies in its ability to transfer information from a

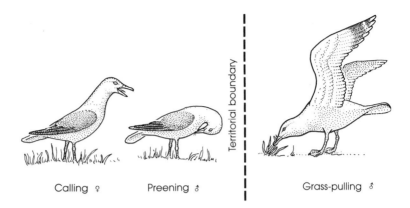

Territorial boundary

Calling ♀   Preening ♂   Grass-pulling ♂

22 **Conflict behavior in herring gulls.** Two males, at the boundary of their territories, are motivated by aggression *and* fear. One bird responds by pulling grass within the safety of its territory, while its opponent preens its feathers rather than attack. The mate of one male calls behind him.

signaler to a particular receiver, then whatever constitutes a less ambiguous message may be selectively advantageous. This selection mechanism may favor individuals that reliably produce a distinctive and unchanging signal in a particular context. The more exaggerated the behavior and the more it involves conspicuous structural traits, the less likely it is to be misunderstood. This is the kind of argument that has been used, for example, to explain the complex dance displays of honeybees (Table 1).

Cumulative Selection
From simple and humble beginnings, an elaborate and complex signal can evolve—gradually. The hypothesis that behavior evolves small step by small step matches what we know about the effects of genetic mutations on development of individuals. The probability that a mutation will have many large (revolutionary) effects, all of which are sufficiently well-integrated to produce a viable individual, seems vanishingly small. It is far more likely that dramatic evolutionary changes are the cumulative product of many minor mutations, each mutant's effects slightly modifying and improving upon what had occurred before, the whole process requiring many generations.

The power of cumulative selection is immense, as Richard Dawkins illustrates with a brilliant analogy [185]. Let us say that a complex current trait is like an English sentence, for example, a line from Shakespeare's *Hamlet*: METHINKS IT IS LIKE A WEASEL. What is the chance of producing this unique combination of letters by chance alone? There are 28 letters and spaces in the phrase and for each of the 28 positions in the sequence one could put any one of 26 letters or a space. So if we generate a random sequence, it might be this: SWAJS EIURNZM MVASJDNA YPQZK. If we kept at it, or had a computer keep at it, the time needed to come up with METHINKS

IT IS LIKE A WEASEL would be enormous, because there are so many possible combinations of letters and spaces (27 possibilities for the first position, times 27 for the next position, times 27 for third . . . times 27 for the twenty-eighth position—you get the idea). The number of possibilities is inconceivably large, and yet only one combination is METHINKS IT IS LIKE A WEASEL.

However, instead of trying to get the "right" combination in one go, let's change the rules so that we start with the random sequence above and use the computer to copy it over and over but with a small error rate built into the program so that every once in a while the computer copies the sequence incorrectly, randomly inserting a new letter into one of the positions. Then we can, as Dawkins did, ask the computer to look over its list and pick the sequence that is closest to METHINKS IT IS LIKE A WEASEL. Whatever "sentence" is closest is used for the next generation of copying, again with errors. The sentence that is most similar to METHINKS . . . is selected for breeding and so on. When Dawkins did this with a number of different starting sequences, he found that it only took 40–70 generations to reach the target sentence, not millions upon millions upon millions of attempts, a few seconds of computer time, not years [185].

Natural selection is cumulative selection. Some mutations induce small random changes in the genetic "sentence" possessed by individuals. Those changed genetic messages that confer higher reproductive success propagate themselves throughout the population. After the new "sentence" is in place, additional small changes that improve the reproductive output of individuals will be incorporated in the same way. The cumulative effect of this process may eventually be large differences between the genetic information possessed by the member of a species and that of its ancestors.

Constraints on Evolutionary Change

The computer program designed by Dawkins "knew" what the goal of the exercise was, the production of the trait METHINKS IT IS LIKE A WEASEL. Natural selection does not have a goal in mind. It is the outcome of differences in reproductive success among alternative genotypes. As George C. Williams notes, as a trait undergoes evolutionary change by natural selection, each transitional state must be adaptive, because each new mutant will either enhance individual reproductive success, or eliminate itself from the gene pool [848]. Natural selection and the other "agents" of evolution do not start from scratch in designing the features of an organism. Instead, changes are layered over past changes. Consequently, from the perspective of a design engineer, there are some rather peculiar features of many characteristics.

Imagine the difficulties human mechanical engineers would have in designing a jet plane if they had to do so by modifying a propeller-driven plane, changing it piece by piece, and having the plane fly in every one of its transitional stages [184]. Yet this is how evolutionary transitions must occur. Stephen J. Gould labels the jury-rigged nature of evolutionary change "the panda principle," in honor of the panda's thumb, which is not a "real" finger at all but a structure built around a highly modified wrist bone [302]. Pandas evolved from carnivorous ancestors whose first digit had become developmentally committed to a foot used in running. Thus, there was no opportunity for selection to favor mutants with mobile first digits that they might use to strip the leaves from bamboo shoots, pandas having become herbivorous bamboo-eaters. Instead, chance resulted in the alteration of a wrist bone that is now used in the fashion of a thumb.

The panda principle can be seen in dozens of cases, as in the skeletons of certain snakes and whales, which clearly show that these now legless animals evolved from four-legged ancestors (Figure 23). Human embryos at a very early stage have gill slits, a feature that only makes sense if we realize that mammals evolved from aquatic ancestors and that we have happened to retain a character that these ancestors possessed. Some key nerve cells regulating flight in the African locust originate from the abdomen, despite the fact that the wing muscles are in the thorax. Over the course of evolution, cells that originally had something to do with abdominal control have taken on a flight control responsibility, leading to a bizarre wiring system [210].

Or consider the persistence of sexual behavior in parthenogenetic *Cnemidophorus* lizards. These species are composed entirely of females; yet if a female is courted and mounted by another female (and females do engage in pseudomale sexual behavior for reasons that are not fully understood), she is much more likely to produce a clutch of eggs than if she does not receive sexual stimulation from a partner (Figure 24) [158]. The fact that courtship has hormonal effects that promote female fecundity in unisexual species is obviously related to the history of the species, which was derived from a sexual ancestor. Even though males are not needed to fertilize eggs, parthenogenetic females retain characteristics that their nonparthenogenetic ancestors possessed, characteristics that a biological engineer would eliminate if he or she could play god in creating a new all-female species.

The lesson of history is that gradual changes can have a large cumulative effect that dramatically reshapes the characteristics of a species—provided each step along the evolutionary pathway is a viable, adaptive trait that serves as a foundation for the next adaptive modification.

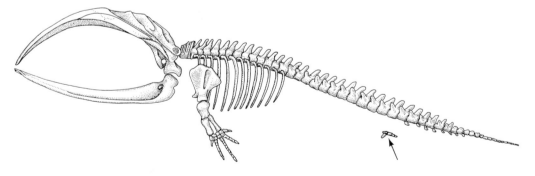

23 **The imprint of history.** The skeletons of some whales include a vestigial pelvis (arrow); these animals had terrestrial ancestors that walked on four legs. The hindlimbs have been lost altogether, but the forelimb bones in the flipper of the whale clearly reveal the five "fingers" of the terrestrial ancestor.

24 **Sexual behavior in whiptail lizards.** On the left, a male of a sexual species engages in courtship and copulatory behavior with a female. On the right, two females of a closely related parthenogenetic species engage in very similar behavior. *Source:* Crews [156].

SUMMARY

1 A fundamental ultimate question about behavior is, What was the sequence of events over evolution that led to the behavior we observe today? Tentative answers to this question are often possible, even though direct evidence on the history of behavior is usually missing or incomplete.

2 The behavior of extinct species may help us understand early stages in the evolution of certain striking modern behaviors, like bird flight or bipedal locomotion in man. Reconstruction of the behavior of extinct animals can

be accomplished in part by comparing the structure of fossil bones with that of modern species for which the relation between bone structure and behavior has been established.

3 When fossil material is absent, it is still possible to conduct historical detective work by using the comparative method. If a cluster of related species exists, one can sometimes trace a pathway leading to the unusual behavior of one of these species by using a set of logical rules. These rules include the following assumptions: (1) if a trait is widespread in a group of related species, the character was likely to be present in the ancestor of those species, whereas (2) if a trait is represented in only one or a few species, it is likely to have evolved relatively recently.

4 Studies of fossils and application of the comparative method to modern species supports the intuitively reasonable argument that behavioral evolution proceeds gradually, by degrees, with changes building upon and limited by past evolutionary events. The cumulative effect of natural selection can slowly produce great changes in evolutionary lineages.

SUGGESTED READING

Andrew Kitchener's paper on the Irish elk [431] is a marvelous example of detective work on how to use fossils to determine the behavior of an extinct species. Hans Bänziger's description of his research on blood-sucking moths [38] is a pleasure to read. Martin Lindauer's classic *Communication among Social Bees* [485] outlines the evolutionary pathways to honeybee dances. Thomas D. Seeley's *Honeybee Ecology* [711] updates the honeybee story. Stephen Jay Gould discusses the panda principle in [302], and Richard Dawkins makes the power of cumulative effects of natural selection clear in his excellent book *The Blind Watchmaker* [185].

DISCUSSION QUESTIONS

1. Imagine that it is your task to determine whether or not an extinct, very ancient species of bat could echolocate. Or whether an equally extinct ancestor of humans could speak a language. What structural characteristics of living bats and humans are linked with echolocation and speech, respectively? How might this help you know what to look for in the fossil material at your disposal? After developing an answer, see [254, 464, 597].

2. I reported the conclusion that workers of some *Trigona* stingless bees do not communicate information about the distance and direction to a food source. Design an experiment to test whether this conclusion was correct.

3. When a worker honeybee has had to detour around a large obstacle to reach a rich food source, its subsequent dance ignores the detour and codes for a direct beeline to the food source—even though this is not the path actually taken by the recruiter. What problems does this finding pose for the hypothesis that the waggle dance is a kind of incomplete leading movement in which the recruiter uses its experience to point others toward a flower patch [296]?

4. Some persons have argued that because evolutionary changes are limited by the traits in place in a population, many characteristics of organisms cannot be "optimal." What must "optimal" mean in the context of this argument? Are the existing traits of a species likely to be maladaptive? After your discussion, see the next chapter.

8

The preceding chapter presented techniques for tracing the historical sequence of events underlying a behavior, especially communication signals. There the key question was, What changes have taken place over time as a behavior evolved? This chapter asks how a behavior might increase an individual's chances of leaving surviving descendants. The adaptationist approach to animal behavior is presented, an approach based on the Darwinian assumption that what an animal does raises its own chances for reproductive success, rather than promoting the preservation of its species. The pleasant problem for the adaptationist is to test specific ideas on the adaptive value of a behavior of interest. This analysis requires an understanding of

The Evolution of Behavior: Adaptation and Behavioral Ecology

the relation between an animal and its ecology to see whether its behavior helps individuals overcome obstacles to reproduction present in its environment. The chapter begins with a case study—mobbing behavior by black-headed gulls—which will be used to show how behavioral ecologists test hypotheses on the adaptive significance of a trait. The discussion turns again to communication signals and uses the approach of behavioral ecology to identify the adaptive value of these behaviors, rather than the historical antecedents of modern signals. Because the adaptationist approach has been sharply criticized, the chapter will conclude with an examination of these criticisms and the response of adaptationists to them.

Behavioral   Persons often become interested in the function or adaptive
   Ecology   value of behavior when they watch an animal deal ingeniously
with the myriad difficulties imposed by its natural world. When I see a
sexually frenzied male bee dig into the soil and miraculously uncover an
emerging female hidden to me, I applaud his skill at finding mates. Or
when I watch a black-tailed jackrabbit dashing across his patch of parched
Arizona desert scrub in July, I marvel that he can find enough to eat in his
barren sun-baked environment.

The patient observer is sure to find that any animal possesses special
abilities that enable it to live and reproduce. Behavioral ecologists search
for these special skills and attempt to understand how they function, how
they help individuals leave copies of their genes in the next generation.
Niko Tinbergen, for example, enjoyed watching black-headed gulls, a common, ordinary gull but one with a complex battery of behaviors that offer
much to admire and much to try to understand in terms of adaptation [772,
773]. One of the things nesting black-headed gulls do is dive-bomb their
predators, and we shall use this activity to illustrate the questions behavioral ecologists ask and how they go about getting answers.

Black-headed gulls have an annual cycle of behavior not unlike that of
white-crowned sparrows. Adults breed in coastal European sites in the
spring and summer, disperse in the fall, migrate south (to North Africa) in
the winter, and then return the next spring, usually to the breeding colony
they used the previous year. Dozens or hundreds of gulls pair off and nest
on the ground in open grassy areas, often on islands (Figure 1). Should a
fox, crow, badger, hawk, or human appear in the colony, breeding adults
respond with a volley of loud cries. If the predator approaches, groups of
gulls will fly toward it, continuing to call raucously and defecating profusely
while dive-bombing the enemy.

Ultimate Hypotheses about Behavior

Before we ask whether *mobbing behavior* is adaptive, we have to know
what we mean by an adaptation. We shall define an ADAPTATION as an
inheritable characteristic that gives an individual an advantage over others
with different inherited abilities, an advantage in transmitting its genes
to subsequent generations. An adaptation is *better* than other alternatives
that exist, better than it would be if it were slightly modified, better at
"helping" individuals pass on their genes [845].

When a chance mutation produces observable effects on an individual,
the allele will spread or disappear, the outcome depending in large measure
on whether the effects raise or lower individual reproductive success. If the
mutant trait raises reproductive success relative to that afforded by the
alternative(s) present in others in the population, it is by definition an
adaptation.

Tinbergen had many opportunities to ask himself why the birds mobbed,
for he himself must have triggered mobbing by gulls on many occasions.
Tinbergen tried to imagine why this action might increase reproductive
success, and he quickly came up with a possible explanation for mobbing,
which is that it helps parent gulls confuse predators who would otherwise

1 **Colony of lesser black-backed and herring gulls.** The small structures are blinds for close observation of the birds. Photograph by Niko Tinbergen and Hugh Falkus.

eat their offspring. Having been beset myself by a mob of gulls, I think the hypothesis is a good one, because I found that having a flock of screaming, defecating gulls swirling around me made it hard to concentrate on anything. Distracted by the chaos, I have not noticed as some gulls slip around behind me to come diving down at my head. When these avian kamikaze bombers pull out of their dive at the very last moment, they send the wind roaring through their wings (Figure 2). The effect is unnerving to say the least, particularly if the gull applies the coup de grâce—a not-so-gentle clip on the top of my head with a trailing foot. But although I find it highly plausible that mobbing deters gull predators, it is a belief that needs to be tested, and *that is the point of developing hypotheses.*

Assumptions Underlying Adaptationist Hypotheses

Before consideration of how one might test the hypothesis that mobbing is an adaptation that enables gulls to deter some predators, a brief discussion on the development of adaptationist hypotheses is in order. A critical element of the adaptationist approach is the recognition that all traits have both negative and positive effects on individual reproductive success (direct fitness). You may think that mobbing is clearly beneficial, but a gull can get killed or injured when it attacks a fox or a hawk, even a human. In addition, the mobber expends a certain amount of time and energy that

2 **Black-headed gulls** mobbing a trespasser on their nesting area.

could be spent in other activities. All of this decreases to some extent the probability that the gull will reproduce successfully in the future. This is the negative aspect or the EVOLUTIONARY COST of the action.

Set against the costs are the EVOLUTIONARY BENEFITS of mobbing a predator—in this case, the potential increase in individual fitness that comes from saving some currently existing offspring. To be adaptive, a behavior must at a minimum have benefits (fitness gains) that exceed its costs (fitness losses). Many of the hypotheses to be examined in the pages ahead use the simple economic approach of behavioral ecology to examine whether it is likely that certain benefits exist, and whether they exceed the costs of the action. For example, if mobbing is adaptive, it should have some beneficial effect for the mobber, as in decreasing the chance the predator will find the mobber's young. Moreover, the costs of mobbing, as in risk of injury, should be modest.

It is possible, however, to follow the cost–benefit approach to its logical conclusion, which is that evolved traits will *maximize* benefits relative to costs because such traits will spread through populations in competition with alternatives whose costs more closely match benefits. If it were possible to measure the costs and benefits of a set of alternative traits, one would predict that the action with the greatest difference between benefits and costs should be exhibited by the members of a species. For example, let us say that we could discover how much reproductive success was improved by diving within 1 meter, 2 meters, or 3 meters of a predator. Presumably the closer a gull comes to a fox, the more distracted the fox becomes and the greater the survival chances of the gull's youngsters. The closer the mobber approaches, however, the more likely the fox is to catch it, and the fewer future chances to reproduce the gull has. There will be a point at which benefits exceed costs by the greatest possible margin, and this is how close we predict gulls should come to foxes (Figure 3A).

Behavioral ecologists have most often used OPTIMALITY THEORY to make quantitative predictions of this sort in conjunction with studies of foraging behavior (Chapter 10). Even when the predictions are qualitative, as is true for the large majority of the hypotheses reviewed in the chapters ahead, the underlying approach usually involves the assumption that evolved traits will be optimal in a cost–benefit sense.

Users of optimality theory often do not consider how social interactions within a species might affect costs and benefits of alternative traits. Proponents of GAME THEORY do factor social effects in their models. They formally treat animal behavior as a game between the members of a population, in which the adaptive solution depends on the strategies employed by other members of the population. For example, in considering the evolution of mobbing behavior, game theorists would look not just at the contest between gull and fox but also at what other gulls are doing in the colony.

Imagine a gull rookery composed entirely of "daring mobbers" that approached predators very closely in the company of their fellows, thereby effectively discouraging searching predators. Suppose that in such a population a "cautious mobber" arose by chance, a gull that hung back a bit while its more daring neighbors rushed toward the predator. The mutant genes for cautious mobbing could become more common, with the cautious mobbers parasitizing daring mobbers. In effect, they would be deriving most of the benefits of communal mobbing while letting the daring individuals take the bulk of the risks. Their presence reduces the fitness of the daring mobbers by reducing the communal intensity with which predators are attacked and by placing the burden of risks more heavily on the daring birds.

At some point, however, the spread of cautious mobbing genes would stop because the cautious mobbers' success depends on being able to take advantage of daring ones. At an equilibrium point daring mobbers and cautious mobbers would enjoy equal fitness (Figure 3B). Thus, instead of one single "optimal" solution to the problem of adaptive mobbing, two quite

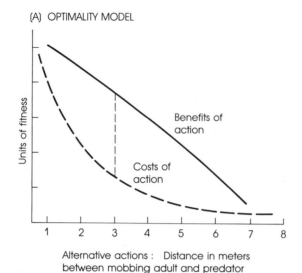

(A) OPTIMALITY MODEL

Units of fitness

Benefits of
action

Costs of
action

1   2   3   4   5   6   7   8

Alternative actions :   Distance in meters
between mobbing adult and predator

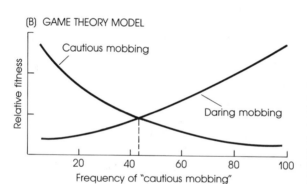

(B) GAME THEORY MODEL

Relative fitness

Cautious mobbing

Daring mobbing

20   40   60   80   100

Frequency of "cautious mobbing"

3 Optimality and game theory models (hypotheses) contrasted. (A) An optimality approach generally employs the assumption that there is one alternative that maximizes the difference between fitness benefits and costs. The graph shows that mobbing gulls enjoy highest fitness if they come 3 meters from the predator. (B) Game theory models can accommodate the coexistence of two alternative behaviors. When practitioners of daring mobbing make up 45 percent of the population they have the same fitness as individuals exercising the alternative behavior (cautious mobbing). The two behaviors will be maintained in the population at this equilibrium point.

different behavioral responses to predators could coexist in the population. If only all the gulls would practice "daring mobbing," they would all benefit maximally, but such an outcome would not be EVOLUTIONARILY STABLE for the reasons we have discussed. The game theory approach makes us aware that natural selection can favor traits that enhance success in the arena of social competition while reducing the effectiveness of individuals in their interactions with the other aspects of their environment.

We shall examine hypotheses based on game theory primarily in the context of competition among males for mates, a competition in which the success of one set of tactics can be strongly affected by the behavioral decisions taken by other males (Chapter 13). But we should remember that there is more than one way to produce adaptationist hypotheses for any given trait, and because this is true, the importance of testing hypotheses becomes all the more critical.

Testing Ultimate Hypotheses   Fortunately there are many ways to test hypotheses about the adaptive function of behavior. All these methods involve using a hypothesis to make a prediction whose accuracy can then be evaluated by (1) additional observation, (2) experiment, or (3) comparative data from other species. The hypothesis that mobbing is adaptive for a gull because it deters some predators generates a number of simple optimality predictions. Benefits will be high relative to costs—if mobbed predators tend to overlook nests and young and if the risk of injury to a mobbing gull is low.

The simplest way to test these qualitative predictions is through direct observation of gulls and predators. During Hans Kruuk's 2-year study of predation at a black-headed gull colony, he saw gulls attack many crows and herring gulls [452]. Carrion crows and herring gulls love gull eggs but cannot capture and eat adult birds, so mobbing black-headed gull adults are reasonably safe—as predicted. Both crows and herring gulls can avoid dive-bombing gulls, but this requires that they continually face their attackers. As a result, they are distracted and ineffective in their search for nests and eggs while under group attack—as predicted.

Kruuk also employed experiments to test the prediction that mobbing by black-headed gulls deters some predators [452]. For each experiment he placed 10 hen eggs, 1 every 10 meters, in a line running from outside to inside the gull nesting area. Carrion crows and herring gulls hunting for food in the vicinity of the colony often discovered the eggs; and Kruuk was able to determine the order in which the eggs were taken. If mobbing gulls deter egg predators, then hen eggs outside the colony should have been more vulnerable than those inside the colony. They were. Moreover, the more the gulls bombarded their enemies, the less success the predators had in finding hen eggs (Figure 4).

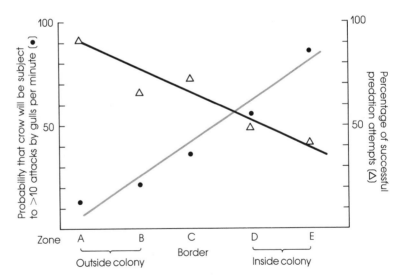

4 **Effectiveness of mobbing** in preventing egg-eating predators from finding eggs. In this experimental test, the frequency of intense attacks (●) on crows increases when these predators are within the borders of a nesting colony. As mobbing increases, the percentage of hen eggs discovered and eaten (△) decreases. After Kruuk [452].

Thus, both observational and experimental tests support the hypothesis that group mobbing is an adaptive response to certain egg-hunting predators. Note that these tests did *not* involve measuring reproductive success by counting the number of surviving offspring produced. There are obvious practical difficulties in securing these data in most studies, including ones on mobbing behavior. Instead Kruuk looked at the number of eggs that survived, on the (reasonable) assumption that the more eggs that go untouched by a predator, the more chances an adult has to rear surviving young.

Behavioral ecologists often have to settle for an indicator or correlate of reproductive success when they attempt to measure it. In the chapters that follow, "reproductive success" is used interchangeably with such things as egg survival, young that survive to fledging, number of mates inseminated, and so on. The reader should keep in mind, however, that the bottom line is the number of offspring that reach adulthood and reproductive age, and that correlates of this measure will be accurate to varying degrees.

The Logic of the Comparative Method

There is still another way to test adaptationist hypotheses—the COMPARATIVE METHOD. In the preceding chapter we showed how comparisons among living species can be used to reconstruct a historical pathway leading to a modern behavior. Comparisons can also be used to test ideas on the adaptive value of a behavior [147, 200, 667]. For an example, let us use the comparative method to test the prediction that mobbing behavior by black-headed gulls is an evolved adaptation that thwarts predators. If this is adaptive, then at least some other birds that have nest predators should have evolved the same solution. However, be forewarned that *some comparisons are valid for testing this prediction, whereas others are not.*

Many gull species that nest on the ground mob egg-eating predators. But we cannot use this information to test our adaptationist hypothesis, because the different species of gulls may behave similarly only because they inherited the mobbing trait from the original proto-gull. Our assumption is that animals with a common ancestor, of the same phylogenetic lineage, will have inherited some genes from this ancestor and so will develop similar nervous systems and similar behavior.

If, however, behavior evolves in response to special ecological conditions, then related species with *different* ecological problems should diverge behaviorally from the pattern they inherited from their common ancestor. If we show that related species with different ecological problems have different behavioral abilities, we support the hypothesis that these particular traits are adaptations to the special problems each species faces.

Therefore, gull species that did not have nest predators should not exhibit mobbing behavior—if our hypothesis on the function of the behavior is correct. The kittiwake gull nests on nearly vertical coastal cliffs (Figure 5) where its eggs cannot be reached by mammalian predators and where swirling sea winds make maneuvering difficult and dangerous for large predatory gulls and hawks. Kittiwakes are relatively small, delicate gulls with clawed feet, and they can land and nest on tiny ledges (Figure 6). As

5   **Nesting habitat** of cliff-nesting kittiwake gulls at Orkney
Island, Great Britain. Photograph by Arthur Gilpin.

6 **Adaptation for cliff nesting.** Kittiwake gulls have evolved the capacity to alight on the narrowest of cliff ledges. This small spot is often used as a perch, judging from the guano stains below it. Photograph by the author.

a result, the kittiwake's eggs and young are not at risk from predators and, as predicted, the adults do not mob predators when the occasional enemy drifts past the colonial nesting site. The absence of mobbing in the predator-free kittiwake is a case of DIVERGENT EVOLUTION, a case that supports the hypothesis that mobbing evolves in response to predator pressure [162].

The other side of the coin is that species from different phylogenetic lineages are expected to behave differently because they have different ancestors that endowed them with distinct genetic-developmental characteristics. If, however, they have been subjected to similar selection pressures, they may have independently evolved similar behavioral traits through CONVERGENT EVOLUTION (Figure 7). If a common ecological factor can be related to an apparent case of convergent evolution, this finding strengthens the argument that the trait represents an adaptation to this environmental pressure.

There are animals unrelated to black-headed gulls that have convergently evolved mobbing behavior. These species experience ecological circumstances that are similar to those affecting black-headed gulls, but not those affecting kittiwakes. For example, many pairs of the colonial bank swallow may nest together in a sand quarry or river bank (Figure 8). Their young, like young black-headed gulls, are vulnerable to predators. As expected, (1) the swallows actively mob only potential enemies, like blue jays, (2) they are very rarely killed while mobbing, and (3) they successfully deter certain of their enemies through communal harassment [383].

Yet another colonial species that engages in group harassment of predators is the California ground squirrel [358, 608]. The squirrel's primary enemies are snakes, which are able to invade nest burrows and capture the young. The adults above ground are relatively safe, and they collectively mob a snake intruder, kicking sand in its face and generally making its life as miserable as possible to "encourage" it to go elsewhere (Figure 9).

Donald Owings and his colleagues found that young, inexperienced squirrels living in places with venomous rattlesnakes are significantly less likely to approach any snake than are naive young squirrels from areas without rattlesnakes. Wild-caught adult squirrels will mob rattlesnakes, but they discriminate between them and nonvenomous gopher snakes—keeping farther away from rattlers when harassing them [608].

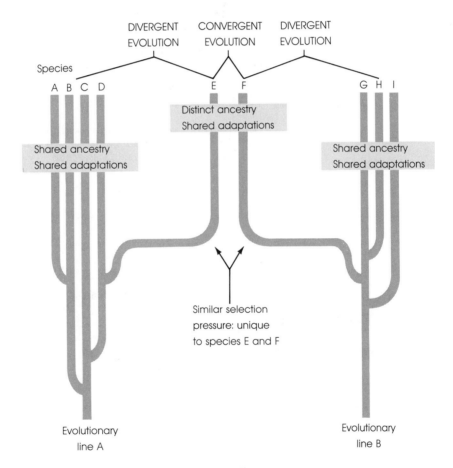

7 **Divergent and convergent evolution.** Species E has experienced different selection pressures from those acting on its relatives (A–D); as a result its traits are distinctive for its phylogenetic line. The same is true for species F, which has diverged evolutionarily from its relatives (G–I). Because E and F are subject to similar selection pressures, they have convergently evolved similar traits even though they are unrelated species.

8 **Convergent evolution: bank swallows.** A colony of bank swallows, another social species in which mobbing of predators occurs. Photograph by Michael D. Beecher.

9 **Convergent evolution: ground squirrels.** Colonial California ground squirrels, like certain ground-nesting gulls and colonial swallows, have evolved mobbing behavior. One squirrel kicks sand at a rattlesnake, while others give a variety of alarm calls. Courtesy of R. G. Coss and D. F. Hennessy.

These findings make sense in the context of a cost–benefit analysis of mobbing behavior. The greater the personal danger in approaching an enemy, the higher the costs of mobbing and the more likely it is that the costs will exceed the benefits of this trait. In places in which mobbing carries with it higher costs, the smaller, less agile, juvenile squirrels do not mob. Although the adults do harass snakes, they do so more cautiously when dealing with a highly dangerous opponent.

Weak Tests and Strong Tests of Adaptationist Hypotheses

Thus far, we have tried to confirm or reject a particular hypothesis on mobbing solely on the basis of tests of predictions derived from this one hypothesis. Although many predictions from the hypothesis have been examined in many different ways, testing a single explanation is not ideal because we run the risk of overlooking other hypotheses that may yield some of the same predictions. As outlined in Chapter 1, if alternative hypotheses A and B both produce prediction X, then if the prediction is true we cannot rule out either A or B. This is why most biologists especially admire research that tests *multiple* working hypotheses in ways that permit clean discrimination between the competing alternatives.

William Shields has used the multiple hypotheses approach to analyze mobbing in the barn swallow [720]. Several pairs of this small bird sometimes nest together in barns or under bridges. The birds attack predators by diving and swirling about them, occasionally even striking them in flight. One hypothesis on the function of active mobbing is the now familiar one, namely, that mobbing increases the survival of the offspring of the mobber by distracting or deterring the predator from a vulnerable nest with eggs or fledglings. But there are alternatives. For example, mobbing may directly increase the survival of the mobber by giving him or her a chance to assess the enemy and perhaps drive him away from the area, or it may increase the survival of the group as a whole by protecting the entire colony from a raiding predator.

The three hypotheses, the parental care, self-defense, and group defense alternatives, yield different predictions. If active mobbing is a form of self-defense, then there is no reason to expect seasonal variation in its frequency, but mobbing is strongly associated with the breeding season of the birds. We can therefore reject this hypothesis. If mobbing is a form of cooperative group defense, then we predict that the age and breeding status of mobbers will simply reflect the composition of the population as a whole. But active mobbers are not a random sample of those available in a cluster of breeding birds (Table 1). When Shields placed a stuffed screech owl near a site occupied by a number of breeding barn swallows, he found that adults that had young in the nest were overrepresented in the active mobbing category, whereas juveniles and nonbreeding birds were especially likely *not* to become involved in dive-bombing the owl. By formally testing a set of alternatives, Shields provided a much stronger case that the adaptive value of active mobbing in barn swallows helps parents protect their eggs or nestlings.

TABLE 1

Mobbing by barn swallows: The percentage of individuals of various types present in the studied colonies, and the percentage actually participating in mobbing a predator

Type of swallow	Representation in population (%)[a]	Active mobbers (%)
Adult: nonbreeding	6	2
Adult: before incubation	9	11
Adult: during incubation	14	10
Adult: with young	51	77
Juveniles	20	0

Source: Shields [720].
[a]Percentages reflect the average for the population as a whole over the 8 weeks of the study.

The Behavioral Ecology of Communication Signals

Having examined how behavioral ecologists have explored the adaptive basis of mobbing, let us turn again to the topic of communication. In the preceding chapter we discussed how evolutionary biologists have tried to answer questions about the history of some communication signals given by empidid flies, hyenas, and honeybees. Here I wish to contrast the phylogenetic approach with the adaptationist approach, which enables us to explore communication in a very different way.

Let us begin at the beginning. How do we know when animals are communicating with one another? When a black-headed gull gives a special call and its mate looks up, then flies to join it in mobbing a predator, we have little doubt that the caller communicated with the other gull. An action by one individual, the signaler, conveyed information that altered the behavior of another, the receiver.

The discovery that an action by one individual affects the behavior of another animal is the first step in determining whether communication is occurring. If you were to observe a whale leap almost all the way out of the water, turn over in midair, and come crashing back into the ocean, sending up an eruption of water, how would you tell whether this was a message from the "breaching" whale (Figure 10)? Breaching may be just a way to dislodge irritating parasites on its back. But there may be more to it than this, because humpback whales are more likely to breach if other individuals are in the neighborhood [832]. Furthermore, the behavior occurs more frequently during the breeding season than outside these months. Finally, if one whale breaches, the action serves to trigger the behavior in other whales within a radius of 6 miles. All this suggests that breaching conveys information from a signaling humpback to others, although just what this information might be remains a mystery.

The next phase in an adaptationist analysis of the transfer of information between individuals is to ask the $64,000 question: Do the benefits exceed the costs for the signaler, the receiver, or both? This question forces us to consider the several different combinations of fitness consequences for individuals involved in a social interaction (Table 2).

10 **A breaching killer whale,** leaping out of the water to come crashing down—perhaps to communicate with other whales in the neighborhood? Photograph by Jeff K. Jacobsen.

There is a difference between the transaction that takes place between a calling gull and its mate and that which takes place when a cuckoo chick begs for food from its foster parent. The cuckoo parasite benefits from the transmittal of information to its host, but the host wastes its parental care on a member of another species instead of expending it on its own offspring. There is also a difference between deceitful "communication" between cuc-

TABLE 2
The possible fitness effects on signaler and receiver
as a result of transferring information from one to the other

| | Fitness effects on | |
Category of behavior	Signaler	Receiver
Cooperative signaling	+	+
Deceitful signaling	+	−
Spiteful signaling	−	−
Incidental signaling	−	+

koo and host and the incidental "communication" that takes place when a sparrow hawk watches a male anole lizard flashing his red dewlap at a female anole and uses the information to nail the male lizard.

The focus on the fitness consequences of signaling and responding to the signal enables us to identify interesting puzzles worth exploring and to make predictions about what we might find when examining communication systems [186]. For example, deceitful signals are a problem, because they require that a receiver react in ways that lower its fitness. How can such responses evolve? Wouldn't individuals that simply ignored deceitful signals outreproduce those that were fooled and harmed as a result?

The Evolution of Deceit
and Countermeasures to Deceit

To answer these questions, we must first establish beyond reasonable doubt that deceit occurs in nature. It does. We have already discussed brood parasites and orchids that sexually deceive certain male bees and wasps (Chapter 2). In Chapter 11 we shall discuss other cases in which edible species that resemble other toxic animals fool their predators into leaving them alone.

The firefly *femmes fatales* studied by James Lloyd provide another famous example of deceit [490]. The females of some predatory fireflies in the genus *Photuris* answer the flashes given by males of certain other species in the genus *Photinus*. Each species of *Photinus* has its own code with a female answering her male's distinctive flash pattern by giving a flash of her own after a precise time interval. Some *femmes fatales* can respond at the correct interval to male signals of three *Photinus* species [492]. If a *Photuris* female succeeds in luring a male *Photinus* close enough, she will grab, kill, and eat him (Figure 11).

Another less certain case of deception involves "pseudoestrus" by female langurs. In Chapter 1, I described how male langurs may kill young infants in a band after they assume control of it. Males sometimes even kill the offspring of females that were pregnant at the time of the takeover, although they do not usually kill their own offspring when these are born to the females with whom they have mated since the takeover.

Female langurs in the early stages of pregnancy at the time of a takeover sometimes actively solicit copulations from the new male—even though they are not ovulating and cannot be fertilized. Sarah Hrdy proposed that these females might be deceiving their partner into treating their future offspring as if they were the new male's progeny, even though they are not.

Although the paternity-confusion hypothesis has not been tested systematically with langurs, it has been used to generate predictions tested with other species. For example, Boel Jeppson found that when pregnant female water voles moved into an area occupied by a male other than their previous mate, the females resumed cycling and mated with the new male (judging by presence of a fresh male-donated *vaginal plug* in the female's reproductive tract shortly after the move). Some of these females subsequently gave birth to offspring so soon after moving that these offspring could not have been fathered by the second male [410].

If males generally tolerate offspring of females they have mated [459], we can better understand why deception *might* work in langurs and water voles (assuming that the deception hypothesis is correct in these cases). Males cannot test the paternity of their offspring directly but must rely on indirect cues that are correlated with paternity. If a male mates with a female, records the experience, and uses it to guide his behavior toward the offspring of this female, he will *usually* behave adaptively. That is, he will often be the father of those offspring, in which case killing them would be disadvantageous. Because in most cases males gain by copulating with estrous females, and receptive ovulating females gain as well, deceptive females can mimic the signals of estrus, and thereby parasitize a system of information transfer for their own benefit, a system that evolved because of its mutual advantages *under other conditions*.

This pattern seems to be generally true of deceitful signalers. In South America, certain bird species routinely flock together. In a mixed-species group, the white-winged tanager-shrike perches and pursues insects flushed

11 **Firefly *femme fatale*.** Females of predatory *Photuris* fireflies attract and eat males of *Photinus* fireflies. Photograph by James E. Lloyd.

up by the other species that forage actively in dense foliage. Because the percher scans its environment, it is usually first to see predatory hawks, and when it does, it sounds the alarm, causing the others to dive for safety or freeze. The tanager-shrike, however, also is especially likely to give this call when it pursues a flying insect that another bird is chasing (Figure 12). The tanager-shrike is usually crying "wolf," because there is no hawk, but the other birds usually abort their chase, thus giving the tanager-shrike an uncontested shot at its prey. Why do the other birds permit this to happen? As Charles Munn points out, most of the time the other birds gain by paying attention to the alarm call, because the price to be paid for not reacting may be death—if the tanager-shrike really has spotted a hawk cruising in for the kill. Under these conditions, the benefits of treating each alarm call seriously will outweigh the costs, permitting the tanager-shrike to take unfair advantage of the relationship from time to time [578].

Similarly, a female *Photuris* firefly that lures a male *Photinus* to his death is able to use deceit because *Photinus* males have more to gain than to lose as a rule by responding to the sexual acceptance signal of their species. Males that completely avoided *Photuris* females would probably also avoid their own females, because the signals are nearly identical, and the nonresponding males would leave few descendants to carry on their cautious behavior. In any event, male *Photinus* have a counterstrategy for dealing with *femmes fatales*. *Photuris* predators catch *Photinus* males on only 1 of 10 attempts, in part because *Photinus* males do not throw themselves upon individuals giving the receptive female signal. Instead, they land some distance away, and then walk *slowly* forward, apparently ready to decamp should anything appear amiss [495].

Displays and Deception

From an adaptationist perspective, deceitful ("illegitimate") signalers impose a cost on an otherwise beneficial communication system [606] (Figure

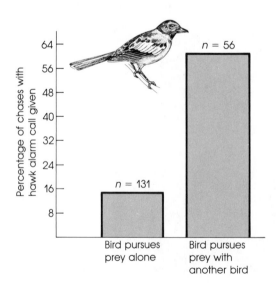

12 **A bird that cries "wolf."** White-winged tanager-shrikes are far more likely to give their hawk alarm call when pursuing a flying insect that is also being chased by another bird than when it is chasing a prey by itself.

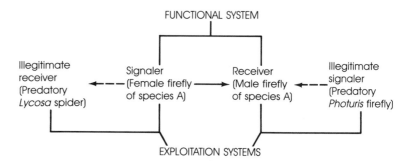

FUNCTIONAL SYSTEM

Illegitimate receiver (Predatory *Lycosa* spider)

Signaler (Female firefly of species A)

Receiver (Male firefly of species A)

Illegitimate signaler (Predatory *Photuris* firefly)

EXPLOITATION SYSTEMS

13 **Evolved communication systems** consist of a legitimate signaler and receiver (such as a male and female firefly of the same species). Such systems can be exploited by illegitimate signalers (e.g., a predatory *Photuris* firefly that lures males of other species to her) and illegitimate receivers (e.g., a spider that detects a signaling female firefly).

13). Illegitimate signalers reduce the net gain enjoyed by the "legitimate receivers" in a communication system. If it were consistently disadvantageous to respond to a signal, then selection should result in the spread of the mutant genes of nonresponding individuals.

A special problem for the adaptationist, given this argument, is to explain why so many animals use highly ritualized visual and acoustical displays to make critical decisions. For example, rather than fight directly, males of many animals settle some conflicts by employing highly ritualized displays that often do not involve the slightest contact between the two rivals, let alone an outright battle. Even fighting appears to have an element of comic opera about it, with the two contestants separating after a few pushes and shoves. After a body slam or two, a subordinate elephant seal generally lumbers off as fast as it can lumber, inch-worming its blubbery body across the beach to the water while the victor bellows in noisy, but generally harmless, pursuit.

Prior to the recognition that "for-the-good-of-the-species" hypotheses have serious logical problems (see Chapter 1), it was common to hear that the resolution of conflicts via harmless threat displays had benefits for the species or group, such as population regulation, or the prevention of injury to the "superior" individuals who would improve the genetic stock of the population. Species-benefit hypotheses imply that the gentlemanly losers belong to an expendable population reserve ready to sacrifice itself for the welfare of the species as a whole. However, self-sacrificing individuals, if they ever existed, would sooner or later be replaced by mutant types that behaved in ways that increased their fitness.

Therefore, the working hypothesis of the modern biologist is that an individual must resolve conflicts in ways that raise his or her reproductive success, regardless of the consequences this behavior has for the population as a whole. But the ritualized nature of animal aggression would seem to open the door to deception, with some individuals mimicking the displays that get others to yield a useful resource like a breeding territory or a mate. Some observers have suggested that this is precisely what happens [186].

For example, males of many birds announce their ownership of a breeding territory by their advertisement call, their song. Other birds may be deterred from challenging a singing male for his valuable territory strictly on the basis of his song. John Krebs suggested that males sometimes concede defeat without actually fighting because they have been fooled by the complex song repertoire of an opponent. The male of the great tit has from two to eight different song types; he tends to change his territorial song when he changes his perch [449]. This switch may help convince a territory-hunting intruder listening in the woods that there are more individuals in the area than there really are, a conclusion causing him to leave the "saturated" woodlot without competing for a territory with the singing male.

The notion that male birds may be trying to deceive others about the number of territorial defenders in an area was named the Beau Geste hypothesis [445] in honor of the hero of P. C. Wren's adventure novel about the French Foreign Legion. Beau Geste deceived enemies besieging an undermanned fortress by propping dead men on the ramparts where they created the false impression that the fort was well-defended. Support for the Beau Geste hypothesis comes from an experimental study of red-winged blackbirds, whose males also employ a complex song repertoire and change song types when they change perches. An empty territory site occupied by a speaker broadcasting a variety of song types was significantly less often invaded by trespassing males (other than neighboring territory owners) than an empty territory with a speaker that played only a single song type (Figure 14) [873].

This result, however, is also compatible with another hypothesis, which is that males signal their territory-holding ability through their songs. For example, if males that have a complex song repertoire are older, or more powerful, or in better condition than males with a less varied song, intruders might avoid the versatile songster, not because they are deceived about

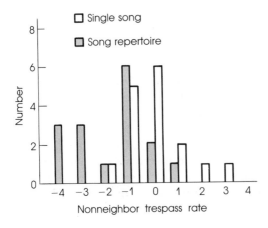

14 The "Beau Geste" effect. The trespass rate by intruding red-winged blackbirds is lower when the intruders (from sites other than neighboring territories) can hear tapes of a repertoire of redwing songs. More intruders enter a territory when tapes from that site broadcast just one song type. *Source:* Yasukawa [873].

the number of occupants in the area, but because they can accurately assess that they would be mauled if they were to trespass against such a singer.

Ken Yasukawa and William Searcy attempted to test the Beau Geste and "male fighting ability" hypotheses by looking at the relationship between male density, number of song types, and rate of territory intrusion by rival red-winged blackbirds under natural conditions. Although there was a significant correlation between the number of song types being sung in an area and the density of males there, the trespassing rate per territory was *not* linked to the density of males in an area [874]. This result violates a key prediction of the Beau Geste hypothesis, which is that males will keep away from areas of high male density, enabling a male in a low-density site to deceive territory-hunting males by singing many different songs.

If we accept the conclusion that a large repertoire is not being used deceptively by male redwings, then why did male intruders stay away from loudspeakers broadcasting large song repertoires during Yasukawa's experiment? In redwings, but not in some other birds, there is a correlation between the age and experience of a male and his repertoire size. If we assume that older, more experienced birds are tougher opponents, at least in redwings, then intruders might do well to avoid a male with a large repertoire, because it is an indirect cue of his fighting ability.

Leaving aside the troubling issue of why repertoire size is *not* correlated with age and experience in other songbirds, let us accept that a male redwing can reduce the number of intruders he has to chase, just by singing a few more song types. The obvious question then becomes, Why do not all territory owners "pretend" they are old and experienced by boosting the number of song types they produce while advertising their territory?

Selection for "Honest" Signals

One hypothesis for the failure of some individuals to produce the kind of display that intimidates rivals at a distance is that they would if they could, but they can't. This hypothesis has been tested with the European toad, *Bufo bufo*, whose males fight for possession of receptive females. When a male finds another male mounted on the back of a female, he may grapple with the other male to pull him from the female. As soon as one male touches him, the mounted male croaks. There is much variation in the pitch of croaks of different males, and Nicholas Davies and Timothy Halliday used an ingenious experiment to show that deep-pitched croaks deterred attackers more effectively than high-pitched ones [176].

The two researchers placed mating pairs of toads in tanks with an attacker for 30 minutes. The paired male, which might be large or small, had been silenced by looping a rubber band under his arms and through his mouth. Whenever the second male touched the pair, a tape recorder supplied a 5-second call of either low or high pitch. Small defender males were much less frequently attacked if the other male heard a deep-pitched call (Figure 15). Thus, deep croaks deter rivals to some extent (although tactile cues also play a role in determining the frequency and persistence of an attack, as one can see from the higher overall attack rate on smaller toads).

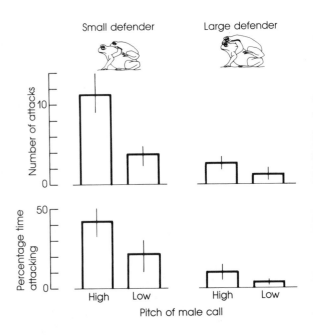

1 5 **Deep croak experiment.** Single male toads make fewer contacts and interact less with silenced mating rivals when a tape of a low-frequency call is played than when a higher-frequency call is played. *Source:* Davies and Halliday [176].

So why don't small males "pretend" to be large by giving low-pitched calls? Because a small male simply cannot produce a deep croak, given that body mass and the unbendable rules of physics determine the pitch of the signal that a male can generate. Thus, toads have evolved a defensive signal that accurately announces their body size. By attending to this signal, a male can determine something about the size of his rival.

A large male that produces a signal that a small male cannot deceitfully mimic can advertise that he is stronger than smaller toads. He gains when small rivals withdraw at once because he does not have to waste his time and energy in an out-and-out fight. Small males gain because they do not waste their time and energy in a battle they are very unlikely to win. Both benefit, and in this sense the communication that takes place between them qualifies as a cooperative interaction.

Imagine two kinds of aggressive individuals in a population, one that fought with each opponent until physically defeated or victorious, the other that withdrew as quickly as possible after an initial check of the rival's fighting potential. There is little doubt that "fight no matter what" types would sooner or later, and probably sooner than later, encounter an opponent that would thrash them soundly. The "fight only when the odds are good" types would be far less likely to suffer an injurious defeat at the hands of an overwhelming opponent.

Further imagine two kinds of superior fighters in a population, one that generated signals other lesser males could not mimic and another that produced threat displays that lesser males could equal. In the second species mimics would arise by chance, and this would favor receivers that ignored the deceptive signal, reducing the value of producing it. This in turn would lead logically to the spread of the genetic basis for the other kind of signal, which could not be devalued by deceitful signalers.

If this argument is correct, we can predict that threat displays of many unrelated species will exhibit convergent properties that honestly inform rivals about the size and fighting capacity of the displayer [623], given that body size, muscle mass, and strength are intercorrelated in many animals [468, 766, 834]. In the red deer, males compete for possession of harems of females during the fall rutting season. Although all-out fighting does occur, typically among males closely matched in physical condition, most inter-actions never even reach the stage of antler clashing. Far more often, one male withdraws after a round of competitive roaring (Figure 16). These acoustical displays are energetically demanding to produce, and only a male in top condition can continue to roar at a high rate for many minutes [145, 146]. Thus, like European toad competitors, challengers and harem-masters gain accurate information about the fighting potential of a rival by listening to his acoustical performance. They can use the information in the display to determine whether the opponent is likely to be much stronger or weaker, and so decide adaptively whether to withdraw or to challenge the rival to a real fight.

Red deer rivals also have a display called the parallel walk (Figure 17). Here, too, males have an opportunity to assess their relative size close up before they ever engage in a direct physical confrontation. Similar displays have evolved in many fish, birds, and insects, in which two individuals stand side by side or head to head, permitting the two contestants to judge how big they are relative to the enemy before fleeing or fighting. In the bizarre stalk-eyed flies, for example, body length is highly correlated with distance between the two eyes. The animal's vision is such that males can readily judge whether they are larger or smaller than an opponent during their ritualized fights (Figure 18), with smaller males decamping in the face of a threat from larger flies [117, 187].

16 **An honest signal.** Only a male red deer in top condition can sustain roaring contests for long periods. Photograph by Timothy Clutton-Brock.

17 **Assessment display: red deer.** Male red deer and many other animals have aggressive displays in which opponents move side by side, thus permitting each other to judge accurately the relative size of the rival. Photograph by Timothy Clutton-Brock.

18 **Assessment display: stalk-eyed fly.** Males of this stalk-eyed fly confront one another head to head, an encounter permitting them to determine the size of the opponent. *Source:* de la Motte and Burkhardt [187].

Coping with Illegitimate Receivers

We have argued that deception in the past has shaped communication signals, favoring honest advertisements that make deception difficult for would-be exploiters of communication systems. Cooperative signalers and receivers can also be exploited by illegitimate receivers. Have communication signals and the ways they are used been affected by this selection pressure as well?

Some predators listen for the acoustical signals of their prey. A now-famous example is the tungara frog and its enemy, the fringe-lipped bat. Male frogs call loudly at night to attract mates; in this they sometimes succeed, but for every benefit there is also a cost, and in this case the cost takes the form of bats that locate calling males, which they sweep from the water and devour (Figure 19). Why call if it could get you killed? The male frogs are in a bind that the bats exploit. If they do not call, they do not mate and leave no descendants. If they do call, they stand a chance of mating and a chance of being tracked down and eaten. The trick is to reduce the costs. The frogs do this in any number of ways, as Michael Ryan, Merlin Tuttle, and their associates have shown. First, they stop calling when they detect a flying bat or when experimenters send a moving cardboard model of a large bat sailing overhead [794].

Essentially the same simple, sensible tactic has evolved independently in a Sonoran Desert katydid, although the insect scans for acoustical instead of visual cues of bat enemies. Males sing their ultrasonic mate-attracting calls from the tops of creosote bushes. But while they are singing,

19 **Predation pressure** can affect the evolution of communication signals. The fringe-lipped bat is an illegitimate receiver that tracks its prey, male tungara frogs, by listening for their calls.

they are also listening for the ultrasonic cries of bats; and when they hear these sounds (either from the real thing flying nearby or from a little hand-held ultrasound generator held by a nearby researcher), they immediately shut up (Figure 20) [740].

The tungara frog sometimes compromises mate attraction in response to predation risk even when it has not detected an approaching bat. The frog's advertisement call has two parts, an introductory whine and one or more terminal chucks (Figure 21). When a male is calling by himself, he uses only the whining component—even though in playback experiments the frog team demonstrated that females almost always come to speakers playing a whine–chuck call rather than to those propagating only the whine call. These same kinds of playback experiments, however, showed that fringe-lipped bats were slightly more than twice as likely to inspect, even land upon, a speaker broadcasting a whine–chuck [685].

The frog team further showed that tungara frogs are safer in a large chorus than in a small one, primarily because a bat has many more victims to choose among in a large group [689]. They concluded therefore that when a male frog was alone and highly vulnerable to a predator, the frog modified his call to make him harder to find, compromising his attractiveness to females but increasing his chances of living through the night.

Bat predation has probably influenced the evolution of the communication tactics of male tungara frogs. Another classic demonstration of the impact of illegitimate receivers on a communication signal involves the

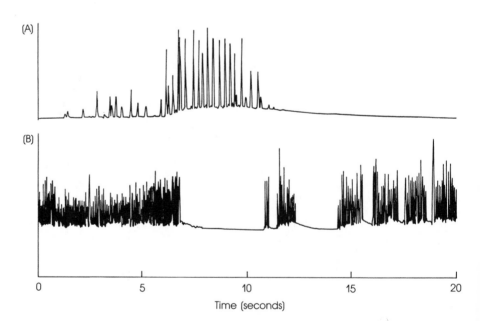

20 **Coping with illegitimate receivers.** Katydids that hear an approaching bat's ultrasonic cries stop calling. (A) A record of bat cries near a katydid. (B) A simultaneous account of the katydid's calls. *Source:* Spangler [740].

(A)

21 Mate attraction versus predation risk. (A) A male tungara frog calling while floating in his pond. Photograph by Michael Ryan [687]. (B) Sonagrams of five calls of increasing complexity. The mate-attraction call of the tungara frog has two components, an introductory "whine" followed by one or more "chucks." Females, and fringe-lipped bats, prefer males giving chucks as well as whines. *Source:* Ryan [686].

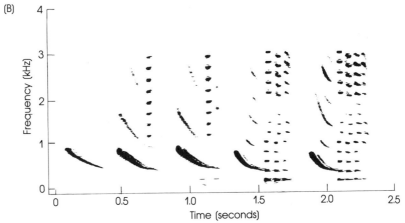

(B)

difference between the mobbing call and the "seet" alarm call of great tits (Figure 22) [518]. These small European songbirds sometimes approach a *perched* hawk or owl and give the mobbing call, a loud signal with most of its energy in the 4.5 kilohertz range. Mobbing signals may attract other birds to the site to join in harassing the predator.

If a great tit spots a *flying* hawk at some distance, it gives the "seet" alarm, which appears to warn mates and offspring of possible danger. This call is much softer and has a higher frequency (in the 7–8 kilohertz range) than the mobbing call. The properties of this signal are such that it attenuates (cannot be detected) after traveling a much shorter distance than the

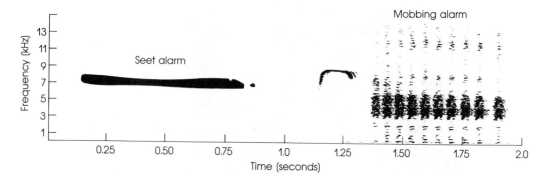

22 **Great tit alarm calls.** A sonagram of the "seet" alarm call (courtesy of Peter Marler) and the mobbing alarm call of this species (courtesy of William Latimer). Note the lower frequency of the mobbing signal.

mobbing signal travels, thus compromising its effectiveness in reaching distant legitimate receivers, but also decreasing the chance that the caller will alert the predator of its presence. The probability of informing the hawk is further reduced because the hawk cannot hear relatively high-pitched sounds very well, whereas the great tit has high sensitivity to acoustical stimuli in the 7–10 kilohertz range (Figure 23). As a result a great tit can hear a "seet" call given by a bird 40 meters away, whereas a sparrow hawk more than 10 meters distant will fail to detect the same signal [433].

The remarkable convergence in the "seet" calls of European songbirds suggests that selection by bird-eating hawks favored individuals of many prey species that produced alarm notes that were hard for hawks to hear, even if they might be somewhat less effective in reaching the intended receivers of the alarmed signaler (Figure 24).

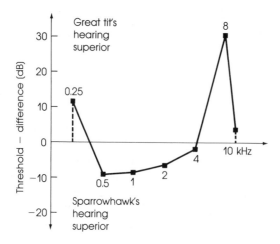

23 **Hearing abilities of great tits and their predators, sparrowhawks.** Great tits are more sensitive to sounds in the high frequency range of the seet call than are sparrowhawks. *Source:* Klump [433].

Channels of Communication

There are many other kinds of questions about communication signals that arise if one starts thinking about the costs and benefits to individuals of alternatives. One of the alternatives for animal species has to do with the channel of communication. Why, for example, do birds use airborne acoustical calls, whereas banner-tailed kangaroo rats communicate by thumping their hindfeet on the ground, creating vibrations in the earth [658], and subterranean mole rats whack their heads against their tunnels to the same end [363]? Why is it that bark beetles employ odors to attract mates, whereas empidid flies and fireflies use visual displays? As indicated in the previous chapter, accident, chance, and especially the constraints imposed by past evolution play a role in determining the nature of an animal's communication abilities. But it is also possible that different channels of communication have been selected for their ability to overcome obstacles to information transfer imposed by a particular environment. To test this hypothesis, we should consider the following factors that influence the fitness value of a signal to a sender:

1. How effectively the signal reaches desired receivers
2. The amount of information encoded in the signal
3. The energetic cost to the sender of producing and broadcasting the signal
4. The ease with which the sender can be located by a receiver, if this is a goal of the sender

These parameters of a message vary, depending on the environment in

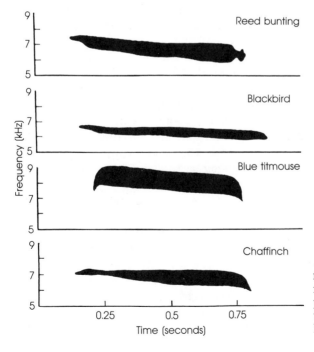

24 **Convergent evolution of a communication signal.** Many unrelated birds give high-pitched "seet" calls when they spot an approaching hawk. *Source:* Marler [519].

TABLE 3

The properties of the major channels of communication

Ability of signal to reach receiver	Channel			
	Chemical	Acoustical	Visual	Tactile
Range	Long	Long	Medium	Very short
Transmission rate	Slow	Fast	Fast	Fast
Flow around barrier?	Yes	Yes	No	No
Night use	Yes	Yes	No[a]	Yes
Information content				
Fadeout time	Slow	Fast	Fast	Fast
Locatability of sender	Difficult	Fairly easy	Easy	Easy
Cost to sender				
Energetic expense	Low	High	Low to moderate	Slow

[a]Except bioluminescent signals.

which they are used (Table 3). Obviously, visual signals, except for self-manufactured bioluminescent flashes, cannot be used at night and will not pass around physical barriers, unlike chemical and auditory messages. Moreover, visual signals may be easily detected by the sender's predators. On the other hand, visual communication enables instantaneous localization of the sender if the receiver gets the message, unlike most other communication modes.

Pheromones, for example, must be tediously tracked by a receiver, and these long-distance chemical signals are dependent upon wind conditions. Too little wind and the signal goes nowhere; too much and the pheromone may be blown erratically hither and yon, a result that makes it difficult for receivers to find the sender. In addition, because a chemical message lingers in the air a long time, it is probably difficult (but not impossible [148]) for an animal to send out a rapid-fire series of pheromonal messages, each conveying a new bit of information. This constraint lowers the rate of information transfer by chemical means relative to acoustical and visual signals, which usually have a rapid fadeout time and can be quickly replaced with new messages [853]. But acoustical messages are energetically demanding to produce, so much so that a calling male frog may exhaust its energy reserves in just a few hours [822].

Thus, the utility of a communication channel depends on the ecology of the species, as we can illustrate by comparing the species-identification signals used by different groups of insects. Because moths can fly to a perch exposed to night breezes, they can use pheromonal messages to attract mates from a great distance. The sex pheromones that advertise species membership tend to be moderately long chain carbon molecules (Figure 25), small enough to be lightweight and volatile, but large enough to be distinctive in structure [851]. The great diversity of pheromones makes it possible for females to attract males of their species whose olfactory apparatus is specifically "tuned" to their females' signal.

The hypothesis that the medium-range molecular weight of sex pheromones is related to the advantages of species specificity can be tested by examining alarm pheromones. There is a trade-off between molecular distinctiveness (which is best served through molecules of high weight) and volatility (which is best served by molecules of low weight). Sex attractants generally sacrifice a certain amount of volatility in order to achieve a unique structure and thereby convey unambiguously the species membership of the signaler [852]. An alarm pheromone is under no such constraint. If a worker ant or termite is to arouse other colony members to repel an attacker, the primary requirement is that the message be broadcast as quickly and widely as possible. Alarm pheromones tend to be extremely lightweight and convergently similar in structure among social insects [852].

Crickets advertise species membership primarily via the auditory channel [9]. Constant chirping is much more energetically expensive than the release of a minute amount of chemical and may also provide some male

Bombykol (silkworm moth)

Disparlure (gypsy moth)

2,3-Dihydro-7-methyl-1H-pyrrolizin-L-one
(Queen butterfly)

Honeybee queen substance

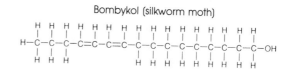

25 **Species-specific olfactory signals.** The sex pheromones of insects tend to be moderately complex hydrocarbons; species often differ in the chemical they use to advertise species membership.

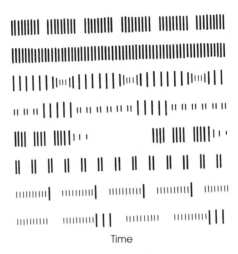

26 **Species-specific acoustical signals.** Species of crickets typically differ in the pattern of their mate-attracting calls. In the eight different calls illustrated here, each pulse of sound is represented by a vertical bar. The height of the bar indicates the intensity of the pulse.

competitors and predators with information damaging to the chirper. (Because pheromones are released in such small quantities and are so distinctive, predators must be highly specialized to use pheromonal messages to track down their prey.) The risk of exploitation associated with a loud calling song may be one reason why male crickets that have succeeded in attracting a female switch to a soft song during courtship.

On the plus side, auditory messages are less subject to environmental vagaries than olfactory signals. Moreover, because crickets live in and under grasses and debris, an auditory signal can usually be broadcast more effectively than an olfactory one. A moth perched high on a tree is exposed to wind currents; a cricket under a log is not. Species specificity of the signal is produced by varying the temporal spacing of the components of chirping or trilling song (Figure 26).

Fireflies are nocturnal insects that manufacture their own visual signals with the aid of photochemical equipment at the tip of their abdomens [491]. Their conspicuous flashes would seem to make them highly vulnerable to bats, more so than signaling moths and crickets. However, fireflies are probably poisonous [229]. Thus, they are able to use an easy-to-locate and economical visual communication channel. The specificity of the message is achieved through the evolution of unique patterns of light pulses, which vary from species to species in duration, intensity, frequency of occurrence, and color [491] (Figure 27).

Overcoming Obstacles to Long-Distance Communication

The broad correlation between the kind of environment in which moths, crickets, and fireflies live and the kinds of species-specific signals they use provides a weak test of the hypothesis that communication signals are adaptive solutions to particular ecological factors. Now let us take a closer look at specific hypotheses on how animals overcome environmental obstacles to long-distance acoustical communication.

The ease with which sounds of different frequencies are transmitted over

long distances depends upon the environment. Acoustical signals may be
absorbed by vegetation quickly, reducing the number of receivers they may
reach. Eugene Morton broadcast sounds of different frequencies in the forest
and grassland habitats of Panama [572]. Tape recorders placed at various
distances from the sound producer made a record of what frequencies
reached them. Morton found that sounds of low and high frequency were
relatively quickly absorbed in the forest, but sounds between 1500 and 2500
hertz were much less severely attenuated. Bird species in the Panamanian

27 **Species-specific visual signals.** Each firefly species has
its own special pattern of flashes. Courtesy of James E.
Lloyd.

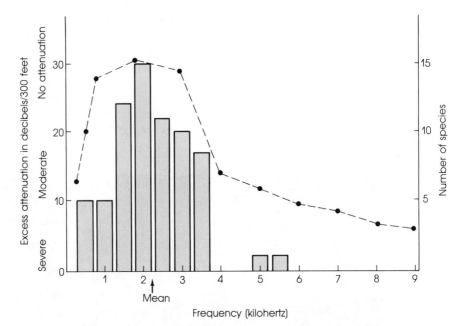

28 **Exploitation of a "sound window."** Sound attenuation in lowland forest habitats is least for those frequencies that birds most often use in their calls. The histograms show the number of species employing various frequencies in their calls; the dotted line shows the attenuation effect of the habitat on sounds of different frequencies. *Source:* Morton [572].

forest typically signal within the 1500- to 2500-hertz band. These songs can be projected the greatest distance with the least energetic expense (Figure 28).

The hypothesis that forest-dwelling birds have evolved songs with frequencies that are transmitted more effectively in forest habitats has been challenged on a variety of grounds [525, 526]. An alternative hypothesis is that the key problem for birds is one of song distortion or degradation rather than of song attenuation. In forests, sounds echo back from leaves and limbs; the echoes create competing noises that may damage the integrity of a song. Forest birds may therefore sing low-frequency songs, not so much to project a message a great distance, but to avoid deflection by small objects such as leaves and branches, which greatly interfere with high-frequency sounds.

Although the relative importance of attenuation and degradation as factors affecting song evolution remains to be resolved, Morton's hypothesis has been tested by Peter Waser and Charles Brown in Kenyan and Ugandan rain forest, where they discovered a "sound window" higher in the trees. Using modified and improved techniques for determining the attenuation of sounds, Waser and Brown showed that low-frequency sounds (125–200 hertz) were much less impeded than lower or higher frequency sounds broadcast from 7 to 8 meters above the ground (Figure 29). They note that

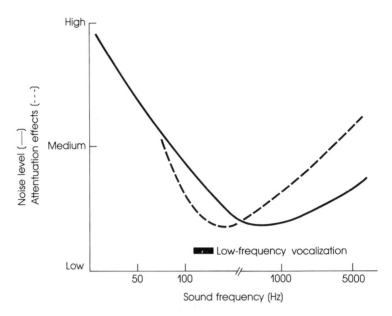

29 **Blue monkey's sound window.** For long-distance communication, this African primate has a special low-frequency vocalization (black bar) that falls in the sound frequency range that is least likely to be masked by background noise and that is least attenuated by obstructing vegetation. *Source:* Brown and Waser [107].

the blue monkey (as well as other tropical forest primates) produces distinctive vocalizations that fall within this range, and that the blue monkey has special acoustical sensitivity to low-frequency sounds. These results suggest that certain primates may pitch their calls to take advantage of a relatively quiet channel available to them in the canopy of the forest [107, 816].

Opportunities for efficient communication can change over the course of a day as various background noises rise and fall. Alex Kacelnik and John Krebs examined the possibility that territorial great tits sing at dawn because this is a relatively quiet time of day [416]. Numerous studies document that wind and air turbulence increase after dawn, producing noises that attenuate the songs of birds. It is plausible, therefore, that territorial males can project their messages farther at less cost in the hour or two around dawn instead of later in the day.

It may also be, however, that sneaky male intruders are most abundant at dawn (perhaps because receptive females mate at this time of day [513]). Further, in the cool air and dim light of dawn, insects are less active and less visible to bird predators (Figure 30). Therefore, territorial males have the most to gain by trying to repel other males with their songs at dawn and least to lose in feeding opportunities then. The dawn chorus in great tits and other birds probably occurs because a combination of ecological factors makes the benefit-to-cost ratio for singing most profitable then.

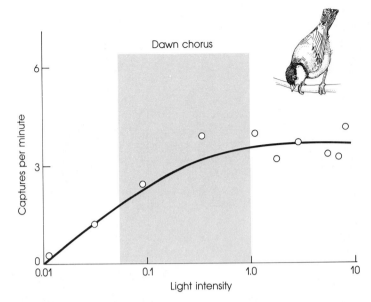

30 **Ecology of the dawn chorus.** Great tits sing during early morning hours when prey capture is reduced by poor visibility due to low light intensity. *Source:* Kacelnik and Krebs [416].

Social Competition and Communication

The environment of some signaling species provides not only physical barriers to signal transmission but also competition for "air space" in the form of calling conspecifics. The question is, Can calling be viewed as an evolutionary game, with males under pressure to use tactics that deal with an environment filled with signaling rivals?

James Lloyd's studies of fireflies provide a catalogue of techniques used by males to reduce the signal effectiveness of opponents while increasing their own odds of mating [494, 495]. In those *Photinus* species in which males cautiously approach a potential mate (because she might be a *Photuris* predator), males sometimes mimic the female signal (which is given at a standard interval after a male has produced his species-specific pattern). Lloyd suggests that this may confuse rival males that have also gathered on the scene, diverting their attention from the signaler that attracted them all, which might be a receptive female of their species.

Competition for broadcast space has also clearly played a strong role in the evolution of calls and their delivery in some frogs. The Puerto Rican coqui treefrog gives its call, "co-qui," in a rain forest filled with loudly calling rivals and members of other species. Peter Narins has suggested that the following features of the coqui's call help males transmit their messages effectively to appropriate receivers [582]:

1. The call is very loud: 90–100 decibels at 50 centimeters from the male.

2. Males call from elevated perches and orient themselves to face skyward, the better to propagate the song in the horizontal plane [584].

3. The song's energy is focused in a band of frequencies not used by other species of frogs in the forest, another case of a species exploiting a "sound window" in their environment.

4. Males are remarkably adept at producing a call almost instantly after a nearby rival's "co-qui" has ended. This enables an individual to inject his signal in the relatively quiet interval between neighbors' calls [583].

The coqui treefrog achieves moderately high densities, but nothing like that of another treefrog, *Hyla ebraccata*, which forms compact choruses of many males, each separated only by a meter or two from its neighbors. When other males are calling, the rate of signals given by individuals increases, males are more likely to give a multinote call than a single-note version, and the frogs synchronize their signals with great skill. Thus, as soon as one male pipes up, others nearby react in a fraction of a second, in effect covering the call of the rival with their own. When many males are calling in a small area, the possibility of injecting a signal in a quiet space may be reduced, favoring synchronous calling (although there are as always competing hypotheses on why synchrony has evolved in some species, and alternation in others). In any case, males do *not* call randomly, and playback experiments show that if a female is given a choice, she will hop toward the speaker whose signal stops last (Figure 31). The results of this experiment are consistent with the hypothesis that males time their calls to reduce the effectiveness of rival signals while enhancing their own chances of attracting a mate [821].

Criticisms of the Adaptationist Approach We have seen how adaptationists pick a behavioral attribute that interests them, like the timing of a mate-attraction signal, which they assume has evolved via individual selection. They then propose one hypothesis or, better still, alternative hypotheses on

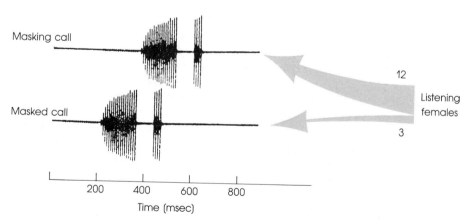

Masking call

Masked call

12

Listening females

3

200   400   600   800

Time (msec)

31 **Competition in acoustical signals.** Males of a *Hyla* frog may try to mask the calls of their rivals. In experiments, when females have a choice of going to a speaker giving the lead call and one that gives a call that covers the end of the lead call, they usually go to the masking call. *Source:* Wells [821].

how the trait might actually have the effect of increasing individual reproductive success. As we mentioned in Chapter 1, some evolutionary biologists, particularly Stephen J. Gould [303, 304], have claimed that this approach is fatally flawed by its Panglossian philosophy, the belief that all components of an organism are perfectly adapted products of natural selection.

We can expect many, if not all, features of living things to be less than perfect, if by a perfect trait we mean (1) a trait that is designed without respect to the historical constraints that restrict evolutionary change in a species, (2) a trait that is designed without respect to the other activities the organism engages in or the other traits that the organism possesses, and (3) a trait that is impervious to developmental "noise," the slight deviations from the norm caused by environmental and genetic perturbations. If you could, for example, build a mouth-brooding cichlid fish from scratch, you would probably not design it in such a way that a parasitic catfish could take advantage of it. And yet in nature, there is a catfish that somehow gets its eggs into the mouth of the mouth-brooding cichlid; once there the eggs hatch into catfish fry that eat the offspring of the host cichlid right in the host's mouth [762]. Brood parasites offer dramatic evidence of the "imperfection" of animal behavior.

A perfect mouth-brooding parent has not evolved in the parasitized cichlid species, because natural selection cannot redesign a species' behavior in a single step but is constrained at each step by the current characters of the animal. You will recall the panda principle from the preceding chapter, the principle that evolutionary changes must be compatible with what already exists in the population. An evolved developmental plan is a marvelously complex, integrated thing and to change it in any major way is unlikely to be beneficial. You might imagine that a cichlid fish capable of sensing the presence of a catfish parasite would be better off than one that accepts these parasites. But the past history of the fish may have so committed it to a particular pattern of mouth-brooding that *possible* modifications of this pattern are not likely to be superior, especially since the current system is exploited by parasites only rarely.

Just as it is possible to invent a more "perfect" mouth-brooder, if you or I were free to draw up plans for perfect nest-building behavior by kittiwakes, we might well make the gull's bill large and broad, the better to scoop up large amounts of mud to take to their cliff ledges. The bigger the bill, the fewer the trips needed, the more energy saved, and the fewer risks of being attacked by a predator while on a mud-gathering trip in an English meadow. But perfection vis-á-vis mud-collecting surely interferes with the utility of the bill in fish-catching, or in aggressive interactions, or in feeding the young through regurgitation. Thus, the bill's actual shape and size must be a compromise controlled by conflicting selection pressures, just as the tungara frog's calling behavior is shaped not just by the problems of attracting females but also by the presence of deadly bats [687].

Gould argues that historical constraints, and the constraints imposed by conflicting selection pressures, must often prevent a particular trait from achieving perfection (in the sense outlined above), and therefore adaptationists are wasting their time in analyzing every trait as if it were a perfectly adaptive product of natural selection. For example, Gould chastises those who have tried to find adaptive value in the pseudopenis of female spotted hyenas, which as you will recall from Chapter 7 may be the developmental product of high levels of testosterone in females of this species. Gould favors the hypothesis that the pseudopenis is an incidental effect of selection for high androgen levels in females because androgens enhanced female aggressiveness and chances for dominance [299].

This is a plausible hypothesis, and it is certainly true that most traits are unlikely to be "perfect" in the sense of being designed in a vacuum for a particular role or task. Does it follow, therefore, that critics are right in suggesting that adaptationists would do well to abandon their approach to evolutionary biology?

Adaptationists have, at least since 1966 when George C. Williams's *Adaptation and Natural Selection* appeared, recognized that adaptation refers to that which is better, not that which is perfect in some idealized design engineer sense [845]. Natural selection operates on existing alternatives; alleles that are better than the competing alternatives at promoting the development of reproductively superior individuals will, by definition, spread through the population. The traits linked with those alleles that produce higher reproduction than that produced by alternative traits are adaptations. The assumption of adaptation employed by adaptationists is therefore not that existing traits are "perfect" but that they contribute more to reproductive success than the alternatives that have occurred in the population in the past.

Furthermore, the goal of an adaptationist hypothesis is to be tested, and not to be taken on faith alone. I can imagine someone using the adaptationist approach to model what the optimal shape and size of a kittiwake bill should be. A realistic modeler might well take into account the conflicting forces on bill design created by mud-collecting, fish-catching, and so on. The goal of the exercise would be to test the model, to see how closely reality matched the predictions taken from the hypothesis that factors X, Y, and Z had influenced the evolution of the kittiwake's bill.

If someone wished to test a hypothesis on the possible adaptive value of a pseudopenis for a female spotted hyena, more power to him. Perhaps if maleness in the past were associated with dominance in the ancestors of spotted hyenas, the possession of a malelike penis by a female might promote her dominance of other females [337]. In other words, the pseudopenis might complement heightened aggressiveness in a mutant female's drive to dominate rival females. Maybe this hypothesis is wrong, maybe it fails to consider historical or developmental or interactive constraints that mean the trait is not adaptive. If so, an accurate test of the hypothesis will reveal that it *is* wrong. This is all to the good, because eliminating one possibility helps focus attention on other alternative hypotheses, such as the incidental-effect hypothesis outlined above. The incidental-effect hypothesis needs to be tested as well, and it would be a useful thing to do.

The Value of the Adaptationist Approach

The pleasure of science comes in dealing with difficult puzzles, which requires a way to identify hard cases and a way to solve them. Adherents of

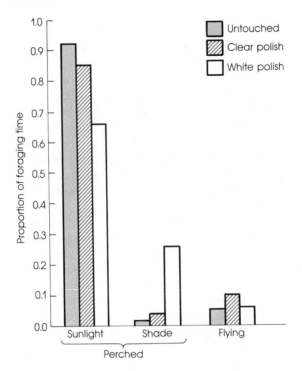

32 **Bill color affects foraging behavior.** When the upper mandible of a willow flycatcher's bill is experimentally whitened, the birds are forced to forage more in shady areas, where reflectance off the upper bill does not interfere with detection of prey. *Source:* Burtt [119].

the theory of evolution by natural selection can find interesting problems by uncovering traits that have obvious costs in terms of individual reproductive success, but no obvious benefits. I am impressed by how often an apparently minor detail of a living thing proves to have probable adaptive value after the trait has been examined via the adaptationist approach. For example, a recent paper showed that the dark color of the upper mandible of the willow flycatcher has as an adaptive consequence: the reduction of glare reflecting into the flycatcher's eyes. This permits the bird to forage in sunny areas, whereas birds whose upper mandibles had been painted a pale reflective color spent more time perched in the shade (Figure 32) [119].

Or to take an example from black-headed gulls, these birds pick up and carry away the eggshells from their nest after the young have hatched. This is a superficially trivial action that occupies only a few minutes out of an adult's lifetime. Only a "Panglossian" adaptationist would insist on wondering how this behavior might contribute to the reproductive success of the eggshell-removing birds. But Niko Tinbergen is such an adaptationist, and he proposed that the benefit of this behavior pattern might be the removal of a conspicuous visual cue that predators use to locate nests with edible eggs. Tinbergen felt reasonably confident that there had to be *some* benefit to the behavior, because in order to get rid of eggshells parents leave their nest and vulnerable young unguarded for a few minutes.

Tinbergen tested his antipredator hypothesis experimentally [775]. He took some intact gull eggs from a colony and scattered them through sand dunes visited by carrion crows, which have a great fondness for this food. By some of the unhatched eggs (which have a highly camouflaged, mottled exterior of tan and dark brown), he placed some broken eggshells whose

TABLE 4

The presence of an eggshell near a gull egg increases
the risk that the egg will be discovered and eaten by crows

Distance from eggshell to egg (cm)	Crow predation		Risk of predation (%)
	Eggs taken	Eggs not taken	
15	63	87	42
100	48	102	32
200	32	118	21

Source: Tinbergen [775].

white interior was visible; by others, no shells were placed. The crows found
and ate a much higher proportion of the eggs that had shell cues lying
nearby (Table 4).

Another test of the antipredator hypothesis involves a comparison of
kittiwake and black-headed gull behavior. What prediction would you make
about the kittiwake parent's response to eggshells after the hatching of its
babies? You are correct. The kittiwake, which has essentially no nest predators, does not even bother to push the shells off the cliff ledge on which
its nest rests [162].

On the other hand, some birds that are unrelated to the black-headed
gull and that also rely heavily on nest concealment for the safety of their
brood do carefully collect and discard broken eggshells far from the nest
(Figure 33). Thus, there appears to be a lawful relationship between the
intensity of nest predation and the existence of eggshell removal. This
finding supports the hypothesis that removing eggshells has evolved in
response to the ecological pressures provided by egg-eating predators.

33 **Eggshell removal
by the bittern.**
Shortly after hatching has
occurred, parent birds carry
off the egg shells. Photograph by Eric Hosking.

My point is that by raising the *possibility* that even this apparently insignificant behavior pattern was adaptive, Tinbergen ultimately contributed to a better understanding of gull behavior. By being willing to test his idea, he was guaranteed to make a contribution. If he had been wrong, his experiment would have told him so. Because he was right, we know that even the small details of animal behavior may sometimes serve specific adaptive functions.

SUMMARY

1   Persons interested in ultimate questions about behavior investigate the current adaptive value of a trait as well as attempting to trace the history of a trait. The underlying assumption of an adaptationist approach is that the ultimate function of a behavioral ability is to contribute to the gene-copying success of an individual. There are many environmental obstacles that stand in the way of individual reproductive success. Adaptationists (or behavioral ecologists) try to discover how behavioral traits overcome these obstacles.

2   All traits have benefits and costs in terms of their effect on the fitness of an individual. The benefits and costs of a trait should vary depending on the environmental pressures acting on an animal. Behavioral ecologists often use an optimality approach to determine whether traits have benefits that exceed their costs by a greater margin than other possible alternatives.

3   A useful hypothesis generates testable predictions, which can be analyzed by gathering additional observational data, by conducting controlled experiments, and by using the comparative approach. Much of this chapter focuses on the comparative method of prediction testing. Differences in the key environmental factors acting on two closely related species should lead to divergent evolution in behavior. Similarities in the ecology of two or more phylogenetically unrelated species should lead to the convergent evolution of a similar behavioral solution to a shared obstacle to reproductive success.

4   The comparative method can be illustrated by examining the conditions associated with the evolution of communal harassment of predators. Mobbing has evolved independently in many species that are faced with effective predators of their offspring. Mobbing does not occur in the kittiwake gulls because these birds nest on cliffs that predators cannot reach.

5   Adaptationist studies of animal communication are based on the proposition that the transfer of information between individuals can have different costs and benefits for the signaler and receiver. The value of this perspective in generating hypotheses is illustrated by an analysis of the evolution of deceitful signaling and the responses of receivers to the risk of being deceived.

6   Different channels of communication have different advantages and disadvantages in various environments. The correspondence between ecological problems and the kind of signal used suggests that selection has pro-

moted effectiveness of signal transmission, detection, and response over evolutionary time.

7 The adaptationist approach has been criticized on the grounds that it is unrealistic to assume that each aspect of the behavior of an animal will advance its fitness. Historical and developmental constraints limit the effects of natural selection on the evolution of animal behavior. However, the assumption of adaptation is used, not as a tenet of faith, but as a means to develop falsifiable working hypotheses. Testing hypotheses produced in this fashion has greatly increased our understanding of the ultimate significance of animal behavior.

SUGGESTED READING

Intensive research on ground-nesting gulls has been done by Niko Tinbergen and his associates [772–775]. One of his students, Esther Cullen, has written a classic article [162] on the kittiwake. The textbook by John Krebs and Nicholas Davies [448] has a helpful chapter on the application of the comparative method to prediction testing in behavioral ecology.

General discussions of the evolution of animal communication can be found in Daniel Otte's review article [606], several chapters in E. O. Wilson's *Sociobiology* [853], and Michael Ryan's *The Tungara Frog* [687].

Stephen J. Gould's criticisms of the adaptationist approach appear in many of his articles [e.g., 303, 304]. For a succinct and pithy reply to Gould, see Jerram Brown's letter to *Science* [111]. Ray Hilborn and Stephen Stearns offer a very different kind of criticism of the multiple-hypotheses approach that focuses on the shortcomings of assuming that a trait can serve only one evolved function [364].

DISCUSSION QUESTIONS

1. If two closely related species possess a similar behavioral trait, the similarity cannot be used to test hypotheses about the adaptive value of the characteristic. But does this mean that trait is not an adaptation in both species?

2. Stephen Jay Gould has argued that if evolution were capable of producing perfectly adapted animals we would observe species that possessed wheels, because these structures clearly promote efficient locomotion. Gould believes that wheels have not evolved in living things because the historical sequence of events leading to the development of locomotory limbs has locked animals into a complex developmental plan that cannot be dramatically restructured into a system capable of producing wheels [300]. How would an adaptationist approach the absence of wheels in living things? What would his or her working hypothesis be and how ideally would he or she go about testing this hypothesis? Order various tests in sequence from the weakest to the strongest possible method for analyzing an hypothesis. See [457 or 196] after developing your answer.

3. In some butterflies, males transfer other substances to their mates in addition to sperm. These other materials have an odor that rival males find repellent. Hypothesis 1: The transfer of these odorants deceives rival males into treating the female as if she were a male and so decreases the likelihood that she will be courted and mated again [284]. If mated females were receptive, what might happen over evolutionary time to the response of rival males to the deceitful antiaphrodisiac? Develop an alternative hypothesis using the "honest signal" approach. What do you need to know to discriminate between the two hypotheses?

The Ecology of Finding a Place to Live

Having introduced the approach of behavioral ecology, I shall in the next several chapters discuss four fundamental decisions that most animals make: where to live, how to gather food, how to avoid predators, and what tactics to use to reproduce. This chapter examines questions related to the first decision, the problem of finding a place to live. In making where-to-live decisions, many species do things that at least superficially appear costly and damaging to individual reproductive success. As I suggested in the last chapter, it is this kind of behavior that attracts the attention of adaptationists because it seems to run counter to the assumption that animals will do things that raise their fitness. For example, some animals invest considerable time and energy in trying to locate certain kinds of habitat while passing up other places. Often animals that are born in or have found a suitable location later abandon the site and disperse to another, even though this requires that they cross dangerous and unfamiliar terrain. In some species dispersers return later to the very spot they left on migratory journeys of immense scope. Finally, individuals commonly defend the spot they eventually select to live in, although territoriality makes many demands on the animals that practice it. This chapter reviews the rich supply of hypotheses on the adaptive value of dispersal, nest site selection, migration, and territorial behavior, all elements of the behavioral ecology of finding a place to live.

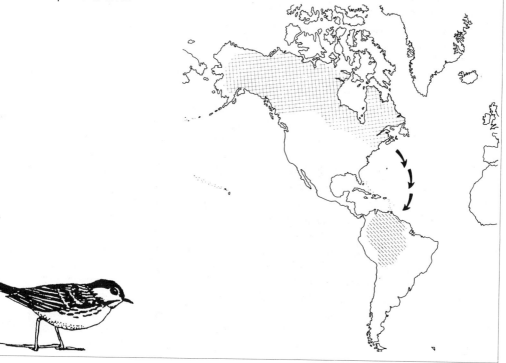

Active Habitat Selection   All naturalists know that certain species are reliably found only in certain habitats. When I was a teenager, I was an avid bird-watcher, and each spring my father and I would set aside one Saturday in May to see how many different species of birds we could find within a few miles of our home in southeastern Pennsylvania. We knew that if we were going to add a grasshopper sparrow to the list we had to check grassy fields; the spotted sandpiper would be down by White Clay Creek; the pinewoods were a must if we wanted some owls; and if we were to see a sedge wren, the flooded meadow in Mr. Boggs's pasture was our only hope. All bird-watchers know that the easiest way to see many species of birds in one day is to visit as many different habitats as possible. Were there amateur insect-listers or spider-watchers, equally keen to exhaust themselves in pursuit of a long list of species, the rule would be the same. The fact that certain species are found only in particular habitats suggests that animals may try to settle in some places while ignoring or avoiding others.

Creative studies of habitat selection have tested hypotheses about the possible evolved function of actively choosing a place to live. A key hypothesis claims that if animals have preferences for certain kinds of living space, then individuals that satisfy their preferences will experience higher fitness than those unable to settle in the favored habitat.

To examine this hypothesis, Linda Partridge studied two very similar, closely related songbirds of Europe: the coal tit, which lives in pinewoods, and the blue tit, which occupies oak woodlands. At the proximate level, this difference between the species is achieved in part because young coal tits innately prefer the foliage of pine trees, whereas blue tits favor oak foliage [624]. But do the bird's proximate preferences for simple habitat cues lead individuals to behave adaptively? If the habitat preferences of coal tits and blue tits promote individual reproductive success, one should be able to show that the birds gather food more efficiently in their preferred habitat [625]. This argument assumes, not unreasonably, that the more efficiently an individual collects food, the more likely it is to survive and reproduce successfully (see also Chapter 10).

Partridge tested the connection between habitat type and foraging success by giving hand-reared birds artificial feeding tasks that required each bird to (1) hack through or pull off a cover that concealed a bit of food or (2) hang upside down to get at a prey item. These are the kinds of things that one can see wild blue tits doing as they hang beneath a broad oak leaf, or rip and pull at a cluster of dead oak leaves for hidden insect eggs and caterpillars. In contrast, coal tits forage on the open clusters of pine needles and pine bark and therefore do not often employ these techniques. As predicted, the hand-reared, untrained blue tits were significantly faster at extracting food during the tests than naive coal tits (Table 1).

Note that we have here an example of a weak test of an adaptationist hypothesis, one that examines a prediction drawn from a single hypothesis, not multiple working hypotheses. Moreover, the hypothesis focuses strictly on a possible benefit of having a habitat preference but does not also evaluate the costs of particular preferences. The foraging results for the two species are consistent with the hypothesis that the proximate mecha-

TABLE 1

Comparison of the foraging skills of naive, hand-reared coal and blue titmice as measured by the time taken to secure food from three artificial dispensers

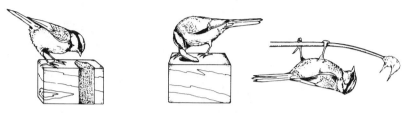

Mean time (in seconds) to secure food

Species	Peck through cover	Pull off cover	Hang upside down
Blue titmice	14	3.8	2.9
Coal titmice	22	5.0	6.1

Source: Partridge [625].

nisms of habitat selection help young birds find good places to hunt for food. This correlation is satisfying and useful; but, as is true for the large majority of studies reported in this and succeeding chapters, additional and more rigorous tests would not be out of order. I leave it to the reader to keep this in mind and to evaluate each case study discussed in terms of whether multiple hypotheses are presented, whether the relation of benefits to costs is considered, and whether the predictions are qualitative or precisely quantitative.

Nest Site Selection by Honeybees

Some insects also actively choose where to live. In the spring, a honeybee colony that has grown sufficiently large will fission, with the old queen and half her worker force leaving the old hive to a daughter of the queen and the other half of her worker retinue. The departing swarm settles temporarily in a tree, the bees hanging from a limb in a mass about their queen (Figure 1). Over the next few days, scout workers fly out from the swarm in search of small openings that lead to chambers in the ground, in cliffs, and in hollow trees. There are often many such sites within range of the waiting swarm, but only some are attractive enough to induce a worker to return to the swarm and perform a dance that communicates information about the distance, direction, and quality of the potential new home (see Chapter 7 on the dance language of honeybees). Other workers attend to a dancing scout and may be sufficiently stimulated to fly out to the spot themselves. If it is also attractive to them, they too will dance and send still more workers to the area. Eventually most scouts will be advertising one location, at which time the swarm will leave its temporary perch and fly to the most popular nest site.

Thomas Seeley discovered that scouts are enthusiastic only about chambers with a volume of between 30 and 60 liters [485, 710]. But the size of

1   **Honeybee scouts dance** on the surface of the swarm, announcing the location of potential homes for the colony. Photograph by E. S. Ross.

the chamber is not the only factor assessed by scout bees. In Bavaria, Germany, where Martin Lindauer did his work, the bees generally choose holes in the ground, with wooden structures and straw-basket hives as second and third choices, respectively. Moreover, if a Bavarian swarm is given a choice between two apparently equivalent hives, one close to the original hive (say, about 50 meters away) and another farther away (about 200 meters off), the bees will choose the more distant of the two [485]. This is true even though the added distance means the queen may become exhausted as she struggles to cover the distance (queens are wonderful egg layers but only marginal fliers).

Lindauer suggested that the preferences of his Bavarian bees provided them with well-insulated nest sites that could keep the bees alive in winter. And by moving a substantial distance from the daughter's hive, a dispersing queen might reduce competition for food between the two colonies, increasing the survival chances of her daughter's colony.

We can test Lindauer's hypothesis that cold winters and feeding competition have influenced the evolution of habitat preferences of Bavarian honeybees by examining habitat selection in Italian honeybees. Both are members of the same species, *Apis mellifera,* but they are considered different subspecies; and the Italian honeybees occur in southern Europe, where the winters are much milder than those in northern Germany [403, 404]. Italian bees do not prefer chambers in the ground to aboveground hives nor do they favor sheltered rather than unsheltered hives. Moreover, they accept smaller cavities and sites closer to a daughter's colony than do northern bees.

All of these differences are correlated with the reduced danger of winter mortality for Italian bees. Because they live in warmer regions of Europe, heavily protected sites are not especially valuable [295]. There is also reduced selection pressure for large colony size, which appears to be adaptive primarily under cold climatic conditions. When the bees form a compact mass within the hive during the northern European winter, the outer layer serves to insulate the others from the cold. In freezing weather, the outermost bees gradually die; so only large numbers of colony members will provide layers of insulation to enable the queen and an adequate worker force to make it through the winter. During the mild winters in southern Europe, large colonies gain no special survival advantage. Smaller colonies not only can occupy smaller cavities, they also can support themselves with food resources from smaller areas. Therefore, Italian bee swarms are not under pressure to disperse great distances from the home hive [295].

Leaf Preferences and Fitness
in Poplar Aphids

Thus far we have relied on habitat selection studies that have not measured reproductive success directly but have shown that a likely correlate of reproductive success, such as feeding efficiency or winter survival, is promoted by active habitat choices. Thomas Whitham has directly tested the hypothesis that habitat preferences raise reproductive success in the tiny poplar aphid [833–835].

Each spring in Utah, aphid eggs begin to hatch in the bark of cottonwood poplar trees. A new generation of tiny black females walks from the trunk to the leaf buds on branches. Each female, and there may be tens of thousands per tree, actively selects a leaf, settles by its midrib almost always near the base, and in some way induces the formation of a hollow ball of tissue—the gall—in which she will live with the offspring she bears parthenogenetically (Figure 2). When her daughters are mature, the gall splits and the aphids within disperse to new plants.

Whitham found that females settling on large leaves produce more numerous and heavier offspring than females on small leaves. If aphids can choose where to settle, they should pick large leaves. They do. In the trees examined by Whitham, there were 35 aphids for every 100 leaves. All the very large leaves on the poplars were occupied, although they made up only 2 percent of the total, whereas the smallest leaves (33 percent of the total) were avoided. Aphids have proximate mechanisms that cause them to prefer the most productive leaves, whose nutrients they will consume while nestled within their protective galls. They are so good at habitat selecting that the average aphid produces more than twice as many offspring as she would if she were to choose a leaf randomly.

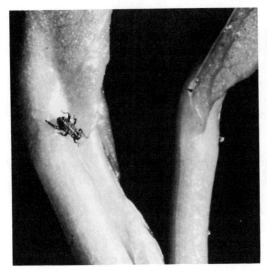 

2 **Poplar aphid on poplar leaf.** The aphid (left) searches for a site to form a gall (right) in which she will produce her young. Photographs by Thomas Whitham.

However, large leaves are in short supply and are quickly taken when the stem mothers emerge. An aphid that encounters a leaf with a gall-forming female can settle with the original aphid or hunt for an unoccupied leaf. A latecomer will have to form her gall farther out on the central rib of an occupied leaf and will not get as much food as the individual nearer the base. Whitham showed that when aphids double up they choose unusually large leaves. The second colonist on a big leaf can do just as well as a single aphid on a smaller leaf, and substantially better than a lone colonist on a little leaf (Table 2). In other words, a female poplar aphid somehow determines the relative abundance of leaves of different sizes and the number of already established colonists on a leaf. She uses this information to do what selection theory predicts she should: select the leaf from among those available that is likely to make the greatest possible contribution to her reproductive success [835].

TABLE **2**
The effect of leaf size and position of the gall on the reproductive success of female poplar aphids

Number of galls per leaf	Mean leaf size (cm)	Mean number of progeny produced		
		Basal Female	Second Female	Third Female
1	10.2	80	—	—
2	12.3	95	74	—
3	14.6	138	75	29

Source: Whitham [835].

Leaving One Homesite for Another   Once a poplar aphid adult finds a spot to live in, the animal will spend the rest of its life at that location. But for other species, like the honeybee, dispersal from an established home base to another living area happens regularly. Moving about is a costly business in terms of energetic expenses and risk of exposure to predators; this observation raises the evolutionary question of why animals are so often willing to pay the price and disperse.

Honeybees are unusual in that a parent *voluntarily* moves while its offspring stay behind. Far more often young family members disperse to new locations while the parents remain in place for another breeding attempt. Persons who have reviewed the extensive literature on the dispersal of young animals from their birthplace have found that as a rule (1) dispersers do not move far, but (2) usually the members of one sex move farther than members of the other [719, 817].

Take Belding's ground squirrels for example. This small mammal, which lives in high mountain meadows in California, spends the first 4 weeks of life largely underground in its mother's burrow. Each adult female ranges aboveground over a circular home base roughly 50 meters in diameter, and when the young animals come out of the burrow they stay within this area for some additional weeks. Males start leaving the natal area when they are 9 to 10 weeks old, never to return. Female squirrels stay put, remaining much closer to their natal burrow. At 12 months of age, male Belding's ground squirrels that are found again have moved an average of about 170 meters from their birthplace, whereas their sisters have moved a mere 50 meters (Figure 3). The little squirrels follow the typical pattern for mammals, in which the more adventurous travelers are males, but even they rarely move more than a few home-range diameters from their birthplaces.

The Inbreeding Avoidance Hypothesis

But why should young Belding's ground squirrels move at all? It seems obvious that a young animal wandering away from a place with which it

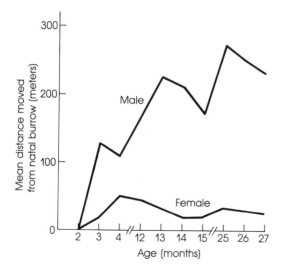

3 **Distances dispersed** by male and female Belding's ground squirrels. Males go much farther from their natal burrow than females. *Source:* Holekamp [371].

had become familiar is putting itself in a difficult situation. It must find food and shelter in strange places where predators wait for easy victims [817].

According to one argument, costly dispersal by young animals helps them avoid INBREEDING DEPRESSION by getting them away from their close relatives. When closely related individuals do breed together the offspring they produce are more likely to carry damaging recessive alleles in double dose than are the offspring produced by less closely related pairs. The risk of genetic problems should in theory reduce the average fitness of inbred offspring, and in a number of species researchers have documented high juvenile mortality in inbred populations [655].

But if the point of dispersing is simply to avoid inbreeding, then one would expect there would be as many mammals in which the females were the longer-distance travelers as species with far-moving males. This is not the case, and therefore Paul Greenwood suggested that although both sexes gain by avoiding inbreeding, female mammals remain at their natal territory because their reproductive success is especially dependent on possession of a territory in which to rear their young (Figure 4). Defense of a territory is presumably improved if one is familiar with an area and if the borders of one's territory are shared with relatives, who may be less aggressive than genetically unrelated strangers (Chapter 15). Male mammals do not attract mates by possessing a territory and instead have much to gain reproductively by searching for willing partners wherever they may be found [309].

Belding's ground squirrels are a case in point. Females that remain near their birthplace enjoy assistance from their mothers in territorial defense

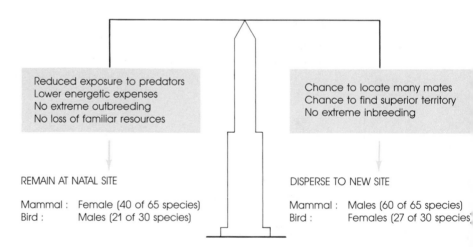

Reduced exposure to predators
Lower energetic expenses
No extreme outbreeding
No loss of familiar resources

Chance to locate many mates
Chance to find superior territory
No extreme inbreeding

REMAIN AT NATAL SITE

Mammal :   Female (40 of 65 species)
Bird :   Males (21 of 30 species)

DISPERSE TO NEW SITE

Mammal :   Males (60 of 65 species)
Bird :   Females (27 of 30 species)

4   **Dispersal of the sexes in birds and mammals.** Nondispersal and leaving the natal site carry different costs and benefits. Paul Greenwood's survey of 65 mammal and 30 bird species showed that female birds and male mammals are much more likely to disperse than are male birds and female mammals. *Source:* Greenwood [309].

and protection of burrows against rival females. Males, on the other hand, travel to sites where receptive females are most numerous at the start of the breeding season and fight for possession of mates as they appear. They have less to gain by staying at home. Thus, the mating system of the squirrels leads males to disperse and permits females to remain in familiar ground [371].

Greenwood pointed out that in birds females usually disperse and males tend to stay at home. Male birds, unlike male mammals, need a territory if they are to attract mates, and therefore males benefit greatly by remaining near the familiar birthplace in the company of male relatives. Female reproductive success will depend on the resources contained in a partner's territory, and this may favor dispersal to "shop around" for the best possible site [309].

The Competition Hypothesis

Jim Moore and Rauf Ali have argued that competition, not differences in the benefits of staying near one's birthplace or avoiding inbreeding, may be responsible for either males or females being forced to move farther than the other sex. For example, as males fight with one another for access to mates (a common occurrence in mammals and many other animals), losers may find it advantageous to find a place where they might win. Females rarely fight with each other for mates (Chapter 13), and this may generally enable them to be more sedentary [568].

Let us apply the competing hypotheses on dispersal to Belding's ground squirrel. In this species, males generally disperse in their first year of life, *before* reaching sexual maturity and at a time when competition for mates is absent. Thus direct competition among males for mates cannot be the immediate cause of male dispersal. The juvenile males may, however, have some means of assessing their future chances with rivals, and they may apply this information and leave areas of high future sexual competition. If this claim is true, it could rescue the competition hypothesis for male-biased dispersal in this squirrel, although at the moment it seems unlikely to apply here [371].

The factors causing dispersal in lions provide an interesting lesson on the complexity of the problem. Lions live in groups, or prides, dominated by one or several adult males. They are typical mammals in the matter of dispersal in that almost all young males leave their natal prides whereas a majority of females spend their entire lives in the area in which they were born (Figure 5) [648]. The nondispersing females benefit from their familiarity with their natal territory. They know where to find food and where to find safe dens when they give birth.

The proximate reason for the departure of young males is usually the arrival of a new mature male or males that displace the previous pride owners and evict, sometimes violently, the subadult males in the pride. So we might think that the aggressiveness of new pride masters to the males in the pride is sufficient explanation for their emigration, and that therefore one need not invoke inbreeding avoidance as a factor in male dispersal. However, if young males are not chased off after a pride takeover, they

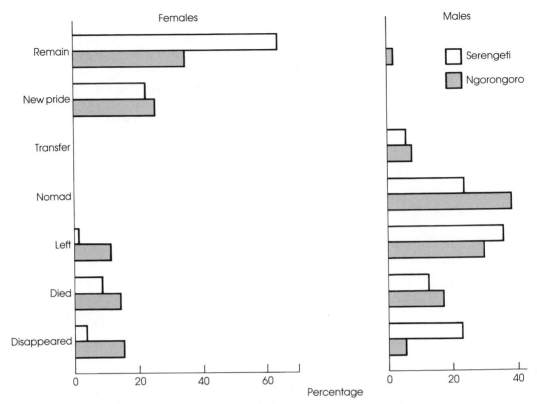

Females   Males

5   **Fate of subadult lions** by 4 years of age. In the Serengeti
    and Ngorongoro Crater, female lions tended to remain with
their natal pride, whereas males left or became nomads. *Source:*
Packer and Pusey [614].

often leave voluntarily. While they remain, they almost never mate with
their sisters and half-sisters. Moreover, mature males that have claimed a
pride will sometimes disperse again, expanding their range to annex a
second pride of females at a time when their daughters in the first pride
are becoming sexually mature. All of this suggests that there are inhibi-
tions against inbreeding in lions and that males disperse to seek nonrelatives as mates—although the timing of their departure from their birthplace is not always under their control [340].

How Far to Disperse?
When an animal moves from its living place, either voluntarily or because
it has been forced to do so, how far should it move? Dispersing female lions
almost always settle in an area immediately adjacent to their natal territory, a finding in keeping with the general rule that dispersers do not move
very far. William Shields has interpreted this to mean that animals attempt
to avoid extreme *out*breeding! His argument in many ways stands the
inbreeding avoidance hypothesis on its head. While accepting that extreme
inbreeding (such as sib–sib mating) may produce inbreeding depression,

Shields points out that extreme outbreeding may also yield reproductive depression.

Shields suggests that producing offspring that have something like the parental genomes would often be advantageous for parents (and their offspring). Extreme outbreeding is genetically disruptive, breaking apart those gene combinations that have proved successful in dealing with local conditions in the past. Moderate inbreeding dampens the production of extreme genetic novelty in progeny, and this is a good thing, so Shields argues, if the offspring are likely to live in much the same area as their parents. According to this view, limited dispersal is a mechanism that, like a mating preference for cousins [45], promotes moderate inbreeding and the retention of useful parental gene combinations in one's offspring [719].

The Competition Hypothesis Again

Peter Waser provides a counterhypothesis to the outbreeding avoidance argument, which parallels the counter to the inbreeding avoidance hypothesis discussed above. According to this view, animals move only as far as they are forced to by tougher competitors. The fact that juveniles usually disperse and adults stay put is seen as a result of the ability of older, larger, more experienced individuals (parents and other adults encountered during dispersal) to force weaker juveniles to seek a home elsewhere [817].

The competition hypothesis can be tested by examining whether there is a relationship between the turnover rate or mortality rate of adults and the median distance moved by dispersing individuals. If a disperser simply moves in a line away from home until it comes across the first homesite made vacant by the recent death of an adult, then the distance it has to travel will be a function of the turnover rate of its population. If adults are subject to a high death rate, a disperser will not have to move far before finding an opening.

If competition and travel minimization explain dispersal patterns, then we do not have to invoke outbreeding avoidance as a promoter of reduced dispersal. Waser has reanalyzed data on dispersal distances and mortality rates in deermice and has found that the observed average distance moved by animals from their natal sites matches that predicted from the competition hypothesis [815] (Figure 6).

There are other species in which dispersers appear to move *farther* than would be expected if they were willing to accept the first vacancy, and these cases could be analyzed for evidence that individuals are attempting to avoid inbreeding. On the other side of the coin, if dispersal data suggested that individuals were dispersing shorter distances than expected from the competition hypothesis, then a case might be made for outbreeding avoidance. Given the number of competing hypotheses, the adaptive value of dispersal is still very much an open question and a challenging invitation to additional research.

Changing Breeding Locations

It is not just juvenile animals that move from one home base to another. In many species, adults that have reproduced in one location may either

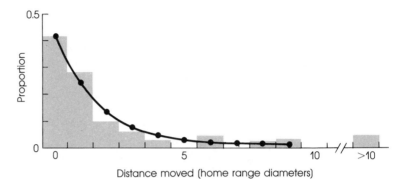

6 **Dispersal of *Peromyscus* deermice.** The predicted (dashed line) and observed (bar graph) dispersal pattern of deermice under the hypothesis that competition causes individuals to disperse. The match is close, supporting the hypothesis that competition drives dispersal in this species. Waser [815].

return to it the next breeding season or shift to a new site. If the move is adaptive, individuals that voluntarily abandon one potential breeding place for another should do better in the new spot than in the one they left behind. This prediction has been tested in a number of animals, among them red-winged blackbirds and female goldeneye ducks.

Red-winged blackbird males defend territories in marshes; females inspect potential nesting areas and pick males on the basis of the resources in their breeding territories. Once having staked out a claim, male redwings are highly site faithful from year to year. In one population more than 70 percent of the males that came back to the marsh reclaimed the same territory they had held before. But what about those that moved? Les Beletsky and Gordon Orians predicted that if these birds were trying to improve their lot (1) they should have had relatively poor reproductive success in the place where their mate(s) had nested the previous year and (2) they should be more successful after moving than before [53]. Figure 7 presents data that support these predictions.

In a similar study of female goldeneye ducks, Hilary Dow and Sven Fredga discovered that only slightly more than 40 percent of nesters returned to the same nest box the next year [205]. Goldeneyes nest in tree cavities in nature, but they will readily accept nest boxes provided for them by Swedish game managers. Those females that were faithful to the same box had a much lower probability of nest failure than the females that moved.

Why then did not all the goldeneyes reuse last year's nest box? Females moving to a new nest box were more likely to have failed in the previous year's breeding attempt than those that stayed put (Table 3). By leaving a nest site associated with failure, females did better; over 80 percent of the females that experienced nest failure in the preceding year succeeded at the new site [205]. Once a predator like a pine marten finds a nest box it is likely to return the next year, a fact that favors ducks that move to a new spot if their nest has been destroyed by a marten.

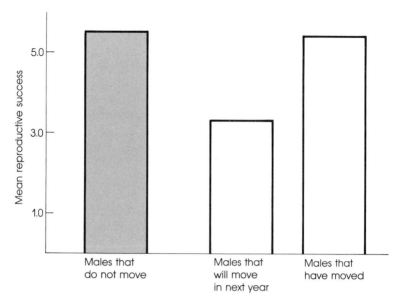

7 **Dispersal and reproductive success** in red-winged black-birds. Males move when they have reared relatively few fledg-lings; after moving they improve their reproductive success in the next breeding season.

Migration The capacity of female goldeneye ducks and male red-winged blackbirds to return to the same nesting area on successive years becomes considerably more impressive when one realizes that for much of the year the birds may be hundreds or thousands of miles from the breeding site. Nor are goldeneyes and redwings at all unusual. Nearly half of all the breeding birds of North America take off in the fall for a trip to Mexico, Central or South America. That translates into something like 12 billion North American birds engaged in a trip that the average human would not take lightly, even with an airline ticket [152].

Tiny ruby-throated hummingbirds weighing a quarter of an ounce fly

TABLE 3

Factors influencing the tendency of female goldeneye ducks to use the same nest box in consecutive years

	Current year	
Previous year	Female moves to new box	Female reuses nest box
Nest fails	16 (89%)	2 (11%)
Nest succeeds	78 (45%)	95 (55%)
G statistic = 13.6 $P < 0.001$		

Source: Dow and Fredga [205].

nonstop across 800 kilometers of the Gulf of Mexico twice a year. An Arctic tern breeding in Canada may complete a nearly 40,000-kilometer round-trip course each year (the equivalent of seven trips across the continental United States) (Figure 8). And North American birds are not the only magnificent migrants; a migrating hoopoe (with the exquisite scientific name *Upupa epops*) was watched by an ornithologist named Lawrence Swan as it hopped up a Himalayan pass over 6,000 meters high [756]!

Among the mammals, wildebeest, caribou, bison, seals, and whales make journeys of thousands of miles each year. Migratory reptiles include leatherback turtles, which cover up to 5000 kilometers in traveling between a breeding area in French Guiana and points near North America and Africa [645]. One population of green sea turtles nests on Ascension Island, a tiny speck of land in the center of the Atlantic Ocean between Africa and Brazil. The adult female turtles visit the island only to deposit their eggs in beach

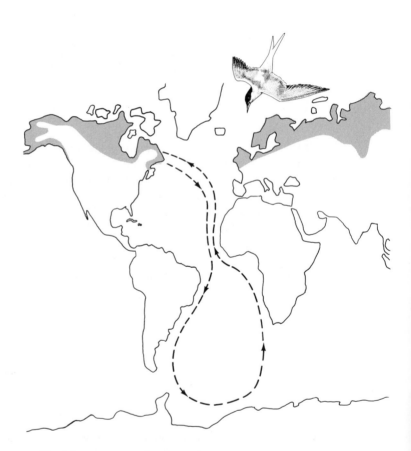

8 **Migratory path of arctic terns** takes birds from high in the Northern Hemisphere to Antarctica and back. Some young birds may spend 2 years circling Antarctica before returning to their northern breeding grounds.

sands. They then swim 1800 kilometers or so to warm, shallow water off Brazil, where they feed on marine vegetation for several years before returning, usually to the same beach, to lay another clutch of eggs [140].

The Costs of Migration

Migration poses a major historical problem, for how could a 40,000-kilometer round-trip journey evolve by degrees? Perhaps in some (or many) cases, long-distance migrations have been derived from short-distance dispersal tendencies that took individuals from one region to another immediately adjacent area. Subsequent changes in climate or geology may have gradually increased the distance between regions with useful resources, favoring ever more "ambitious" dispersers [152]. For example, some have suggested that the amazing Atlantic migrations of green turtles have grown longer and longer as the continents of Africa and South America have drifted slowly apart.

Migration also creates many puzzles for persons interested in the adaptive basis of the behavior. The evolutionary costs of migration are obviously major. Not only must the migratory individual invest in the development of the complex physiological systems (whatever they might be) that make navigation possible, but it may spend weeks or months each year on its energetically demanding journeys. Moreover, one can hardly overestimate the risks taken by many migrants. We mentioned earlier that one species of hawk, Eleanora's falcon, makes a living off exhausted songbirds that have crossed the Mediterranean Sea in the fall. Many other predators also have a field day with migrants.

Many migrating individuals take action to reduce the costs of their trip. Thus, it is common for migrants to travel in groups, perhaps to dilute the risk of being captured by a predator (see Chapter 11). In Europe a host of songbirds travel to central Africa by way of Spain and Gibraltar in order to cross the Mediterranean at its narrowest point [823]. This route is longer but has a shorter over-water component, perhaps decreasing the risk of drowning at sea. In light of this hypothesis, however, it is paradoxical that some North American songbirds, including several tiny warblers, migrate to South America from eastern Canada by way of the Atlantic Ocean (Figure 9) [849]. At first glance, this seems positively suicidal because to reach South America by a transoceanic route requires a nonstop flight of more than 5,000 kilometers. One would think it would be far safer to travel along the coast of the United States and down through Mexico and Central America rather than to invite death by drowning. But there is no question that migratory songbirds do appear regularly on islands in the Atlantic and Caribbean, so that some, perhaps most, do survive their ocean crossing.

There may be special advantages for the blindly courageous blackpoll warbler that attempts this trip. First, the sea route from Nova Scotia to Venezuela is about half as long as a land-based trek. Timothy and Janet Williams estimate that a blackpoll warbler flying under good weather conditions can go from Maine to South America in 50–90 hours of continuous flight [849]. Second, there are very few predators lying in wait in midocean or on the Greater Antilles chain of islands that blackpolls hope

9 **Transatlantic migratory path** of blackpoll warblers from southeastern Canada and New England to their South American wintering grounds. Courtesy of Janet Williams.

(unconsciously) to reach. Furthermore, easterly winds in the northern Atlantic that blow strongly after the passage of cold fronts give the birds a tail wind at the start of their journey, while westerly breezes in the southern Atlantic blow the birds to their island landfall.

For all their navigational and meteorological skills, migrant birds still pay a steep price for their trip, especially if they are forced into the water during storms over the Atlantic. The general point is that although migrants may reduce the costs of their travels, they cannot eliminate them. What ecological conditions can possibly elevate the benefits of migration enough to outweigh the dramatic costs of this behavior?

The Benefits of Migration
Migration is a special form of dispersal and can be investigated in a similar manner. The fact that so many unrelated species have convergently evolved

similar patterns of behavior makes use of the comparative method potentially productive. The search for ecological correlates of migration by means of the comparative method suggests that migratory behavior has no single function but has evolved in response to several different ecological pressures. First, individuals of some species move to an area when that region is rich in resources and then leave it when the place becomes less suitable. Temperate-zone birds typically follow a migratory pattern in which they move from a warm wintering area with acceptable food supplies to temperate regions that experience a spring and summer surge of insect production that may support a large brood of young [110].

In any number of unrelated fishes, individuals migrate from fresh water to the sea and back during their lifetime, or from the sea to fresh water and back. In northern latitudes most migratory species, including the familiar salmon, spend the bulk of their lives in the ocean. In the tropics migrants are far more likely to move into fresh water from the sea. This pattern reflects the fact that food production in the northern oceans exceeds that in bodies of fresh water, whereas just the reverse is true for tropical latitudes. Fish, like birds, appear to migrate to reach distant locations rich in food [315].

The other major migratory pattern involves movement from a feeding area to a protected breeding or birthing location. When an Adélie penguin female waddles onto the Antarctic mainland, she moves into an area a long walk from the nearest feeding grounds (Figure 10). The windy, rock-strewn slopes chosen by the Adélie penguin offer the best of a poor situation for rearing young in the Antarctic region [232]. Likewise, marine mammals, fishes, and reptiles may abandon rich feeding grounds and swim without access to food hundreds or even thousands of miles to sites suitable only for reproduction [592]. Seals that mate and bear their young on land, as well as marine turtles that lay their eggs in sand, tend to choose areas that are isolated and well sheltered. The green sea turtles that nest on isolated cove beaches in Ascension Island were free from predators (prior to human settlement) [140].

Whales and fishes are well adapted to bearing young and eggs in the water; however, some areas are very much better than others. A stream in which a salmon was born has proved itself suitable for reproduction. A mature salmon will therefore endure a long and difficult journey from oceanic feeding grounds to freshwater spawning grounds, risking death from exhaustion or predation, in order to have offspring in or near its birthplace. Whales exploit the extremely food-rich cold waters of northern and southern oceans, converting the tons of food they catch into tons of blubber. Then in fall in the northern Atlantic and Pacific and in spring in Antarctic seas, the animals move away from their frigid feeding grounds to warmer birthing sites in tropical or temperate waters.

The Migration of the Monarch Butterfly
Some insects, as well as whales and songbirds, migrate. Each fall millions of monarch butterflies move from as far away as southern Canada and the northeastern United States to small patches of fir forest high in the moun-

10 **Breeding grounds of Adélie penguins** in coastal Antarctica. Two hundred thousand pairs nest here in Cape Crozier, where they have open water near the rookery late in the season, good landing beaches, and bare ground (covered for the moment by snow) for nest sites. Photograph by David H. Thompson.

tains of central Mexico where they spend the winter [515] (Figure 11). One marked individual was recaptured after a 2500-kilometer journey, although most individuals make somewhat less monumental trips [796]. Migration occurs only in monarch populations confronted with seasonal changes in the availability of the milkweed plants upon which females lay their eggs. In the eastern United States, milkweeds die off in late autumn and grow again in spring. Thus, there is a strong seasonal fluctuation in the resource base needed by monarchs if they are to reproduce, and the butterflies living there migrate. In some areas of the western United States milkweeds are present year-round, and here the movements of monarchs are much reduced.

But why should some eastern monarchs fly hundreds (or thousands) of miles to spend the winter in the mountains of Mexico? Surely there are suitable places far closer to the spring and summer breeding grounds? Perhaps migrating monarchs are so selective in choosing overwintering sites because so few areas promote individual survival and subsequent reproductive success. One possible reason they travel so far is to find places without the nighttime freezes that occur during most winters throughout

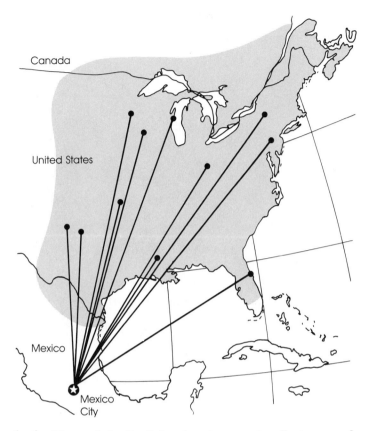

11 **Monarch butterfly's migratory routes.** Eastern populations move to a few small locations in the mountains of
central Mexico. Lines connect mark–recapture points. *Source:*
Malcolm [515].

the eastern United States. In the high mountain fir forests on the Mexican
wintering grounds, freezes are very rare with temperatures staying within
a narrow range (4° to 11°C during the coolest winter months). Occasionally,
however, snowstorms do strike the mountains, and when this happens
mortality among the monarchs can be great. Over 2 million monarchs are
estimated to have died after a night in which temperatures dropped several
degrees below freezing. There are many sites in Mexico in which this risk
could be completely avoided. But William Calvert and Lincoln Brower note
that in warmer areas, the monarchs' metabolism would be activated, with
consequent depletion of the energy reserves the animals will need to reproduce in the spring. By remaining cool (but not frozen), the butterflies
conserve their fat supplies [125].

Butterflies are also sensitive to water loss, and the high mountains in
which they overwinter receive more moisture than alternative lowland
sites, thereby helping the insects avoid desiccation. This benefit also comes
with a cost, because higher humidity raises the temperature at which
monarch tissues can be damaged by cold.

12 **Cluster of monarch butterflies** overwintering in a coastal Californian location. Photograph by E. S. Ross.

The hypothesis that central Mexican mountains possess a uniquely favorable mix of climatic features that keep the insects cooler and moister than they would be elsewhere can be tested by examining the other major overwintering sites in coastal California. Here, too, immense numbers of individuals gather from regions in the west where the foodplant is absent in winter (Figure 12). These locations are close to the Pacific Ocean and so subject to the moderating influence of the ocean and a moist airflow off the water. It seems probable that monarchs have evolved a migratory pattern that enables them to wait out the nonproductive months with reduced risk of death from freezing, exhaustion, and desiccation, and that the locations that offer this combination of benefits are sufficiently rare that long-distance travel to reach them is favored.

Territoriality   This chapter has examined why animals show preferences for some living space over others and why animals might move from one location to another. We now turn our attention to yet another problem facing some animal species: Having settled temporarily or permanently in one place, should the individual be territorial and defend the location against others? Just as there is diversity among animals with respect to habitat or nest site preferences, readiness to disperse and to migrate, so too there are species that are extraordinarily aggressive in competing for a living area while others ignore or even tolerate their fellows in the area in which they live. Many of the species mentioned in this chapter are territorial defenders for part of the annual cycle and nonterritorial occupants of a HOME RANGE (an undefended living space) for the remainder of the year.

Just as dispersal from a birthplace and migration over thousands of miles are obviously expensive, so too territoriality is a trait with some clear disadvantages. Consider the tiny poplar aphid, a highly territorial creature despite its small size. A female settling on a young poplar leaf continuously probes and feeds on tissue at the base of the expanding leaf, creating a depression that eventually grows into a gall. But if a second female attempts to occupy the leaf, the initial resident and the newcomer devote their time and energy to a kicking and shoving match that may last 2 days (Figure 13). While engaged in aggression, females cannot probe the leaf, delaying gall formation and reducing the ultimate size of the gall. The combatants may even die during their fight [836]. Wouldn't it be to their advantage to skip the fighting and move to a spot that could be claimed without the costs of territoriality?

If the expense of territorial defense is worth paying, then there must be reproductive compensation from the defended resource. As we noted earlier, sometimes two females "share" the same leaf, one forming a gall right at the base of the leaf, the other gall maker moving slightly farther out on the leaf. The reproductive success of females on 14-cm^2 leaves that had the leaf all to themselves was greater than that of females unable to monopolize a leaf of this size (Figure 14). Furthermore, on one leaf with two galls, the female in the position closest to the base of the leaf had more offspring than the female in the more distal position. The advantages of sole ownership explain why females fight to keep others away from "their" leaves, and why failing this, they try to be the occupant of the basal position.

Territoriality and Reproductive Success
Poplar aphids that can keep other females from settling on their leaves produce a relatively large number of offspring. The prediction that owning a resource-containing territory advances the owner's reproductive success

13 **Territorial dispute between two poplar aphids.** Females may spend hours kicking one another to determine who gets to occupy a preferred leaf or the superior location on a leaf. Courtesy of Thomas Whitham.

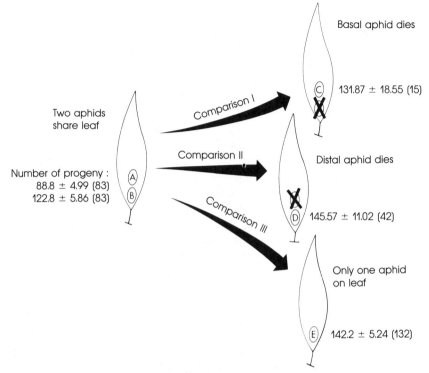

14 **Number of progeny** produced by aphids that succeed in monopolizing a poplar leaf versus those that are forced to share a leaf with a rival. Figures shown are averages ± one standard error (sample size). *Source:* Whitham [836].

has also been examined in a few vertebrates. Many birds defend breeding territories from which they gather a large proportion of the food needed for their offspring. If suitable breeding sites really are in short supply, then one should be able to find nonterritorial, nonbreeding individuals in populations of territorial birds.

In her study of a population of the rufous-collared sparrow, Susan Smith captured many individuals in a mist-net (a fine-meshed, black, nylon net that can be strung between poles in areas traveled by birds). She gave each bird a unique color band combination [734] so that she could plot the location of known individuals and their breeding activity, if any. As is true for most songbird studies, her plot was divided into territories, each one owned by a breeding male and his mate. There were also nonterritorial "floaters," who had well-defined home ranges within one or more territories of other birds.

The nonterritorial birds remained quiet and inconspicuous, and so "lurked" within territories permanently. They could not breed, however, because a singing male floater would not be tolerated by a resident nor would a nesting female intruder be permitted by the female territory holder. Instead, members of the underworld population waited for an opportunity to acquire a territory when a territorial bird of the appropriate sex disappeared. Immediately upon the demise of the male resident, the male floater in a territory would claim the site and the female in it. Female floaters did the same when the resident female died or moved away. Subordinates from outside the territory had little chance to secure the area in competition with an established floater, who, if able to make the transition, would then be able to reproduce [734].

Smith's study supports the hypothesis that territory owners will outreproduce individuals that cannot claim a territory. The same thing is true for a European songbird, the great tit, in which floaters sometimes pair up and attempt to breed on another pair's territory without ever claiming a territory of their own. These pairs essentially sneak into an established territory, build a nest as quickly as possible, and behave as inconspicuously as they can. If detected by the resident territoral male, they may be evicted; but they sometimes succeed in nesting successfully, especially if they enter a territory after the host birds have passed the highly aggressive territorial establishment phase and are incubating eggs. The fact that territorial pairs pay the price of aggression, singing, and patrolling suggests that they must do better reproductively than the sneaky intruders, *if* territoriality is adaptive in this species. The Belgian ornithologists Andre Dhondt and Jeannine Schillemans found that intruding birds laid a clutch of eggs only slightly smaller than that of territorial birds, but more of their nests failed (largely because of interference from the territory owners). As a result, intruders fledged an average of four young, whereas territorial pairs fledged an average of eight young [194].

The Belgians studied a woodland population of the great tit, but sometimes this species nests in hedgerow habitat as well. In an English population of the bird studied by John Krebs, males compete first with songs, displays, and fights for territories in woodland habitat [444]. The study area became completely filled with color-banded individuals at a time when other birds still searched for breeding sites. Some of these individuals eventually settled in hedgerows adjacent to the woodland, established territories, and attempted to breed.

Because great tits prefer woodland territories to hedgerow territories, Krebs predicted that great tits in woodland habitat would produce more fledglings than birds in hedgerows. This proved to be the case, as only 2 of 9 nests (22 percent) succeeded in producing fledglings in hedgerows whereas 54 of 59 woodland nests (92 percent) yielded fledglings.

To test whether the hedgerow birds really were excluded from superior breeding territories that they would have preferred to occupy, Krebs "removed" (a euphemism for "shot and killed") six pairs from their woodland real estate. Within a few days, most of the territories were claimed by new defenders—most of whom were banded birds from the hedgerow area, which they had abandoned as soon as superior habitat became available (Figure 15), another example of dispersal to improve reproductive chances.

Territoriality and Calories
We have looked at a sample of studies that directly test whether possession of a limited resource yields more descendants for the territory owner. It is

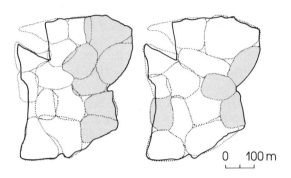

15 **Territories of great tits** before (left) and after (right) a removal experiment. When six pairs were removed (gray territories on left), the vacated area was occupied in part by neighbors that expanded their territories and by replacement pairs (gray territories on right) that arrived within 3 days. *Source:* Krebs [444].

0   100 m

rare, however, that direct measurements of fitness can be secured. More often, researchers have to assume that something they can measure is linked with fitness, in which case they can test indirectly whether territorial possession enhances reproductive success. For example, many animals compete for feeding territories with winners enjoying exclusive rights to the food in the defended area. Presumably acquiring food promotes survival and thus, reproductive success. Therefore, we can indirectly test whether animals that defend feeding preserves gain reproductive advantages by examining whether the calories they gain exceed the calories expended in defense of the site.

For part of the year, African golden-winged sunbirds compete for possession of patches of a flowering mint (Figure 16). Frank Gill and Larry Wolf conducted an economic study to see whether territorial birds did, as predicted, get more calories than they spent in defense of their sites [285–287]. Wolf and Gill counted the number of flowers in a patch of the sunbird's foodplant. They next measured the rate of nectar production per flower by covering a number of flowers (to prevent animals from removing the nectar). At the end of the day they inserted a glass micropipette into the flower to take up the nectar by capillary action. The micropipette was then sealed and taken to a lab to determine its sugar (and thus caloric) content. Daily nectar production in a patch was estimated by multiplying the number of flowers there by the quantity of nectar produced per flower per day. If a bird could monopolize these flowers, then it alone enjoyed the nectar produced by them. This calculated value represented the caloric benefits of territorial defense of a patch.

But there are special costs for territorial sunbirds as well, particularly the calories lost when chasing intruders from the area. Because physiological studies have shown what it costs for a bird of the sunbird's size to perch, forage, and pursue, one can quantify the calories spent while defending a territory and compare these with the calories gained by having a private foraging preserve.

Because it is the nonbreeding season when golden-winged sunbirds defend their flower territories, we can assume that the birds are trying to secure their minimum daily energy requirements as quickly as possible. Territoriality will promote this goal only if the energy available in a de-

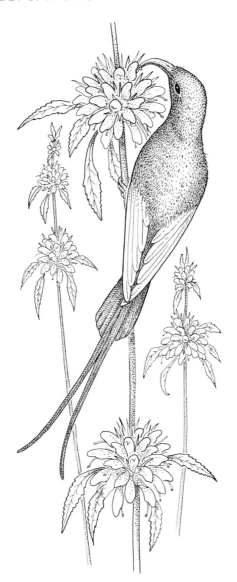

16 **Golden-winged sunbird** on a nectar-producing mint. These birds sometimes defend patches of mint.

fended patch exceeds that available in other nondefended sites. As Table 4 shows, a bird that owns a territory with flowers producing nectar at the rate of 2 microliters per bloom per day potentially saves 4 hours of foraging time if alternative sites are generating nectar at only half this rate. In caloric terms, the added time spent perching instead of foraging represents 2400 calories saved, from which must be subtracted the calories expended to defend the rich patch against intruders. Each hour of defense flight burns up 2000 more calories than would be spent if the bird had abandoned territoriality and were foraging nonaggressively elsewhere. An hour's defense would be worthwhile if the bird were holding a 2-microliter area, while forcing other sunbirds to forage in 1-microliter patches (net caloric

TABLE 4

Energetic costs and benefits of territoriality by sunbirds under different conditions

Activities	Expenses/hour (calories)
Resting on perch	400
Foraging for nectar	1000
Chasing intruders from territory	3000

Nectar production (microliters per blossom/day)	Hours required to collect sufficient nectar for 1 day
1	8
2	4
3	2.7
4	2

Nectar Production

In undefended site	In territory	Hours of resting gained	Calories saved[a]
1	2	8 − 4 = 4	2400
2	3	4 − 2.7 = 1.3	780
4	4	2 − 2 = 0	0

Source: Gill and Wolf [286].
[a]For each hour spent resting instead of foraging, a bird expends 400 instead of 1000 calories, which equals 600 calories saved per hour. Total calories saved = 600 × number of hours spent resting instead of foraging.

gain = 2400 − 2000 = +400 calories). But territoriality becomes disadvantageous if the bird spends an hour chasing intruders to protect flowers producing at the 3-microliter rate when other undefended patches are producing at the 2-microliter rate (net loss = 750 − 2000 = −1250 calories).

If sunbirds are adaptively territorial, they should be sensitive to the relative rates of nectar production in various patches and the time required to deal with intruders. As predicted, when nectar production is uniformly high, the birds are not territorial. But even during a single day, if nectar productivity in different patches begins to diverge, some individuals will begin to take up territorial residence in the rich areas.

As the density of birds increases, so does the cost of repelling intruders. In response to increased defense expenditures, birds tend first to contract their territories, reducing the area from which they repel conspecifics. If the rate of intrusion continues to rise, the birds will abandon territoriality altogether. Thus, sunbirds have the capacity to switch from foraging in a home range to defending a territory, the choice depending upon variable ecological factors that influence the economic return of their behavior [286].

Nor are sunbirds unique in this regard. John Craig and Murray Douglas have found that the totally unrelated New Zealand bellbird defends nectar-producing trees under some circumstances while feeding pacifically in flocks under other conditions [153]. In one patch of forest there was a single *Vitex* tree, a plant that produces high-quality nectar in large quantities per flower. The tree came into flower when few other sources of food were available. It therefore attracted such large numbers of birds that no one individual attempted to defend it in its entirety from intruders.

Subsequent to the flowering of the *Vitex* tree, another more abundant and widely distributed species of tree (*Dysoxylum*) began to bloom. Male bellbirds began to abandon the *Vitex* tree and to establish territories in areas with patches of *Dysoxylum*. If males switched for economic reasons, those that became territorial should have secured more resources than by feeding on *Vitex* nectar. The New Zealanders established that the maximum daily caloric intake for any bird feeding at the *Vitex* tree was about 100 kilojoules of energy, with most individuals securing less than half this amount. Territorial males on *Dysoxylum* gained more than 125 kilojoules a day.

We can also predict that if economics influence an individual's territorial decisions, then the more valuable the resources in an area, the more a territory holder will be willing to invest in territorial defense. Paul Ewald tested this prediction in an ingenious experiment in which he tricked black-chinned hummingbirds into fighting for possession of feeders that the two contestants valued differently [249]. He placed identical feeders at 10 meter intervals and allowed hummers to establish territories, each centered on a single feeder. After marked individuals had claimed their feeders, Ewald then changed the rate of nectar flow provided by his pump-driven feeders so that one bird had a "poor" feeder (giving fluid at 0.6 ml/hour) while the other enjoyed a "rich" feeder (which offered 0.8 ml/hour). Over the provisioning period, the owners of poor feeders would have been able to secure about 80 percent of their daily requirements at their feeders, while the territorial defenders of rich feeders took in 130 percent of daily energy expenditures.

After the birds learned what quality nectar was offered by their feeder, Ewald then gradually moved the feeders closer and closer until the birds came into conflict. Eventually, one hummer came to defend both feeders. If past feeding experience influenced the birds' decision to fight for possession of a territory containing their feeder, then those hummers that owned the rich feeders should have been willing to invest more in aggressive efforts to control their sites than owners of poor feeders. In 13 of 17 staged contests the winner was the bird that owned the rich feeder. The birds that had fed on the low-return feeder had less to gain by maintaining possession of their resource and so were less motivated to defend it.

Long-Term Effects of Territoriality on Fitness

Behavioral ecologists have used nectar-feeding birds to test the hypothesis that animals will be territorial when this enables them to meet their daily caloric requirements most efficiently. The same hypothesis has been applied to the pied wagtail, a European songbird, which also defends feeding territories during the nonbreeding season [173]. The wagtail, however, eats insects not nectar. In southern Britain, its winter territories are 600-meter stretches of riverbank. Territory owners consume the aquatic insects that are constantly being washed up by the river. Wagtails, like sunbirds, have the ability to switch back and forth between territoriality and nonterritorial behavior. When insect renewal rates fall within an area, the territory owner often temporarily joins flocks of nonterritorial birds that forage widely over the countryside. When the river edge is productive, a territorial bird will remain at its site and vigorously defend it against outsiders.

But there are two puzzling aspects of their behavior that violate the prediction that individuals are territorial in order to monopolize a rich food resource. First, residents sometimes tolerate another bird on their territory for a time, in effect letting another individual reduce the food supply at the site [177]. Second, even on the best of days during Nicholas Davies's study in the winter of 1974, the food-collecting rate for a territorial wagtail never exceeded 20 items per minute, whereas flocking birds collected 21–29 prey each minute during the same period [173]. Examination of the feces of the birds (one of the less glamorous activities of a behavioral ecologist) showed that the size of the insects consumed by territorial and nonterritorial birds was the same, and therefore territorial birds were not eating fewer but larger prey.

Owners tolerate a satellite only on days of high food abundance. This reduces the cost of having a satellite share one's territory. Moreover, on these days intruders are especially likely to invade the territory. The satellite assumes 20–50 percent of the defense of the area against these outsiders, and as a result the resident actually increases its net energy gain by associating with a satellite under these conditions (Figure 17). End of puzzle number one.

But how can we account for the territorial birds' persistent return to areas that yield fewer prey than nondefended regions visited by flocking birds? Why bother to be territorial when the hunting is better elsewhere? Davies suggests that when snow covers the meadows, the river edge may

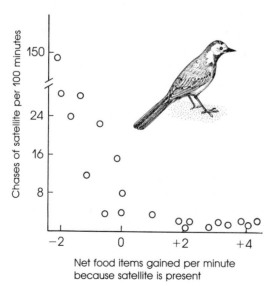

17 **Pied wagtails chase satellites** from their riverbank territories when the presence of another bird reduces their intake of food. Satellites are tolerated when their help in driving off other intruders results in a net gain in food eaten for the resident. *Source:* Davies and Houston [177].

be the only place with a predictable and constant supply of food. To maintain its body weight, a wagtail must feed 90 percent of the time and collect about 7500 food items per day. A single day of starvation can kill it. Thus, territorial ownership may provide an element of insurance against exceptionally bad weather. It would be useful to test the hypothesis that animals consider not only the relative productivity of different foraging areas, but also the *reliability* of production, in deciding whether to be territorial [173].

How Large Should a Territory Be?

Territory size varies enormously across species, reflecting in part the size of the species (the minute poplar aphid defends a territory a few millimeters long whereas the larger wagtail defends 600 meters of riverbank). In addition, the function of the territory is correlated with its size. There are many species in which territorial males defend a space that they use only for the performance of mate-attracting displays (see Chapter 13). Large birds and mammals in this category may defend display sites that are only a few square meters in size. We have restricted our discussion of territoriality in this chapter largely to feeding or all-purpose territories in which the resource defended is or includes food that the owner exploits. Not surprisingly, these territories tend to be much larger than display territories.

Even within a food-territorial species, however, the size of the defended area may vary considerably. Consider a rufous hummingbird on migration through the mountains of California on its way to wintering grounds in Mexico. In the course of its journey, the bird stops for a number of days in rich feeding areas where it behaves in a highly territorial fashion. But the number of flowers of Indian paintbrush that a bird defends changes from day to day. If the bird is attempting to maximize its net energy gain per day, we can predict that when hummers adjust their territory size they will gain weight more rapidly. But how do you get a rufous hummingbird to reveal its daily starting and ending weight without having to subject the bird to traumatic captures at the beginning and end of each day (and without having to subject researchers to the trauma of having a prized "hummer" disappear after capture and release)?

Lynn Carpenter and her colleagues invented an effective solution [138]. Within a territorial patch, they placed a highly attractive perch on a very sensitive scale that could be read through binoculars from a distance. Territorial birds like to be where they can survey their domain easily and some accepted the experimental perch happily, thereby weighing themselves at the convenience of the observers.

Most of the birds in the study adjusted their feeding territories in the course of their stay in the mountain meadows. These trial-and-error changes resulted in an increase in weight gain so that the birds increased the daily rate at which they were storing fat on one or a few days before they resumed their southward journey. A representative case appears in Figure 18. Note that this bird first expanded its territory substantially. It then gave up some of the flowers it had been defending, thereby reducing its caloric expenses in aggression and gaining weight more rapidly as a result.

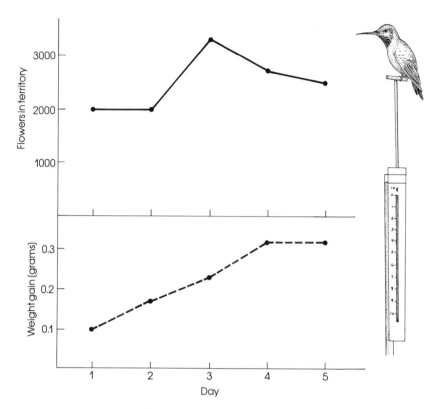

18 **Record of weight gain** by one rufous hummingbird that adjusted its territory daily. *Source:* Carpenter et al. [138].

A prediction from the resource maximization hypothesis is that if resources increase within an area, the size of defended territories will not decrease because the costs of patrolling the area will remain about the same but the gains available will increase, thus increasing the net resource intake. In contrast, the cost minimization hypothesis predicts that with more resources available, a territory owner could afford to decrease the size of the area defended, saving time and energy while still meeting its basic resource needs.

In an observational test of these hypotheses, David Paton and H. A. Ford measured territory size of two Australian nectar-feeding birds in relation to the number of nectar-producing *Banksia* inflorescences in a study plot [626]. These highly territorial birds defend feeding sites all winter long, with the size of the territory varying greatly (Figure 19). Paton showed that natural territories of whatever size always contained enough inflorescences to yield about 18 kilocalories per day. The birds defended larger areas when the nectar sources were diffusely distributed and smaller areas when the *Banksia* flowers were tightly clustered.

The wagtail offers another opportunity to test whether birds adjust territory size in ways that keep costs at a minimum while still providing enough food to keep the defender alive. The territories claimed by wagtails provide them with a renewable resource, insects washed onto the riverbank.

The rates of renewal can vary; and it can be shown mathematically that at times of high influx of insects on the riverbank a contraction in territory size would yield a higher rate of energy gain per unit time, thus permitting a bird to meet its daily requirements faster than by continuing to defend a full 600 meters of riverbank.

Wagtails, however, do not respond to short-term changes in food renewal rates with adjustments in the size of the territory defended. Instead, they continue throughout the winter to maintain a territory of about 600 meters. Again, long-term considerations apparently favor this inflexibility of behavior. Data on variation in rates of prey renewal show that a wagtail with a territory of 300 meters would secure insufficient food to stay alive on 14

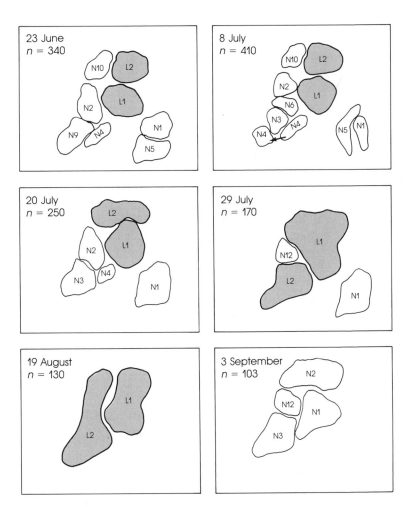

19 **Change in territory size** in relation to food resource concentration in Australian honeyeaters. Each circle represents the size of a territory of a single, identified little wattlebird (L; shaded territories) or New Holland honeyeater (N; open territories) on a particular date. *n*, Number of banksia flowers in the area on that date. *Source:* Paton and Ford [626].

out of 42 winter days. In contrast, territories 600 meters or more in length will receive sufficient insect flotsam to keep a bird alive on the territory on 38 out of 42 days. Therefore, birds may consistently defend a full 600-meter territory to reduce the risk of facing a starvation day [386]. The differences in the territorial behavior of wagtails, honeyeaters and hummingbirds illustrate the significance of the effect of time frame and insurance benefits on the fitness gains from defending territories of different sizes.

The Evolution of Interspecific Territoriality

There are many other questions about resource-based territories that have been explored from an evolutionary perspective. For example, some species treat certain other species with about as much nastiness as they reserve for conspecifics. Several workers have proposed that interspecific territoriality will occur to the degree that individuals of different species exploit the same resources, and thus come into direct competition. Figure 20 shows that in winter when mockingbirds defend berry-laden trees, they are more aggressive toward those species that primarily eat fruit [567].

Another bird that practices discriminating interspecific territoriality is the bell miner, an attractive green honeyeater that lives in the eucalyptus woodlands of southeastern Australia. When I was on sabbatical leave at Monash University in Melbourne some years ago, I learned about bell miners because they are social and they produce clear, loud, bell-like calls that are said to get on the nerves of people living near a bell miner colony. Only after leaving Australia, however, did I discover that these colonial honeyeaters have an exotic diet of psyllid lerps (Figure 21). Lerps are thin shields that little psyllid bugs form with carbohydrate secretions from their bodies; the lerp covers and protects the bug from ants. Bell miners carefully pluck these edible covers from psyllid bugs, which then secrete new ones, only to have them eaten later by foraging honeyeaters. The birds sometimes show a lack of restraint by eating the whole bug, but typically they do not consume the psyllid that makes the golden lerp [509].

Bell miners appear, therefore, to be managing a food resource for the long haul; and wherever these honeyeaters occur, psyllids flourish while the eucalyptus trees that the bugs are draining of sugary fluid do poorly.

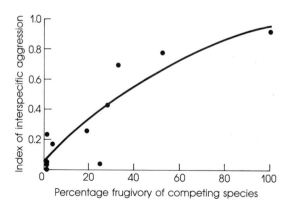

20 **Interspecific territoriality.** Mockingbirds, which defend fruit-producing trees in winter, are more aggressive to other species that also feed on fruit. *Source:* Moore [567].

Observers of bell miners have long noted that the birds are exceedingly aggressive, not just to other bell miners, but also to a wide variety of other unrelated species *that feed on psyllid bugs*. To test whether interspecific territoriality helped bell miners protect the lerp-producing psyllids, a group of Australian researchers trapped and removed a colony of 34 bell miners from their group-defended territory. Shortly thereafter, large numbers of psyllid-eating birds invaded the area, and the number of psyllid bugs declined to practically zero in just 2 months [509].

Bell miners practice interspecific territoriality in ways that permit them to monopolize a resource that they husband for its long-term yield. But defense against sitellas and pardalotes costs bell miners some of the gains they derive from protecting a psyllid farm. In theory, the level of competition from other species should affect decisions made by bell miners on what to defend and where to be territorial. The general point has been explicitly tested by Paul Ewald and Raymond Bransfield in their study of two competing hummingbirds. They showed that when Anna's hummingbird was present, these birds forced black-chinned hummingbirds to invest more time and energy in defense of rich feeding sites. Under interspecific pressure,

21 **A bell miner** holding a lerp removed from a psyllid bug in the group's territory.

black-chins gained greater caloric payoffs by switching to poorer food sources that were less attractive to dominating, rival Anna humming-birds [250].

SUMMARY

1 Many animals actively select certain places to live from among a set of alternative sites. Individuals able to occupy preferred places should have higher fitness, a prediction that has been tested with affirmative results for a number of species.

2 The selection of living space often occurs in the context of leaving one spot for another, as when juvenile animals abandon the place where they were born to go in search of a new home. There are many competing hypotheses to account for the fact that dispersers rarely move far, and that usually one sex moves farther than the other. Dispersal might occur because some individuals are forced by more powerful rivals to move, or dispersers may choose to move various distances voluntarily in order to avoid inbreeding with close relatives or to avoid outbreeding with genetic strangers.

3 Migration is a special form of dispersal in which the migrant may return eventually to the place it left. The costs to individuals of migratory journeys appear to be high; counterbalancing benefits may arise if migrants can exploit seasonal bursts of food productivity during their reproductive season or if migrants can secure safe sites for breeding or for giving birth.

4 In choosing living space, some animals invest additional time and energy in defense of the site. Territorial behavior is strongly correlated with the occurrence of valuable resources in small, economically defendable patches. Owners of defended breeding sites typically outreproduce those that fail to secure a territory; when some breeding territories are preferred to others, owners of the preferred locations enjoy greater reproductive success. When feeding territories are established, either owners gain caloric benefits in excess of their caloric expenses incurred during defense of the site or else they are able to secure their minimum daily caloric requirements more quickly, saving time that then can be spent resting rather than foraging.

5 Territory size is related to the density of resources present in an area. In keeping with an economic analysis of territoriality, owners expend energy in defense of sites in relation to their value, and they attack other species in relation to the overlap in their resource needs.

SUGGESTED READING

The papers by Thomas Whitham [834–836] and Linda Partridge [623, 624], as well as Martin Lindauer's book, *Communication among Social Bees* [485], offer modern analyses of habitat and nest site selection.

For competing views on the evolution of dispersal, see the papers by Peter Waser [817], William Shields [719], and Paul Greenwood [309].

Robin Baker has written a huge book on animal migration [34].

A classic exposition of the cost–benefit approach to territoriality is by Jerram Brown and Gordon Orians [113]. The papers by Nicholas Davies and his colleagues on wagtail territoriality are especially well done [173, 177, 386].

DISCUSSION QUESTIONS

1. Using the various hypotheses on animal dispersal, derive a set of predictions that will permit you to discriminate among them and develop a research proposal on Belding's ground squirrel that would get the data you need to test your predictions.

2. Some lizards defend feeding territories. Let's say we found that juvenile lizards, despite their smaller size, were guarding territories larger than those defended by most adults. Two hypotheses for this difference are that (1) juveniles are forced into suboptimal habitat and therefore need to defend a larger area to obtain sufficient food, and (2) juveniles need to defend more space because they are growing actively and need to eat more than adults, which need only to maintain their body weight. How might you test the competing ideas experimentally (see [741] after preparing your answer)?

3. Males of many animals guard territories that contain food needed by their females. The males do not consume the food but use it to attract females with whom they mate. What are the costs and benefits of this kind of mating territoriality for males *in terms of females contacted*? Use the list to predict what ecological factors should lead to this kind of mating territoriality.

4. Let's say that we want to use a game theory model to analyze the decision of some great tits to employ the nonterritorial route to breeding. Develop such a model and show what predictions follow from it.

10

The Ecology of Feeding Behavior

Even a cursory survey of the animal kingdom reveals a wealth of techniques used by members of different species to locate and consume food. Almost everything organic is grist for somebody's mill. Even the most extraordinarily well-hidden, or elusive, or dangerous, or poisonous animal, even the most repellent substances (from a human perspective) contain calories and nutrients that some animal exploits. But collecting food is not a simple task. Nutritious items are often in short supply, and they may be more than mildly reluctant to be consumed. In order to survive and reproduce, animals must be adept at making decisions about what food items to feed upon, where to search or wait for meals, whether to defend a food patch, how to overcome a prey's defenses, and so on. Behavioral ecologists have begun to analyze these decisions in some species. The chapter's first section employs the comparative method to test ideas on how predators overcome the specialized defenses of certain prey. I then shall examine an area of current controversy in feeding ecology, the hypothesis that the dietary differences among some animals are evolved responses to competition for food in the past. The last section examines how optimality theory can be formally used to analyze adaptation in foraging. This chapter provides some case histories of the approach and its utility (and limitations) for an evolutionary understanding of feeding behavior.

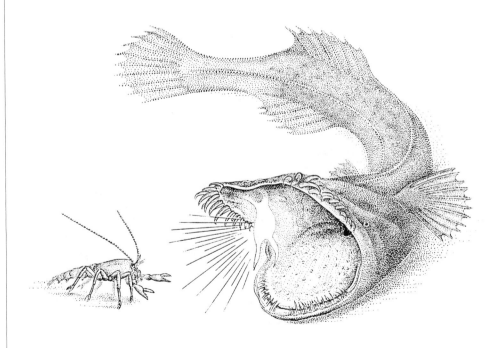

The Diversity of Prey Capture Techniques   Animals are usually experts at finding and securing their favored food. Within 15 minutes of hitting the African plain, a mound of elephant dung may be a seething mass of dung beetles, hundreds of which have flown up an odor trail to this impressive resource. Once in the dung, the beetles set about consuming it immediately or carting it off in round balls that they form as quickly as possible and push over the ground away from the competition (Figure 1) [356]. The far more modest droppings of small birds are just as eagerly consumed by certain insects that treat bird excreta with the enthusiasm that humans reserve for a meal at Chez Louis [660].

Inanimate dung does not fight back when dung beetles show up, but many living prey do resist their consumers (as bat-detecting moths illustrate). The next chapter will offer many more examples of sophisticated antipredator techniques. As prey evolve superior defenses, they create selection pressure on their predators, pressure favoring those with still better countermeasures that defeat the prey's adaptation. The result is a kind of arms race, with food and consumer locked in a spiral of evolutionary changes.

For a possible example of one such arms race, consider a jet-black darkling beetle ambling across the desert with its abdomen tip pointing upward at a slight angle. Touch an *Eleodes* and the beetle stops to elevate its abdomen higher. Continue to molest the beetle and it discharges a mix of irritating and poisonous chemicals. Although this may protect some beetles against some enemies some of the time, grasshopper mice counter this defensive maneuver by quickly grabbing the beetle before it fires and stuffing the tip of its abdomen into the sand. The mouse then crunches its way through the insect from the head down [227]. Skunks use a different tactic, rolling the beetle vigorously on the ground with their forepaws until the glands are completely emptied, after which the now-defenseless tenebrionid can be eaten with impunity [728].

1 **An Arizonan dung beetle.** Here a male and female cooperate in rolling away a ball they have formed of grade A cow dung. They may consume it themselves, or the female may lay an egg on the dung ball (after the pair have buried it underground), and the resulting grub will feast on the food at its leisure. Photograph by the author.

Prey Detection

Any number of attributes of predators have been shown to (1) increase the probability that a foraging animal will detect something edible or (2) increase the probability that a detected food item will be converted into a consumed food item. Here I shall present some examples of characteristics that fall into each category, beginning with probable adaptations for prey detection.

We earlier discussed how fringe-lipped bats locate the next meal by listening for calling frogs. These bats discriminate among potential prey on the basis of their acoustic signals [688]. Caged, hungry bats, given a choice between flying to a speaker playing a novel call that had a certain resemblance to a poisonous toad's call and flying to one that played a call similar to an edible frog species, nearly always avoided the speaker playing the toadlike signal but visited the other (Figure 2).

For predators faced with less noisy prey, there are still some things that can be done to help the hunter encounter a meal. Some oceanic "flashlight" fish possess a bioluminescent light organ just under each eye, which they can "switch" on and off voluntarily. These animals hunt in darkness at night or at great depths; they activate their light organ to produce enough illumination to spot their prey, which would otherwise be invisible [189, 540].

SEARCH IMAGE FORMATION provides another special mechanism for enhanced visual perception of prey. Luuk Tinbergen suggested that birds learned from experience just what key characters were associated with common prey, and they formed a search image specifically for these prey [770]. Alexandra Pietrewicz and Alan Kamil used operant conditioning techniques to test whether search image formation occurred in blue jays. They trained their birds to respond to projected images of cryptically colored moths positioned on an appropriate background. When a slide flashed on a screen, the jays received a short time to react; if the jay detected the prey, it pecked at a key, received a food reward, and was quickly shown a new slide. If the bird pecked when shown a slide with no cryptic moth present, it not only failed to secure a food reward but it had to wait a minute for the next chance to get some food [635].

In the course of their tests, the jays were shown a series of 16 slides, half of which had a moth present. In one experiment, the eight moth slides were all of the same species; but on another 16-slide trial, two species of moths appeared in the series in random order. When the birds were given repeated experience with just one species, they apparently formed a search image, because they made fewer errors as the series proceeded. But when they had encounters with two prey species on the projected slide, they did not improve, a result suggesting that they either failed to form a search image or were prone to use the search image for one species when the other was on the screen, thus reducing their prey detection accuracy under conditions when both prey were at equal frequency (Figure 3).

A similar study conducted in the field with European blackbirds demonstrated that these birds also improve in finding camouflaged prey after

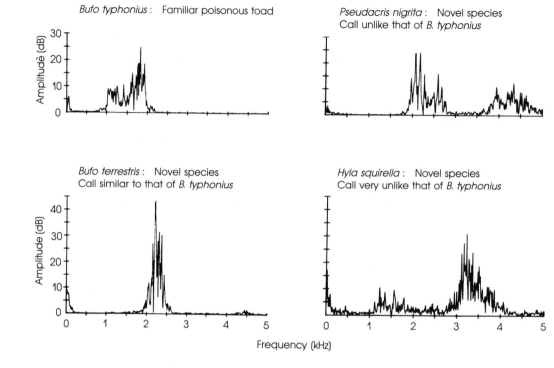

| | *Bufo typhonius*: Familiar poisonous toad | *Pseudacris nigrita*: Novel species Call unlike that of *B. typhonius* |

| | *Bufo terrestris*: Novel species Call similar to that of *B. typhonius* | *Hyla squirella*: Novel species Call very unlike that of *B. typhonius* |

Frequency (kHz)

Species	Number of trials	Number of approaches
B. terrestris	47	2 (4%)
P. nigrita	52	16 (31%)
H. squirella	59	42 (71%)

2 **Prey detection.** Predatory frog-eating bats discriminate among potential prey on the basis of their calls. Bats familiar with *Bufo typhonius* almost never approach a loud-speaker broadcasting the calls of a novel species from a different region, a species with a call whose energy is largely in the 2-kilohertz range, just like that of *B. typhonius*. The bats were more likely to approach a speaker playing the calls of two other novel species that have signals unlike that of *B. typhonius*. *Source:* Ryan and Tuttle [688].

they have experience with the food item. In this case the "prey" was a flour and lard cylinder, dyed green or brown and placed on trays covered with green- or brown-painted gravel. The birds were familiar with the food because they had been given opportunities to forage for undyed pastry "worms." When hunting for the dyed prey on the background that made them inconspicuous (i.e., a green gravel background for green pastry cylinders), the birds found the last four items more quickly than the first four.

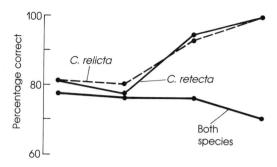

3 **Search image formation in blue jays.** When given the task of finding an underwing moth image that appeared in 8 of 16 slides, blue jays did not improve if there were four slides of one species and four of another intermixed in the series. But if all eight slides showed just one species, the jays became better at finding the moth over the test, suggesting that they had learned to search for the key cues associated with the one species. *Source:* Pietrewicz and Kamil [635].

This result is consistent with the notion that prey perception improves when the bird has the chance to form a search image of a particular prey type [467].

Prey Capture

Locating food is the first step; ingesting it is the goal, however, and this is often easier said than done because some food items resist capture. As a result, many animal species have evolved apparent aids for the capture of unwilling victims. The orb webs of orb-weaving spiders are a familiar example. Other less familiar spiders use silken lines to entangle and subdue their prey in many different ways. For example, a theridiid spider of the western United States appears to specialize on *Pogonomyrmex* harvester ants, a highly social animal with a potent sting. The spider dashes up to a walking forager and applies a line of silk to its body, which it attaches to the ground, nailing the ant to the spot. Then the spider bites the ant on a leg, and retreats for a few minutes while the ant struggles on its leash before succumbing to the poison. The spider soon retrieves its prey, which it transports to a safe feeding spot [642].

The famous little arthropod *Peripatus*, which is thought to have retained many of the "primitive" characteristics of the extinct ancestor to modern arthropods, uses a similar tactic when it comes across a cricket or spider. The caterpillar-like *Peripatus* has a modified kidney that produces a gluey protein; it fires the glue from a pair of modified limbs near its mouth, entangling its victim before injecting it with lethal saliva [279].

Just as there has been convergence in the use of entangling restraining substances, so, too, the use of deceitful food lures to entice victims within capture range has evolved in ambush predators ranging from small spiders

4 **Lure-using predator: angler fish.** An angler fish with a "minnow" appendage on the front of its head (top) can wave it about (bottom) to lure its prey within striking range. Photographs by David Grobecker.

to giant alligator turtles [223, 307, 647, 837]. For example, an angler fish has a bizarre projection from the front of its head, a projection that consists of a thin rod from which is suspended a bit of tissue with an uncanny resemblance to a small fish (Figure 4). This bait is waved about seductively by the anglerfish, who explosively engulfs any small predatory fish that comes to investigate [636].

Still other predators employ mimetic sexual lures, rather than feeding lures, to secure their prey. I have already mentioned firefly *femmes fatales* (Chapter 8). The bolas spider's sexual mimicry is equally devious. The spider releases a scent that mimics the sex pheromone of certain female moths. Males of these moths fly upwind in search of a mate but may find a bolas spider waiting with a sticky globule on the end of a silken thread (Figure 5). If the spider swings the blob and hits the moth, the predator will feast on the captured insect. Because the spider makes its living throw-

5 **Lure-using predator: bolas spider.** A bolas spider swinging its lure—a ball impregnated with the sex pheromone of certain moths. Males of this moth species approach the spider and are captured by the sticky ball. Photograph by William G. Eberhard.

ing a ball, William Eberhard, an entomologist who also happens to be a dedicated sports fan, gave the name *Mastophora dizzydeani* to one species of bolas spider [218].

Tool-Using Animals

There is a tendency to think that because humans are intelligent and use tools, other tool-using creatures must also be exemplars of intelligence and adaptability. This is not true. For example, ant lions (larval Neuroptera) and worm lions (fly larvae) have convergently evolved trap-building and tool-using behavior (Figure 6) [782]. These animals construct pits in sandy areas frequented by ants. When an ant topples over the lip of the trap and begins to slide toward the bottom where the predator awaits her, the ant lion or worm lion will speed the descent of its victim by tossing sand grains at it.

Other animals that make good use of tools in feeding are the archer fish [510], the woodpecker finch [461], the Egyptian vulture [798], the sea otter [327], the chimpanzee [798], and us. The archer fish maneuvers close to the surface of water in mangrove swamps and spits out water droplets capable of knocking insects and spiders out of overhanging vegetation as much as

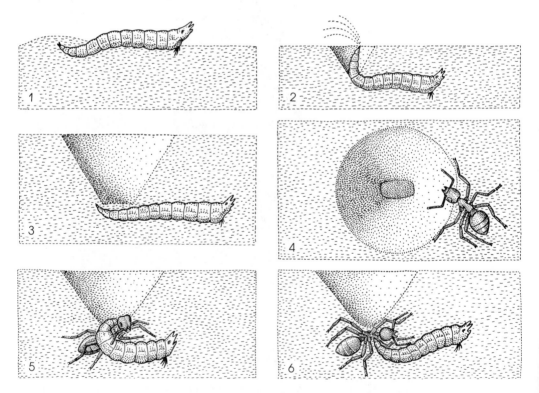

6 **Convergent evolution in trap-using predators.** The worm lion (a fly larva) shown here uses the same techniques to build a trap for ants as does the ant lion (a member of a different order of insects, the Neuroptera).

7 **Convergent evolution in tool-using animals,** all of which use tools to obtain prey that are difficult to secure. A woodpecker finch (top left) inserts a spine into a tunnel made by a wood-boring insect. Photograph by Iräneus Eibl-Eibesfeldt. An Egyptian vulture (top right) smashes a thick-shelled ostrich egg with a rock. Photograph by Hugo and Jane van Lawick-Goodall. A sea otter (lower left) uses one clam to crack open another clam, which is resting on the otter's chest. Photograph by Karl Kenyon. A chimpanzee (lower right) inserts a grass stem into a termite gallery; termites that cling to the probe will be eaten when the stem is withdrawn. Photograph by Leanne Nash.

4 feet away. Although the fish can leap some distance into the air in pursuit of these food items, the use of water droplets permits the fish to capture prey out of reach of its most acrobatic leaps. The other four tool-users either insert probes into narrow crevices or use rocks to open thick-shelled prey (Figure 7).

What ecological pressures do tool-using species share that might lead to the evolution of the behavior? In each case, a tool-using individual gains access to food that is inaccessible to conspecifics without tools. Ants can scramble out of pits; insects perched above the water or in tiny tunnels in wood or in hard-packed termite and ant nests are difficult prey to capture for fish, finches, and chimpanzees, respectively, and hard-shelled mollusks and ostrich eggs are not easily opened by otters and Egyptian vultures. Manipulation of a tool means more calories captured for less energy expended for all these creatures.

Cooperative Prey Capture

Cooperative hunting by terrestrial carnivores has evolved independently in three families: cats (Felidae), dogs (Canidae), and hyenas (Hyaenidae). In each case, the social cats, dogs, and hyenas take prey that weigh from 6 to 12 times as much as any one adult hunter [453, 550, 700]. Solitary species of cats, dogs, and hyenas forage for much smaller victims as a rule. Michael Earle has pointed out that the mass of a foraging unit (which is the weight of a single animal for solitary hunters, but the combined weight of the hunting pack for cooperative hunters) is strongly correlated with the weight of the largest prey regularly taken by the predator [217]. In effect, group hunters combine forces to achieve the weight needed to bring down prey that would otherwise be vulnerable only to a much larger individual.

The correlation between cooperative social hunting and the killing of unusually large prey suggests that a primary function of group foraging by these unrelated animals is the capture of massive and dangerous herbivores. A giraffe can crush the skull of a lion with one kick, and a moose will gladly do the same to a wolf. But through cooperation, lions or hunting dogs overwhelm adult zebra or wildebeest by providing the communal mass needed to subdue the prey quickly (Figure 8).

8 **Social prey capture.** A pack of wild dogs have cooperatively caught a wildebeest that they are about to kill and eat. Photograph by Norman Myers.

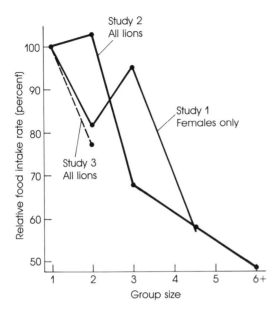

9 **Group size and foraging success in lions.** In three different studies, solitary lions gained as much or more food per individual as lions hunting in larger groups. *Source:* Packer [612].

The comparative test of the hypothesis that sociality is an adaptation to promote the capture of large prey (relative to the body weight of an individual predator) relies on a correlation between social life and prey size. The hypothesis can and should be tested in additional ways as well, and this has been done with instructive results in the case of lions. If social hunting promotes individual reproductive success, then an increase in group size (up to a point) should be correlated with an increase in the food intake *per individual.* When Thomas Caraco and Larry Wolf analyzed George Schaller's data on this point, they found the food intake per lion per day was higher for lions hunting in pairs than for lions hunting alone. However, groups of three or more did more poorly than groups of two (Figure 9) [131]. This observation is damaging to the group-foraging benefit hypothesis, because lions often associate in hunting groups greater than two. Craig Packer has therefore proposed an alternative *multifactor* hypothesis to account for the evolution of social living in lions [612]. Lions, unlike solitary cats, are large, live in open country, and occur in relatively high density. They can and do kill large prey on their own, but a dead zebra or wildebeest cannot be consumed immediately by one individual. A dead zebra lying on an open plain is visible to many other lions in the neighborhood, so a solitary hunter is likely to lose a portion of her kills to conspecific thieves. By tolerating her daughters and female relatives, however, a female converts that loss into a gain, because the individuals who benefit by exploiting her kills are her offspring or others who share genes with her (see Chapter 15 on the evolution of aid given to relatives).

Packer's hypothesis can be tested by examining the rate of loss of prey to conspecifics and other scavengers in solitary felids, especially the tiger, cougar, and leopard, all of which can kill large prey. Because these species live and hunt under dense cover, it is highly likely that the risk that others will spot their kills is reduced. Discovery is also less likely because the density of these carnivores never approaches a high level, whereas many lions live together in the short-grass African savannah, where vast herds of wildebeest roam. Additional data on this point are required, but Packer's analysis illustrates again the need to consider alternative hypotheses when testing any one idea on the adaptive value of a trait.

Social Foraging by Gulls

Many species of birds forage socially (Figure 10); and in the case of black-headed gulls, several hypotheses have been advanced on why they so often forage together. Perhaps the flocks are composed mainly of birds that are trying to follow successful foragers back to a productive spot. To the extent that followers exist, they might raise the fitness of the birds they follow when they reach the foraging ground, or have no effect, or have a negative effect. Let's consider this last case (the parasitism hypothesis) in which followers deplete a food source that a leader has found. The parasitic followers gain by taking advantage of a successful forager to locate food that they would probably not have found on their own [813].

However, both leaders and followers might gain by flying together to a food source, because fish may be caught more easily when subjected to group assault than when confronted by a solitary predator (the cooperative foraging hypothesis). One might imagine, for example, that when a flock of gulls plunges repeatedly into a school of fish, some of the frantic prey, in attempting to escape from one gull, might fail to see another predator and so be more easily captured. Thus, the feeding rate might be higher for all the members of a cooperative hunting flock than for a solitary hunter.

A prediction from the parasitism hypothesis is that highly successful foragers should be followed out from the colony on their way back to a distant, rich, food site. This hypothesis has been subjected to an experimental test in which an abundance of dead fish was placed on a raft out of

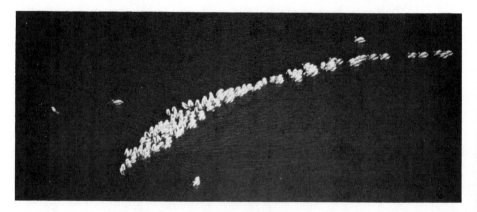

10 **Cooperative foraging** in white pelicans foils the escape maneuvers of their prey—schooling fish. Photograph by Ralph and Betty Ann Schreiber.

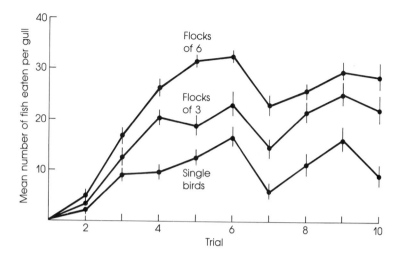

11 **Group size and foraging success in captive black-headed gulls.** A gull in a flock of six consistently captured more prey per 3-minute trial than single gulls or those in groups of three. *Source:* Gotmark et al. [291].

sight of the colony a kilometer or two offshore. Wandering gulls soon found the food, however, and gorged themselves. An observer near the raft radioed information about the identity of a successful gull back to a colleague by the colony, who could then record what happened upon the return of the bird. The gull landed by its nest with conspicuously distended crop and often with fish in its beak to feed its chick. Although a successful forager could have been identified, no neighbor actually flew behind the forager when it headed back to the raft. Moreover, when the next gull did leave the colony, there was absolutely no tendency for it to fly in the same direction as the successful fish-finder [21]. The failure of the data to support the parasitism hypothesis is useful. It sharply decreases our confidence that black-headed gulls exploit birds that have found food out of sight of the colony.

A group of Swedish researchers have recently tested the parasitism hypothesis again, in conjunction with the cooperative foraging hypothesis for flocking in gulls [291]. Happily, black-headed gulls will accept captivity, a fact that made the following experiment possible. Captive gulls were introduced into a large indoor aviary with a fishing pool, where they could hunt for living prey. They were permitted to forage alone, or in groups of three or six. If food-finders are subject to parasitism by followers, then birds that were hunting alone without companions should have done better than those that were swimming about the tank in groups of three or six (the number of prey was the same for each experimental run). Figure 11 illustrates that, contrary to this prediction, birds in groups caught more fish per unit time than solo hunters. Social birds improved their foraging success by confusing their prey; a much higher percentage of attempted captures resulted in success for the flocking birds, because the fish in fleeing from one gull often swam toward or near another hunter, whereas single birds had to catch fish swimming away from them. The result of flocking was a much higher percentage of fish grasped by their head or body.

To the extent that we believe that these findings have relevance for field conditions, they enable us to eliminate the possibility that gull flocks tend to form because some individuals are attempting to take advantage of others. The possibility that flocking represents a form of cooperation with increased food-capturing success enjoyed by all participants remains viable for black-headed gulls.

Diversity in Cooperative Feeding by Ants

Let us apply the cooperative foraging hypothesis to ants. The entire spectrum of foraging group sizes is represented among these insects, from species that gather food individually to others that form huge groups when collecting food. The North African ant *Cataglyphis bicolor* is typical of solitary foraging ants, which hunt for small, immobile items that one ant can easily collect and transport to the nest [819]. Other species, like *Leptothorax* ants, usually forage alone but sometimes form groups of two [373] when one worker has found a relatively large, but immobile, food source (a large dead insect too heavy for one ant or a colony of aphids to be milked, for example). The discoverer returns to her nest and regurgitates a food droplet to colony mates. Then she turns around, raises her abdomen in the air, extrudes the stinger, and releases a chemical signal. Another worker may approach her and touch her body with her antennae. The leader starts off with the follower close behind, running in tandem, all the way to the food site (Figure 12).

The species that are most similar ecologically to the social carnivores and the cooperative fish hunters are the ants that practice MASS RECRUITMENT to a food source [852]. Some of these ants show the kind of flexibility that supports the hypothesis that cooperation evolves in ants when there are large prey to subdue (Figure 13). A giant *Paraponera* ant of Central America recruits others only when it has found a large insect prey, not a small bug or a source of nectar or other food that does not fight back [93]. Far more famous and sophisticated mass recruiters are found in the army and driver ants of the tropics. When a worker has discovered a large insect or small vertebrate near a foraging column, she recruits dozens or hundreds

12 Tandem-running in an ant. The lead ant has recruited a follower to go with it to a food item too large for one ant to retrieve. Photograph by Michael Möglich.

13 **Cooperation in prey capture.** A group of weaver ants carries a much larger ant, which they have killed, back to their home nest. Photograph by Bert Hölldobler.

of her fellow ants to help her subdue the victim by laying down a scent trail from the prey to the column. Similarly, by means of mass recruitment, the tiny fire ants introduced into the southern United States are able to prey upon large insects that happen to pass close to a colony's nest. An ant that discovers a potential victim rushes back to her nest, laying an ephemeral odor trail that is gone in less than 2 minutes. The pheromone arouses other workers, who hurry out along it; if they are excited by the discovery, they, too, run back, applying more trail pheromone. This process can lead to an exponential increase of individuals at the food source, and the capture of prey far larger than any one foraging ant could handle (Figure 14).

Competition and Animal Diets The list of gastronomic novelties and ingenious techniques for securing them could be expanded almost indefinitely, but the point is that there appears to have been an astounding radiation in diets, with each animal species diverging from its ancestors to consume things or combinations of foods not eaten by other animals. This point is further reinforced by an examination of some closely related clusters of species, each member of which has its own specialty. The classic example involves Darwin's finches [88, 461]. These superficially drab little birds, which occur only on the Galapagos Islands off Ecuador, feed on everything from tiny

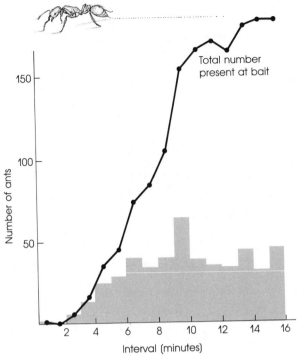

14 **Mass recruitment in the fire ant.** Ants that have found prey lay a pheromone trail (above) from the prey back to the nest, stimulating others to join them. The number of trail-laying ants increases rapidly (bar graph), leading to a rapid buildup of ants arriving at the prey (line graph). In this experiment, a cockroach was used as the prey. *Source:* Wilson [850].

insects to huge, thick-hulled seeds (Figure 15). There are ground-feeding seed eaters with powerful, crushing beaks, cactus consumers with beaks adapted for probing and slicing, tree-bud and leaf eaters with cutting beaks, and insect eaters with delicate, forceps-like bills. When more than one of these species exploits the same category of foods, as in the case of the seed-eating ground finches, their beaks are specialized for one component of the total resource (e.g., large, medium, and small bills capable of crushing large, medium, and small seeds, respectively).

Traditionally, the favored hypothesis for cases of dietary differences has been that competition among different species conferred a reproductive advantage on those individuals that happened to reduce overlap in food preferences. By exploiting food supplies not tapped by another species, an individual should have access to more energy and therefore have more opportunities to produce young than those members of its species that struggle to secure the same resources required by another species. Because the diets of even closely related species are different, many species can coexist in the same region, each supported by its own food supplies, so the argument goes [e.g. 461, 571].

More recently, however, a number of ecologists have challenged the traditional view [188, 705, 839]. A key criticism can be phrased as a question: What is the probability that by chance alone two species would happen to share identical or very similar diets? Given that different species have separate ancestry and distinct genetic makeups from the moment of their reproductive isolation from one another, the probability must be low indeed.

The different diets of animals could therefore be an incidental effect of their separate histories and have little or nothing to do with a reduction of ecological overlap in food demands.

In the case of the Galapagos finches, an alternative to the competition hypothesis was suggested early on by Robert Bowman, who pointed out that the plant life differed markedly on the various islands in the Galapagos Archipelago. If the species arose on different islands, their different bill shapes may have evolved as adaptations to the local flora and not in response to competitive pressures among species [88].

Peter Grant has summarized the ways to test the competition versus the noncompetition hypotheses for the divergence in bill shape and diet in the Darwin's finches [305]. Consider one example that involves the concept of COMPETITIVE RELEASE. Assume that competition has led a species to restrict its diet in order to avoid overlap with the feeding preferences of other

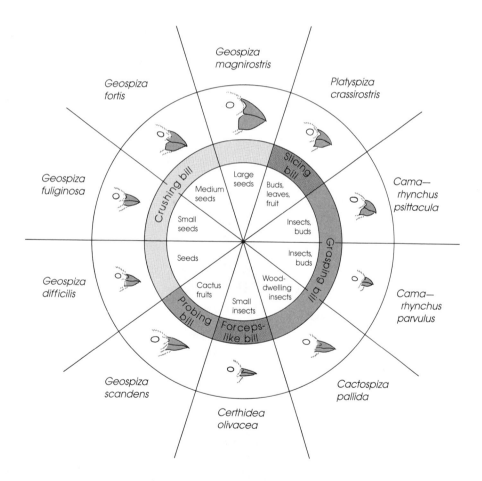

15 **Diet diversity in the Darwin's finches.** Each species has a distinctive beak suited for its distinctive diet. *Source:* Bowman [88].

species. If so, in the absence of the putative competitors, the species in question is released from the pressure of competition and should have a broader diet than when competitors are present. The large cactus ground finch, *Geospiza conirostris*, occurs on islands in the Galapagos with and without another large finch, *Geospiza magnirostris*. On an island in which it is the only *Geospiza* present, *G. conirostris* has a bill size that is more or less average for *Geospiza* as a group and its diet is very broad, including a wide diversity of seeds. But on an island where the other large finch also occurs, the bill size of the cactus ground finch has shifted away from that of its potential competitor and its diet is narrow and different from that of the other finch.

A similar test of the competition hypothesis using totally different birds

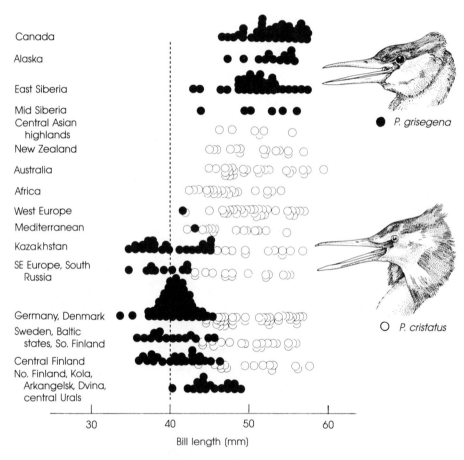

16 **Competitive release in grebes.** In areas in which the red-necked grebe (*Podiceps grisegena*) lives without the great-crested grebe (*Podiceps cristatus*), its bill is much longer than in areas in which the similar sized great-crested grebe is present. *Source:* Fjeldså [262].

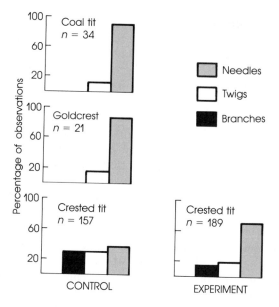

17 **Competitive release in coal tits.** Experimental removal of coal tits and goldcrests from a winter flock resulted in the expansion of the areas foraged by the remaining crested tits, which utilized spruce needles much more in the absence of the smaller competitors. *Source:* Alatalo et al. [5].

comes from a recent study of grebes. In four different species of grebes that occur alone in lakes with no other grebes present, the birds have evolved a general-purpose bill and a broad diet of fish and insect prey. But when these same grebes live and breed in areas that support more than one species of grebe, their bills have become more specialized, either a thin, small bill specialized for insect prey or a larger, longer bill that facilitates the capture of fish. Thus, individuals in solitary populations of the silver grebe have the intermediate, general-purpose bill. But on Lake Junin in Argentina, which they share with the long-billed Junin grebe, they possess a short bill and consume insects [262]. And in Europe where the red-necked grebe and the long-billed great-crested grebe live together, the bill of the red-necked grebe is much shorter on average than in regions of Siberia and Canada in which the great-crested grebe is absent (Figure 16).

Rauno Alatalo and his colleagues have devised an experimental test of competitive release with still another group of birds, European titmice. In the winter, willow tits, crested tits, and coal tits flock with goldcrests. As the mixed-species flocks forage through the woodlands, the little goldcrests and coal tits concentrate on the outer twigs and needles of conifers while the larger willow tits and crested tits hunt on the more substantial branches and trunks of pine and spruce trees. Alatalo and his fellow researchers selected one flock to be the experimental group from which all goldcrests and coal tits were removed, and another flock was left as the control. In the absence of coal tits and goldcrests, crested tits foraged much more on spruce needles (Figure 17) and willow tits shifted to twigs in pines. This finding suggests that the presence of the other species provided competition that kept the larger titmice from exploiting certain sources of food in their environment [5].

Rodents, Seeds, and Competition

The various demonstrations of competitive release indicate that the competition hypothesis has merit, at least for some groups. However, to illustrate the point that competition need not be responsible for diversity in feeding behavior among species, let us examine the case of kangaroo rats and pocket mice. There are several species of each of these seed-eating rodents that live together in the southwestern United States. They differ, however, in size, with the smallest kangaroo rat weighing at least three times as much as the largest pocket mouse. If the example of the Darwin's finches were to be our guide, we might expect that the larger kangaroo rats would take larger seeds and the smaller pocket mice would gather smaller seeds, thus apportioning food resources between them and avoiding direct feeding competition.

The first study on the diets of these animals in areas of geographic overlap reported that there was segregation between species in seed preferences [109]. This report was taken to support the competition-avoidance hypothesis. But later research documented that in some areas at least the two groups of species were consuming exactly the same kinds of seeds [729]. Two other hypotheses have been offered for the coexistence of kangaroo rats and pocket mice. First, the two species appear to have different microhabitat preferences: the pocket mice stay close to cover, the kangaroo rats travel through more open, sparsely vegetated locations in the same general area [674]. The spatial segregation of the species would mean that they would not come into contact frequently and therefore would not often compete directly for food. Second, the two species have different modes of foraging. Kangaroo rats use their large powerful hindlegs to propel them rapidly through their nocturnal world. Pocket mice have much shorter legs and travel more slowly. Thus, kangaroo rats can afford to hunt for the rare clusters of wind-blown seeds in the open desert, traveling economically in search of these patches, whereas pocket mice are gleaners that pick up each seed they encounter [662]. Once again the species might avoid competition, because, although they eat the same seeds, they forage for them in different ways and so should utilize different sources.

But are the differences in body size, locomotion, microhabitat selection, and foraging tactics the product of selection for avoidance of food competition? One might argue that the two kinds of animals first evolved different antipredator tactics: the kangaroo rats relying on rapid escape dashes and the pocket mice hiding in dense cover to avoid detection by their enemies. These divergent antipredator adaptations would impose constraints on the foraging behavior of the two kinds of rodents with the associated, but incidental, result that they do not collect food in an identical manner.

If the constraints hypothesis is correct, then we can predict that other rodents that employ the locomotory mode of kangaroo rats will also forage in the open and utilize the kind of foods that kangaroo rats prefer. There is another fleet, hopping rodent in western deserts—the kangaroo *mouse*, a creature with elongate hindlegs but with a body size that is about the same as that of pocket mice. John Harris showed that this creature, like the larger bipedal kangaroo rats, preferred to forage in the open but showed a preference for scattered seed [345] (Figure 18). These results show that bipedal, hopping locomotion permits foraging in the open, presumably because the kangaroo-like rodents are safer from predators thanks to their speed, but does not lead necessarily to specialization on clumped food sources.

Harris proposed that the kangaroo *rats* use their large body size to monopolize rich patches of seeds, relegating smaller species to leftovers, namely, isolated seeds. If he is right about the large rats' ability to outcompete the smaller mice for patches of food, then the removal of kangaroo rats should result in competitive release for the little fellows, permitting an increase in their foraging range and numbers.

An experimental test of competitive release has been done with kangaroo rats and pocket mice in the following manner [577]. Eight square plots (50 meters × 50 meters) were fenced off; four were assigned to be controls and four were experimental exclosures from which all the resident kangaroo rats were removed. In both kinds of plots, holes were placed in the fence at intervals; but in the controls the holes were big enough for kangaroo rats to pass through, whereas in the experimental plots the passages were large enough only for small rodents. For 3 years, the populations within the exclosures were monitored (Figure 19). The removal of the large kangaroo rats did result eventually in a small, but significant, increase in the number of smaller seed-eating rodents of which the pocket mice were a component. In contrast, there was no evidence of competitive release for the small, omnivorous rodents that should not be primary competitors with kangaroo rats.

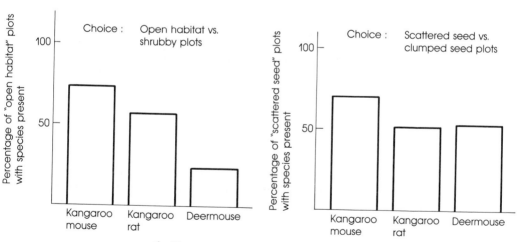

18 **Foraging niches:** foraging behavior of kangaroo rats, kangaroo mice and deermice (animals similar in size to pocket mice). The bipedal, hopping kangaroo mice and rats prefer to forage for seeds placed by an experimenter in open habitat, whereas deermice stay near shelter. Kangaroo mice tend to collect scattered seeds, while kangaroo rats and deermice gather clumped and scattered seeds equally. *Source:* Harris [345].

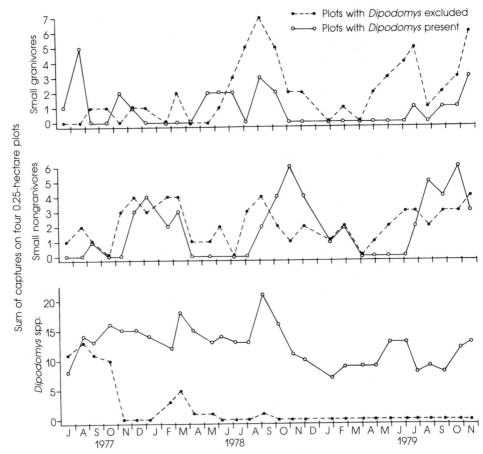

19 **Experimental test of competition.** Exclusion of seed-eating kangaroo rats (*Dipodomys*) from exclosures in the desert resulted in an increase in the numbers of small seed-eating rodents there over time relative to plots with kangaroo rats (top). Removal of kangaroo rats had no such effect on the numbers of other rodents that do not feed on seeds (middle). *Source:* Munger and Brown [577].

This experiment indicates that some species may be in direct competition for a limited food resource. But many issues in the competition debate continue to be explored, thus ensuring that a fundamental question—why do different species eat different foods—will occupy many ecologists in the future.

Optimal Foraging Behavior We have seen that animals employ a diversity of methods to secure the food they consume and that their resulting diets are probably shaped by competition from members of their own and different species, as well as by many other ecological pressures, including predation. Given that there are many influences on what an animal eats, we can ask whether the way an animal makes its foraging decisions within these constraints advances its fitness more than any other alternative. One way to approach this question is with optimality theory, which we briefly discussed in Chapter 8. This approach recognizes that alternative traits that deal with a given ecological problem will have different costs and benefits; the theory assumes that natural selection will favor whatever attribute maximizes the net gain in fitness. If so, the attribute that yields the best return should spread through a population. An example of how the theory can produce a specific hypothesis with quantitative predictions follows [880].

A familiar sight on the northwestern coasts of North America are crows dropping prey onto rocky beaches. A beach-combing crow spots a clam, snail, or whelk, picks it up, flies into the air, and then drops its victim. If the mollusk's shell shatters on the rocks, the bird plucks out and eats the soft body of its victim. The adaptive significance of the bird's behavior seems straightforward. It cannot use its beak to crack open the extremely hard shell of certain mollusks. Therefore, it breaks its prey by dropping them on rocks. This seems adaptive. Case closed.

But we can be much more ambitious in our analysis of this component of a crow's foraging behavior. When a hungry crow decides to hunt for whelks, it has many choices to make: which whelk to pick up, how high to fly up before dropping the prey, and how many times to keep trying if the whelk does not break on trial 1, 2, or 3. One optimality model assumes that crows should choose whelks that will yield the greatest possible energetic return for the time and energy invested in opening them. The point of feeding behavior is to ingest sufficient energy to maintain one's body and to reproduce. Animals that consistently invest more energy in foraging than they gain in collected food obviously are destined for a short life and low fitness. When they perish, so will their genes.

Selection should eliminate not only totally incompetent foragers but also those that forage less efficiently than their genetically different rivals. We assume that a crow that maximizes its net energy gain when searching for food will usually leave more descendants than a less efficient individual. The less productive crow will either accumulate lower energy reserves (reduced benefits) or be forced to spend more time foraging (higher costs). In either case, its fitness is likely to be reduced through its relatively early demise or inability to produce large numbers of surviving offspring.

Reto Zach tested whether northwestern crows really do maximize the difference between foraging benefits and costs [880]. (1) He observed that crows only dropped large whelks 3.5–4.4 centimeters long. (2) They flew up about 5 meters to drop their selections. (3) They kept trying until the whelk broke, even if many flights were required. If their ultimate (unconscious) goal were energy maximization, (1) large whelks should be more likely to shatter after a drop of 5 meters than smaller ones, (2) drops of less than 5 meters should yield a much reduced breakage rate whereas drops of much more than 5 meters should not improve the chances of opening a mollusk a great deal, and (3) the chance that a whelk will break should be independent of the number of times it is dropped.

Zach tested each of these predictions in the following manner. He erected a 15-meter pole on a rocky beach and outfitted the pole with a platform whose height could be adjusted and from which whelks of various sizes could be dropped. He collected samples of small, medium, and large whelks and dropped them from different heights (Figure 20). He found that large whelks required significantly fewer 5-meter drops before they broke than either medium-sized or small whelks. Second, the probability that a large whelk would break improved sharply as the height of the drop increased up to about 5 meters. After this height, the improvement in breakage rate was very small. Third, the chance that a large whelk will break is always 25–30 percent on any drop. A crow that abandoned a "recalcitrant" whelk would do no better with a new one and would have to invest time and energy to find the replacement.

Zach went one step further by calculating the average number of calories required to open a large whelk (0.5 kilocalories) and subtracting this amount from the food energy present in a large whelk (2.0 kilocalories); net gain 1.5 kilocalories. In contrast, medium-sized whelks, which require many more drops, would yield a net loss of 0.3 kilocalories; and trying to open small whelks would have been an even poorer investment. Thus, the crows' rejection of all but large whelks is clearly adaptive, and their selection of dropping height and persistence in the face of failure enabled them to reap the greatest possible energy return for their foraging activities [880].

Subsequent to Zach's study, two other Canadian researchers, Howard Richardson and Nicholas Verbeek, looked again at the feeding behavior of the northwestern crow, this time considering how the birds selected littleneck clams [666]. A crow cannot tell how big a clam is until it has excavated it from the sand. Often, when a crow digs up a littleneck, it leaves it lying

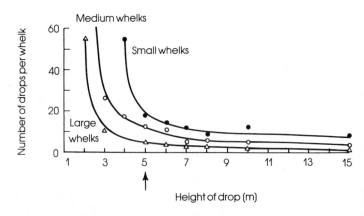

20 **Optimality model tested.** The number of drops at different heights needed to break whelks of different sizes. Northwestern crows drop large whelks only and from a height of about 5 meters, thereby minimizing the energy they expend in opening whelks. *Source:* Zach [880].

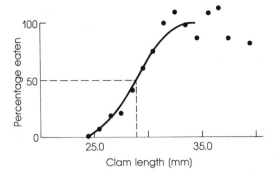

21 Prey size and foraging preferences. Northwestern crows usually leave uneaten any clams they dig up that are less than 28 millimeters in length. By selectively feeding on only the larger clams, crows increase the rate at which they secure calories. *Source:* Richardson and Verbeek [666].

on the beach and hunts for another one. Sometimes, however, it flies up and drops the clam it has found and consumes the interior flesh after the clam breaks when it hits the stony beach. Small clams are left behind; large ones are consistently eaten (Figure 21).

The Canadians determined that, unlike whelks, small, medium, and large littlenecks are equally prone to crack open when dropped onto rocks. Thus, there was no time penalty for trying to break smaller clams. They further established that the larger the clam, the more time it took to extract the meat when the clam shell was fractured. Despite the time cost of feeding on large clams, the crows rarely left them behind.

By figuring out how much energy was available in the tissues of clams of different sizes, Richardson and Verbeek showed that the bigger, the better—in terms of energy profit gained per unit time of food preparation. Therefore, a preference for larger clams increases the energy gained for a foraging bird. But crows do not consume only the very largest clams they find by digging into tidal flats. Instead, they accept about half the clams of about 29 millimeters long, with the consumption rate increasing to 100 percent for clams 32–33 millimeters long.

In order to calculate the maximum energy gain for a particular diet, one must include the search costs for prey of given sizes. Although a 33-millimeter clam contains more energy that a 29-millimeter prey, a crow that has found the smaller clam will have to spend some time (and energy) to locate a larger replacement. As it turns out, the math indicates that the optimal strategy for a crow in terms of maximizing net energy gained per unit of foraging time is to take clams 28.5 millimeters or larger and to reject clams smaller than the 28.5-millimeter minimum. As Figure 21 shows, this is essentially what the birds do [666].

Minimizing Foraging Costs

The northwestern crow is remarkably good at doing the things needed to maximize the difference between energy expended and energy gained when searching for whelks or littleneck clams. Is the bird an exception or are there other creatures like it? Malte Andersson applied an optimality approach to the foraging behavior of nesting whinchats, a small European songbird. Like Zach, Andersson assumed that these birds should be energy maximizers, because when collecting prey for their young in a nest, whinchats ought to be under pressure to gather food as efficiently as possible. The more food supplied to the young, the greater the likelihood that they will survive to fledge.

Andersson further assumed that the food in a whinchat's territory is likely to be more or less uniformly distributed. Using these assumptions, one can construct a mathematical model that generates these intuitively reasonable predictions: (1) foragers should spend more time close to the nest rather than far away—so as to keep travel costs down; (2) the higher the food density, the more time the bird will spend in the sectors nearest the nest—because it need not travel far to get the food that its offspring require; and (3) the higher the food density, the smaller the total area searched for food by a foraging bird.

By observing foraging parents and recording how far from the nest they were while hunting, Andersson collected data that supported the first prediction, namely, that the birds devoted most of their searching effort to the area closest the nest. He then performed an experiment to test predictions 2 and 3 by adding food evenly throughout an area with a radius of 90 meters about the nest. The added food consisted of mealworms (beetle larvae) placed in 225 petri dishes. Each of his five subjects changed its foraging behavior to concentrate still more on those sectors closest to the nest and to reduce the total search area. The mean distance a male foraged from his nest declined from 44 meters to 29 meters during the first 2 hours of the food supplement experiment. The effect of the birds' decisions was to enable them to provide their young with more food per unit time spent foraging than had they hunted farther from the nest site [16].

Yellow-eyed juncos provide a final example of a test of an optimality model based on the assumption that foragers are attempting to maximize their net energy gain. When these sparrows have just fledged, they handle large prey clumsily, taking a long time to crush a big grub, for example, before swallowing it. In contrast, very small food items can be picked up and swallowed without much handling. Kim Sullivan has demonstrated that because of the long handling times associated with large mealworms, recently fledged juncos actually gain more energy per unit time feeding by eating small mealworms [753]. Experienced older foragers have learned to manipulate insect larvae skillfully and can prepare a large mealworm in less than half the time it takes a novice junco. This means that they do not pay a great time penalty for selecting large mealworms, and indeed the caloric payoff per second of foraging time is the same for large and small mealworms for adult juncos. If one were to give fledgling and adult juncos free access to an ample supply of small and large mealworms mixed together, how should their choices differ (Figure 22)?

Energy Maximizers or Cost Minimizers?

The optimality models discussed to this point have been based on the assumption that animals are out to maximize their net intake of energy. Although this is a reasonable assumption for something like a breeding whinchat with a nest full of hungry babies, it may not always be a reason-

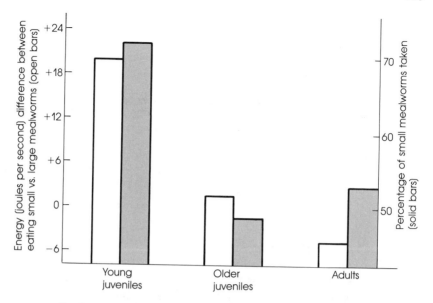

22 **Handling time and diet selection** in yellow-eyed juncos. Young juveniles take so long to prepare large mealworms for consumption that they actually gain more energy per unit time by feeding on small mealworms, and they prefer this size class of prey. For older juveniles and adults, handling time for large prey decreases, so there is little difference in energy gained per time spent feeding on large versus small mealworms, and the birds feed on both types equally. *Source:* Sullivan [753].

able assumption. The African sunbirds, whose feeding territories we examined in the preceding chapter, are a case in point. There we suggested that because the birds were simply trying to keep alive during the winter, they might well be attempting to get what they needed as quickly as possible (minimizing the time invested in foraging and the exposure to predators) rather than trying to maximize the net energy gained per day (net gain = gross caloric intake − total daily caloric expenditures).

Graham Pyke constructed two different optimality models, one based on the assumption of cost minimizing, the other assuming that the birds were energy maximizing over the course of each day. He then tested his models using data gathered by Frank Gill and Larry Wolf. The actual caloric costs and gains vary from individual to individual and depend on the size and quality of the territory and the number of intruders the area attracts. But the average difference between calories gained and calories expended by sunbirds each day is close to 0 (−190 calories). This result is not what one would predict from an energy maximization hypothesis (Table 1) [651]. Because nonbreeding sunbirds do not have to forage for their brood or build up energy reserves for reproduction, their unconscious goal is just to meet their daily energy requirements as quickly as possible. They can then spend the rest of the day safely perched in a sheltered location.

TABLE 1

Comparison of predicted and observed values for various aspects of sunbird territorial behavior: A test of two hypotheses

Hypothesis	Net energy gain (calories)	Flowers in territory	Hours spent		
			Foraging	Defending	Resting
Energy maximization	28,000–48,000	6300–9600	5.7–8.1	1.9–4.3	0
Cost minimization	0	1540–1600	1.7–2.6	0.2–0.4	7.2–7.9
Observed[a]	−160	1600	2.4	0.3	7.3

Source: Pyke [651].
[a]Mean values are given for "Observed."

Optimality Models: Constraints on Foraging Efficiency

People sometimes criticize optimality theory on the grounds that animals will not always behave as efficiently as possible. This criticism is reminiscent of the complaint that not all behavior is perfectly adaptive, and, like that complaint, it misses the point. Optimality models are not constructed to make statements about perfection in evolution, but to make it possible to test whether one has identified the conditions that influence an animal's behavior. As we have just seen, if a sunbird is assumed to be maximizing its total daily energy intake it is predicted to behave differently than if it is assumed to be trying just to acquire the food it needs to stay alive until the next day. This is useful because it enables us to see which model is closer to reality, which assumptions are justified and which are not.

Marmot foraging behavior is influenced by the risk of predation as well as by the advantages of gathering food efficiently. These animals live in burrows in rocky talus slopes, where they are relatively safe from their enemies, which include coyotes and golden eagles. But they have to leave their burrows behind when they head out into the grassy meadows to harvest the vegetation that they consume. Warren Holmes tested the prediction that animals will compromise foraging efficiency (defined strictly in caloric or nutritional terms) to the extent that they are at risk to predators. In the world of marmots, as in many other species, young animals are smaller and less experienced than older ones and therefore juveniles are favored prey of predators. If a juvenile animal's goal were simply to stuff itself with as much food as possible, it would spend as much time as possible feeding, and it would travel away from the well-grazed areas near the talus slope in search of rich feeding areas in the meadow. In fact, juveniles spend more time looking about for danger than do adults, even though this leaves them less time for feeding. Moreover, juveniles travel shorter distances from their burrow retreats than do adults, at the cost of staying away from some highly productive food patches [376].

Predation, Competition, and Foraging

Some ingenious experimental studies with birds and fish confirm the point that individuals will opt for a less profitable, but safer, feeding behavior when the risk of predation is high. Steven Lima placed a feeding tray with sunflower seed bits at 2, 10, or 18 meters from dense cover. On some trials, he offered visiting chickadees large fragments of sunflower seeds that required about 30 seconds to break up and consume. On other trials he gave them smaller bits that could be prepared and eaten in just 10 seconds or so. If the birds' behavior was designed to maximize food intake, they should simply have stayed at the feeder, working through one food item after another, no matter what the size of the seeds or how far they were from cover. In actual practice, they often picked up a seed and flew back with it to the relative safety of pine foliage, where they could prepare and consume the seed out of view of predators (Figure 23). They were much more likely to do this for large food items that were in feeding trays close to cover, that is, when the benefits were high (there would be 30 seconds of decreased exposure to predators, once in shelter) and the costs were low (the flight time and energy expended to reach shelter were low) [484].

Lima also examined the effect of increasing the risk of predation, by sending a flying model hawk sailing by when a chickadee was feeding at the feeder. Chickadees exposed to a simulated predator became more likely to take their sunflower seeds to cover. In other words, they were (sensibly) more willing to sacrifice a degree of foraging efficiency when the danger of being killed was high than when it was low.

A fundamentally similar study was conducted in the laboratory with

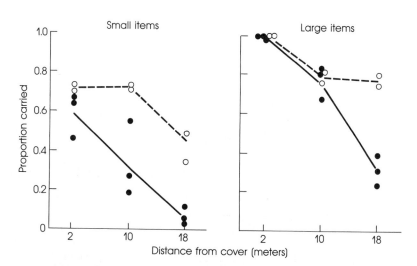

23 **Predation risk and foraging behavior in chickadees.** After a simulated hawk has passed near chickadees (open circles), they were more likely to carry seeds, small and large, from an exposed feeding site to safer cover before preparing and eating the food. In the absence of a predator (solid circles), chickadees tended to take large seeds, not small ones, to cover, and they were more likely to seek cover when it was nearby. *Source:* Lima [484].

very small salmon. In this study the fish were given a chance to rise to the surface to retrieve a floating prey [199]. Under one experimental situation, a photograph of a salmon predator, a trout, was manipulated to drift in and out of view of the salmon. Under these conditions, the juvenile salmon waited longer to leave their retreat to feed than under the condition of "no predator present" (Figure 24).

On the other hand, young salmon were quicker to attack when they were permitted to view a mirror image of themselves, a simulated competitor, at intervals during the feeding trials (not shown in Figure 24). The salmon were willing to take greater risks given an apparently higher probability that a competitor would beat them to it. This example suggests that taking social competition into account may be essential for understanding at least some elements of feeding behavior. Game theory (Chapter 8) may therefore provide useful alternative hypotheses to those taken from optimality models that do not deal with the competitive factor in foraging.

The possibility that an animal's intrinsic competitive ability might also influence his selection of a diet has also been examined in a laboratory study of stickleback fish [555]. These fish are often victimized by certain parasites that change the fish's appearance while feasting on the fish's tissues. Manfred Milinski found that parasitized fish preferred small water flea prey, even when large water fleas were available. Because the larger prey had three times the dry weight of the small prey, the fish "should"

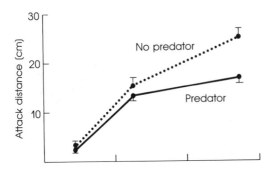

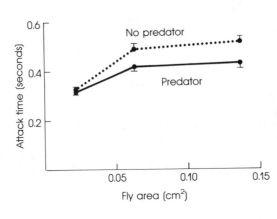

24 **Predation risk constrains foraging** in captive young coho salmon. Larger flies (those with greater "area") are more quickly attacked; but when the fish are exposed to a simulated predator, they are more reluctant to leave their hiding place to capture even large attractive flies floating toward them. *Attack distance* is the distance traveled by the fish from its resting place to the prey; *attack time* is the time taken to capture the prey. *Source:* Dill and Fraser [199].

have chosen the large water fleas if they were selecting their food strictly on the basis of its caloric content. But parasitized fish cannot compete well with unparasitized sticklebacks. Under these circumstances the sicker fish may do better to take food items that their healthy rivals will leave untouched, rather than compete directly and unsuccessfully for the food with the highest caloric value.

Nutritional Constraints on Foraging

There is still another reason why the simple prediction that animals should maximize their caloric intake over a time period will often fail. Individuals almost always require more than energy from their diets if they are to reproduce successfully. Northwestern crows do not eat whelks all day long. In fact, after consuming one or two, they generally leave the beach and go off to hunt a different food elsewhere. Perhaps the advantages of having a varied diet may affect their selection of food. This has been documented for another bird, the European bee-eater.

Despite their skill in catching honeybees, bee-eaters bring their nestlings a mixed diet. To test whether a diversity of foods was beneficial to juvenile bee-eaters, John Krebs and Mark Avery hand-reared nestlings on one of three diets: pure bees, pure dragonflies, or a mixture of bees and dragonflies. They found that the birds given a mixed diet consistently gained more weight per gram of food eaten than nestlings receiving one-item diets [447].

A very different animal, the immense moose, also gains by consuming a mix of plant foods. Although you might expect moose to select only those items that enabled it to maximize net energy intake, because it needs to take in so much, on Isle Royale in Lake Superior these mammals do not feed exclusively on the high-calorie leaves of deciduous trees and assorted weeds. Instead, Gary Belovsky discovered that they also eat a great many aquatic plants, which are relatively low in energy per unit weight. But aquatic plants contain much higher sodium concentrations than terrestrial ones. Belovsky hypothesized that this critical element was a limiting resource for the moose and that the need to consume some minimum amount per day constrained the moose's food selection. He constructed a mathematical model based on this assumption and from it predicted what a moose should eat if it were trying (unconsciously) to consume as many calories as possible, subject to the constraint of ingesting a certain amount of sodium. The predicted diet closely matched the observed one [55].

Another kind of nutritional constraint on foraging behavior is imposed by the defensive properties of certain potential foods, not all of which are actually safe to eat. Thus, plant-eating animals as different as tassel-eared squirrels [255] and leaf-cutter ants [387] select foods that are low in toxic terpenoids, a poison that many plants incorporate in their leaves or cortical tissues to repel consumers.

This kind of selectivity is also exhibited by the herbivorous howling monkeys that Kenneth Glander studied in a riverbank forest in Costa Rica (Figure 25) [289]. Glander discovered that the howlers did everything backwards *if* their goal was simply to ingest as much plant material as possible quickly. First, the more common a tree species, the less likely the monkeys

25 **Selective feeder.** Howler monkeys, although surrounded by leaves, forage very carefully and avoid toxic leaves and leaves low in nutritional value. Photographs by Kenneth Glander.

were to feed on its leaves. Instead they spent considerable time searching out the scarcer species. Second, even with the less common tree species, they were selective, refusing to eat from most individuals available to them. For example, the monkeys took plant material from only 12 of 149 specimens of one "acceptable" tree species growing in the forest. Third, the monkeys preferred the scarcer, smaller, new leaves to the more abundant, larger, mature leaves. Fourth, they often fed "wastefully," eating only the petiole and dropping the larger leaf blade.

Glander hypothesized that the seemingly perverse feeding choices of the howlers actually were adaptive responses to the chemical defenses employed by many trees. Far from being a Garden of Eden, tropical forests are filled with plants whose luxuriant vegetation is poisonous or of very low nutritive value to many consumers.

In fact, Glander showed first that the tree species avoided had leaves with high alkaloid or tannin concentrations. Alkaloids poison howlers; tannins bind with plant proteins and make it costly and difficult for the monkeys to digest the useful proteins. Second, individual trees of the same species varied in their toxicity and tannin concentrations. The accepted trees were, as predicted, relatively low in toxins and tannins. Third, new leaves have more water and less nonnutritive fiber than mature leaves. When the monkeys did eat mature leaves, they selected specimens whose leaves had a higher (12.4 percent) protein content than the mature leaves of trees they rejected (which averaged only 9.4 percent protein). Fourth, "wasteful" feeding occurred because the monkeys were eating the leaf part (the petiole) lowest in toxins while discarding the more poisonous leaf blade.

The conclusion that emerges from these and similar studies is that foraging decisions are often influenced by considerations other than the number of calories that can be ingested during a period of feeding. Although some animals, like whelk-eating crows, may be energy maximizers in the short run, others cannot do this because of the nature of their environments. Tests of optimality predictions have been helpful in showing when energy maximization is not the ultimate goal of animal foraging behavior. Modified optimality models will almost certainly become more precisely predictive in the future and so contribute to a further understanding of the effects of multiple influences on foraging efficiency.

SUMMARY

1 Animal species possess a remarkable array of tactics that appear to overcome the defenses of prey species that are difficult to locate or capture. Several unrelated ambush predators have convergently evolved food (or sex) lures that entice certain elusive and mobile prey within easy capturing range. A small group of predators employs tools to extract energy from some well-armored or well-protected food items. A third category of animals forages cooperatively (1) to overcome the defenses of schooling fish or (2) to capture and subdue large, mobile, and potentially dangerous prey.

2 Typically each species has its own diet and unique set of foraging tactics. Competition among species may currently shape the dietary preferences of some animals, although questions remain about the generality of this conclusion.

3 Optimality theory is the use of evolutionary theory to make specific predictions about what an optimal solution to an ecological problem should be. (Optimal solutions are those that raise an individual's personal fitness more than any alternative would.) The prediction can then be tested against reality.

4 The optimality hypothesis that animals should maximize energy gained per unit of foraging time in order to maximize their reproductive success has been used extensively in the study of feeding behavior. Although some animals do behave in ways that match predictions derived from this hypothesis, many others compromise energy maximization to deal with such things as the need for a balanced diet or the need to avoid predators while foraging. It is possible to construct and test optimality models that incorporate these constraints on the foraging behavior of animals.

SUGGESTED READING

A general review of the literature on predatory behavior is provided in Eberhard Curio's *The Ethology of Predation* [163]. Bernd Heinrich's book, *Bumblebee Economics* [353], is a good companion to this chapter as it deals clearly and simply with competition theory, plant–pollinator interactions, and optimality theory as applied to bumblebees. Reto Zach's article [880] on optimal foraging in the northwestern crow is a model of clarity. D. W. Stephens and J. R. Krebs have written *Foraging Theory*, a book that thoroughly reviews the goals and accomplishments of optimal foraging research [743].

DISCUSSION QUESTIONS

1. Black-headed gulls regularly rob lapwings and golden plovers of the worms they have captured in agricultural fields. How would you develop a simple optimality model that would enable you to make predictions about (a) which of the two species, lapwings or golden plovers, gulls should prefer to attack, and (b) the distance a gull will fly to launch an attack? Remember that the average energy gained in a thievery attempt will equal the energy in a stolen worm times the probability that the attempt succeeds minus the energy expended in flight during the attack. See [763] after you develop your model.

2. Using the comparative tests of adaptationist hypotheses given in this chapter, list examples in which cases of divergent evolution were used to test predictions; then make a separate list of comparative tests based on cases of convergent evolution.

3. Small bluegill sunfish are at greater risk of attack from bass than are large bluegills. If the food items consumed by bluegills are more abundant in open water than in dense aquatic vegetation, what prediction would you make about the foraging habitats selected by small versus large bluegills and the growth rates of the two size classes in ponds with and without bass predators? See [825] after developing an answer.

4. Many small birds flock together when foraging for seeds on the ground. The more birds present, the less often any one bird interrupts its foraging to look around for danger [e.g., 130]. (Why?) But the more birds present, the more competition there is for food, and the longer it takes to find each food item. How might you incorporate these conflicting pressures into a model that predicts what the optimal flock size is? How would changing food abundance or predator pressure affect optimal flock size?

11

The Ecology of Antipredator Behavior

The previous chapter documented that predators often skillfully overcome the defenses of their prey. Once eaten, a dead animal has great difficulty reproducing, a fact that favors individuals with the means to defeat their predators. The defensive behavior of prey is as varied and ingenious as the hunting behavior of their consumers.

Earlier chapters presented hypotheses on the possible antipredator functions of many behavior patterns ranging from eggshell removal by black-headed gulls to the escape dive of bat-pursued moths and lacewings. This chapter categorizes these behaviors in terms of their possible adaptive value, which is either to make it hard for a predator to detect a prey or to make it difficult for a predator to capture a prey that it has spotted. Under these general headings, the chapter examines how behavioral researchers have analyzed cases of cryptic behavior, techniques that deflect attack, chemical repellents, and the spectrum of antipredator effects of group living. The goal is to employ comparative and experimental approaches to test hypotheses about the possible survival value of an animal's behavior.

Making Prey Detection More Difficult   I sometimes find insects whose resemblance to their background is so detailed that it takes my breath away (Figure 1). I could have included in Figure 1 my picture of a bark-mimicking insect, a creature my wife found on an Australian eucalyptus tree. Its mimicry of eucalyptus bark is so perfect, however, that the insect cannot be seen in the photograph. But had that insect been resting on any tree other than the eucalyptus species it had selected, the bug would have been far less well hidden and would have seemed far less marvelous.

The point that the behavior of a camouflaged animal is critical to the success of its color pattern can be illustrated by H. B. D. Kettlewell's classic study on industrial melanism in moths [424]. British lepidopterists knew that a completely black form of the moth *Biston betularia* had become much more numerous in Britain after 1850, largely replacing the typical salt-and-pepper form of the moth in many areas near big cities. Kettlewell's explanation for the spread of the melanic form was that they were less conspicuous to bird predators in woodlands blackened with soot from urban factories. The salt-and-pepper form remained abundant in nonpolluted forests, because in this environment whitish wings blended in nicely against lichen-covered tree trunks.

Kettlewell tested his hypothesis in an experiment in which he placed samples of the two forms of moths on the same tree trunk (Figure 2) and observed the reaction of insectivorous birds from a blind. Birds found whitish forms on sooty tree trunks much more quickly than they found melanic individuals; the reverse was true when the experiment was conducted with a lichen-covered tree as the background.

Kettlewell's study has been replicated several times to permit researchers to measure precisely the selective advantage enjoyed by the two forms on different backgrounds. With these measures, it is possible to determine mathematically the expected rate of change in the frequency of two forms as woodlands change from polluted to nonpolluted under the effect of England's new antipollution regulations. These models, when tested against reality, prove to be inaccurate; the black forms have remained more common longer than predicted.

1 Cryptic coloration and selection of a resting place. A Costa Rican moth whose wing colors give its wings the appearance of a fallen leaf. The degree to which this camouflage is successful must depend on the ability of the moth to find appropriate resting places during the day, when birds hunt for insects. Photograph by the author.

2 **Cryptic coloration of salt-and-pepper moths.** The typical form of the moth, which survives better in nonpolluted woods with lichen-covered trees, is shown on the left. On the right is the melanic form of the moth, which is more cryptic when resting on soot-blackened trunks and limbs. Photographs by H. B. D. Kettlewell.

As it turns out, part of the problem with the measurements is that they are based on the incorrect assumption that the moths typically perch directly on tree trunks. In fact, they more often select the shaded patches just below the junction of a branch with the trunk. R. J. Howlett and M. E. N. Majerus glued samples of frozen moths to open trunks *and* to the underside of branch joints. This more recent experiment confirmed Kettlewell's original findings (Figure 3), but this study also showed that melanic moths were particularly likely to be overlooked by birds if the moths were placed in the shade of a branch joint [392].

Salt-and-pepper moths apparently actively select the kind of resting site that complements their color pattern most effectively. Many other animals also have evolved a strong preference for the "right" background [696]. For example, two closely related Australian lizards are light pinkish yellow and dark reddish brown, respectively. In the field they live in the same general area but consistently perch on rocks that are appropriate for their body hues. When taken to an enclosure with light quartzite rocks and dark reddish limestone resting sites, the yellowish species perched on the pale rocks most of the time whereas the reddish lizard greatly preferred the red limestone. Because a small lizard-eating hawk is abundant in the area where these reptiles live, their ability to pick the matching background is likely to be adaptive [281].

Finally, complete concealment from visually hunting predators can be achieved by going *under* the background, a practice of the Egyptian plover,

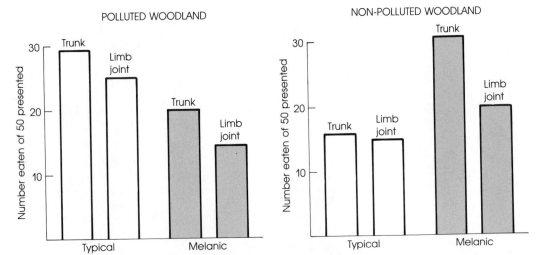

3 **Background selection and predation risk.** Specimens of typical and melanic forms of the salt-and-pepper moth were pinned to tree trunks or the places where a limb joins the trunk. Melanic forms were discovered by birds less often in polluted (blackened) woods; typical forms "survived" better in nonpolluted (lichen-covered) woods. But in every case, specimens on limb joints were less likely to be taken than those pinned directly onto trunks. *Source:* Howlett and Majerus [392].

which nests on sand bars in African rivers. Adults always keep their eggs covered with sand. Should a predator approach young, recently hatched chicks, the parent plovers hurriedly flip sand with their beaks over the crouched youngsters, burying them lightly enough so that they can breathe but cannot be seen [391].

Background Selection and Prey Conspicuousness

If background selection is adaptive for cryptic species, then their predators should have a harder time finding prey on the preferred background than on alternative substrates. Alexandra Pietrewicz and Alan Kamil used the operant conditioning techniques described in the preceding chapter to test whether North American blue jays have trouble finding cryptically positioned prey [634]. Jays were rewarded for detecting a moth in an image projected on a screen before them and punished for making mistakes. The experimenters used a great many different kinds of slides, some with and others without a moth in the photograph. The species of moths were varied, as was their orientation on backgrounds of different sorts. Pietrewicz and Kamil found that when a moth species was perched in its normal orientation on a background appropriate for its color pattern, it was less likely to be detected by a jay.

For example, *Catocala relicta* rests head up with its whitish forewings over its body on white birch and other light-barked trees (Figure 4). When given a choice, it more frequently selects birch bark rather than darker backgrounds as resting places [697]. As predicted, the jays failed to see the moth 10–20 percent more often when examining photos of *C. relicta* pinned to birch bark than when it was pinned to other backgrounds. Moreover, when the moth was oriented vertically on a birch trunk, the birds overlooked it more often than when its long axis was horizontally oriented on the tree trunk. These results confirm the adaptive value of the moth's preference for birch bark.

Removing Telltale Evidence

Some cryptically colored prey advance their concealment with actions other than matching their color pattern to the right background. When a moth larva feeds on its food plant, it creates a cue—a chewed leaf—that a visually hunting predator might use to find a prey. A small bird's hunting time per caterpillar could be enormously reduced if it could productively restrict its search to damaged leaves rather than having to hunt through an entire tree for its victims.

Bernd Heinrich and Scott Collins conducted feeding experiments with captive birds to demonstrate that chickadees can learn to inspect artificially or naturally damaged leaves in preference to intact ones when food is associated with the damaged leaves [357]. Wild-caught chickadees were permitted to enter an enclosure containing ten fresh-cut birch or chokecherry tree tops. Two of the ten trees had damaged leaves and only these trees contained food rewards in the form of impaled mealworms or small edible caterpillars. At first the birds spent some time hopping about in all the trees. But after a few trials, they began to go directly to those trees with damaged leaves (Figure 5), even though the position of trees was changed before each test.

Heinrich also discovered that a number of common caterpillars of the Minnesota north woods either (1) removed themselves from their feeding

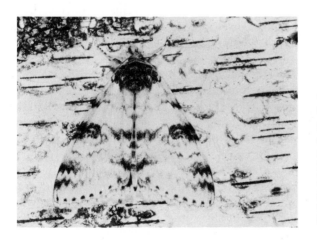

4 **Cryptic coloration and resting position.** The orientation of a resting *Catocala* underwing moth determines whether the dark lines in its wing pattern match up with dark lines in birch bark. Photograph by H. J. Vermes, courtesy of Theodore Sargent [697].

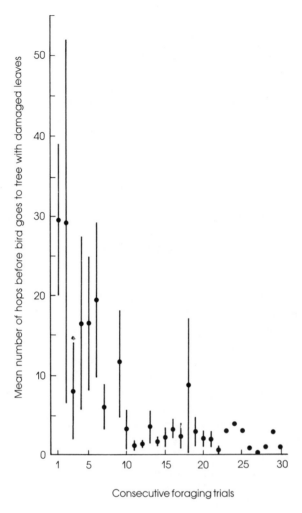

5 **Telltale evidence.** Chickadees learn to visit damaged leaves if these leaves have prey on them. The mean number of hops before captive birds went to a tree with damaged leaves in an enclosure declined rapidly as the chickadees learned to associate this cue with food rewards. *Source:* Heinrich and Collins [357].

sites during the daytime or (2) eliminated evidence of their foraging [354]. In the first category are the caterpillars of *Sphecodina abbotti*. When dawn comes, these animals abandon the grape leaves they have been consuming at night to go to resting sites some distance away on the vine stem, where they are beautifully camouflaged. A bird that looked near damaged leaves for this prey would not find any. In the second category are the larvae of another *Catocala* moth, *C. cerogama*, which neatly snips through the petiole of a largely eaten leaf, dropping what is left to the forest floor, where it can offer no clue about the location of the caterpillar in the tree.

In order to test the hypothesis that these edible caterpillars are trying to hide from their predators, Heinrich predicted that moth larvae that are protected by spines, hairs, and toxins will not invest in efforts to conceal themselves at their feeding sites. Protected species rely on warning coloration and behavior to repel their predators directly (see p. 350) and they do remain on the leaves on which they feed [354].

Making Capture More Difficult Producing a beautifully complete camouflage, selecting precisely the right resting site, and cutting away a partially eaten leaf all carry costs as well as benefits. However, the hypotheses tested to date on these attributes have focused on their possible benefits without examining their costs. It is sometimes possible, and often desirable, to consider how the costs of a particular action might influence its utility. You may recall that in the previous chapter we noted that marmots, chickadees, and baby salmon compromise foraging efficiency when feeding exposes them to predators. This response is probably the most common antipredator behavior of all. The general point is that maximizing lifetime reproductive success—the ultimate currency of evolution—usually requires juggling several competing goals. For example, one might imagine that a prey that has been seen by a predator should immediately run away, making its capture more difficult. More mature reflection, however, reminds us that running away consumes time and energy and prevents the animal from doing other potentially useful things. There are trade-offs between avoiding predators by running from them and staying where they are in order to feed or mate.

A hypothesis drawn from a cost–benefit approach is that the greater the risk of capture, the more willing the animal will be to sacrifice its other fitness-enhancing activities, and the sooner it will flee from a detected predator. Anoles run more slowly when they are cooler, and therefore they are more likely to be captured when they have low body temperatures. Data for one West Indian species show that cool lizards do not permit their enemy (an approaching human) to approach as close as warmer anoles do before dashing to safety. Warm anoles are capable of running faster and therefore can stay put longer in the face of an approaching enemy [875] (Figure 6).

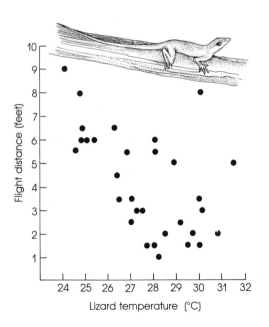

6 Risk and flight distance. The body temperature of an anole affects how fast it can run and how soon it begins to run from an approaching human. *Source:* Ydenberg and Dill [875], after Rand [657].

This admittedly somewhat artificial experiment involving simulated predators suggests that the "flee-as-soon-as-predator-is-seen" hypothesis cannot apply to all species and all situations. Indeed, it has long been known that the response of African antelope and zebras to cheetahs or lions they can see is usually remarkably blasé because they know precisely how close these enemies have to be before they are likely to charge [700].

Running off each time a predator appeared on the Serengeti would be both energetically expensive (for predators are everywhere there) and dangerous, because in running from detected predator X you may run head first into undetected predator Y. One element of the escape behavior of an African antelope, the Thomson's gazelle, has been interpreted as an adaptive response to this risk. The gazelle sometimes stotts in the presence of a predator; *stotting* involves jumping about a half-meter off the ground, with all four legs held stiff and straight and with the white rump patch fully everted (Figure 7). T. Pitcher proposed that by leaping up in this fashion a fleeing gazelle could scan better for predators hiding in ambush. He suggested that it might pay an alarmed gazelle to sacrifice some speed initially in order to determine what the safest escape route might be [637].

The antiambush hypothesis has considerable plausibility, but it is only one of *eleven* alternative explanations for stotting compiled by Tim Caro [135, 136]. We shall confine ourselves to just five of the more likely possibilities in this excellent example of the importance of using the method of multiple working hypotheses to analyze the function of an antipredator behavior.

Let us begin with the antiambush hypothesis, which predicts that stotting will *not* occur on short-grass savannah but will be reserved for tall-

7 Stotting behavior. A Thomson's gazelle leaps high with its legs held stiffly, perhaps to signal to a cheetah that the predator has been seen.

grass or mixed grass and shrub habitats where predator detection could be improved by jumping into the air. However, gazelles feeding in short-grass habitats do stott regularly. Moreover, gazelles can look where they are going by jumping high even while fleeing at full speed, without stotting, which does slow them down [136]. The antiambush hypothesis fails its test, despite its intuitive appeal.

Let us move on to the alternatives, which include the (1) alarm signal, (2) social cohesion, (3) confusion effect, and (4) communication with predator hypotheses.

1. Stotting might warn conspecifics, particularly offspring, that a predator was dangerously near. This could increase the survival of offspring and relatives, thereby improving the representation of the signaler's genes in subsequent generations (see Chapter 15 on indirect selection).

2. Stotting might enable groups of gazelles to form and to flee in a coordinated manner, increasing the difficulties for a predator that was attempting to cut a victim out of a herd.

3. By stotting, individuals in a fleeing herd could confuse and distract a following predator, keeping it from focusing on one animal.

4. Stotting is an action that communicates to the predator that it has been seen and that the gazelle is prepared to flee or is fleeing from it. Predators may choose to abandon chases with alerted prey because they are hard to catch.

Table 1 lists the predictions that are consistent with these four hypotheses. Because the same prediction sometimes follows from two different hypotheses, we must consider multiple predictions from each hypothesis if we are to discriminate among them.

Caro learned that a solitary gazelle will sometimes stott when a cheetah approaches. This helps eliminate the alarm signal hypothesis (if the idea is to communicate with other gazelles, single gazelles should not stott) and the confusion effect (because a predator will become confused only if several

TABLE 1

A comparison of the predictions taken from four alternative hypotheses on the adaptive value of stotting behavior in Thomson's gazelles

Hypothesis	Prediction			
	Solitary gazelles stott	Grouped gazelles stott	Stotters direct white rump toward predator	Stotters direct white rump toward other gazelles
Alarm signal	No	Yes	No	Yes
Social cohesion	Yes	No	No	Yes
Confusion effect	No	Yes	Yes	No
Predator communication	Yes	No	Yes	No

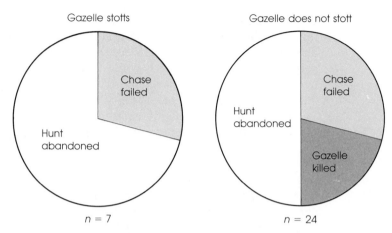

Gazelle stotts   Gazelle does not stott

Chase failed   Chase failed

Hunt abandoned   Hunt abandoned

                            Gazelle killed

n = 7   n = 24

8 **Stotting: a sign of vigilence?** Test of the hypothesis that stotting is a signal to a predator that it has been seen. Cheetahs abandon hunts more often when gazelles stott than when they do not. *Source:* Caro [136].

animals are bobbing up and down in a mass of fleeing gazelles). We cannot rule out the social cohesion hypothesis on the grounds that solitary gazelles stott, because there is the possibility that the solitary individuals try to attract distant gazelles to flee together for the mutual benefit of all. But if the goal of stotting is to communicate with fellow gazelles, then stotting individuals, solitary or grouped, should direct their white rump flash to other gazelles. Stotting gazelles, however, orient their rump to the predator.

The only hypothesis that is still standing is the one that claims that gazelles stott to tell cheetahs that they have been detected. It is often in the predator's interest to give up on prey that have seen them, because the odds of capture decline sharply then. Caro found that cheetahs were significantly more likely to abandon hunts when the gazelles they were stalking or following stotted than when the potential victims had not stotted (Figure 8). Often a cheetah slowly stalking a feeding gazelle made the mistake of moving when the gazelle had its head up instead of when it was grazing. Under these conditions, the gazelle often stotted, and the cheetah gave up (in disgust?) and lay down on the ground. Although stotting does not increase a gazelle's detection of its enemies, it does let its would-be killers know that they are dealing with a vigilant individual [136].

Misdirecting an Attack

Even vigilant animals may be surprised by a hunter. If this happens, the prey may still be able to protect itself to some degree. One tactic is to try to cover vital body parts, particularly the head. Predators typically assault an animal's head because damage to the brain quickly renders a prey totally helpless. A convergent response to this danger is to conceal the head when threatened by tucking it under harder, less vulnerable, body parts. Turtles are adept at this maneuver, and so too are animals as different as crustacean pill bugs, rattlesnakes (Figure 9) [214], and mammals like the armadillo, the hedgehog, and the Australian spiny echidna.

An alternative way to protect one's head is to induce a predator to strike at some other, expendable body part. A particularly ingenious trick of this sort may be practiced by Antarctic krill, small marine crustaceans that gather together in large groups to feed on plankton. As krill get bigger, they shed their "skins" from time to time. But when scuba divers startled masses of krill by diving toward them, great numbers of krill suddenly and simultaneously molted, leaving their cuticular exoskeleton floating conspicuously in the water while they "tail-flipped backwards like tiny lobsters" [339]. This raises the (still untested) possibility that predators sometimes may be fooled into attacking the molted exoskeletons.

Another tactic whose function has been examined in more detail is to display false heads to predators. False heads tend to be on posterior portions of the animal's body, parts that can be sacrificed without incurring a mortal wound [837]. Hairstreak butterflies often have false heads (Figure 10), and it is common to find hairstreaks alive and well with their expendable false heads neatly snipped out of their hindwings, presumably by birds [668].

Mark Wourms and Fred Wasserman conducted an experimental test of the hypothesis that false heads on butterflies work by misdirecting the attack of bird predators [870]. They painted false heads of various sorts on the wings of living cabbage white butterflies, which were then presented in a feeding dish to captive blue jays. The researchers recorded whether the jays pecked the third of the body and wings that included the actual head, or the middle portion, or the back third of body and wings. Although the birds did in fact peck at the hindwings more often in the experimentally altered specimens, the difference in their treatment of the two groups of butterflies (altered and unaltered) was not statistically significant.

9 **Rattlesnake hiding its head,** a highly vulnerable portion of its body, from a potential predator. Photograph by Michael King, courtesy of David Duvall.

10 **Hairstreak butterfly** with its false head (arrow) on one wing snipped out by a bird. Photograph by the author.

They then gave several aviary-bound blue jays the chance to hunt for living cabbage whites, some of which had been experimentally endowed with false heads and some of which were unaltered controls. They found that after a jay had taken a butterfly in its bill, it usually transferred it from bill to foot, preparatory to killing it and removing its wings. Cabbage whites with false heads escaped at this stage almost three times as often as controls (Figure 11) because the jays mishandled the experimental butterflies by pecking at their false heads. This experiment supports the hy-

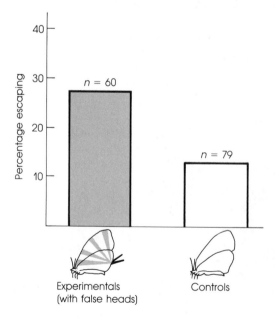

11 **Misdirecting an attack.** In an experiment, blue jays mishandled captured cabbage butterflies with artificial false heads, permitting them to escape more often than unaltered controls. *Source:* Wourms and Wasserman [870].

12 Distraction display by a desert gecko. The gecko raises and waves its tail as an expendable lure when threatened by a potential predator (left). Photograph by Justin Congdon. The same species; an individual has lived to regenerate its tail after losing it to a predator (right). Photograph by the author.

pothesis that false heads induce life-saving mistakes, but when prey are already captured rather than before [870].

More work needs to be done with living butterflies that naturally possess false heads in order to find out when these structures come into play in predator–butterfly encounters. But the use of a moving body part as a lure to deflect an *initial* strike is well documented for some lizards, which twitch their tails when threatened by a predator, distracting their enemies from their heads (Figure 12). Lizard tails may be brightly colored, as in the blue tails of young skinks. Nine of 19 blue-tailed skinks escaped when attacked by a captive snake, whereas only 1 of 15 skinks had similar good fortune when their tails were painted black to blend in with the rest of their body [149].

Escape for the blue-tails came about because the snakes tended to attack the base of the conspicuous tails, and not the body or head of the lizard. When a snake grabs a young skink's tail, it breaks off—thanks to the tail's fragile connection to the rest of the body. As a final touch, the autotomized tail may thrash wildly on the ground after breaking off, a response that may help keep the predator busy while the lizard escapes. In an experiment, captive snakes were offered a skink's tail that was either still twitching or had been allowed to exhaust itself into immobility. The snakes took 40 percent longer to "subdue" and consume the thrashing tails than to eat the nonmoving tails. The behavior of the sacrificed appendage can make a big difference in the time available to a tailless lizard for escape [195].

Making a Predator Hesitate

Some deflection lures work by occupying a predator while the prey, or what is left of it, escapes. Another way to buy time for escape is to startle an enemy sufficiently to reduce the efficiency of its attack. In Chapter 1 we met one of the more spectacular representatives of the many moths that possess false eyes, which they expose suddenly when jabbed in the thorax (see Figure 1, Chapter 1). There we presented the hypothesis that a sudden change in the moth's appearance persuades some predators to hesitate in attacking, as has been shown experimentally for another lepidopteran [71]. This might provide the moth with a few seconds to prepare for flight, or the bird might simply decide to go elsewhere for a meal.

The function of brightly colored hindwings that are concealed when a moth is at rest has been examined in Theodore Sargent's studies of *Catocala* underwing moths. In flight, the brightly patterned hindwings draw the attack of captive pursuing birds (supporting the deflection hypothesis). When a resting underwing moth is grabbed by a blue jay, the moth struggles, suddenly exposing its hindwings, which typically have orange, red, or yellow bands on a dark background. The sudden flash of color appears to startle jays, at least initially; the birds gape involuntarily, sometimes allowing the moth to escape [697].

The startle hypothesis has been rigorously tested by Debra Schlenoff, who trained blue jays—those most cooperative of predators—to pick up artificial moth prey that had a pinyon nut seed attached underneath them [702]. Blue jays like pinyon nuts, and they soon learned to grab the moths that were presented to them. These artificial moths were so designed that plastic hindwings popped out into view when the model was removed from its presentation board.

In one experiment some blue jays were trained on moth models that had gray hindwings. After a while, the birds showed no hesitation in removing the experimental models and consuming the pinyon nut reward. Then the birds were offered a series of models on the presentation board, one of which had brightly patterned hindwings like *Catocala* moths. When the birds picked up these models, they were very prone to drop them or cry in alarm (Figure 13).

Calls and Startle Responses

STARTLE RESPONSES can be induced by sudden, loud noises as well as by sudden changes in visual appearance. A deep growl or a sharp hiss may frighten a predator because these sounds are also made by large animals that are potentially dangerous to the predator [573]. Sometimes even animals that are usually silent, such as rabbits, will produce surprisingly intense piercing calls when in the clutches of a predator. Are these "fear screams" adaptive because they startle the predator? In a paper entitled "Adaptation unto death," Goran Högstedt used the multiple hypothesis method to examine the function of these vocalizations. He noted that fear screams are loud, as required by the startle hypothesis, but the calls are persistent rather than sudden and unexpected, and their persistence should reduce their potential to startle [369].

Because fear screams contain both high and low frequencies of sound, features that enable others to locate the source of the sound easily, perhaps they warn others of danger. Högstedt points out that, if this is so, one would

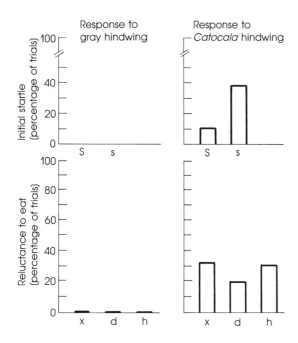

13 **Test of the startle hypothesis.** Wild-caught blue jays trained to remove nuts from artificial models of underwing moths with plain hindwings often are startled when first exposed to a model with brightly patterned hindwings, and they are less likely to eat the nut. S, intense startle; s, weak startle; x, nut not eaten; d and h, forms of hesitation in eating the nut. *Source:* Schlenoff [702].

expect conspecifics to respond to the cries by taking some action; yet they apparently largely ignore the calls, a reaction that makes a certain amount of sense, because a predator with a captured prey will be occupied with that victim rather than hunting for uncaptured specimens nearby.

There are still other possible functions for the screams. It could be that they are given by young animals to enlist the aid of their parents. But if this were the only function of the behavior, one would expect fear screams to be given only by dependent young, and this is not true. Adult birds captured in mist-nets are just as likely to scream as juveniles when handled by a human being (Figure 14).

The evidence points to a fourth hypothesis, namely, that the desperate

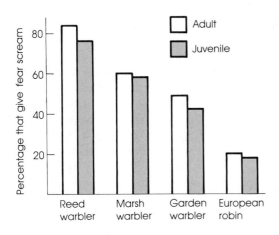

14 **Fear screams and age** in four European birds. For many species, juveniles are no more likely to give fear screams when handled by a human than are adults. This finding suggests that the function of the call cannot be to enlist parental assistance. *Source:* Högstedt [369].

prey's call benefits the victim by attracting other predators to the scene, predators that may turn the tables on the prey's enemy or at least interfere with it, sometimes enabling an about-to-be-dispatched animal to escape in the confusion. This hypothesis requires first that predators be attracted to fear screams—and they are, as Högstedt showed by broadcasting taped screams of a captured starling, a tape that had hawks, foxes, and cats hurrying to the recorder.

Furthermore, the attract-competing-predators hypothesis produces the prediction that birds living in dense cover will be more prone to give fear screams than birds of open habitats. In areas where vision is blocked, a captured bird cannot rely on other nearby predators to see it in its hour of need. Therefore, as a last ditch effort to avoid death, it can attempt to attract competitor predators to the scene with calls. When handled, mist-netted birds that live in dense cover are more likely to give fear screams than species that occupy open habitats [369].

Fighting Back   A starling in the grip of a sparrow hawk is reduced to a kind of doomsday device, calling for other killers on the very outside chance that this will create enough confusion so that the starling can slip away. Before reaching these desperate circumstances, many animals attempt to deter capture by physically resisting their enemies, a potentially effective option even for a small animal like the black widow spider.

Black widows abound in my tool shed, and I once found a big female with an elaborate web living comfortably under our kitchen table. I, like most Arizonans, treat these spiders with a mixture of fear and respect because of their toxic venom, but other animals are far less intimidated. It takes time for the spider to maneuver itself into position to use its fangs, and a small mammal, like a deer mouse, can immobilize a maneuvering spider with its own quicker bite. Deer mice find black widows completely edible, and in the laboratory they have been shown to capture this prey safely. But about half the interactions in an experimental arena ended with the mouse repelled and the spider alive [799]. In a large majority of these cases the cause of the spider's survival was traced to the rapid secretion and deployment of a strand of silk, which the spider produced as soon as it was jostled (Figure 15). The silk strand was adorned with droplets of a profoundly adhesive substance. With its hindlegs, the spider applied the strand to its attacker's face. When this happened, the mouse typically recoiled and attempted to clean itself of this material, sometimes by rolling frantically on the ground. In nature, the spider would have easily escaped during this time by scrambling from its web to a safe crevice.

Rich Vetter tested the importance of the sticky silk strand by blocking the spinnerets of some spiders with wax. When attacked by mice, these individuals were unable to defend themselves with adhesives and were three times as likely to die as unclogged spiders. Although the glue is not toxic (consumption of the viscous substance did not affect the mice), it is sticky and may require immediate removal if it is not to interfere with the protective and thermoregulatory properties of the mouse's pelt.

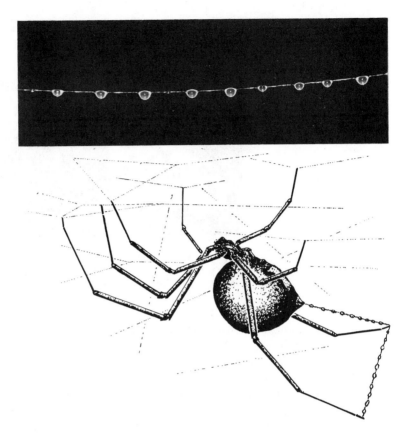

15 **Black widow spiders** (below) apply a sticky strand of silk (above) to predators. Courtesy of Richard Vetter.

Chemical Defenses

Insects as well as spiders practice chemical warfare by spraying, injecting, wiping, or regurgitating toxic, sticky, irritating, or bad-tasting chemicals on predators [228]. Consider the larva of a notodontid moth that lives in Western Australia. I spotted the big caterpillar as it was feeding on the leathery leaf of a banksia tree. When I went over to take a closer look, the caterpillar suddenly dropped its head and abdomen from the leaf and formed an inverted U. Undeterred I touched the animal, at which point it everted a complex, red, antler-like sac from the underside of its body near the head; with this device, called an osmeterium, the larva proceeded to spray me with a strongly scented aerosol of acetic acid (Figure 16).

I do not know what natural predators the moth larva douses with its acetic acid spray, although I guess that the chemical defense works against some birds and perhaps also against ants, which are major enemies of their fellow insects. Ant predation seems responsible for the chemical-behavioral defenses of an Asian relative of the honeybee, *Apis florea*, which builds an exposed honeycomb on a limb in low vegetation (Figure 17). Worker bees

16 **Chemical warfare.** This moth larva can spray acetic acid from the osmeterium (arrow) on the underside of its body. Photograph by the author.

coat the branch on which the nest is supported with rings of a chemical substance so sticky that ants cannot walk to the colony without becoming entrapped [712].

Other insects with ant enemies have convergently evolved similar sorts of ant guards and ant-off substances (Figure 17). For example, a chemical defense system occurs consistently within a group of wasps whose females start nests by themselves and must leave the nest unguarded at times when they are off foraging. In closely related species whose females found a new colony with large numbers of workers in attendance, the use of ant guards has not evolved, presumably because the many aggressive workers directly deter ants from trying to invade the nest [406, 440].

Chemical deterrents are also well-represented in vertebrates, as anyone who has had the misfortune to irritate a skunk knows all too well. Salamanders are less famous for their chemical defenses, but some species employ adhesive repellents that are functionally similar to those used by the black widow. When certain salamanders are grasped by a garter snake, they writhe and thrash while releasing secretions from their tail and body. Because the secretions are sticky, the salamander's tail tends to adhere to the body of the snake. The goal of the salamander may be to frustrate attempts to be eaten long enough to induce the snake to release it, somewhat the worse for wear, but still alive [30]. Occasionally, a salamander may even succeed in coating its enemy so thoroughly with tar-baby glue that the snake is rendered helpless (Figure 18), a most desirable outcome from the salamander's perspective [30].

Still other salamanders have evolved toxic skin secretions, rather than adhesives [100]. The familiar red eft of moist woodlands can cover its brilliantly crimson body with a deadly tetrodotoxin exudate [643]. Birds appear to be sensitive to this poison and quickly learn to avoid the salamander [390].

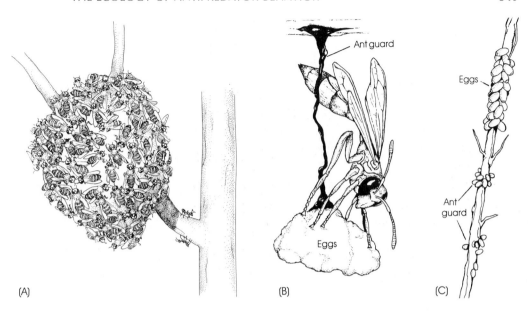

17 **Convergent evolution in ant guards.** (A) An Asian honeybee coats the approaches to its exposed nest with a sticky ant trap. (B) Paper wasps rub the pedicel of their nest with a chemical ant repellent. (C) Owl fly females ring the grass stalk on which their eggs are glued with egglike objects that guard the eggs from ants. After Seeley et al. [712], Jeanne [405], Henry [359].

18 **Salamander wins.** A small garter snake has been glued together by the sticky secretions of a salamander it attempted to consume. Photograph by Stevan Arnold.

Warning Coloration

Warningly colored animals often behave in ways that enhance their already conspicuous appearance (Figure 19). Striking displays that show off the reds, oranges, yellows, and black patterns might remind experienced predators of past unpleasant experiences with other noxious, similarly colored prey. This reasonable idea has been tested by seeing whether young chickens learn to avoid conspicuous prey faster than cryptic ones of equal noxiousness. After having been fed for a few days on food crumbs dyed blue or green, some chicks were then offered a mix of bitter, quinine-soaked crumbs on either a blue or a green arena floor. An aversion to green was more quickly formed on a blue background, and fewer blue baits than green ones were taken when the substrate was green [288]. If it is generally true that animals can learn more quickly to avoid things that are conspicuous, the adaptive value of warning coloration becomes more obvious.

Warning coloration may also promote survival simply because predators are innately more cautious in dealing with bright color patterns [841],

19 **Warning coloration and behavior.** A tropical wasp that combines bright warning coloration with social defense (left). Photograph by W. D. Hamilton. Ladybird beetles (right) are familiar examples of brightly-colored, noxious animals that often cluster together, thereby enlarging the effect of their conspicuous appearance. Photograph by the author.

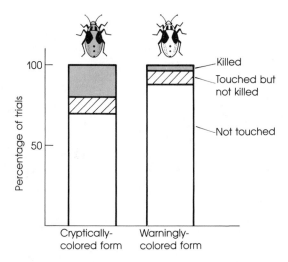

Percentage of trials

100

50

Cryptically-colored form   Warningly-colored form

Killed

Touched but not killed

Not touched

20 **Advantage of warning coloration** for a true bug. Birds were more reluctant to touch brightly colored red forms of a true bug than cryptically colored grey forms. Moreover, they killed far fewer warningly colored types. *Source:* Sillén-Tullberg [722].

perhaps as a result of the fact that warningly colored animals as a group tend to be dangerous prey. This hypothesis has been tested by Birgitta Sillén-Tullberg, who took advantage of the fact that a species of a lygaeid bug comes in two color forms, one with red spots and the other without these bright warning colors [722]. The flavor of both types is equally unpleasant, a fact established by grinding up the warningly colored and cryptic types, and offering birds gelatine pellets that looked the same but contained a mash of either one or the other bug. Birds rejected both types of pellets at the same rate.

But even though the two types of bugs tasted equally bad, hand-reared great tits were considerably more likely not to touch the brightly colored form (when it was offered to them intact rather than in a nondescript pellet). Even when they did seize the red-spotted bug, they were more likely to drop it before it had been killed than the cryptic form (Figure 20). Therefore, bright coloration predisposed inexperienced young birds to avoid the bugs altogether and to release them alive should they get up sufficient courage to attack.

Batesian Mimicry

Many experiments show that birds can learn to avoid certain prey after having an unhappy experience with them. The ability of predators to learn to reject bad-tasting prey on the basis of their appearance has been exploited by a host of deceptive species [837]. These animals, although perfectly edible, look like a protected species. This is Batesian mimicry, named after its discoverer Henry Bates, a nineteenth-century English naturalist and explorer. Probably every organism with a powerful predator repellent has one or more edible species that mimic it and so deceive some predators.

Batesian mimicry can also work if animals have an innate fear of cues associated with potentially lethal animals (see Figure 18, Chapter 2 on motmots and coral snakes). For example, consider the case of the snake-caterpillars. These large, 2- to 6-inch larvae of several moths respond to being touched by forming a triangular head complete with snake eyes

21 **Mimicry of a predator.** A snake-mimicking caterpillar (right) resembles the Mexican vine snake (left) when the caterpillar lowers the anterior part of its body and changes its body shape to create a triangular snake head complete with realistic eyes. Photographs by James D. Jenkins and Lincoln P. Brower.

(Figure 21). Moreover, the "snake" will strike accurately at the object that touched it [837]. You can bet that this secures the attention of a would-be predator. These mimics may be especially designed to repel lizards, because a transformed larva closely resembles the head of a certain vine snake that is fond of lizard meals. I suspect that lizards do not have to learn through experience that they should stay away from lizard-eating vine snakes.

The effectiveness of Batesian mimicry in color pattern and visual appearance has been demonstrated experimentally many times now [e.g., 101, 102]. But what about acoustical Batesian mimicry? For example, some harmless snakes, like the gopher snake, produce sounds that have a passing resemblance to the rattle of a rattlesnake. Once in the Arizona desert I closely approached a snake I knew perfectly well to be a gopher snake, but when it made a sound like a rattler I suddenly found myself treating it with vastly greater respect.

No one has carefully examined the possibility that gopher snakes enjoy antipredator gains as a result of their vocalization. But a well-studied case of acoustical Batesian mimicry does exist [677]. Burrowing owls (Figure 22) in their nest tunnels give a call that sounds very much like an aroused rattlesnake's rattle. Rattlesnakes also often spend the day in underground burrows, and therefore it is plausible that an owl that sounded like a lethal snake might persuade an unwanted intruder to go elsewhere.

As a comparative test of this hypothesis, Matthew Rowe and his co-workers pointed out that the burrowing owl is the only member of its family that nests underground (i.e., in rattlesnake habitat) and is the only owl in its family that possesses a rattling call.

For an experimental test of the acoustical mimicry hypothesis, the researchers tested whether Douglas ground squirrels discriminated between rattlesnake rattles and other burrowing owl vocalizations. Douglas ground squirrels are burrowing rodents that live in the same areas as burrowing owls and are thought to compete for nest holes with the owls. If burrowing owls produce "rattlesnake" signals, then ground squirrels should treat the mimetic signal with the same respect they show the real McCoy. When captive ground squirrels were given an opportunity to enter an artificial burrow, they were equally reluctant to enter when they heard tape recordings of rattlesnake rattles and burrowing owl hisses, but much less hesitant if they heard an owl's scream-chatter or white noise coming from the burrow (Figure 23). These results support the argument that the owl is an acoustical Batesian mimic of rattlesnakes.

Associating with a Protected Species

Batesian mimics derive protection by looking like a dangerous, poisonous, or noxious species. Other animals use the defenses of well-protected species

22 **Acoustical Batesian mimicry.** The ground-nesting burrowing owl, whose hissing call may mimic a rattlesnake's rattle. Photograph by Stephan Schoech.

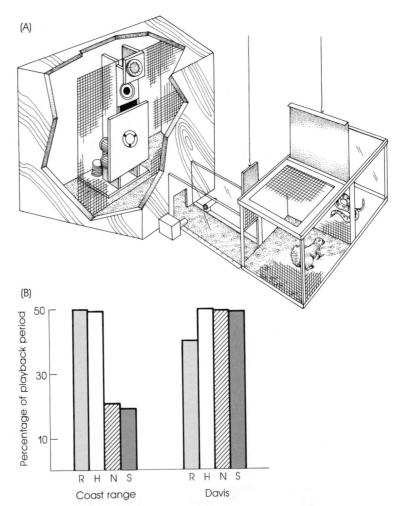

23 **Test of acoustical mimicry hypothesis.** (A) The test apparatus. During acoustical trials, the ground squirrel could not see the rattlesnake but instead heard various sounds coming from the enclosed burrow. (B) Ground squirrels from the coast range area (where rattlesnakes occur) spend more time in the runway leading to the burrow during trials when rattlesnake rattles (R) and burrowing owl hisses (H) are played back to them from the burrow than when undifferentiated white noise (N) and burrowing owl scream-chatters (S) are played. Ground squirrels from Davis, where rattlesnakes do not occur, make no such discrimination between the various playback tapes. *Source:* Rowe et al. [677].

in a different way. For example, some marine slugs (nudibranchs) eat the stinging tentacles of various jellyfishes and corals. The tentacles are covered with cells that fire poisonous hairs and barbs into most of their enemies, but the nudibranchs are immune. In fact, they store any untriggered cells in special sacs and use them against their own predators [882]. Similarly, hedgehogs are able to eat toads with no ill effect despite the extreme poisons in toad skin (which have felled many a naive dog). Hedgehogs often partially skin their victims. One might think that this was done to reduce contact with the toxins in the skin. But after polishing off a toad, a hedgehog may munch on the deadly parotid glands in the skin and then froth at the mouth, licking the combination of saliva and toad poison onto its spines (Figure 24). For good measure, it may also wipe its body with the skin of its prey [99].

The use of a protected species by a prey need not involve the death and recycling of the noxious species. Caterpillars of a lycaenid butterfly are generally attended by a retinue of ants, which feed from "honeydew" glands on the larva's back. At least two hypotheses have been offered to account for the ant–butterfly association: (1) the butterfly feeds the ants to keep them from killing it, and (2) the butterfly feeds the ants to enlist their aid in repelling enemies [633].

The lycaenid does have enemies, particularly small parasitoid wasps and flies, which lay their eggs on butterfly caterpillars. A parasitoid's offspring will eventually consume the caterpillar before it reaches reproductive age. To examine whether being with ants confers protection from these enemies, Naomi Pierce and Paul Mead prevented ants from tending some larvae while permitting other caterpillars to retain their keepers. In order to create a group of untended larvae, they placed a sticky ring of Tanglefoot about the base of a plant on which one or more caterpillars were feeding. Ants are unable to cross this viscous barrier. After establishing the two groups of caterpillars, each larva was inspected daily in order to collect it when it had reached full size and was ready to pupate. The caterpillars were then held until they pupated or until they died from the emergence of parasitoids from their bodies. In two populations, many more parasitic wasps and flies emerged from untended larvae than from lycaenids that had ant guards

24 **Recycled chemical defenses.** Here a hedgehog anoints itself with a foam containing toad toxins, the toad having recently been eaten by the hedgehog. Photograph by E. D. Brodie, Jr.

TABLE 2

The effect of experimentally preventing ants from guarding lycaenid butterfly larvae on the success of parasites of the larvae

Site	Larvae without ants		Larvae with ant guards	
	Percentage parasitized	Sample size	Percentage parasitized	Sample size
Gold Basin	42	38	18	57
Naked Hills	48	27	23	39

Source: Pierce and Mead [633].

with them (Table 2). This experimental evidence, in conjunction with direct observations of ants driving off or killing parasites, shows that *Formica* ants are helpful protectors of these butterfly larvae [633].

Social Defenses   Some animals gain protection by associating with members of another species; many other animals enhance their survival chances by living with members of their own species and communally repelling predators (Figure 25). Nowhere is social defense more fully developed than among the ants and termites. These social insects have a specialized soldier

25 **Cooperative defense.** Social defense by muskoxen (top), which form a ring when threatened by wolves. Photograph by Ted Grant (Information Canada Phototheque). Australian sawflies (bottom) rest in a ring, facing outward. They apply defensive regurgitates on their enemies if disturbed. Photograph by the author.

caste whose members are generally larger than other workers and are armed with formidable mandibles. Their sole function is to protect the colony against dangerous invaders, such as predatory ants or raiding vertebrates. Soldiers communicate with one another via alarm signals, particularly chemical scents, that attract large numbers to a point of combat. The reaction of soldiers to intruders varies among species: they may chop their opponents in half with shearing mandibles, or stab them with piercing jaws, or snap them away with the peculiar "finger-snap" mandibles possessed by some termite species.

Alternatively, a colony's defenders may entangle enemies in the manner of nasute termites, whose soldiers manufacture and store resinous materials in liquid form in a huge gland in its head. When alarmed, the blind soldier somehow manages to point its "nose" toward the enemy and, by contracting powerful head muscles, the soldier fires a liquid entangling thread at the foe, often an intruding ant. The defensive spray also contains alarm substances that draw still more soldiers to the area, where they, too, may engage the invaders [852] (Figure 26).

26 **Social chemical defense.** Nasute termites spraying a fruit fly (in the center) with strands of sticky thread. Photograph by E. Ernst.

Several species of ants practice a variant on this theme. The ant's abdomen consists largely of a gland filled with an entangling glue. In effect, the ant is a living grenade, for it rushes at an enemy, simultaneously constricting its abdominal muscles so violently that its body wall bursts and the gland explodes. The released glue entraps the intruders and prevents them from harming the colony [527].

Alarm Signals

In social insects, defensive substances often act as alarm signals as well as deterrents to attack. Warning other individuals of impending danger could have advantages for the alarm giver if (1) the signal aided the animal's offspring, or other close relatives, as is true for nasute termite soldiers and the like (see Chapter 15), or if (2) the signal enabled the members of a group to coordinate their escape, offering mutual benefits for all concerned as they try to thwart a predator.

Belding's ground squirrels, which live in high densities in mountain meadows, offer a probable case of mutually advantageous alarm calls. When these small mammals see a hawk or falcon sailing in for the kill, they give a specific call, a high-pitched whistle, as they dash for cover. When other squirrels hear the alarm call, they, too, instantly sprint for the nearest shelter, thereby creating a brief moment of pandemonium.

Whistling squirrels are almost never captured, a fact that enabled Paul Sherman to dismiss the hypothesis that callers warn others at risk to their own safety (Table 3). Noncalling animals attempting to escape were significantly more likely to be killed. If callers are safe, why do they call? And if noncallers are at risk, why do they run when they hear the alarm whistle? Sherman argues reasonably that if the first animal to see a incoming hawk failed to call as it ran for safety, it might be targeted for attack, because it was especially conspicuous while moving. By calling, it can get lost in the chaos created as large numbers of squirrels dash in all directions. Noncallers are under selection to run in response to an aerial signal, because if they do not, they will be sitting out in the open with a goshawk or prairie falcon headed their way, not a happy prospect for a ground squirrel. The caller may be safest if he can get others to run, but the noncallers are more safe if they react to the call than if they ignore it, producing benefits for all participating in the signal-giving and signal-responding community.

TABLE **3**

The effect of giving an alarm call to an aerial predator on the survival of Belding's ground squirrels[a]

	Number escaped	Number captured	Percentage captured
Alarm callers	41	1	2
Noncallers	28	11	28

Source: Sherman [716].
[a]$P < 0.01$.

TABLE 4
Effect of group size on the response of caged starlings to a simulated hawk

	Group size	
Behavior	1 Bird	10 Birds
Mean number of times per minute each bird stops foraging to look around	23.4	11.4
Percentage of foraging time spent in surveillance	47	12
Mean time (seconds) from appearance of hawk model to flight of starling	4.1	3.2

Source: Powell [644].

Improved Vigilance

A squirrel in a breeding "colony" or a bird in a foraging flock may be able to flee from a predator that it had not personally detected by reacting to the alarm calls or escape behavior of its companions. In an experimental test of the hypothesis that social animals may gain from improved vigilance, G. V. N. Powell released an artificial hawk model that traveled along a line over a cage in which lived (1) a single starling or (2) a group of 10 of these despised, but adaptable, birds. The average reaction time of the solo bird was significantly slower than that of a bird in a group, despite the fact that single birds spent about four times as much time looking around for danger as a bird in a flock [644] (Table 4).

R. E. Kenward took this test one step closer to reality by releasing a hungry, trained goshawk at a standard distance from flocks of pigeons [422]. As predicted, the larger the flock, the sooner the prey took flight and the lower the probability that the goshawk would make a kill (Figure 27). Although this experiment provides strong support for the vigilance hypothesis, it should be noted that it does not prove that flocking pigeons are more safe in big groups than in small ones. If being in a large flock attracts disproportionately more predators, the benefits gained by improved vigilance could be negated by a greatly increased frequency of attacks.

If animals can gain by associating with members of their own species, perhaps they can also improve their communal vigilance by associating with members of other species. For example, the dwarf mongoose is a highly social little mammal that travels in open African country in search of its prey. Some individuals take turns acting as lookouts that sound a warning to their fellow foragers should they spot a hunting raptor. There may be as many as six mongooses putting in a 15- to 45-minute stint of guard duty at any one moment [659].

Guard duty is risky business because the guards stand in the open on elevated lookout perches, a position that makes them especially conspicuous to predators. Moreover, while watching for hawks, guards go hungry. Anne Rasa noticed that dwarf mongoose often travel with a couple of species of hornbills, strange big-billed birds that feed on flying insects flushed out by foraging mongooses. Hornbills have many of the same predators as their

(A)

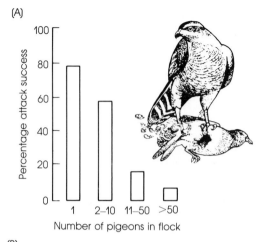

Percentage attack success

Number of pigeons in flock

(B)

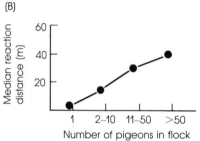

Median reaction distance (m)

Number of pigeons in flock

27 Group size promotes vigilance. The larger the group of woodpigeons, the sooner the flock detected and flew away from an approaching goshawk, thereby reducing the predator's chance of making a kill. *Source:* Kenward [422].

mammalian traveling partners, so when a hornbill flies up in alarm, the mongooses are quick to dive for cover, taking advantage of the cues of danger provided by the birds. Rasa showed that the number of mongoose guards declined when the troops were accompanied by many birds (Figure 28). Hornbills in effect take the place of mongoose guards, thus enabling the members of the band to spend more time feeding and less time risking their lives to announce the arrival of killer hawks.

The Selfish Herd

Mongooses and hornbills appear to warn each other with special alarm calls, with the hornbills going so far as to give their alarm cries when they spot certain predatory hawks that are too small to be a threat to them but big enough to threaten their mongoose companions. Although cases of this sort seem to please people, who take them as evidence of harmony and cooperation in nature, it is entirely possible for groups to form even though individuals never risk a thing for others. There is no indication that starlings or woodpigeons attempt to warn each other of danger. Instead, the birds in a group simply respond to cues provided by their frightened fellows. Thus, what looks like a coordinated flight response of an alarmed flock of birds may be the result of individuals that are trying (unconsciously) to avoid being left behind to face a predator alone.

To emphasize the possibility that what looks like a cooperative group may actually be composed of reproductively competitive (selfish) individuals

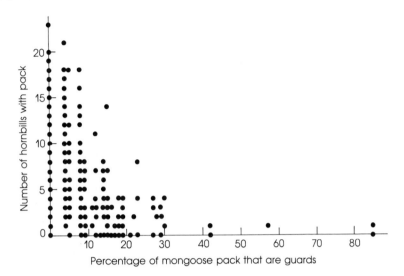

28 **Defensive mutualism.** When dwarf mongooses are accompanied by many hornbills, a smaller proportion of the pack adopts the guard role at any given moment. *Source:* Rasa [659].

that are always trying to keep others between themselves and potential danger, W. D. Hamilton coined the phrase THE SELFISH HERD [332]. Hamilton used a game theory model to make the point that the fitness of any one competing animal is dependent on what the rest of the population is doing. If one animal uses another as a living shield, this favors "shields" who get in position to use others as their protection. The result will be a clumping of individuals, a selfish herd whose members may be no safer than they would be if they all agreed to spread out. But as long as some selfish shield-seekers exist, spreading out cannot evolve.

When Adélie penguins leave their breeding groups to go out to sea to feed, they often gather in groups on rocky ledges by the ocean and then jump into the water together to swim out to the foraging grounds. The value of this becomes clearer when one realizes that by every jumping-off point there is likely to be a leopard seal lurking in the water. The seal can only capture and kill a certain small number of Adélies in a short time. By swimming out in a group through the danger zone, many penguins will escape while the seal is engaged in dispatching one or two unfortunate ones (the DILUTION EFFECT; see later). If you had to run a leopard-seal gauntlet, you would probably do your best to be neither the first nor the last one into the water. Adélie penguins appear to have a similar goal. By going out in the middle of a wave of paddling penguins, an individual can use his companions as a living shield to protect him against the predator, in which case penguin flocks would qualify as selfish herds.

In a previous edition of this book, I suggested that a penguin might go so far as to try to lure another individual into leaping first by lunging toward the water as if he were going in. At that time I was unaware of a claim made in 1914 by M. G. Levick that penguins would sometimes try to push a fellow penguin into the water, presumably to test for the presence of a leopard seal [476]. Levick wrote, "When they succeeded in pushing one of their number over, all would crane their necks over the edge."

Selfish herds need not always be composed of murderers, but in such a group we should at least be able to detect competition for the central positions. Support for this prediction comes from various sources [130, 398], including research on bluegill sunfish [317]. These animals breed in colonies, with males defending nesting territories against rivals. Females visit a colony to lay their eggs. There is intense competition among males for central territories, and larger (generally older) males are more likely to win this competition. Females prefer to lay their eggs in these territories, which are far safer from predators than peripheral nests. It is the peripheral male that is first to confront a foraging bullhead or cannibalistic bluegill, and as a result his eggs are twice as likely to be eaten by a predator as those of a male with a central nest.

If peripheral bluegills are at special risk, why do they accept their inferior position? If a younger or weaker individual has no reasonable probability of forcing a more powerful rival to yield his superior position, then the subordinate animal has two options: to nest on the outskirts of a group and incidentally shield the central animals or to nest as a solitary individual. In the bluegill, solitary nesters (which sometimes occur in this species) experience higher snail infestation rates and chases with predators than peripheral, colonial males (Table 5). Note also that although mobbing of a predator is more likely to occur in the center of the colony, peripheral males also sometimes practice communal defense of their nests, an option unavailable to the solitary nester. A subordinate fish may benefit slightly by nesting in a group even though more dominant males will use him as a living shield against their enemies [317].

The Dilution Effect

Perhaps the simplest antipredator advantage of living in a group is to overwhelm the consumption capacity of the local predators. We can express the dilution hypothesis in terms of Adélie penguins and leopard seals. If a seal can kill one penguin every 3 minutes, then a penguin in a group of

TABLE 5
Predation pressure on bluegill nests

Position of nest	Mean number of snails per nest	Mean number of chases[a] per nest	Group response[b] (%)
Central	6.9	1.5	50
Peripheral	13.7	8.7	8
Solitary	29.7	10.4	0

Source: Gross and MacMillan [317].
[a]Number of times a nest defender is forced to chase a potential egg predator per hour.
[b]The percentage of chases at a nest in which two or more males simultaneously pursued the predator.

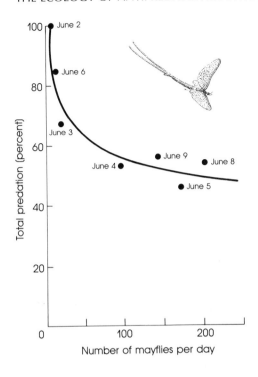

29 **Dilution effect: mayflies.** The more female mayflies emerging on a June evening, the less likely any one mayfly is to be taken by a predator. *Source:* Sweeney and Vannote [757].

100 passing through the predator's foraging zone has a survival chance 10 times better than that of a penguin in a group of 10 (provided each takes about 3 minutes to get through the danger zone).

The dilution effect has been clearly demonstrated for mayflies, tasty aquatic insects that are especially vulnerable to predators during a transition to adulthood, when they are emerging from the water to mate and lay their eggs. Mayflies tend to emerge en masse during a few hours on a few days each year. Bernard Sweeney and Robin Vannote tested the hypothesis that synchrony in emergence was an antipredator adaptation [757]. They placed nets in streams to catch the cast skins of mayflies, which molt on the water's surface just as they change into adults, leaving their molted cuticle to drift downstream on the current. Counting these molted skins gave them an idea of how many individuals emerged on a given evening from the stretch of stream they monitored. The nets also caught the bodies of females that had died after laying their eggs; females expire on their own immediately after having dropped a clutch of eggs in the water, provided a nighthawk or a whirligig beetle does not end their life prematurely.

Because there was day-to-day variation in the number of mayflies emerging, Sweeney and Vannote could measure the difference between the number of cast skins of emerging females and the number of intact corpses of spent adult females that washed into their nets. The more females emerging, the better their chance of surviving to lay their eggs (Figure 29).

It is possible, however, that synchrony of emergence is adaptive because it gives individuals a better chance of finding a mate. But if this alternative hypothesis is correct, then parthenogenetic mayflies composed entirely of

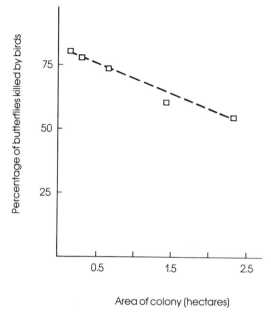

30 **Dilution effect: monarch butterflies.** The larger the group of monarch butterflies in an overwintering cluster in Mexico, the lower the total percentage of those eaten by predators. *Source:* Calvert et al. [126].

females should show less synchrony of emergence than sexual species. They are, however, just as likely to emerge in a swarm over a few days as sexual species, so the dilution hypothesis appears valid for mayflies generally.

Monarch Butterfly Defense Systems   The dilution effect may be responsible for the extraordinary density of monarch butterflies at their overwintering sites in coastal California and mountainous central Mexico (Figure 12, Chapter 9). If thermoregulatory and metabolic considerations were the only factors involved in the formation of monarch aggregations, we would expect to find individuals more broadly distributed than they are. Instead, they are clustered in almost incredible densities (tens of thousands may rest upon a single tree). Because chilled monarchs are incapable of flight, they cannot fly away from their predators. Orioles and grosbeaks are estimated to have killed about *2 million* monarchs in a Mexican colony of something over 20 million butterflies during one winter [104]. Lincoln Brower, William Calvert, and their associates have shown that the risk of dying for any one monarch is inversely proportional to the number of butterflies clustered within an aggregation site (Figure 30) [126]. Moreover, monarchs on the less densely populated periphery of a colony, large or small, have a much higher chance of being eaten than those in the denser core [104].

It is more than mildly surprising to find birds preying so heavily on monarch butterflies, because the butterflies exhibit classic warning coloration (orange and black stripes). Moreover, some monarchs contain toxic cardiac glycosides in sufficient quantities to be dangerous even to a human consumer [103]. In dealing with monarch butterflies, therefore, you would be wise not to follow the practice of the famous lepidopterist E. B. Ford, who wrote, "I personally have made a habit, which I recommend to other naturalists, of eating specimens of every species which I study" [264].

Monarch snacks are also not recommended for some birds. A blue jay that eats a poisonous monarch becomes ill and vomits 15–30 minutes after its meal (Figure 31). In the laboratory, a single encounter that leads a blue jay to vomit is sufficient to educate the predator to avoid the butterfly thereafter. But why is it advantageous to carry toxic chemicals in one's tissues if one must be eaten to enable the toxins to cause illness? This is a general problem that applies to the many other organisms that are warningly colored and cause delayed vomiting or internal injury only when consumed. A vomited monarch does not fly off into the sunset. However, as we saw earlier (Figure 20), warningly colored, bad-tasting animals may sometimes be released by a predator while still alive. In a laboratory experiment using quail as predators, monarchs regularly survived attacks because they were usually quickly released [842]. If the same holds true in nature, selection may have favored toxic individuals simply because they taste bad and not because they cause emesis when swallowed. Vomiting and the learning it promotes may be protective adaptations of birds, which rid themselves of ingested toxins and avoid second mistakes.

Monarchs and Their Food Plants

But we are still without an explanation for why some Mexican birds can feast on monarch butterflies without vomiting. The solution to this puzzle lies in the recognition that there are costs and benefits to any trait, includ-

31 **Effect of monarch toxins.** The blue jay that eats a toxic monarch vomits a few minutes later. Photograph by Lincoln P. Brower.

32 **Recycled plant poisons.** When threatened by a predator, Australian sawflies regurgitate sticky aromatic oils sequestered from the eucalyptus leaves they have eaten. Photograph by the author.

ing being protected by toxins from predators. Monarchs do not manufacture their own chemical deterrents but simply incorporate some poisons that may be present in their foods [103], a trick that has independently evolved in a number of other insects (Figure 32) and even some vertebrates (see Figure 24). Monarch females lay their eggs primarily on milkweeds (Figure 33), despite the fact that some members of this family have evolved powerful toxins (cardiac glycosides) that interfere with many basic cellular activities [103]. Monarch larvae not only consume poisonous milkweeds safely, they store and retain the plant's poisons in their tissues when they undergo metamorphosis to adulthood. Proof that the butterfly acquires toxins from its foods came when Lincoln Brower and his associates "persuaded" some monarch larvae (through artificial selection) to eat cabbage. When the adults reared on this harmless food plant were fed to jays, the birds did not vomit. In contrast, jays that consumed butterflies reared on the milkweed *Asclepias curassavica* always vomited a short time later (Figure 31) [106]. Chemical analysis showed that the glycosides in the milkweeds and in the butterflies reared on them were identical.

Thus, the monarch that feeds on a toxic foodplant benefits from the evolved defenses of the plant. But it probably pays a price as well. The larva must devote some of its metabolic energy to gather the toxins and to store them safely. If it did not have these expenses, it presumably could grow faster or larger, reaching maturity sooner or as a stronger individual. Larger size, greater strength, and earlier maturation could affect fitness favorably.

In fact, monarch females in nature oviposit on a variety of foodplants, ranging from the very poisonous to the completely nontoxic [209]. The selection of a foodplant in areas with several possibilities constitutes an

33 **Monarch butterfly laying an egg** on a milkweed. Photograph by Fred A. Urquhart.

evolutionary game between competing females. The best option for a female will depend on several factors, including what the other females in her population are doing. If many are laying on toxic foodplants, those females that lay eggs on edible food plants will produce edible offspring that may nevertheless be avoided on sight by predators familiar with conspecific toxic monarchs.

But this brings us back full circle to the Mexican monarch eaters. These birds apparently have learned that there is variation among monarchs in palatability. They overcome "automimicry" by throwing away some (toxic) monarchs after pecking at them while eating others after checking their taste. Moreover, they may avoid whatever toxins are present in some mildly poisonous monarchs by eating just the abdomen and the muscles of the thorax, which contain the lowest concentrations of poison [260]. No defensive system is perfect. Endless evolutionary "wars" between foodplant, herbivore, and carnivores have produced complex tactics and countertactics of exploitation and defense—and many entertaining problems for modern observers of these battles.

SUMMARY

1 Antipredator behavior is a dominant feature of the repertoire of most animals. Antidetection adaptations help individuals avoid interactions with their predators; anticapture adaptations help individuals reduce the probability of being captured, once detected.

2 A common form of antidetection behavior is cryptic behavior in which an animal enhances the effectiveness of its camouflaged color pattern by selecting the appropriate resting background or by removing cues that predators might use to locate it.

3 The spectrum of anticapture responses include (a) vigilant behavior, (b) misdirecting a predator's attack to an expendable body part, (c) responses that make a predator hesitate during an attack, and (d) mechanical or chemical repellents. Examples of convergent evolution are numerous within these categories.

4 Individuals often use other animals to improve their chances of survival. This may involve exploitation of the warning coloration and behavior of a toxic species by an edible one (Batesian mimicry), a strategy that may deceive some predators. Alternatively, the members of a less well protected species may join or use a portion of a member of another species for their own defense.

5 Members of the same species may form groups that have antipredator benefits including (a) group attack on a predator, (b) cooperative flight from a predator, (c) increased vigilance at lower individual cost, (d) the use of fellow group members as a physical shield against predation, and (e) the union of forces to swamp the feeding capacity of local predators.

6 The monarch butterfly's antipredator responses illustrate that all techniques for dealing with predators have costs as well as benefits, that no tactic works equally well against all enemies, that antipredator behavior can have effects on other elements of an animal's behavioral repertoire (such as oviposition site choices), and finally that what works best for any one individual may depend on the behavioral characteristics of other members of its species.

SUGGESTED READINGS

Malcolm Edmunds' *Defense in Animals* [222] and Wolfgang Wickler's *Mimicry in Plants and Animals* [837] contain useful reviews of many amazing examples of behavioral defenses.

Thomas Eisner's article on chemical defenses of arthropods [228] and Lincoln Brower's articles on butterflies [103, 105] document the remarkable antipredator adaptations of these creatures.

Bernd Heinrich's article on how predation has affected the evolution of caterpillar feeding behavior [354] and Tim Caro's papers on stotting [135, 136] offer good examples of how to test hypotheses on the possible antipredator value of behavior.

DISCUSSION QUESTIONS

1. Most North Americans have seen white-tailed deer flip up their tail (tail-flagging), exposing the conspicuous white underside of the tail. Construct a list of hypotheses on the possible antipredator function for tail-flagging and a list of predictions that would enable you to discriminate among the various candidates. After completing your answer, check references [65, 365].

2. Bombardier beetles earn their name by firing a boiling mixture of quinones and hydrogen peroxide in the face of ants and other predators that molest them [228]. Use Ydenberg and Dill's cost–benefit approach [875] to develop testable predictions on when the beetle would *not* fire upon being contacted by a potential predator. (Keep in mind that the beetle has only so much defensive material stored in its glands, and that it must metabolize the substance from foods it consumes.)

12

The Ecology of Sexual Reproduction and Parental Care

However skillful an animal is at finding good living space, foraging efficiently, or repelling predators, these abilities only count in the long run if the individual succeeds in passing on its genes. If an animal fails to do this, it will have no lasting influence on the evolution of its species. Therefore, reproductive behavior is the central focus of natural selection. Typically, animals propagate their genes by making copies of them and donating the copies to an offspring. But there is a bewildering variety of ways to do this, and the goal in Chapters 12 through 14 is to show how one can make some sense of this variety. This chapter begins by describing the reproductive behavior of the satin bowerbird to illustrate some of the major questions that will be covered in the three chapters on reproduction: Why do animals reproduce sexually? Why is there variety in the parental tactics of animals? Why do males and females differ in so many aspects of their reproductive behavior? Why do mating systems vary from species to species? This chapter explains why modern biologists have trouble understanding why sexual reproduction ever evolved. The focus then shifts to an examination of the adaptive basis for parental care in some sexually reproducing species, why females more often than males provide care for their offspring, and why there are exceptions to this "rule." The chapter concludes with an analysis of some parental attributes that appear to violate the expectation that animals will behave in ways that promote their reproductive success.

The Puzzle of Sexual Reproduction The most basic question of all for any sexually reproducing species is, Why bother with sex? Look how complicated it is for Australia's satin bowerbird, whose sex life has been studied thoroughly by Gerald Borgia [82–86]. Male satin bowerbirds construct a display bower without which they cannot hope to attract a mate. Although not nearly so elaborate as the bowers of some of its relatives, the satin bowerbird's bower is an attractive piece of work, a carefully interwoven platform of twigs from which rise two parallel walls of twigs tightly aligned and arching over to create a little avenue (Figure 1). Males spend hours making sure that the bower is in good condition, painting a mixture of saliva and charcoal or plant material on the twigs, and gathering blue parrot feathers and yellow blossoms from the surrounding forest with which they ornament the entrance to the bower.

I saw my first male satin bowerbird at his bower in a forest preserve near Canberra, Australia, having been told where I could find a male accustomed to prying people. The satiny blue-black male soon flew into view—carrying a rubber band to add to his collection of decorations, which included blue parrot feathers, blue ballpoint pen caps, and other colorful plastic items. The man-made detritus made the bower less attractive to me, but it highlighted the oddness of the tendency of males to gather objects purely for the decoration of their display sites. In fact, many decorations are stolen from the bowers of other males, even though the birds are

1 **A male satin bowerbird** at his bower. Photograph by Bert and Babs Wells.

extremely aggressive and territorial. Given a chance, males will also wreck other birds' bowers totally. Intruders will even interrupt courtships and attack copulating pairs.

When a female cautiously comes to a bower, she is treated to a dramatic dance by the male, who bounds about her singing noisily while holding a prized decoration in his beak. The extraordinary courtship song of satin bowerbirds consists of various buzzes and other mechanical sounds followed by a bout of mimicry in which the male imitates the songs of other species, including the hooting of laughing kookaburras.

The elaborate show of visual and vocal displays notwithstanding, most females fly away before the courting male can copulate. Females reject male after male before finally settling on a partner that is permitted to mate when the female crouches in his bower and tilts her body forward. After a single copulation, she goes off to nest in an area that may or may not be in her mate's territory. In either case, he does nothing to help the female while she single-handedly incubates the eggs, feeds the young for many days, and shepherds them about for a time after they have fledged. While the female cares for her brood, the male continues his displays. In one exceptional case, a male succeeded in mating with 33 females in a single breeding season, although most other males copulated with few or no females.

The sexual behavior of satin bowerbirds is unquestionably strange in its details, but it exhibits the same basic elements of sexual reproduction shown by many other animals. There are two sexes in bowerbirds. Males, by definition, are those individuals that produce sperm, the smaller gametes, tiny in fact, barely more than a set of genes from the male in a package just large enough to contain the energy needed to drive the genes to an egg. Females, by definition, are those individuals that produce eggs, the larger gametes, which in birds and most other animals are enormous relative to male sperm (Figure 2). A male bowerbird expends much time and energy in attempts to introduce his sperm into females, where these sperm have a chance to fertilize a few eggs. A handful of males succeed in mating often in a breeding season; many males fail to copulate even once. All or almost all mature females in the population mate, and they invest primarily in the care of their offspring.

The key evolutionary question is, Why do female satin bowerbirds bother with males? How much simpler it is for animals that reproduce asexually (Figure 3). If the point of life is to inject one's genes into the next generation, what more direct way to do so than to run off photocopies of one's genome and place them in offspring? Imagine a mutant bowerbird that reproduces parthenogenetically. All of her offspring will be capable of reproducing on their own (and so can be considered egg-producing females), whereas the sons of a sexual female must mate with sexual, egg-producing females in order to propagate their mother's genes. If we assume that the sons and daughters of a sexually reproducing female require the same amount of maternal resources per gamete, then for every son produced, a sexual female loses one daughter. If one half of the progeny of the sexual female are sons and the other half daughters (which is the typical case), then

2 **Hamster egg being fertilized** by a hamster sperm. The photograph (magnification ×1640) illustrates the size difference between male and female gametes. Photograph by David M. Phillips.

3 **Asexual reproduction.** A parthenogenetic female aphid giving birth to an offspring that is sexually identical to the mother. Photograph by E. S. Ross.

parthenogens should have twice as many surviving daughters as sexual females [532]. (We assume that a parthenogenetically produced daughter has the same survival chances as a sexually produced one.) Each of the parthenogenetic female's daughters should also have twice as many female offspring as the genetically distinct sexual females in the population. The proportion of parthenogenetic females will therefore rapidly increase (if our assumptions are correct), eventually leading to the extinction of the sexually reproducing females and males (Table 1).

Now, although most sexual species are similar to bowerbirds in that the male's contribution to his offspring consists of little more than a set of genes, there are animals in which the two "sexes" produce gametes of equal size. (In this case, there are no males or females, but there can be sexual reproduction if the gametes are produced by means of meiosis [Chapter 3] and fused to produce new individuals.) In many other species, males offer parental care or nutritional benefits to their offspring. In this case, when a female accepts sperm from a male that she uses to fertilize her eggs, she is receiving more than just genes. The other benefits could more than double her production of offspring and so enable her to outreproduce a parthenogenetic competitor.

But as G. C. Williams points out, this situation would still be vulnerable to invasion by a mutant parthenogen that would accept the large gamete or other material goods from a donor "male" but would not incorporate the genes of the donor in her offspring [847]. Such an individual would double her production of offspring and also double the propagation of her genes by avoiding meiosis and the reduction in chromosome number in each gamete that this entails. A mutant allele whose female carriers accepted assistance from male partners while rejecting their genes should spread throughout the population, leading to a parthenogenetic species.

Thus, even when males offer females more than their genes, there would seem to be great advantages for females that reproduce asexually [846]. Yet, as we are all very much aware, sexual reproduction is still with us. Why is this so?

TABLE 1

Advantage of parthenogenesis if sexual and parthenogenetic females produce the same average number of offspring

| Generation | Sexual individuals | | Parthenogenetic females | Proportion of females that are parthenogenetic |
	Males	Females		
1	49	49	1	1/50 = 0.02
2	49	49	2	2/51 = 0.04
3	49	49	4	4/53 = 0.08
4[a]	45	45	7	7/52 = 0.13
5	43	43	13	13/56 = 0.23
⋮	⋮	⋮	⋮	⋮
n	0	0	100	1.00

[a]To keep the total population of this species about 100, across the board cuts of 10 percent and 5 percent were made in generations 4 and 5, respectively.

Individual Selection and Sexual Reproduction

It is well known that sexual reproduction results in the continual formation of novel genotypes. In the making of gametes during meiosis, the genetic information in the original reproductive cell is thoroughly scrambled because of recombination. As a result, an egg or sperm that is produced may carry any one of a huge number of possible combinations of genes. The genotypes that are produced by the union of eggs and sperm are virtually guaranteed to be unique (except in identical twins). The fact that almost every individual has a different combination of genes in a sexually reproducing species ensures a substantial amount of diversity in the visible characteristics of the members of the species.

Thus, sexual reproduction produces variant individuals that may, in times of rapid environmental change, survive and reproduce while most others die. This could have the effect of perpetuating populations that would otherwise have become extinct had the parental generation been reproducing asexually and thus not generating new variation in each generation. Indeed, many biologists have suggested that sexual reproduction evolved because it ensures the future survival of species. This *group selection* argument, however, supposes that individuals are gratuitously providing a certain number of variant offspring for the sole purpose of establishing a genetic bank for the long-term advantage of the population. If such foresighted individuals did exist, logic suggests that they would soon be eliminated by other genetically different types that acted ("selfishly") to maximize their immediate reproductive success (Chapter 1).

There are more sophisticated group selection hypotheses on sexual reproduction [532, 846], but let's consider the possibility that sexual reproduction confers a fitness advantage to individuals. If under some circumstances, sexually reproducing offspring survive better or reproduce more than asexually produced ones, then we would be able to understand why sex is maintained in populations despite the apparent benefits of parthenogenetic reproduction.

There are a number of hypotheses on why individuals might benefit by reproducing sexually. We shall consider two: the *lottery hypothesis* and the *coevolution hypothesis*. George C. Williams has argued that sexual reproduction is analogous to the strategy one employs in a lottery [846]. You do not usually win a prize in a lottery by buying a large number of tickets with exactly the same number. Instead, lotteries favor the possession of diverse tickets. The environment of some species, but not all, acts like a lottery in that the environment changes from place to place, from season to season, and from year to year. If one's offspring disperse to find relatively unexploited areas, they may not find a place with conditions identical to those a parent experienced. Therefore, an individual that reproduced asexually might very well have progeny beautifully suited for conditions that were impossible to find or that no longer existed (even in the place where the parent had lived).

For species that live in variable habitats, the vacant slots in the environment that potentially can be exploited by offspring represent the different prizes in a lottery. Winning tickets are those offspring that find and are capable of filling these variable slots. By reproducing sexually, an individual places different samples of his genotype in a variety of different progeny, who will have a chance to distribute themselves over a range of habitats rather than all being highly dependent on the same set of ecological conditions. And the offspring of one individual will by virtue of their differences be less likely to compete with their siblings for limited resources than would asexually reproduced offspring, all of which share the same genes and same adaptations for the same environmental conditions.

The Coevolution Hypothesis

The lottery hypothesis focuses on the unpredictably variable *physical* environment of some species. In contrast, the coevolution hypothesis proposes that sexual reproduction is favored by the evolutionary interaction between a species and the other species that it seeks to exploit or that seek to exploit it. An animal's *biotic* environment can coevolve with it as the changes in species A have selective consequences for species B.

For example, imagine the following evolutionary game between a parasitic species and its host. If the host were to reproduce asexually, then those genotypes in a parasitic species that best attacked the dominant asexual genotype would spread through the parasite species. Should a sexual form arise in the host species, however, it might produce some offspring whose distinctive combinations of genes conferred resistance to the common form of the parasite. These different forms would enjoy high reproductive success, but as the new successful combinations became common, selection would favor sexual parasite parents that happened to produce some offspring whose gene combinations enabled them to exploit the now common form of the host. Sexual reproduction by parasites in turn favors parents in the host species that have offspring with uncommon gene combinations that escape the increasingly common parasite type [335].

The host and parasite exert FREQUENCY-DEPENDENT SELECTION on each other (Figure 4). As any gene combination becomes frequent in either the host or parasite population, it loses its fitness advantage. Whatever is rare enjoys a reproductive edge. Sexual reproduction, by constantly breaking up gene combinations and creating new ones, could be favored over asexual reproduction, which conserves a given gene combination, in an environment

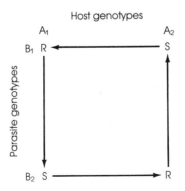

4 **Host–parasite interactions.** Whichever host or parasite genotype is rarer will enjoy a selective advantage. The arrows point in the direction that selection will move the population. A_1, A_2, host genotypes; B_1, B_2, parasite genotypes; S, susceptible; R, resistant. *Source:* Clarke [144].

dominated by frequency-dependent selection. Coevolutionary races between parasite and host, or between predator and prey, could be responsible for the maintenance of sexual reproduction.

The lottery hypothesis and the coevolution hypothesis can be tested by examining the predictions they make about where we would expect to find sexual and asexual species. If unpredictably variable physical environments favor sexual individuals because they produce variable offspring that may find a special niche for their unique traits, then the less stable the environment, the more likely we are to find sexual species. For example, consider the following aquatic habitats: temporary rock pools that form after a heavy rain, seasonal ponds and streams that dry up over a period of months, permanent streams and rivers, and seas and oceans. Temporary rock pools provide conditions that are surely more unpredictable, variable, and ephemeral than those found in a river or ocean. The lottery hypothesis predicts that sexual species will predominate in ephemeral and seasonal aquatic habitats and that asexual species will be more common in the larger, more permanent bodies of water.

In contrast, the coevolution hypothesis predicts that sexual species will be found in environments in which biotic interactions are most numerous, thus leading to some that are complex and intense. In an ephemeral rock pool the number of species is far fewer than in a river; the number of different parasites and predators with which an individual of any one species is likely to interact is therefore much reduced. Under these conditions, there should be a reduction in the advantage to producing genetically distinctive offspring to escape coevolving enemy species; asexuality should prevail.

Asexual species do dominate ephemeral habitats, whereas in more stable environments like large lakes, tropical forests, or the open ocean, one is far more likely to find species that reproduce sexually. In general, asexual species occur in environments that have reduced species diversity and that undergo great fluctuations in physical conditions, like desert rock pools, or mountaintops, or tidal zones. These facts contradict the predictions of the lottery model while supporting the coevolution hypothesis [54].

What about those species that alternate sexual and asexual generations during the year? The lottery hypothesis requires that the production of offspring by sexual means should occur when the resulting offspring will face the least predictable and most variable physical conditions. For some species this seems to be the case. For example, in a species of freshwater *Hydra* that lives in the northern United States, individuals reproduce asexually as newly hatched polyps in the spring and throughout the summer [180]. When fall arrives, the hydras begin a sexual phase, producing eggs and sperm (Figure 5). When an egg is fertilized, an embryo is formed; this embryo detaches itself from the parent and attaches to the substrate. It then spends the winter in a quiescent state before hatching out in the spring and beginning the phase of growth and asexual budding. Thus, an individual hydra produces variable offspring at just that time when the conditions its progeny will face are least predictable (because after a winter has passed the aquatic environment may be very different from its state the previous year).

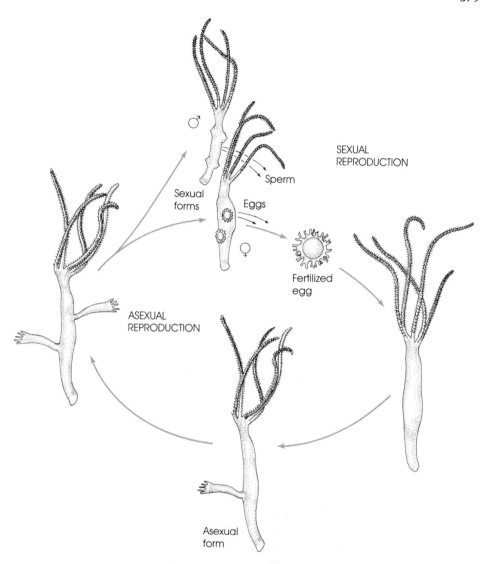

5 **Annual cycle of a *Hydra*** exhibits both sexual and asexual generations.

Although this one case is consistent with the lottery hypothesis, a wide-ranging survey of species that have both asexual and sexual phases of their life history reveals that the general rule is that sexual reproduction occurs at peak population densities. It is then that parasites are likely to be most numerous, thus conferring the greatest advantage on individuals able to produce genetically novel offspring that may escape the parasites that have evolved a capacity to exploit the common genotype in the parental generation [787].

A New Zealand snail has both parthenogenetic and sexual members simultaneously intermixed in its populations, but the proportion of sexually reproducing individuals, as measured by the proportion of males, is greater

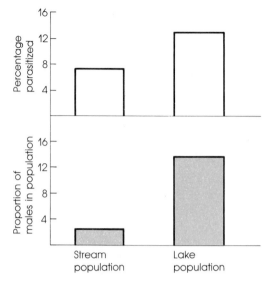

6 **Parasite pressure** and the frequency of sexual reproduction in a New Zealand snail. More males (and thus more sexual reproduction) occur in lakes where parasites are more numerous. *Source:* Lively [489].

in lakes than in streams. The percentage of specimens that are infected with parasites is greater for snails from lakes than for stream snails (Figure 6), an observation suggesting again that biotic complexity, particularly as represented by parasites, favors sexual reproduction [489].

Currently, the coevolution hypothesis appears stronger. Environmental unpredictability per se is not likely to be the selective factor that has favored the evolution of sex, but the last word has not been said on this fundamental problem in evolutionary biology.

Parental Investment and the Reproductive Strategies of the Sexes   Whatever the advantages of sex, there are many profound evolutionary consequences linked to this mode of reproducing. The heart of the matter is that male and female reproductive strategies are basically different. The essence of femaleness is the production of large gametes, each one of which contains far more energy than is present in a sperm made by a male of the species. A female bowerbird allocates the energy available to gamete production over her lifetime to a handful of mature eggs; males make vast numbers of tiny sperm with the resources they devote to gamete production. In birds it is not uncommon for a single egg to weigh 15–20 percent of the weight of the female, and some go as high as 30 percent [463]. With the same amount of resources, a male could produce trillions of sperm. Male birds often do make and expend millions of sperm in a single insemination attempt, but the male that used sperm equal to 5 percent of his body weight in a single season would be a sexual athlete of the first degree.

Among mammals, the same correlation between gamete size and numbers produced holds true. A human female makes only a few hundred cells that can develop into large eggs [165]. In contrast, a single male could theoretically fertilize all the women in the world because he can produce billions of minute sperm in his lifetime.

The difference between males and females of a species in their allocation of resources per gamete can be expressed as a difference in their PARENTAL INVESTMENT per offspring [784]. Parental investment is whatever increases the probability that an existing offspring will survive to reproduce at the cost of the parent's ability to generate additional offspring. Males and females probably have access to equal amounts of energy to invest in reproduction. Therefore, the female's "decision" to allocate relatively more energy per gamete means that she sacrifices opportunities to produce additional gametes in favor of increasing an egg's chances of developing after it is fertilized. The typical male makes a sperm that contains little or no resources that will help the zygote develop. Thus, the male's parental investment per offspring may be zero, whereas the female's parental investment per offspring will usually be substantial.

The concept of parental investment was invented by Robert Trivers, who noted that there is more than one way in which a parent can sacrifice for an existing offspring [784] (Figure 7). The female bowerbird not only allocates considerable energy to each gamete, she also goes to the trouble of making a nest, incubating the eggs, and finding quantities of cicadas and other insect food for each offspring. The energy spent, time lost, and risks taken on behalf of one fertilized egg, at a cost to the female of future chances to reproduce, all constitute parental investments on her part.

One of the clearest examples of the gametic cost of helping an existing offspring comes from those tropical frogs whose females feed their tadpoles unfertilized eggs. In so doing, they forfeit some potential offspring to assist already living progeny [650]. Female animals often make these kinds of postfertilization investments to promote the welfare of their progeny, but males do so much less frequently. In many mammals, for example, the male impregnates a female and then departs. The female nourishes the embryo(s), and then cares for and feeds her offspring with milk after they are born. These activities place a heavy energy burden on females, and they may lead to her death, either through increased physiological stress or through heightened exposure to predators. The maternal care given to an embryo or dependent offspring is often the major component of a female's parental investment.

Why Is Parental Care More Often Provided by Females?

Although there are interesting and useful exceptions, surveys of the animal kingdom reveal that in species in which only one sex provides parental care, females are more likely than males to be the care providers. One might be tempted to think that this stems from the fact that females make a larger parental investment in each gamete and are therefore likely to invest more to prevent the loss of what they have already tied up in each fertilized egg. This, however, is what has been called the CONCORDE FALLACY in honor of the British and French governments' decision to throw good money after bad in the development of the Concorde aircraft. These governments would have saved money if they had aborted their investment in the Concorde in midstream, even after having already spent billions, when

7 **Parental investment takes many forms.** A male frog carries his tadpole offspring on his back (top left). Photograph by Roy McDiarmid. A crocodile mother transports her babies to water after they hatch from her nest (top right). Photograph by Jonathan Blair. A female cicada-killer wasp struggles with a cicada she has paralyzed with a sting (bottom left). She is transporting it to a burrow, where she will feed it to her offspring. Photograph by the author. Male danaid butterflies feeding on a plant containing toxins, which they may store and pass with their sperm to their mates (bottom right). Females may incorporate the toxins in their eggs to protect them against predators [76]. Photograph by Michael Boppré.

it became apparent that the Concorde could only lose money after it became operational.

For a parental animal, the central question is a calculating one: Will an additional investment in an existing offspring yield a higher lifetime fitness return than a new investment in a future attempt at reproduction [694]? This choice can be viewed as a problem in optimality, and the solution will not necessarily be to invest more in the existing offspring simply because the parent has already committed resources to that individual. Imagine that a parent bird abandons a dependent nestling at a stage when the baby bird has 1 chance in 10 of surviving to reproduce. If by leaving one youngster to die, the adult bird saves sufficient parental investment to ensure a future breeding attempt that will yield an offspring with 2 chances in 10 of surviving, its "callous" behavior is adaptive.

Thus, the fact that a female has already invested much energy in making a large egg does not mean that she should automatically give parental care to the offspring that results from fertilization of that egg. The problem that an individual of either sex must solve (unconsciously) is whether parental care increases the survival chances of the assisted offspring sufficiently to compensate for the loss of future offspring that the parental care demands. Note that this argument assumes, realistically, that animals do not have access to unlimited supplies of time and energy; consequently, helping one individual subtracts from the animal's ability to reproduce in the future.

The "greater past investment" hypothesis cannot account for the prevalence of female parental care, but fortunately it is not the only hypothesis on this matter. Mart Gross and Richard Shine helpfully reviewed and tested three other alternatives [320]. The first of these is the "low reliability of paternity" hypothesis, namely, that females are more likely to be the parent of the offspring they assist than are males. If a female lays a fertilized egg or gives birth to an offspring, this progeny will definitely have 50 percent of her genes. A male has no such assurance, especially if the species practices internal fertilization, in which case a male's partner may have already mated with another individual and have used his sperm to fertilize her eggs. The argument continues—to the extent that a male runs the risk of caring for progeny other than his own, the benefit of parental care falls for a male, thus reducing the benefit-to-cost ratio of male parenting and making its evolution less likely than maternal care.

It is true that in teleost fishes (Figure 8) and amphibians, mode of fertilization is strongly correlated with the nature of parental care, with males rarely providing parental care in cases in which there is internal fertilization (Table 2). This result, however, is also compatible with competing hypotheses (see below) and so cannot be used to discriminate among alternatives. Moreover, the "low reliability of paternity" hypothesis, although intuitively appealing, cannot apply if the reliability of paternity is the same for parental and nonparental males within a population, as it must often be [826]. Imagine a species with a mix of males, mostly nonparental types, but with a few parental mutants. The fact that mutant males take care of some offspring will not in itself affect the "fidelity" of their mate(s). Thus, the reliability of paternity will be the same for the two kinds of males in the population, and its effect on the evolution of the parental behavior by males will cancel out. Paternal care can spread through the population, even if the reliability of paternity is low, provided the paternal male improves the survival rate of those eggs he does fertilize enough to compensate for any reduction in the number of mates he can find.

A hypothetical numerical example may help make the argument clearer. We shall set the proportion of eggs actually fertilized by a male when he copulates with a female at 0.40, a low figure. Because he is largely occupied

8 **Paternal care in two teleost fishes.** Male sticklebacks defend the eggs laid in the nests they build (left). Photograph by William Rowland. Both males and females of the cichlid fish secrete a nutritious substance that their young can glean from the parent's body (right). Photograph by A. van den Nieuwenhuizen.

with paternal care, the average paternal male copulates with only two females (each with an average of 10 eggs), whereas the nonpaternal types acquire five mates (giving them access to a total of 50 eggs). The survival of protected offspring is 50 percent, but only 10 percent for eggs that lack a paternal guardian. The reproductive success of the paternal male will be $0.40 \times 20 \times 0.50 = 4$ surviving offspring that bear his genes; the average fitness of the nonpaternal males is less ($0.40 \times 50 \times 0.10 = 2$). Thus, a low reliability of paternity is not in itself an absolute barrier to the evolution of paternal care by males.

Furthermore, if one uses the logic of this hypothesis alone to predict the occurrence of *maternal* care in relation to mode of fertilization, then all other things being equal, one would predict that the probability of maternal care should be unaffected by the mode of fertilization. Whether a female's eggs are fertilized internally or externally, by one or by three males, the offspring that arise from her eggs are her offspring, and they carry her genes. In fact, however, female parental care is significantly more likely in

TABLE 2

Relation between mode of fertilization and the sex
providing parental care in teleost fishes and amphibians[a]

Sex providing parental care	Internal mode of fertilization			External mode of fertilization		
	Fish	Amphibian	Total	Fish	Amphibian	Total
Male	2	2	4	61	14	75
Female	14	11	25	24	8	32

Source: Gross and Shine [320].
[a]The numbers are the numbers of families with one or more species in a given category. A single family may appear in more than one category.

species that practice internal fertilization than in species practicing external fertilization. This result, and the logical difficulties outlined above, suggest that the reliability of parenthood is *not* the key factor in the evolution of parental care by one sex or the other.

The Order of Gamete Release and the Association Hypothesis

Still another hypothesis is that uniparental care is determined by the order in which eggs and sperm are released. Whichever sex can leave first after "mating" is predicted to do so, deserting the other to care for the offspring, if parental care is to be given. By this argument, internal fertilization should be linked with female parental care, because after donating his sperm internally a male is free to desert his mate, leaving her with the decision to make about whether or not to provide parental care. In cases in which such care was advantageous, the female would be "forced" to provide it. But when fertilization is external, females often deposit their eggs before males shed their sperm, giving the females a brief moment in which to flee, leaving males holding the eggs.

The data in Table 2 are consistent with the gamete order hypothesis, but our confidence in this explanation is weakened because it makes a prediction that is *not* correct. There are many parental species of fish in which males and females release their gametes simultaneously. In such cases, the chance that the male will be the parental sex should equal that of the female, because both are present when gamete release is complete. In a sample of 46 species of this sort, however, males were the parental sex in 36 of the species, significantly more than the 23 predicted by the gamete order hypothesis. Moreover, there are many frog species in which paternal males release their sperm into a nest they have constructed *before* the female places her eggs in the nest.

The association hypothesis for the prevalence of maternal care is perhaps the most straightforward of all [320]. Its simple point is that because females carry the eggs, mothers are often in a position to do something helpful for the offspring when they emerge. The father has no direct control over the birth of offspring by his mate and may not be in the neighborhood when this happens. Even if he could provide helpful parental care and even if it would be in his genetic interest to do so, in many cases his physical separation from his progeny make paternal behavior impossible.

The relation between internal fertilization and female parental care (Table 2) is consistent with the association hypothesis, a correlation rein forcing the point that this result cannot be used to identify which alter native is more likely to be true. There are, however, some fish in which fertilization is internal but whose females lay their eggs immediately after receiving sperm. In such cases, males will still be present when their offspring are released. The association hypothesis predicts that males of these species will be just as likely to be parental as males of species whose females lay their eggs just before they are fertilized externally. This pre diction is correct.

This is not to say that the association hypothesis is certainly true. Note that the association hypothesis also predicts an equal number of cases in which males and females are the parental sex in fishes whose males and females simultaneously release their gametes. The fact that male care is more prevalent in this group suggests that other, yet to be considered forces play a role in the evolution of parental care.

Why Do Male Fishes So Often Provide Parental Care?

As we have seen, the rule that females are the sex that takes care of offspring is a rule with many exceptions, none more helpful in testing competing hypotheses on the evolution of parental care than the exception provided by many male fishes. Mart Gross and Craig Sargent have analyzed the varieties of parental care in fishes to determine what conditions beyond simple association with the young favor uniparental male care [319].

The benefit of parental care is increased survivorship of the young. Any gain of this sort is likely to be the same for a parental male or female Gross and Sargent reasoned that if one sex is parental and the other is not the *costs* of parental care should be unequal. Two categories of cost may well differ for males and females. There is a potential *mating cost* to parental care. By remaining with their offspring, males typically incur a higher mating cost than females, for their reproductive success can be enhanced by finding many females and fertilizing their eggs. A female's reproductive success is limited by the number of eggs she produces, not the number of mates she acquires. For females, parenting with its time and energy demands imposes a *fertility cost*, a reduction in the future production of eggs.

Paternal male fishes are almost without exception territorial, defending a site that females visit to mate and lay their eggs. In a mating system of this type, the mating cost of parental care all but disappears for males. By guarding a territory, they also guard the eggs they have fertilized; the territory and the presence of eggs make the males attractive to additional females, increasing male mating success. This outcome creates a very fa vorable benefit-to-cost ratio for male parental care.

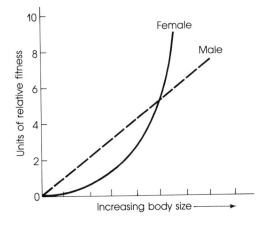

9 **Female fecundity may increase exponentially** with body size in many fishes. If male fitness rises linearly with increases in body size, this difference between the sexes could account for the evolution of exclusive male parental care in fish.

For female fish, in contrast, parenting may involve a steep fertility cost. By devoting herself to one brood, a female is in effect reducing her growth rate because she cannot feed effectively while caring for her current offspring. That loss in body size may be especially damaging for her if her fecundity increases exponentially with increasing body size, as is thought to be the typical case for fishes (Figure 9). In other words, for each lost unit of growth resulting from being parental, a female pays an especially heavy price in the loss of eggs produced in the future. Males that are parental also grow more slowly than they would otherwise, but since to attract mates they must be territorial anyway, the decrease in growth *resulting from parental care* is slight or trivial. Thus, in fishes, male parental behavior is probably the evolutionary result of males paying a lower price for parenting than females, not because they gain more from this behavior in terms of improved survival of their offspring [319].

How to Maximize the Parental Payoff We have reviewed hypotheses on parental care that are based on the assumption that there are trade-offs for animals that are parental, gains to be made and expenses to be paid. By considering what these gains and expenses were and how they might vary between the sexes and in different environments, progress has been made in explaining the patterns of parental behavior that exist in nature. We can extend this approach to still other aspects of parental care to determine whether parents behave in ways that help them maximize the difference between benefits and costs.

Consider, for example, the simple hypothesis that parents will be selective in the distribution of parental care, avoiding offspring that are not their own and investing exclusively or primarily in individuals that carry their genes. A prediction from this hypothesis is that parents will recognize their offspring when the chance exists that they might misdirect parental care to genetic strangers. Because ground-nesting gulls nest in colonies and because their young become mobile after they are a few days old, parents may be confronted with juveniles that are not their offspring. But

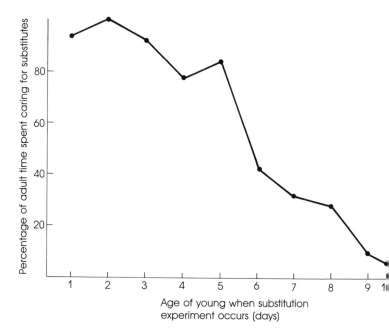

10 **Chick recognition in the ring-billed gull,** a colonial ground-nesting gull. Parent gulls tend to accept other gull's chicks for the first 5 days after their own brood has hatched. But the level of acceptance declines sharply thereafter, as shown in the sharp decline in the percentage of time the adults spend in parental (brooding or feeding) activities with substitutes that are older than 5 days. Adult gulls were checked 20 times per day at five nests and their activity (brooding, feeding young, or standing) was recorded. *Source:* Miller and Emlen [556].

they rarely feed young gulls other than their own, because once their chicks are 4 or 5 days old, parents recognize them and will not accept substitutes (Figure 10).

The fact that chick-recognition is standard practice for many ground-nesting colonial gulls [774] could arise because these species have a common ancestor. Therefore, behavioral similarities among gulls cannot be used to test the hypothesis that recognition of biological offspring is adaptive for a particular species because it reduces the risk of misdirected parental care. One can, however, legitimately test this hypothesis by predicting that any gull species in which there were no opportunities for mistakes in parental care would have diverged from the standard gull pattern and would not recognize its offspring.

We turn to the cliff-nesting kittiwake for a test of this prediction [162]. Because the birds nest on small cliff ledges, often with sheer drops hundreds of feet to rocks and waves below, there has been strong selection favoring chicks that stay put. Kittiwake chicks crouch within their nest, pointed away from the precipice until they are several weeks old and have fully feathered wings. Only then do they leap from the ledge to glide down to

11 **Kittiwakes fail to recognize their young.** Adult kitti-wakes (left) will even adopt the chicks of cormorants (right), which are utterly different from their own offspring in appearance. Photographs by the author.

the ocean, where they begin the next phase of their development. Although in a ground-nesting gull, the parents almost never care for transferred young older than 5 days, and indeed may even kill (and eat) them, kittiwake adults calmly accept strange chicks much older and larger or much younger and smaller than their own. In fact, one can substitute a cormorant chick (the very antithesis of a young kittiwake; Figure 11) and the parent gull will feed the newcomer as if it were its own progeny.

We could further test the hypothesis that offspring recognition evolves when there is a chance that a parent will adopt a stranger by testing the prediction that parental kittiwakes do learn who their offspring are in the *postfledging* phase when the young birds are still partially dependent on their parents for food and protection. At this time the young birds travel about but may return to their nest to be fed by their parents [116]. It is precisely under these conditions that the risk of feeding someone else's begging offspring would be high, favoring parents capable of discriminating behavior. This test remains to be done.

Divergent Evolution
in Fledgling Recognition in Swallows

The hypothesis that the risk of misdirected parental care is linked to the evolution of offspring recognition has also been tested comparatively by looking at two very closely related bird species, bank and rough-winged swallows. These birds not only resemble one another in appearance but also share the habit of nesting in burrows in clay banks and sand quarries. They differ, however, in that the bank swallow is a highly colonial species, whereas solitary roughwing pairs nest by themselves. As it turns out, young bank swallows produce far more distinctive vocalizations than do young

Bank swallow　　　　　Roughwing

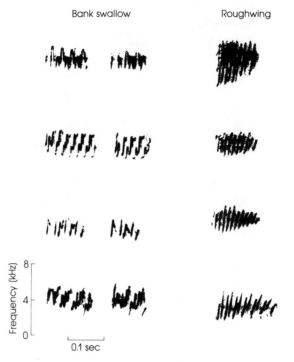

Frequency (kHz)

8

4

0

0.1 sec

12 Parental recognition of young in bank swallows is based on the adult's ability to learn the distinctive properties of the vocalizations of their young. *Source:* Beecher [48].

rough-winged swallows (Figure 12). By the time the fledgling bank swallow emerges from the nest burrow, its parents have learned the key vocal characteristics of their progeny [48, 51]. They use what they have learned to feed just their own offspring while treating strange fledglings that happen to land at their burrow entrance in an unkind manner. In a large colony with many closely spaced entrances, young birds often wind up in "wrong" nests.

The rough-winged swallow, which through its evolutionary history has never had the chance to feed another's fledglings, should not show the same parent–offspring recognition. To test this prediction experimentally, Michael and Inger Beecher transferred fledglings between roughwing burrows and found complete acceptance of the transplants. Rough-winged swallows will even act as foster parents for bank swallows [48].

The divergence between ground-nesting gulls and cliff-nesting kittiwakes and between colonial bank swallows and solitary rough-winged swallows provides multiple lines of support for the argument that offspring recognition evolves when there is a chance that parents will care for someone else's offspring.

Apparent Nondiscriminating Parental Care

There are exceptions to the rule that parents will provide care only to their genetic offspring. In fact, the claim that ground-nesting gulls learned to recognize their 5-day-old young was first made by Niko Tinbergen in his studies of herring gulls. Tinbergen's study and most others like it with ground-nesting species involved the experimental transfer of chicks from one nest to another. It now appears that much of the aggressive, nonparental behavior of adult birds may have stemmed from the "frightened" behavior of the displaced youngsters [306]. Older juveniles that *voluntarily* leave their natal territory (when, for example, they are very hungry, having lost one or both parents) do not flee from potential adopters; instead they beg for food and crouch submissively. Several herring gull researchers have found that these young gulls, even when 35 days old, have some chance of being adopted [375].

Given that these observations fail to support the adaptationist hypothesis that colonial herring gulls should exhibit discriminating parental care, let us consider some alternatives, something that ideally we should have done at the outset. One proximate possibility is that herring gulls use a simple "rule of thumb" to guide their parental behavior, one that goes something like "Feed any chick that begs appropriately and is matched in age with the other chicks in my nest." It is known that most herring gull adoptions occur when the strange chick is able to make its way right to the nest of its host adults and when it is the nearly the same size and age as its hosts' biological offspring [306, 375]. The question of the adaptive value of such a rule of thumb versus the alternative tactic of learning to recognize one's young as individuals remains to be examined in this case. Such a rule, however, would probably be highly effective in herring gull colonies with widely scattered nests. If low-density colonies have been typical for most of herring gull evolution, the risk of misdirected parental care would have been slight, and the advantage of a sophisticated mechanism for offspring recognition reduced. Herring gulls are known to have experienced a population explosion recently, and the adoption studies have occurred in high-density colonies.

There is still another alternative, namely, that herring gulls have only recently become a ground-nesting gull, having previously nested on cliffs where chick recognition confers no fitness benefits [375]. There are even today populations of cliff-nesting herring gulls [60]. According to this non-adaptationist hypothesis, chick recognition would be adaptive for herring gulls, but their historical legacy is such that they have not yet had time to evolve optimal parental behavior. This interesting idea awaits rigorous testing, but its existence reminds us once again of the possibility that not every attribute will be adaptive and of the need to consider more than one competing explanation for a behavior.

For another case of apparently unselective parental care, we turn to the Mexican free-tailed bat. Pregnant females of this bat migrate to certain caves in the American Southwest, where they form colonies of millions. After giving birth to a single pup, the mother bat leaves her offspring clinging to the roof of the cave in a cluster of dozens or hundreds of other youngsters (Figure 13). When a female returns to nurse her pup, she is besieged by babies. Given the shoulder-to-shoulder packing of hundreds, even thousands of pups, early observers believed that females could not possibly relocate their own offspring, and they claimed that females nursed communally. In the group selectionist climate of the times, it was not thought surprising that females would give their milk on a first-come, first-serve basis without regard to the identity of the pup.

13 **A crèche** of Mexican free-tailed bat pups left together by their mothers, who are foraging outside the cave. Photograph by Gary McCracken.

Communal nursing, however, violates the prediction that mothers will be discriminating in providing parental care, and to test this prediction, Gary McCracken decided to reexamine the behavior of free-tailed bats [541]. He captured and took blood samples from females and the pups nursing from them. Using starch-gel electrophoresis, a technique for identifying variation in specific enzymes, he was able to establish whether mothers and pups had the same enzyme variants. For example, there were six forms of the enzyme superoxide dismutase represented in the population from which he took blood samples. This means that the gene coding for this enzyme was present in six different alleles.

If female bats are indiscriminate parents, the enzyme variants of females and the pups they nurse should not be correlated. But if females tend to nurse their own offspring, then they will tend to share the same alleles and thus the same form of superoxide dismutase and the other enzymes. McCracken compared the actual correlation of genotypes of adult–nursing pup pairs with that expected if nursing is random. Despite the chaos of the bat rookery, females found their own pup about 80 percent of the time [541], even though we have little idea how they do this.

Tolerance of Siblicide: A Parental Puzzle

We conclude this chapter with an examination of a strange parental decision, namely, the tolerance of some parent birds of lethal aggression among their offspring, or siblicide as Douglas Mock calls it [562]. In the great egret, for example, brothers and sisters fight for possession of fish that their parents bring to the nest. The dominant individuals in a brood may dispose of their sibling rivals by bludgeoning them to death with bill strikes or forcing them out of the nest to die (Figure 14). Much of the violence takes place under the placid gaze of parent birds, which do not intervene. How can it possibly be to a parent's advantage to lose one or even two members from a brood of three or four as a result of siblicide?

One hypothesis is that parents do *not* gain from these actions, which have evolved because of the advantages enjoyed by winning siblings, not because of selection on parental behavior. There are situations in which we expect the fitness interests of parents and offspring to conflict with one another, when, say, an offspring can gain more genetic representation by receiving a parental investment that will yield the parent more if it is distributed in other ways. As Robert Trivers has explained, we can then expect to see parents trying to do what will aid them the most, and their offspring opposing this endeavor [784].

In the case of the killer egrets, however, we do not see opposition between parents and their offspring. Indeed, the potential for lethal sibling interactions is facilitated by an earlier parental decision, which is to incubate

14 **Siblicide in the great egret.** Young egrets fight while their parent stands at the nest without interfering. Photograph by Douglas Mock.

the eggs as they are laid. Because 1 or 2 days separate each egg laying, the young hatch out asynchronously, with the firstborn getting a head start in growth. The big differences in body size in a brood of three or four enable the larger birds to dominate the smaller ones. The dominant ones get to eat more by aggressively monopolizing the small fish their parents bring to the nest. This imbalance tends to exaggerate the size differences among siblings, and the runt of the litter is typically the victim, when siblicide occurs.

Parent egrets that delayed incubation until a clutch was complete would have synchronously hatching young of the same size, making it much more difficult for any one chick to kill a nestmate. Perhaps parental interests are served by having the chicks eliminate those members of the brood that are unlikely to *survive to reproduce*. Perhaps parents can only hope to rear three or four offspring in exceptional years when food is abundant, in which case sibling rivalry would be reduced. In typical years, when food is moderately scarce, parents cannot hope to provide enough food for all offspring, and so a reduction in the brood accomplished by siblicide saves the parents time and energy.

The adaptive sibling rivalry hypothesis has been tested [563]. In cattle egrets, relatives of the great egrets, the normal interval between egg layings is 1.5 days. Doug Mock and Bonnie Ploger created three categories of broods by shuffling newly hatched chicks from nests in a cattle egret colony: synchronous broods (all had hatched on the same day), normal asynchronous broods (birds that had hatched at 1.5-day intervals), and exaggerated asynchronous broods (birds that had hatched 3 days apart).

If the normal interval is optimal in promoting efficient brood reduction, then the number of offspring fledged per unit of parental effort should be highest for the normal asynchronous brood. This prediction was confirmed. Members of synchronous broods not only fought more and survived poorly, they required more food per day than control broods, resulting in a low parental efficiency score (Table 3). The exaggeratedly asynchronous broods experienced the same mortality rate as normal broods, but they, too, demanded and received a higher daily rate of feeding, an outcome that reduced parental efficiency relative to those with normally asynchronous broods.

Thus, cattle egret parents know (unconsciously) what they are doing

TABLE 3

The effect of hatching asynchrony on parental efficiency in cattle egrets

	Mean survivors per nest	Food brought to nest per day (ml)	Parental efficiency[a]
Asynchronous experimental	2.29	65.1	3.52
Control	2.33	53.1	4.39
Synchronous experimental	1.90	68.3	2.78

[a]Parental efficiency = (surviving chicks produced/volume in milliliters of food brought to the nest per day) × 100.

when they employ an incubation and egg-laying schedule that induces moderate differences in size and fighting ability in the chicks in their nests. Sibling rivalry and siblicide actually help parents deliver their care only to offspring that have a good chance of surviving to reproduce, and enable parents to keep their food delivery costs to a minimum.

SUMMARY

1 Asexually reproducing females need not produce sons and should enjoy a great advantage over sexual females as a result. The persistence of sexual reproduction in many species suggests that sexual females enjoy some fitness gains that more than compensate for the costs of producing sons.

2 Two hypotheses have been advanced to account for sexual reproduction in terms of individual selection. The lottery hypothesis argues that sexual females gain by producing genetically variable offspring, some of which can cope with the unpredictable variation they will inevitably encounter in their physical environment. The coevolution hypothesis proposes that parasites and other elements of the biotic environment, which coevolve with the species they exploit, favor sexual reproduction in the host species as means of producing some offspring whose novel gene combinations help them escape exploiting species.

3 Males and females of sexually reproducing species employ fundamentally different tactics of reproduction. Males make many small gametes (sperm) and usually exhibit little or no parental care, but instead try to fertilize as many females (and eggs) as possible. Females make many fewer, larger gametes and invest in parental care more often than males. Any investment in an existing offspring at the expense of future reproduction constitutes parental investment.

4 Parental care, a major form of parental investment, involves trade-offs in which increased survivorship of existing offspring is weighed against various costs of parenting, including reduced fecundity and opportunities to mate. Females are more likely than males to engage in parental care when ecological conditions are such that the benefit of increased offspring survival outweighs the disadvantages of parenting. This may be because females more often than males are in a position to identify their offspring and provide useful care for them.

5 Fishes are a major exception to the rule, in that males often provide uniparental care. In the fishes, however, males may pay only a small penalty for caring for eggs laid in their territories, whereas the cost to females of sacrificing growth to care for young may be extreme, given an exponential relationship between body size and fecundity in females.

6 An evolutionary analysis of parental care has produced an optimality hypothesis that states that individuals will make their parental investments in ways that maximize their benefit-to-cost ratio. This hypothesis is examined in the light of two puzzling cases, apparent indiscriminate nursing by a colonial bat and parental indifference in some birds to lethal interactions among their nestling offspring.

SUGGESTED READING

Adrian Forsyth's *A Natural History of Sex* [265] offers an entertaining introduction to the topics discussed in this and the next two chapters.

Gerald Borgia and his colleagues have greatly advanced knowledge on the intriguing reproductive behavior of bowerbirds [82-86]. The lottery hypothesis on sexual reproduction is presented in G. C. Williams's *Sex and Evolution* [846]. W. D. Hamilton's review of this book is a useful guide to its key argument [333]. The coevolution hypothesis is clearly contrasted to the lottery hypothesis in Robert Trivers's *Social Evolution* [787].

Trivers also wrote an extremely influential paper on parental investment [784]. Papers by Mart Gross and his colleagues have helped explain patterns of parental behavior in vertebrates [319, 320]. Douglas Mock's studies of siblicide provide beautiful examples of how to do comparative and experimental tests of hypotheses on a grimly fascinating phenomenon [562, 563].

DISCUSSION QUESTIONS

1. In Gary McCracken's study of Mexican free-tailed bats, he found that although females usually fed their own pups, they made a substantial number of "mistakes," feeding pups of other females about 20 percent of the time [541]. Does this mean that the parental behavior of this species is not optimal (see Chapter 8)? How many hypotheses can you devise to account for the "mistakes"? A suggestion: What *disadvantages* might a female encounter from leaving her pup off by itself where she could find it 100 percent of the time, so that she could deliver milk only to her offspring?

2. In some insects, males feed their mates at the time they are transferring sperm to them (see Chapter 13). In these species, females often mate with several males, receiving and storing sperm from all their partners. Under what conditions would the nuptial feedings provided by males be properly categorized as parental investment? When wouldn't they be parental investment? See [693] after preparing your answer.

3. Barn swallows belong to the same family of birds as rough-winged and bank swallows. In terms of coloniality, barn swallows are more similar to rough-winged swallows than to bank swallows. They sometimes nest in small, loose colonies, but they may also nest alone. After the young have fledged, parents and their offspring remain separate from other family groups. What prediction do you make about parental recognition of offspring in this species? (See [551] to check whether your prediction is supported or not.)

4. What factors might favor egg recognition by parent birds? What prediction might you make about the occurrence of this ability in the guillemot, a seabird that incubates a single egg placed on rocky cliff ledges (guillemots nest in astonishing densities—up to 34 birds per square meter)? (See [67] after developing your answer.)

13

The Ecology of Male and Female Reproductive Tactics

As the satin bowerbird illustrates, the behavior that male animals use to woo females is often strange and wonderfully puzzling. No less intriguing is the apparent choosiness that female bowerbirds and so many other animals exhibit. Why is it that males so often take the initiative in locating mates and courtship, and why is it that females so often reject their suitors? The great differences in the reproductive tactics of males and females may reflect the differences in their parental investments per offspring. These differences may have helped create a category of natural or individual selection called sexual selection. An understanding of sexual selection helps us understand the remarkable diversity of mating tactics, especially the differences between males and females in how they go about getting a chance to reproduce. This chapter explores what sexual selection is and shows that it is composed of two components, one arising as a result of the competition among individuals for mates, the other arising as a result of mate choice that favors attributes preferred by the choosy sex. The text examines hypotheses on how the competition-for-mates element of sexual selection may have affected the evolution of aggressive behavior and alternative mating tactics of males. The chapter then analyzes mate choice and the possible effects of this form of sexual selection by first considering how mate choice might operate in species in which the chosen sex has something of practical value to offer the choosy sex. The final topic will be the controversial question of how mate choice might work in species in which males offer only genes, nothing more, to their partners.

Sexual Selection   Like almost everything else important in evolutionary theory, Charles Darwin discovered sexual selection, which he defined as the effects of "a struggle between the individuals of one sex, generally the males, for the possession of the other sex" [171]. Darwin recognized that males usually fight among themselves for opportunities to mate with females, which appear to prefer some males over others. In the typical species, the number of descendants a male had would depend in part on how well he did in competition with other males for females. In addition, females might prefer some males over others, and if so, this, too, would affect a male's reproductive success. The result could be the sexual selection of traits that enhanced male success in the reproductive struggle, even though they might well be a burden in the competition to stay alive.

The possibility that natural selection for traits that promote survival might be opposed by sexual selection for traits that were useful in acquiring mates seemed likely to Darwin, given the obvious costliness of male ornaments and structures that were apparently used to attract or compete for mates (Figure 1). Darwin had read about bowerbirds, whose exaggerated sexual behavior was precisely the kind of thing that caused him to propose the distinction between sexual and natural selection. A male bowerbird's bizarre antics cannot be useful in any other context than the sexual one. Investments in building bowers, gathering decorations, and fighting with other males seem certain to reduce the survival chances of individuals. This cost, however, might be offset by the sexual success of males with elaborate bowers and dramatic displays.

Both natural selection and sexual selection operate in fundamentally the same way, occurring when individuals differ in the number of surviving offspring they produce. We can be specific about this. Sexual selection requires the same conditions that produce natural selection. Individuals must differ in their attributes, and the differences must affect the number of surviving offspring produced (either because individuals differ in their ability to compete with others of the same sex for access to mates or in their ability to attract members of the opposite sex). If the differences are heritable, sexual selection will lead to the spread of the reproduction-enhancing attributes.

Although sexual selection is merely a subcategory of natural selection, the distinction Darwin made between the two processes usefully focuses attention on the selective consequences of sexual interactions *within* a species. Sexual selection helps explain phenomena that do not make evolutionary sense in the context of dealing with other aspects of the "environment" (including climate, predators, diseases, difficulties in finding food, and so on).

Bateman's Principle

After Darwin, the concept of sexual selection was largely ignored until 1972, when Robert L. Trivers [784] resurrected the topic by drawing attention to a study by A. J. Bateman [43]. Bateman reasoned that sexual selection should produce different degrees of variation in the reproductive success of males and females. Unlike a female, whose fitness does not usually

1 **Sexually selected male "ornaments."** Darwin believed that sexual selection via female choice was responsible for the evolution of elaborate plumage and remarkable displays of males in some species of birds of paradise (upper left) and grouse (upper right). Photographs by Crawford Greenewalt and Haven L. Wiley. Darwin argued that the strange horns and snouts of certain beetles (lower left) must also arise via female choice, although it is now known that males fight for mates with this structure. Darwin recognized that the enlarged claws of some crustaceans, like this fiddler crab (lower right) evolved as aids to combat among males. Photographs by the author.

increase with the number of mates she has, a male's fitness will generally rise with each additional female he inseminates. Therefore, males can be expected to compete for access to females and their eggs. Some males may mate with several females, and this means that other males will fail altogether (females tend to mate just once or twice, for they will receive all the sperm they need to fertilize their relatively few eggs from one or two males).

The reproductive output of males will be highly variable, therefore, whereas almost all females will mate and produce about the same number of offspring, the number being limited by the number of eggs each can produce.

Bateman's experiments with fruit flies demonstrated that reproductive success was more variable for males than for females, and most similar studies since Bateman have confirmed this point (Figure 2). Recall that satin bowerbird males secured from 0 to 33 copulations in one breeding season in Gerald Borgia's study [82]; because almost all females laid just two eggs, female reproductive success per breeding season varied from 0 to 2.

Bateman argued that the differences in reproductive variance between the sexes were caused by their different gametic investment in each offspring, with females making a relatively large contribution of resources in the egg and males making a tiny contribution in the sperm that fertilized the egg. Because females and their scarce eggs are a limited resource from the male perspective, whereas sperm are available in abundance, males compete for females and females choose among males.

Trivers expanded this argument by noting that females not only make larger gametes, they also are more likely to invest in parental care for their offspring, a fact that makes them an even more limited resource for males [784]. But if males happened to make a larger parental investment than the females of their species, they should be the limiting sex even though the male's sperm had little to add to the nutrients contained in the egg. Trivers predicted that in a species with greater male parental investment, females would compete to secure the parental care and other valuable assistance that males were willing to offer. As we shall see, this prediction is supported by considerable evidence. In a species showing sex-role reversal, variance in female mating success would be expected to be greater than that of males.

As is generally true, an alternative explanation exists for the observation that males generally show greater variance in reproductive success than females do. William Sutherland showed mathematically that males will exhibit greater variance than females do *by chance alone* if males and females differ in "mating time," that is, the time spent with a partner in various activities, *plus* the time given to parental care and replenishment of gametes spent in one mating [755]. So, random mating, not competition for mates or mate choice, can lead to differences in the sexes in variance in mating success—given the appropriate conditions [395]. The appropriate conditions often do apply, because females more often than males spend extra time following a mating with parental care. Female fruit flies, for example, copulate and then search for oviposition sites for several days before they are ready to mate again. Thus, one cannot automatically conclude that high variance in male mating success stems from interactions among males or the choosiness of females.

The direct way to determine whether sexual selection is at work in a species is to search for evidence that one sex is competing for access to the other sex, producing the COMPETITION FOR MATES component of sexual se-

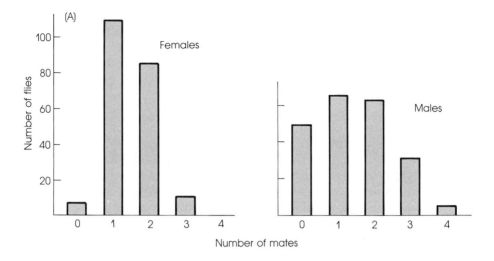

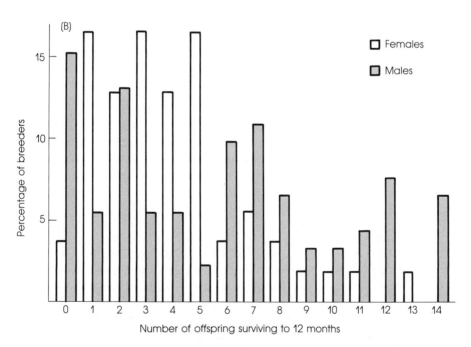

2 **Males and females and variance in reproductive success.**
(A) Bateman's experiment. Male fruit flies exhibit higher variance in reproductive success (as measured by number of mates) than females. Note that a fairly high proportion of males do not secure a single mate. *Source:* Bateman [43]. (B) Bateman's principle also applies to natural populations, including lions in the Serengeti. Note again that many males fail to reproduce, while a few have exceptionally high lifetime reproductive success (here measured in terms of surviving offspring). *Source:* Packer and Pusey [614].

lection [165, 853]. Many elements of the behavior of male bowerbirds appear to have evolved through this kind of sexual selection, as males aggressively attempt to prevent other males from mating [84]. Sexual competition is most obvious when one male assaults a rival who has mounted a female at his bower. In addition, bowerbirds aggressively protect their bowers and their decorations, while stealing feathers from and destroying the bowers of nearby males. Males whose bowers are depleted of feathers and whose walls are pulled apart lose chances to mate because females favor individuals with intact, elaborately decorated bowers. Thus, aggressive interactions among males determine which individuals have a chance with visiting females, which take an active role in selecting a partner.

Female preferences create the MATE CHOICE component of sexual selection. If males differ in some attribute (for example, ability to build and defend a bower) and if females use these differences to discriminate in favor of one type over another, they affect the fitness of males and the evolution of male attributes (assuming that there is or has been a genetic basis for the relevant differences among males).

Thus, there are two possible kinds of sexual selection, and in the remainder of this chapter we shall ask how might competition for mates and mate choice have shaped the evolution of male (or female) reproductive behavior. We begin with the evolution of aggressive behavior.

Competition for Mates and Aggression among Males   Females are a scarce and valuable resource in most species, and a common but not universal male response is to fight for mates. The focus in this section will be on these aggressive species, like bowerbirds, in which males fight for mates. We earlier discussed the direct link between male territorial ownership and mating success in certain birds (Chapter 9). Winners of territorial contests also have more mates than losers in fishes, lizards, and insects [765, 786, 814]. But is aggression adaptively employed in species whose males live in groups? There may be relatively little overt fighting among males in a "social" species, but peace in the group typically reigns because each individual knows its place in a dominance hierarchy, a social ranking established by aggressive interactions in the past. If social competition for high rank is the product of sexual selection, then there should be a positive correlation between being a "top dog" and reproductive success.

Many studies, like the one on cowbirds described in Chapter 4, have documented that higher ranking individuals copulate more frequently than males of lower status [23, 388, 508]. The mammalogist T. S. McCann conducted a representative test of this prediction with southern elephant seals of South Georgia Island in the Atlantic Ocean [539]. He ranked a set of 10 individually recognized males on the basis of the outcome of fights among them on the breeding beach. These impressively massive animals posture chest-to-chest and then try to bite each other about the neck. Losers can be readily identified when they drop their aggressive stance, often retreating with the winner in pursuit. (Most interactions, however, do not reach this stage but are resolved with a threat or two.) The number 1 male had from 14 to 157 encounters with the other top nine males and won them all.

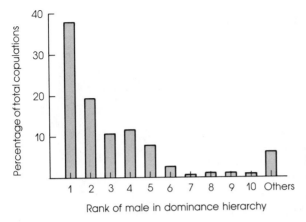

3 **Dominance and mating success.** High-ranking males of the southern elephant seal copulate many more times than low-ranking individuals. *Source:* McCann [539].

The number 2 male elicited submission from all but the top male, and so on down the line. McCann recorded the number of copulations each individual secured over the breeding season (Figure 3). These data clearly support the hypothesis that dominance (and thus fighting ability) leads to sexual success.

But when Glen Hausfater counted copulations in a troop of baboons with a clear dominance hierarchy, he found that males of low and high status were equally likely to copulate with females [349]. This result exposes an initial assumption made by McCann and Hausfater, namely, that the number of copulations is correlated with the production of surviving offspring. Hausfater suspected that this assumption was false for his baboons. He therefore modified his original hypothesis and tested it again with new data. He proposed that high-ranking males copulated with females when their eggs were especially likely to be fertilized. Dominant males generally guarded estrous females during the third day before the female's sexual skin diminished in size. On this day, females usually ovulate. In other words, dominant males monopolized females when their eggs could be fertilized (Figure 4) [349]. (Much of the statistically significant correlation

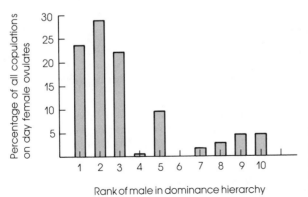

4 **Dominance and fitness.** Dominant male baboons tend to monopolize copulations with females on the day their mates ovulate. *Source:* Hausfater [349].

5 **Sexual dimorphism in seals.** A highly sexually dimorphic
species, the elephant seal (top); the male (here attempting to
copulate) is vastly heavier than the female. Photograph by Burney
LeBouef. In contrast, harbor seals (below) show little sexual dimorphism in body size. The top seal in this photograph is an adult
male; the lower seal is a pregnant adult female. Photograph by
Deane Renouf.

between dominance rank and number of effective copulations in the baboons studied by Hausfater arises because young males that have just begun to mate have very low dominance and very few effective copulations [59].)

In baboons and elephant seals, large males generally dominate smaller ones. The relationship between large body size and superior fighting ability holds for everything from minute insects like aphids [836] and thrips [154] to elephant seals [539]. It is also striking that adult male baboons and elephant seals are bigger—much bigger—than adult females (Figure 5). Not all species are SEXUALLY DIMORPHIC in this way; and when there are differences in the weights of males and females, the degree of the difference varies greatly.

Richard Alexander and his co-workers realized that differences in the degree of sexual dimorphism in body size were a possible measure of the investment males made in fighting capacity. A costly investment in body growth and maintenance could only be sexually selected *if* there were exceptional reproductive rewards to be gained by good fighters. Therefore, the degree of sexual dimorphism should be greatest in species in which males can monopolize many mates and least in species that were limited to one or two mates per breeding season [13]. Data on the ratio of body lengths of males to females, a standard measure of sexual dimorphism, of several mammalian groups support the prediction (Figure 6). Thus, an elephant seal male may have as many as 100 mates in his harem, and

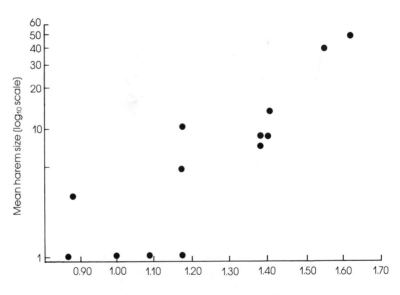

6 **Sexual dimorphism and male mating success.** Opportunities for mating success and investment in male body size are positively correlated in seals and other pinnipeds. When some males can acquire large harems, sexual selection has resulted in extreme sexual dimorphism. In monogamous species, males and females weigh about the same. *Source:* Alexander et al. [13].

males are about 60 percent longer than females and much, much heavier. Monogamous pinnipeds, in contrast, show little or no sexual dimorphism in body length (and weight).

Sexual Selection and Alternative Mating Tactics

In the competition for mates, mating territories, or high social status, there are of necessity both winners and losers. What is surprising at first glance is that losers are apparently so quick to give up. Only 4 percent of 4000 fights between southern elephant seals involved actual contact between the two males [539]. As a rule, a large dominant male need merely bellow noisily to have a subordinate retreat or cringe submissively in acceptance of his inferior status. Throughout the animal kingdom noncontact threat displays rapidly settle most disputes among males.

Chapter 8 explained that competing animals use threat displays to assess the fighting ability of a rival. Upon judging that an opponent would defeat them if they were to engage in combat, subordinates (wisely) withdraw. In some cases, subordinates may, *if* they live long enough, achieve the body weight or experience that will enable them safely to assume dominant status and the reproductive rewards associated with high rank.

But subordinates may also do some mating while they wait, for they are not as reproductively silent as they were once assumed to be. One of the tricks used by "loser" males is to pretend to be a female, a tactic that may give them access to females that might otherwise mate with another male (Figure 7) [28, 202, 269]. I recently watched male rove beetles pretend they

7 **Female mimicry in salamanders.** (Top) A female has straddled the male's tail to signal him to move forward and deposit a spermatophore. (Bottom) A male has interrupted the courting pair and now mimics female behavior, while the real female moves away. Photographs by Stevan Arnold.

8 Satellite behavior. Two noncalling male spadefoot toads (arrows) wait near a calling male, perhaps to grasp females attracted by the caller. Photograph by Brian Sullivan.

were females in the mountain forests of Costa Rica. In this species large males successfully defend small territories about a bit of mammal dung or on the carcass of a dead sloth or kinkajou. These none-too-aesthetic materials attract flies, which the beetles catch and eat. Because females come to dung to feed, territorial males have a chance to mate with them, and they often do. Large males employ their huge jaws to bite smaller males, forcing them from areas where females gather.

Occasionally, however, I watched a courtship that seemed to be going nowhere. The courting male kept tapping at the tip of the abdomen of the other individual, as is the practice in this species. However, instead of standing still and permitting the courter to mount, the other beetle kept moving slowly about the area. Sometimes in the course of these meanderings, the courted beetle encountered a female, at which point to my initial surprise and subsequent delight, he, for it was a he, began to court the female. If she was receptive, the male that had been pretending to be a female copulated with the real female literally under the nose of his thoroughly duped territorial rival.

Female mimicry is by no means the only alternative mating tactic of males. For example, some male toads, frogs, and crickets practice SATELLITE BEHAVIOR, by waiting silently and nonaggressively by a loudly calling territory owner (Figure 8). A satellite sometimes intercepts females that were heading for the caller [123, 629, 820]. Satellites of a different sort exist in bighorn sheep, whose dominant males wait in areas where herds of females gather and then court and directly defend receptive, ovulating females [366, 367]. In these same areas, Jack Hogg found other males that lurk near alpha males and their females and battle dominant rivals for

9 **Alternative mating tactics in bighorn sheep.** A group of coursers (above) wait their chance on a hillside near a tending dominant male who is with a female. (Left) Here a courser has broken through the defenses of a dominant male and has mounted a fleeing female. Photographs by Jack Hogg.

access to females. From time to time, satellite males try to rush past a dominant male and onto a female with whom they might copulate (an act that lasts only a few seconds in bighorns) literally on the run, for females ran away from these sneaky intruders (Figure 9).

Still other males stayed away from dominant males and their slippery satellites and instead tried to find and keep a female from going to areas with dominant individuals. By blocking her path whenever she turned to go in an unwanted direction, a male was sometimes able to monopolize an estrous ewe.

Alternative Mating Tactics: Evolutionary Maintenance

If two (or three) ways of acquiring mates exist in a species, we can ask, Why hasn't selection eliminated all but the most reproductively successful of the alternatives? In many cases, one option (for example, being a dominant male) is clearly superior reproductively to the other tactics used by males [203, 436] (Figure 10).

Game theory, a topic introduced in Chapter 8 and championed by John Maynard Smith [531] and Richard Dawkins [183], has proved especially useful in analyzing alternative mating tactics. The essence of game theory is that the fitness payoff to one individual depends on what the other members of the population are doing. Thus, social interactions determine what option in the game of reproductive competition will yield highest fitness for a participant in the game. That STRATEGY (a set of behavioral rules) that cannot be replaced over evolutionary time by an alternative set is labeled an EVOLUTIONARILY STABLE STRATEGY or ESS.

Like optimality theory, game theory provides a way to make precise quantitative predictions about what evolution should produce, predictions that can then be tested against reality [41]. Unlike optimality theory, however, game theory takes into account the dynamic, interactive aspects of social competition. Use of this theory has helped demolish the idea that there can be only one set of behavioral traits exhibited by a species. Given the appropriate conditions, two or more distinct techniques for dealing with competitors can coexist indefinitely.

In game theory, strategy does *not* mean what it means in everyday speech, namely, a plan of action that the individual consciously adopts. Instead game theory defines strategy as an inherited program that influences what actions an individual takes to achieve the ultimate goal of reproducing [203]. A strategy could in theory provide for just one TACTIC, or response; the difference between male rove beetles that use the female-mimic tactic and those that use the territorial tactic might arise because each type had inherited a different strategy. If so, the rule "always behave like a female when confronted by a rival male" would be opposed in the population by the alternative strategy "always try to maintain a territory aggressively." If one strategy outperformed the other, even by a small degree, over evolutionary time the strategy that yielded the higher repro-

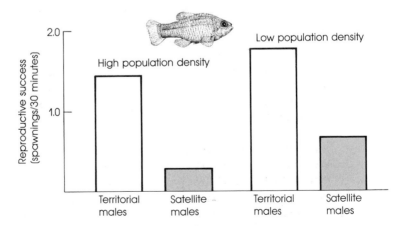

10 **Differential reproductive success** of territorial and satellite male pupfish. Under two different conditions, territorial males clearly spawn more often than satellite males. *Source:* Kodric-Brown [436].

ductive payoff to individuals would replace the other. Therefore, for both strategies to persist in the population, they would have to confer equal fitness on the individuals that possessed them. This is possible, but it generally requires frequency-dependent selection of the sort we discussed first in the hypothetical case of a gull colony with "daring" and "cautious" mobbers (Chapter 8). In other words, when practitioners of two different strategies are at certain frequencies in the population, they may have equal fitness. But if one type becomes more common, it will have lower fitness, a situation temporarily favoring the alternative strategy and leading the population back to its equilibrium point where the two strategies will coexist.

However, a strategy can also specify that an individual perform one of several different tactics, the choice depending on the conditions it encounters. For example, the rule for rove beetles might be "behave like a female *if* you encounter a territorial male larger than you are, but otherwise aggressively defend a territory." Such a strategy is called a CONDITIONAL STRATEGY, a term chosen to emphasize that the tactics an individual employs are dependent on the variable environmental conditions that the individual encounters. Here the set of rules that makes up the strategy provides for flexibility of response. It is entirely possible that all male bighorn sheep have the same strategy, with the three alternative mating tactics arising because individuals experience different conditions. If a male bighorn sheep is small and unable to defeat a large dominant male, it may simply switch to the copulation-on-the-run alternative or try to block an isolated female from reaching places where she will encounter a dominant male.

For a particular conditional strategy to be an ESS, individuals with the flexible set of rules provided by the strategy must have higher fitness than others with different strategies, be they conditional or single-tactic rules. Certain conditional strategies are likely to be advantageous because they help males salvage some reproductive opportunities even when these males are excluded from the most profitable method of acquiring mates. In Richard Dawkins's terms, they help subordinate, younger, or weaker males make the best of a bad job [183]. Imagine a population consisting of some males with the conditional ability to adopt a subordinate role and other genetically different males that persist at all costs in trying to become a territorial or dominant male. If, as would often be the case, the persistent male exhausts itself in futile attacks on stronger males, the conditional strategy is superior.

Two Strategies of Bluegills

With this background in mind, let us analyze two cases of alternative mating tactics. Our first goal will be to ask how we can determine whether the alternatives stem from two different strategies or from a single conditional strategy. Once this is established, we can use game theory to make certain testable predictions about the reproductive consequences for males that use the alternative tactics.

During the breeding season, some bluegill sunfish males defend a nesting

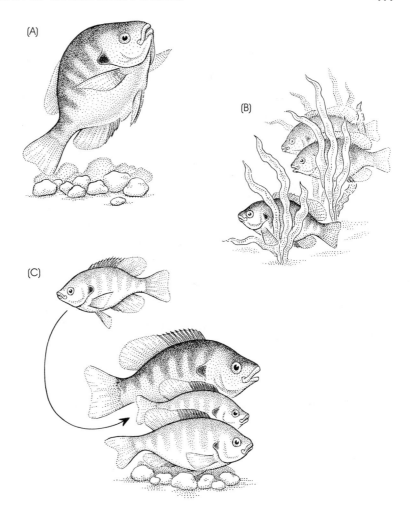

11 **Alternative reproductive tactics of male bluegill fish.** (A) A territorial male guards a nest that may attract gravid females to his nest site. (B) Little sneaker males wait for an opportunity to slip between a spawning pair, releasing their sperm when the territory holder does. (C) A slightly larger satellite with the body coloration of a female hovers above a nest before slipping between the parental territorial male and his mate when the female spawns. *Source:* Gross [314].

site where they will guard any eggs laid in the nest by their mates. Other males are not territorial but try to "cuckold" nest guarders by slipping into another male's nest when a female is spawning there (Figure 11) [316]. There are big size differences between the two categories of males; only large males use the territorial tactic. Furthermore, territorial males out-reproduce cuckolders by a wide margin in any given breeding season.

These results are completely consistent with a conditional strategy hypothesis. For example, males might practice the sneaker-satellite tactic if they failed to grow large, or young males might start reproducing as small cuckolders and then mature eventually into large territory holders. But this is not the case. An experimental study by Mart Gross and David Philipp showed that cuckolder males were far more likely to produce sons committed to the sneaker-satellite role than were territorial types (Figure 12) [318]. Thus, there is a genetic basis for the behavioral difference between the two kinds of males, which by definition are exercising two different strategies, not two tactics in one conditional strategy.

Gross and Philipp identified sons as cuckolders by examining male testes in 1-year-old offspring of the crosses. By this time, some males already have large gonads and mature sperm, in order to embark on a career of stealing fertilizations from territorial males when they are just 2 years old. In contrast, those individuals that are destined to become territory holders 5 or 6 years later have relatively immature testes with no sperm present. They invest all in body growth in order to reach the size needed to compete for a nest in the breeding colony. We can predict that if the sneaker-satellite males and the territorial males possess different strategies, they must confer equal lifetime reproductive success for their practitioners. This may be true, largely because many males that commit themselves to the territorial strategy do not live long enough to be able to get a territory, whereas the small sneaker males can start reproducing much sooner and so are less likely to die without leaving any descendants [316]. Given the appropriate frequency of the two types of males, their *average* reproductive successes can be equal (Figure 13).

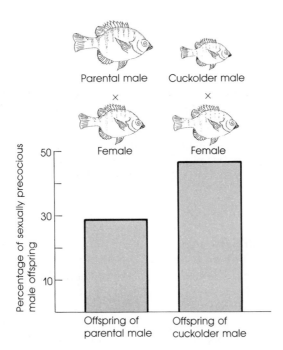

12 **The genetic basis of alternative mating behaviors** of bluegill fish. The male offspring of cuckolder sneaker-satellite males are more likely to exhibit sexually precocious development of the testes than are sons of males that use the territorial-parental strategy. Data courtesy of Mart Gross.

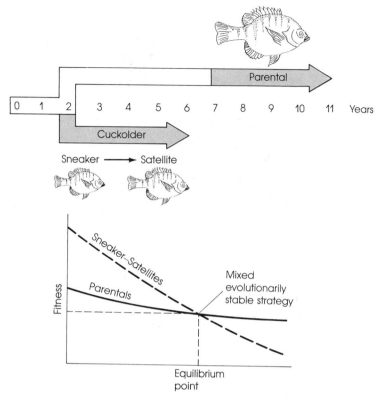

13 **Fitness of alternative mating behaviors.** The two behavioral alternatives in bluegill fish might yield equal fitness for their practitioners. Parental types enjoy high reproductive success—if they live long enough to become large enough to hold a territory. Sneaker-satellites begin reproducing early—but do not experience high annual reproductive success. Nevertheless, when sneaker-satellites make up a certain proportion of the male population, their *average fitness* will be the same as the average fitness of parentals. At this equilibrium point, a mixed evolutionarily stable strategy will exist. Courtesy of Mart Gross.

Three Tactics of Scorpionflies

Randy Thornhill showed that *Panorpa* scorpionflies use three different tactics for mating: (1) some males defend dead insects that attract receptive females, which feed upon the carrion; (2) others secrete salivary materials on leaves and wait for females to come to consume this nutritional gift; and (3) still others offer their mates nothing at all but force them to copulate (Figure 14).

Thornhill placed groups of 10 male and 10 female *Panorpa* in cages with two dead crickets. Some of the male *Panorpa* were large, others medium-sized, and still others were relatively small. They competed for the crickets, and the larger males won both the carrion and 60 percent of all matings,

14 **Forced copulation** in *Panorpa* scorpionflies occurs when a male without a nuptial gift grabs a female and will not release her until she mates with him. Photograph by Thomas E. Moore.

with individuals averaging nearly six copulations each. Medium-sized males generally attempted to attract mates with salivary gifts but were much less successful (gaining about two copulations per male). Small males were unable to claim crickets and appeared incapable of generating sufficient saliva to be attractive to females. They employed the forced copulation route but were least successful of all (averaging only about one copulation per male). These results show that the different tactics do not produce equal fitness gains.

Thornhill therefore predicted that if three tactics were part of one conditional strategy, males using a low-payoff option would switch to a tactic yielding higher reproductive success, if given the suitable conditions. To test these predictions, he removed the males that were able to defend carrion from an enclosure [765]. Other males promptly abandoned their salivary mounds and moved to claim the dead crickets as these became available. Males that had not possessed a cricket or a salivary mound took the abandoned salivary secretions. Thus, male *Panorpa* are able to adopt whichever of the three tactics returns the highest possible rate of copulations, given the nature of the competition they face at that moment.

Forced Copulation

Males of many species besides scorpionflies appear to force females to mate with them. For example, a study of the crab-eater seal showed that males search for and remain with females that are resting on ice floes with an unweaned pup [726]. At intervals, males attempt to copulate despite the fact that females do not become receptive until after the pup is weaned and independent. Nor is the male gentle in his approaches; he frequently bites the female after being rebuffed (and is bitten in return). A research team headed by D. B. Siniff saw males and females covered in blood as a result of their sexual disputes. Although perhaps an extreme example, nothing could illustrate more clearly that sexual reproduction is not a gloriously cooperative enterprise designed to perpetuate the species. A male crabeater seal gains by inseminating a female as soon as possible, because the longer he waits the more likely he is to be supplanted by a rival [726]. Moreover, the sooner he copulates, the sooner he can get on with the search for a second mate. The female, on the other hand, has an interest in seeing that her pup is not prematurely pushed into independence. Sexual conflict is the result.

Although there is good reason to believe that male scorpionflies and crab-eater seals that force copulation on females sometimes father offspring as a result, is forced copulation an adaptation? There is an alternative hypothesis for rape, which is that it occurs as an incidental, even maladaptive, side effect of those proximate mechanisms that make sexual arousal exceptionally rapid for males. These mechanisms are responsible for adaptively motivating males to mate, but they also lead to nonproductive indiscriminate sexuality of the sort that induces male wasps and bees to "mate" with flowers, toads to couple with rubber boots or fingers, and beetles to attempt to inseminate beer bottles (Figure 15). No one argues that copulating with a beer bottle is adaptive per se. By the same token, forced copulation could be another costly side effect for males with a strong sexual drive.

We can test whether rape is a maladaptive or an adaptive reproductive tactic by examining whether females whose eggs can be fertilized are the special object of rapist male birds. The white-fronted bee-eater is representative of the many bird species in which forced copulation occurs [240]. This African bird nests in tunnels in banks. When females leave their nests, they are subject to assault by males that assemble near the colony. The males attempt to force the female to the ground, and once there they press their cloaca on the female, attempting to transfer ejaculates to a female who is not their mate and who strongly resists insemination (by pressing her tail and cloaca firmly on the ground).

Stephen Emlen and Peter Wrege documented that females that were laying eggs, and thus had ova to be fertilized, were the special target of the assaulting males (Figure 16). The same is true for the lesser scaup, a

15 **Indiscriminate sexual behavior** is common among males. A male toad (left) clasps a finger as if it were a female of his species. Photograph by Tony Allan. An Australian buprestid beetle (right) attempts to copulate with a beer bottle. Photograph by David Rentz and Darryl Gwynne.

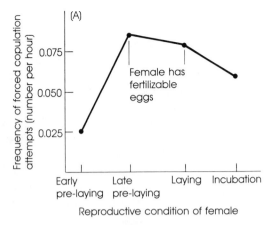

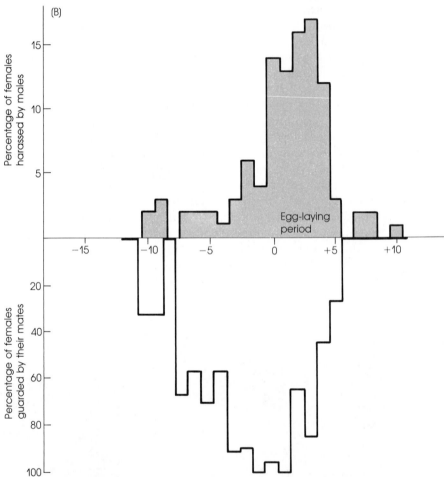

16 **Forced copulation attempts** occur more often around the egg-laying period when females have eggs to fertilize in (A) the lesser scaup and (B) the white-throated bee-eater. In (B), the x-axis shows the days before and after the first egg is laid by a female, with the egg-laying period shown in cross-hatching. Sources: Afton [3]; Emlen and Wrege [240], and unpublished data courtesy of Stephen Emlen and Peter Wrege.

duck whose males sometimes force females to mate (Figure 16) [3]. Thus, in at least some species, male birds force copulation on females that might produce offspring for them [561].

Sperm Competition and Mate Guarding

The occurrence of forced copulation is only one of many events that leads potentially to SPERM COMPETITION between the ejaculates of rival males. When two or more ejaculates are present in a female, then any one male's success in fertilizing eggs is compromised. To the extent that the risk of sperm competition is high, we can expect to see males investing less time and effort in trying to mate with as many females as possible and, instead, guarding a current partner against other males, thereby increasing the likelihood of fertilizing her eggs.

In bee-eaters, scaup, and some other birds, rapist males usually have mates of their own [561], so these individuals have a conditional strategy with two options: a pair-bonding tactic used with one female and the forced-copulation tactic, which is exercised on as many females as possible. There is a trade-off between the two tactics, because those who would engage in forced copulation with nonmates must leave their main partner unguarded to some extent, thus increasing the probability that she will be forced to copulate with rival males. In fact, bee-eater males typically guard their mates *when they are in the egg-laying stage* (Figure 16). When an egg-laden female leaves her nest tunnel, she calls to her male companion and he almost always flies to her and stays with her while she flies from the colony. The presence of a guarding male reduces by 10-fold the chance that a rival male will harass a female [240].

The same tactics have evolved convergently in another colonial bird, the bank swallow [50]. A female swallow traveling to and from her tunnel nest may be mobbed by a group of males and forced to mate. But males usually guard their females just before and during egg laying (when a female's eggs may be fertilized) (Figure 17). Prior to this time and afterward, males sometimes attempt to locate other females that appear to be in the egg-laying process [50]. The switch in use of alternatives occurs at times when the benefit-to-cost ratios of the two tactics change sharply.

In contrast, David Leffelaar and Raleigh Robertson have shown that male tree swallows do *not* guard their mates at any stage of the reproductive cycle [469]. This behavioral difference with bank swallows is correlated with a key difference in nesting ecology: tree swallows are not colonial, but instead nest in scattered tree holes. The risk of cuckoldry is therefore lower for tree swallows while the risk of nest usurpation is higher, because tree holes are scarce. When their mates leave the nest, male tree swallows stay behind to guard the nest rather than flying with a female to guard her. Costly mate guarding advances male reproductive success only if unprotected mates are likely to mate again.

Mate guarding is common in insect species with multiple mating females because, as G. A. Parker showed first, the last male to copulate with a female before she lays her eggs usually wins the sperm competition contest [622]—and this is true for some birds as well [883]. The mechanisms that

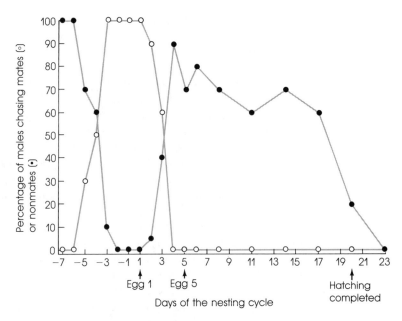

17 **Mate guarding and forced copulation.** Male bank swallows switch reproductive tactics when fertilization gains are likely to be increased by a change in behavior. Males fly with their own mates at times when their mates are fertile but pursue other females at other times. *Source:* Beecher and Beecher [50].

achieve fertilization success for the last male are varied in insects. Jon Waage has discovered perhaps the most spectacular device in the common black-winged damselfly of the eastern United States (Figure 18). Most observers, charmed by the jewellike beauty of the insect, are unaware that the life of this creature, like that of most sexually reproducing species, is marked by violence and competition [805]. Some males are territorial, claiming a stretch of stream bank 1 to 3 meters long from which they repel all other males. Females come to these territories to lay their eggs in barely submerged rootlets and other underwater vegetation. Before she oviposits, the resident male courts and mates with her. He first grasps the front of her thorax with specialized claspers at the tip of his abdomen. The male then pulls the end of his abdomen close to the point where his abdomen joins the thorax, forming a loop. This maneuver involves transfer of sperm from the sperm-producing organs at the tip of his abdomen to a "penis" at the base of the abdomen [806].

After this bizarre transfer has taken place, a receptive female swings her abdomen under the male's body and places her genitalia over the penis (Figure 18). The male then rhythmically pumps his abdomen up and down, during which time his penis acts as a scrub brush (Figure 19), catching and drawing out the sperm stored in the female's sperm storage organ, or *spermatheca*. After a male has removed roughly 90–100 percent of the competing sperm, he releases his own gametes to flow into the female, where they will be stored in the emptied spermatheca for use when she fertilizes her eggs.

Once sperm removal and replacement are complete, the male releases the female and returns to his territory, where he will guard his mate from rival intruders while she oviposits in his territory. If he fails to protect her adequately, she may be forced to fly up from the stream and mate with another male, who will carefully remove all the stored sperm before providing his partner with his own gametes.

Sperm competition may be responsible for certain characteristics almost as remarkable as the black-winged damselfly's use of its sperm-removing penis. William Eberhard has proposed that some sharks use their penis to give a female a contraceptive douche prior to passing sperm to a mate. To this end, a shark penis has two tubes, one through which sperm are transferred to the female, the other being a muscular siphon containing seawater. The male can spray seawater from the siphon at great force, and so might wash out the female's reproductive tract before ejaculating sperm into her [219].

Male birds do not have much in the way of a penis, certainly nothing nearly as elaborate as a shark's or a damselfly's, but males of the dunnock, a drab European songbird, are able to induce a partner to void sperm received from a rival [174]. Males achieve this by pecking at the female's cloaca just prior to copulating, an action that stimulates the female to expel a small droplet of fluid. Nicholas Davies actually succeeded in collecting a cloacal droplet. Sure enough, the fluid contained active sperm.

The sperm expelled by female dunnocks may well have been donated by males other than their cloaca-pecking companions of the moment, because dunnock females regularly mate with males other than their "primary" partner (see Chapter 14). The number of cloaca pecks given by a male

18 **Black-winged damselfly.** A territorial male (left) and a copulating pair (right) in which the male has grasped the female with his terminal claspers while she twists her abdomen forward to make contact with his sperm-transferring organ. Photographs by the author.

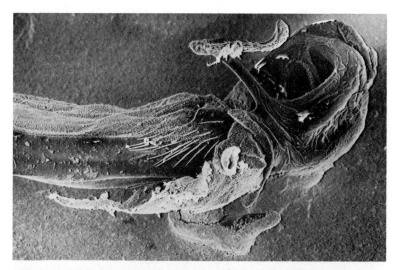

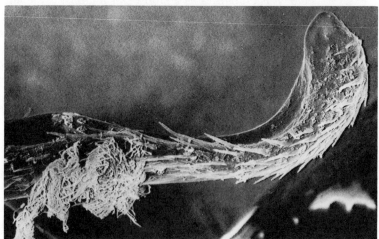

19 **Sperm competition in the black-winged damselfly.** The male's penis (top) has lateral horns and spines that enable the male to scrub out the female's sperm storage organ before passing his own sperm to his mate. A close-up of a lateral horn (below) reveals rival sperm caught in its spiny hairs. Photographs by Jonathan Waage.

before mating increases when he has recently seen another male close to the female (Figure 20). Moreover, when another male is around, a male dunnock copulates more often with his mate, perhaps because ultimately this may swamp any rival sperm not expelled by the female.

Mate Choice by Females: Males Differ in Material Offerings   We have seen that many male attributes, such as their attempts to become socially dominant, to force females to mate, and to guard their mates, may have evolved because these tactics help males fertilize more eggs in competition with other males. Now we shall turn our attention to the possible evolutionary effects

(A)

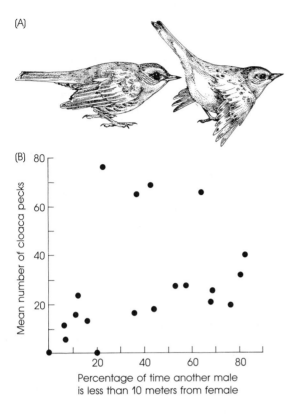

(B)

Mean number of cloaca pecks

Percentage of time another male
is less than 10 meters from female

20 **Sperm competi-
tion in the dun-
nock.** (A) Males sometimes
peck at the cloaca of their
mates, apparently causing
females to expel stored
sperm. (B) Males are more
likely to perform this action
when their mate has re-
cently been near another
male from whom she may
have received rival sperm.
Source: Davies [174].

of sexual selection via mate choice. The theoretical reasons for believing that females may often exert sexual selection on males are straightforward. As noted earlier, a female's reproductive success is not often increased by having more mates. Instead, female fitness is primarily a function of how many eggs she can produce *and by what happens to the eggs after they have been fertilized.* If males differ in their effects on offspring quality and survival, then females should discriminate in favor of those individuals that will contribute the most—and this act of discrimination will create mate choice selection.

The question is, Are females choosy in practice (as opposed to theory) about who gets to fertilize their eggs? No one doubts that females could have some say in this matter, given that they usually can resist forced copulation. Moreover, the rarity with which females produce hybrid off-spring suggests that they surely discriminate between males of their own and foreign species [534], usually on the basis of species-identifying signals given by males of their species (Figure 21).

The key problem, however, has been to demonstrate unequivocally that females prefer some males to others *within their own species.* It is not enough to show that some males enjoy higher reproductive success than others, for as we have seen, this may stem from differences among compet-ing males in their ability to monopolize females and not from female pref-erences. Although it is theoretically possible for female choice to be neutral to females and still greatly affect the evolution of male traits [430, 465], if

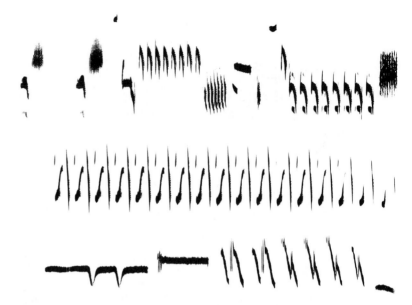

21 **Species-specific songs** of (from top to bottom) a male song sparrow, swamp sparrow, and white-crowned sparrow. Females of different sparrows can use their males' song to identify members of their own species and avoid mating with other species. Courtesy of Peter Marler.

we wish to determine whether mate choice is adaptive for females, we must show (1) that females do not mate randomly, (2) that any nonrandom patterns are not merely the result of male competition for mates, and then (3) that choosy females secure a reproductive advantage [330]. These are demanding requirements.

As an example of the difficulty in unraveling the potential effects of the two kinds of sexual selection, consider the alternative hypotheses available for the following cases. After a new stallion has taken a band of wild mares from another male, the pregnant mares may abort after having been forced to copulate by the new harem owner [61]. After a similar takeover in certain mice, exposure to the odor of the new male's urine causes pregnant females to resorb their embryonic offspring [458, 707].

At the ultimate level, these effects might arise either from male–male competition *or* female choice [707]. It could be that the new males force their females to terminate pregnancy as a way to speed sexual cycling, thereby accelerating the opportunity to father offspring with them. Or it could be that females that have lost or been abandoned by a male prefer to produce offspring with a newcomer that may offer the parental care the previous male failed to provide. If this is true, then pregnancy termination is a form of active mate choice by females. However, it could also be that females abort their pregnancy because the new male will probably kill their offspring should they be born. Thus, infanticide that evolves as a result of male–male competition for access to mates (Chapter 1) could make it advantageous for females to put at end to a doomed pregnancy to save energy for future reproductive effort. If this is the case, the female's response enables her only to make the best of a bad situation, a "choice" imposed on her by infanticidal males.

Female Mate Choice and Nuptial Gifts

Pregnancy termination may or may not stem from female choice. Unequivocal evidence that females do exert adaptive mate choice comes from those species whose males offer their mates a material benefit, such as food or parental care, in return for copulation. Here females may prefer males whose offerings are better than average, creating sexual selection in favor of "generous" males.

Females of the black-tipped hangingfly choose males that give them a large, edible NUPTIAL GIFT when they mate (Figure 22). Randy Thornhill showed that the size of the dead insect gift and its nutritional quality can vary. Females promptly reject males that offer unpalatable ladybird beetles. Even if the food item is edible, and copulation begins, the duration of mating is dependent on the size of the prey gift [764]. If the gift is small, and the meal lasts less than 5 minutes, a female will leave without having accepted a single sperm. But if the prey cannot be polished off in less than 20 minutes, the female will depart with a full complement of sperm (Figure 23).

Not only does the duration of copulation regulate how many sperm a female will take from a mate, it also determines whether or not she will

22 **Nuptial gift.** Male hangingflies offer captured prey to their mates. Photograph by the author.

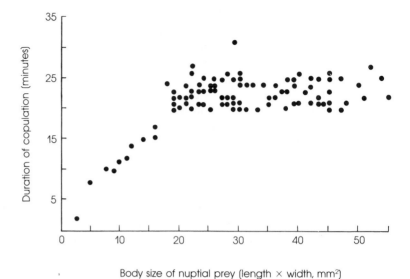

23 **Sperm transfer and the size of nuptial gifts.** The larger the gift, the longer the mating, and the more sperm the male hanging fly transfers to the female. *Source:* Thornhill [764].

24 **Spermatophores are nuptial gifts.** Female katydids, like this Mormon cricket, eat much of the proteinaceous spermatophore they receive from their mates. Photograph by Darryl Gwynne.

become unreceptive and start laying a number of eggs. If the female has been given only a 12-minute meal, she will leave her mate prematurely and, to add injury to insult, she will seek out another male. She then feeds upon his gift and accepts his sperm, which she presumably uses to fertilize the eggs she lays during her nonreceptive phase. In this way females exercise active mate choice, creating sexual selection that favors males that provide large, nutritious, nuptial presents [764].

Spermatophores as Nuptial Gifts

In the katydids, relatives of grasshoppers, males transfer up to 40 percent of their body weight to a female in the form of the highly proteinaceous spermatophore that contains his sperm. Because females feed upon the spermatophore after copulation (Figure 24), we can ask whether females gain reproductively by treating the spermatophores as a nuptial meal. In Darryl Gwynne's laboratory study of an Australian katydid, he arranged to remove spermatophores that had been transferred to some females in order to give them to others [324]. By collecting the eggs produced by females that had eaten different numbers of spermatophores, he found that these donations increased female fecundity and the weight of the eggs they manufactured (Table 1). In nature females of this katydid discriminate against males that do not offer them a spermatophore by not permitting the partner's sperm to enter their sperm storage organ.

Females of the decorated cricket use the same mechanism to discriminate against males that pass a smaller than average spermatophore [692]. The cricket male attaches a two-part spermatophore to his mate's genital opening during copulation. Shortly after separating from the male, the female reaches back and removes one part of the device, leaving the component that contains the sperm in place. While the female feeds on part 1, sperm migrate from the other section into the female's body for storage. The larger the nuptial present, the longer she feeds, and the more time the sperm

TABLE 1

The effect of spermatophore consumption on the weight and number of eggs produced by female katydids (*Requena verticalis*)

Number of spermatophores eaten	Mean number of eggs laid[a]	Mean egg weight (mg)[a]
0	31.9	9.7
1	45.8	10.0
3	59.8	10.3
7	70.7	10.6

Source: Gwynne [324].
[a]Mean values based on the performance of 12–14 females per group. All individuals were maintained on the same amount of rolled oats and carrots after having been mated except that some females received 2 or 6 supplementary spermatophores to consume over a two-week period. Females in the 0 spermatophore group were prevented from eating the spermatophore they received from their mate.

have to exit from the *sperm ampulla*. But as soon as she is done with part 1, the female turns to part 2, the ampulla, munching her way through this edible structure and demolishing any sperm that remain behind. Males that pass a skimpy spermatophore fail to transfer the maximum amount of sperm to a mate.

Access to Monopolized Resources

Some male hangingflies, katydids, and crickets gather, process, and transfer food directly to a discriminating partner, in effect "bribing" females into accepting a full complement of sperm. Nuptial gift giving, however, is only one of several ways in which males exchange material benefits for copulations. A more familiar technique is for males to defend a territory visited by receptive females that mate with the owner and exploit the territory's resources.

We have already (Chapter 9) presented information on the importance of territorial possession for many birds that pair and nest within a food-producing area. Given that females appear to avoid non-territory holders and voluntarily accept certain territorial males, this would seem to be a case of female choice that has obvious benefits for selective females. Here we shall discuss an additional case of mate choice in relation to a resource-based territory.

Bullfrog males call competitively from places in ponds where water temperature and vegetational characteristics are best suited for egg development. Females come to the chorus and move toward a male. Amplexus usually does not occur until a female actually touches a male, an observation that suggests that a female can mate with the individual of her choice [389]. Females prefer to mate and lay their eggs in the territories of large males—probably because large males win control of the best egg deposition spots. Richard Howard has shown that embryo mortality from overheating and leech predation is lower in territories controlled by larger males than in sites monopolized by their smaller rivals (Figure 25). Thus, by mating preferentially with large males, female bullfrogs gain material benefits.

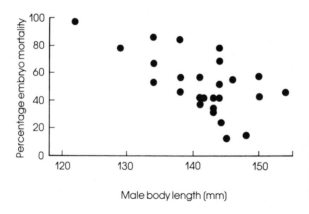

25 **Adaptive mate choice in bullfrogs.** Females prefer large males, and embryo mortality is lower in territories of large bullfrogs. *Source:* Howard [388].

Variation in Male Parental Care and Female Choice

In katydids and bullfrogs, males do not provide parental care for their offspring even though they assist their mates in various other ways. But there are species in which males actively provide for existing offspring; and in these species, variation in the potential paternal investment per offspring could be a relevant factor for female mate choice. The problem is, How can a female tell what a male will do when his young are available to be helped?

Birds often feed their mates during courtship. Females might therefore be able to assess how well a male can collect food through the frequency or quality of his courtship feeding. The first question is, Does courtship feeding correlate with the frequency with which males bring food later to their young? Here there are conflicting results, with the answer being "No" for the pied flycatcher [480], and "Yes" for common terns and herring gulls [589, 840]. No one, however, has yet demonstrated that variation in courtship feeding is used by female birds as a basis for rejecting one male in favor of another.

Females of some species might, however, use subtle indicators of male parental ability, like male body size, to pick a superior caretaker for their offspring. Katherine Noonan examined mate choice in a cichlid fish whose territorial males defend their free-swimming brood against predation [590]. If a female preferred large males, her offspring would probably get a more aggressive, protective male parent. Noonan placed each female cichlid in the central compartment of a three-chambered aquarium. There the female saw a large male and a small male in the two end compartments, but two black Plexiglas barriers in the central unit prevented the males from interacting in any way (Figure 26). Females watched the visual courtship displays of both males, and in 16 of 20 trials, they spawned in the nest near the large male.

Another example of female choice based on male body size comes from a study by Marion Petrie, who documented that female moorhens prefer *small* males [630]. Males of this chunky water bird perform much of the egg incubation duties. The larger the fat reserves of a male, the more days he can incubate, freeing the female to feed and produce another clutch of eggs. The smaller the male moorhen, the higher *proportion* of fat in his body. Petrie observed females fighting for males and found that the winners (the large females) paired off with small, fat males, which have the potential to be good parents.

Finally, observers of some songbirds have argued that perhaps the quality of a male's song offers yet another kind of indirect indicator of male parental ability. If a male collects food for his nestlings and fledglings, his foraging skill affects the survival of his young. Male foraging ability may be related to his size and physiological condition, which in turn might be signaled by the intensity, persistence, or complexity of his courtship displays. Clive Catchpool believes that this is the explanation for the preference of female sedge warblers for males that employ a greater number of

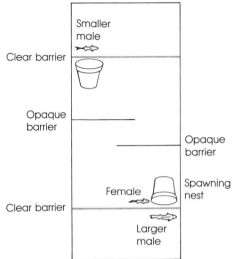

26 **Mate choice in a fish.** The figure shows the aquarium apparatus used to test female preferences in a cichlid fish. The female in the central compartment can observe the males in the end compartments, but the males cannot see each other or interact in any way. In these experiments, females preferentially spawned near the larger male. *Source:* Noonan [590].

syllable types in their complex songs [142]. Whereas moorhen females find small fat males irresistible, female sedge warblers quickly pair off with males that sing songs with a greater repertoire size, leaving late-coming rival females with the males singing less varied songs (Figure 27). Catchpool notes that male sedge warblers defend such small territories that both parents must forage off the nesting territory a good deal, especially when providing for their progeny. Therefore, females cannot assess their reproductive chances by analyzing a territory's qualities directly and so must judge males on the basis of their songs.

Note that female cichlid fish, moorhens, and sedge warblers apparently

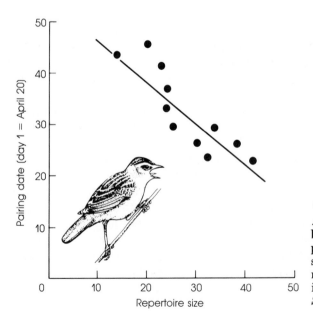

27 **Mate choice in a European warbler.** Females of this species prefer males with large song repertoires. Such males acquire mates sooner in the breeding season. *Source:* Catchpool [142].

favor large males, small males, and tuneful males, respectively. Why females of different species use such different criteria in selecting mates needs additional investigation, and there are many other unanswered questions about the evolution of female mate choice in species with "helpful" males.

Mate Choice by Males

The logic of the argument that females will be choosy when males differ in the material benefits they can offer can be used to predict when *males* will exhibit mate choice. The hypothesis that males are indiscriminate in picking partners is based on the assumption that copulations are not costly for males and that females do not differ much in attributes that affect male fitness. Often these assumptions are correct; sometimes they are not. For example, male butterflies pass a substantial quantity of materials to their mates in addition to sperm. That these secretions are expensive to produce is shown by the difficulty males have in supplying a full complement of materials to two females in quick succession (Table 2). If copulation commits a male to transfer valuable goods to a mate, it may be to a male's advantage to give them to a fecund female [193].

An example is provided by the Mormon cricket, which despite its common name has no religious affiliation and is not a cricket. Males of this large, flightless katydid transfer an enormous spermatophore to their mates along with their sperm [323]. The spermatophore is evidently nutritious because the female consumes it after copulation and sperm transfer are complete. Donation of a spermatophore reduces a male's body weight by about 25 percent; therefore it is reasonable to assume that the transferred materials cannot be quickly and easily replaced. This limits the number of times a male can copulate during his lifetime and presumably favors individuals that choose superior mates.

Sometimes bands of Mormon crickets march across the countryside eating farmers' crops and mating as they go. In these high-density populations, males have access to many potential mates. When they begin to stridulate from a perch, announcing their readiness to mate, females come running. (Note that this sex role reversal is in keeping with the prediction that when one sex provides a limited resource during copulation, the other sex will

TABLE 2

A cost of copulation: Males of the alfalfa butterfly cannot provide two females in quick succession with large amounts of ejaculate secretions

	First copulation (Mean values)	Second copulation (Mean values)
Duration of copulation	55 min	> 4hr to 10 hr
Secretions passed to female		
Weight (mg)	5.8	2.3
Volume (μl)	5.0	1.6
Minutes since first copulation	—	29

Source: Rutowski and Gilchrist [682].

compete for chances to mate.) In order to copulate, a female must mount the male, who then inserts his genitalia and transfers sperm and the spermatophore. But males refuse to transfer a spermatophore to some females. In Darryl Gwynne's study, the average weight of rejected females was significantly less than those that were "permitted" to copulate (Figure 28). By mating with heavier females, males transferred their sperm to more fecund individuals. A male that rejected a 3.2-gram female in favor of a 3.5-gram mate gained about 50 percent more eggs as a result of his choice [323].

Note that just as female cichlid fish cannot measure male parental care directly but must do so indirectly through a correlated character (body size), male Mormon crickets cannot measure female material benefits (egg number) directly but instead use female body weight as an indicator of fecundity. Males of a number of other animals, including stickleback fish, also use body size as a cue to discriminate among females of differing value. Male sticklebacks make a large parental contribution by building a nest and defending the eggs their mates deposit in the nest. If an aquarium-held male is given a simultaneous choice between two females of different size, he shuttles back and forth between the two, distributing his courtship time in a manner that is directly proportional to the eggs to be gained from the two females. If the larger female has 65 percent of the total number of eggs contained in the two females, the male stickleback will spend about 65 percent of his courtship time with her, devoting 35 percent to the less fecund individual [695].

Mate Choice by Females: Males Make No Parental Investment   We have reviewed evidence on the general hypothesis that when one sex has a valuable resource to offer another, and the other sex is variable in some fitness-affecting attribute, mate choice selection is likely to occur. Now there are many species in which males do not do anything to help their mates or offspring above

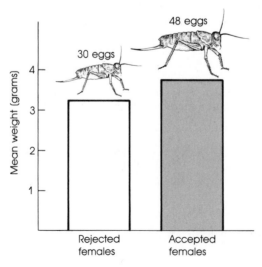

28 **Mate choice by katydid males.** Males of the Mormon cricket reject females that on average weigh less than those they accept. As a result of their choosiness, they gain an average of 18 extra eggs to fertilize. *Source:* Gwynne [323].

and beyond the transfer of sperm. In these species females are on their own; there are no nuptial gifts, no resources to take from the male's territory, no male protection for the offspring, nothing at all except the genes in the male's sperm.

It is among these species that we are particularly likely to see extravagant male characters, extreme sexual dimorphism, and the extraordinary courtship activities that piqued Darwin's interest in sexual selection in the first place. Males of some Australian bowerbirds, for example, defend an all-purpose territory and participate in parental care, whereas others like the satin bowerbird leave the parenting strictly to females. The males of the parental bowerbird species look pretty much like the females of their species, and their courtship routines are nothing special. The species in the non-paternal category are the ones with the remarkable bowers, the unusual displays and PDQ Bach vocalizations.

The same pattern occurs in the South American cotingas. In some species of these birds, males and females form cooperative, parental pairs; and these cotingas exhibit modest courtship behavior and reduced sexual dimorphism. But then there are species like the three-wattled bellbird [735]. The utterly nonparental males spend their days perched on top of forest trees producing ringing calls of stupendous volume that advertise their ownership of a display territory. Females fly from one male to the next, alighting on a limb of the calling tree. The beautiful white and chestnut red male bounces toward the drab female, inflating the three wormlike wattles that hang from about his beak and swinging the pendulous wattles seductively. Barbara Snow writes, "As the male opens his beak wide to utter the call he leans right over the [female] visitor. The latter leans as far away as possible, clinging precariously to the very end of the branch. The delivery of the call, with the visitor flinching at its tremendous noise, is an amazing sight and can only be described as an ordeal by sound which few visitors withstand" (Figure 29). Almost always the female departs rather than mate, just as female satin bowerbirds rarely permit a displaying male to inseminate them.

These observations are highly suggestive of female choice, but as we noted already, convincing evidence of nonrandom mating based on female preferences is needed if one is to claim female mate choice with authority. It is possible, for example, that the extravagant characters of species with no male parental investment evolved solely in the context of male–male competition and that females have not and do not currently discriminate among males of their species on the basis of these characters [24].

What Do Females Prefer?

Are the extravagant displays and bowers of male satin bowerbirds really used by females to choose among males? In order to study mate choice in these birds, Gerald Borgia had to find out who was mating with whom [82]. Other observers had seen bowerbirds copulating, but it was a long wait between matings. Females mate just once per breeding season, and most interactions between the sexes end with the departure of the unreceptive female. For a single bowerbird watcher to accumulate data on the relative

29 **Acoustical display** of the three-wattled bellbird. The male vocalizes at full volume in the ear of a female that has come to his calling perch territory.

mating success of a suitably large number of males would have required thousands of man-hours of observation. Borgia circumvented this problem by placing cameras at a set of bowers, cameras that were activated when a moving object, like a bowerbird, broke an invisible infrared beam passing through the bower avenue. Whenever a female entered the bower, she turned on the camera; and if the male joined her to mate, the event was faithfully recorded on film.

In 1981 Borgia's cameras and team of observers produced 212 records of matings at 22 bowers. The number of decorations, particularly snail shells and blue feathers, and measures of bower symmetry and the density of sticks used in its construction were highly correlated with male mating success. When Borgia and his crew removed decorations on a daily basis from 11 bowers, but left 11 others alone, significantly more matings occurred at the intact bowers than at those that were experimentally depleted [82].

Malte Andersson has conducted another experiment on the effects of male ornaments on female preferences. He studied an African widowbird, whose robin-sized males are endowed with an overgrown tail, about 1 meter in length, which is prominently employed in the bird's courtship displays. Males of this species defend nesting territories with useful resources for females, so it seems likely that females usually select mates indirectly by evaluating their territories. In order to perform his experiment on whether females are attracted to males with long tails, Andersson placed males randomly in one of four groups to eliminate any average difference in territory quality among groups. Having done this, Andersson picked one group of males to be transformed into super long-tailed males. He cut large portions from the tails of other males he captured and Superglued the stolen tail segments to the selected males, artificially lengthening their plumes. Other males had their tails cut and then reglued to the same length, a control group, while others were left with depleted tails. Males of all three groups continued to hold their territories successfully, but about four times as many females subsequently settled on territories of the super long-tailed birds as on the areas defended by the artificially short-tailed birds (Figure 30) [17].

Bowerbirds and widowbirds apparently can use elements of visual courtship by males to reach mating decisions. There are other species whose nonparental males emphasize sound signals in courtship. Do females use these cues to discriminate among males? Ann Hedrick found that females of one cricket species prefer those males able to trill for relatively long

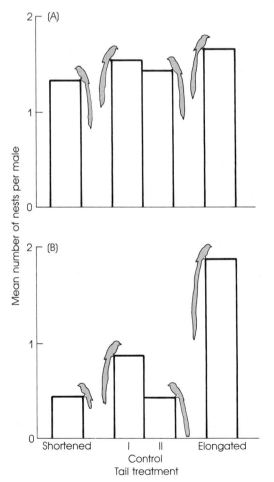

30 **Female preference for long-tailed males** in the widowbird. (A) The mean number of nests in a territory was the same for four groups of nine males prior to the tail alteration treatment. (The numbers at the base of the bars are the values for individual males.) (B) After the experimental treatments were performed, more nesting females were found in the territories of male widowbirds whose tails had been experimentally lengthened than in the territories of birds whose tail length was shortened or left unchanged. *Source:* Andersson [17].

periods [352]. As she points out, the beauty of examining the effects of acoustical signals on females lies in the ability to take the rest of the male away from the signal by using a tape-recording of his call to test its attractiveness. Customarily this is done experimentally by placing a female between two speakers, one broadcasting one song type and the other a different one. The female crickets studied by Hedrick cooperated by marching toward tapes of uninterrupted trilling call in 23 of 25 trials in preference to tapes of short bouts of song. The experimental procedure eliminates the possibility that female preferences that appear to be based on acoustical signals are actually affected by other attributes of males or by subtle male–male effects on each other (when two males are present with the same female).

This technique has been used instructively in the study of acoustical mate choice by frogs and toads. Brian Sullivan showed that in nature female Woodhouse's toads are far more likely to mate with a male giving many calls per minute than one singing at a low rate [752]. This is a toad whose males cluster in ponds and river pools forming a chorus of singing males; females listen to many males and then move to the mate of their choice. To test whether female preferences exhibited under natural conditions could have been influenced by the acoustical properties of male calls, Sullivan took females into the laboratory, where they were placed in an arena with two speakers that played identical calls but at different rates. In 36 trials, females always hopped over to the speaker with the higher call rate.

In the population of Woodhouse's toad studied by Sullivan, there was no correlation between body size and the ability to call many times per minute, and therefore females did not confer a mating advantage on large males. Females of a Panamanian frog do prefer large males [687]. The tungara frog forms choruses of males floating in a pond, filling the air with their two-part calls, a whine usually followed by one or more "chucks." Females enter the aggregation and move about before finally permitting one male to grasp them. Large males have a far higher probability of mating than small ones do, and not because they interfere with small males (Figure 31). Michael Ryan determined that the calls of large males differ from those of small ones only with respect to the sound frequency pattern of the chuck. Large males are able to put relatively more energy into the lower frequency components of this part of the call, and in choice experiments of the sort already described, females consistently went to the speaker broadcasting calls with lower pitched chucks [684].

The Adaptive Significance
of Female Preferences

Thus, there is ample evidence that even when the male contribution to his mate is minimal, females have definite preferences that affect the mating success of different males. But do females gain a reproductive advantage from their choices? If males provide only sperm to their mates, females cannot use male courtship behavior and physical attributes as indicators of the material benefits they will receive from potential mates. So what is it that they *might* gain?

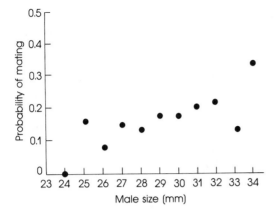

31 **Female preference for large males** in the tungara frog means that the probability that a male will mate is positively correlated with male size. *Source:* Ryan [685].

There have been two major categories of hypotheses on this question, the GOOD GENES hypothesis and the competing views of R. A. Fisher, one element of which is called the RUNAWAY SELECTION hypothesis [31, 347]. Proponents of the good genes argument believe that the traits males exhibit provide accurate information on the utilitarian value of the males' genes [437, 881]. In other words, a male's extravagant characters provide information to the female on the value of the male's genes in helping her produce viable offspring. Just as females of species with male nutrient investment or parental investment might assess the value of a male's services through indicators such as body size or singing behavior, so, too, in species without these material benefits, females might assess male genetic quality indirectly through his courtship performance [766]. For example, the volume of a male bellbird's "bonk" might be correlated with his age, body mass, or physiological condition. To the extent that a male could pass on alleles that would permit the female's offspring to survive longer, feed more efficiently, or grow larger on less food, the female would gain.

Proponents of runaway selection, however, argue that "over-elaborate" male courtship can evolve even though females are not assessing male genetic quality. They point to mathematical models developed by Russell Lande [465] and Mark Kirkpatrick [430], models that develop the runaway selection argument of Fisher [261]. According to these scenarios, females in an ancestral population might have had a slight preference for certain male characteristics, perhaps *initially* because the preferred traits were indicative of some survival advantage enjoyed by the male. Females that mated with preferred males would produce offspring that carried the genes for the mate preference from their mother *and* the genes for the desired male character as well. Male offspring that expressed the trait preferred by a plurality or a majority of females in the population would enjoy higher fitness in part simply because they possessed the key cues that many or most females found attractive. Female offspring that found these traits attractive would gain by producing preferred sons.

The fact that mate choice genes and the preferred male trait genes are inherited together is the foundation for the runaway process because, as the models of Lande and Kirkpatrick show, it is entirely possible in theory for ever more extreme female preferences and more extreme male characteristics to evolve and spread together. The runaway process only ends when natural selection against too costly or too risky displays balances sexual selection in favor of traits that are appealing to females. Thus, if at one time female three-wattled bellbirds preferred males that could give loud calls because such males could deter intruders from a greater distance, they now favor exaggeratedly noisy males because the mating preference has taken on a life of its own, a preference ultimately enabling the females to produce exceptionally attractive sons.

Even more dramatically, the Lande–Kirkpatrick models demonstrate that right from the start of the process there is no need for female preferences to be directed at male traits that are utilitarian in the sense of improving survival, feeding ability, and the like. Traits opposed by natural selection for improved viability can spread through the population due to the runaway process. Instead of mate choice for genes that promote the development of useful characteristics in offspring, runaway selection can then yield mate choice for *arbitrary* characters that are a burden to individuals in terms of viability, a disadvantage in every sense except that females mate prefentially with males that have them!

Testing the Good Genes
and Runaway Selection Hypotheses

The development of good genes and runaway selection arguments has run away from the few attempts to test these alternatives. It has proved very difficult to derive *competing* predictions from the two hypotheses that everyone agrees would provide a way to settle the issue. Instead, what we have for the most part are attempts to see whether existing evidence is compatible with one or the other hypothesis.

Advocates of the good genes approach note that females can make utilitarian choices among males on the basis of genetic quality in species whose males have *not* evolved exaggerated display and ornaments. For example, in some species females actively refrain from mating with their brothers and fathers [382, 546, 611]. Typically, in these cases females learn from experience to identify males they have grown up with, and they use the proximate cue of familiarity to avoid males that will usually be very close relatives. In so doing, they gain an ultimate genetic benefit by not producing highly inbred offspring (Chapter 9).

Moreover, in house mice, females favor males that do not carry the t allele, an allele that is lethal in double dose [473]. This is true even though males with the $+/t$ genotype tend to be dominant to males with the favored $+/+$ genotype. Here the distinguishing cue is an olfactory one, as shown by the greater frequency with which females entered a side chamber containing bedding from a $+/+$ male rather than entering a chamber with bedding from a $+/t$ male. By avoiding such males, females with a $+/t$ genotype avoid the loss of one-quarter of their litters (those with the lethal t/t genotype).

In addition, Malte Andersson has developed a genetic model to be applied to species with elaborate ornaments or displays; the model shows that if preferred characters develop *only* in individuals in top condition, then such traits and the preferences for them can spread in the population via female choice [20]. The idea here is that if sexual ornaments and elaborate courtship displays truly reflect the health or nutritional status of a male, females that pick top males produce more viable young and top males gain by advertising their condition in ways that weaker, diseased, or food-deprived males simply cannot match.

The extreme displays and ornaments of some males are probably hard to match by others in their population. For example, in satin bowerbirds, only older dominant males can provide the things that females like, including complex vocalizations and well-built bowers with many scarce decorations (Figure 32) that must be defended or stolen from aggressive rivals [84-86]. A female preference for attributes that males might be hard-pressed to produce has also been observed in guppies, whose females favor males with large tails (Figure 33), active courtship displays [68], and considerable amounts of orange in their body coloration [435]. Heavily parasitized males infected with gut nematodes or skin parasites are incapable of sustaining a high display rate, so females are unlikely to mate with partners susceptible to damaging parasites [421]. Moreover, Astrid Kodric-Brown points out that the preferred orange pigments are carotenoids, scarce substances that guppies must accumulate from high-protein foods, such as small insects. By favoring males with large amounts of orange color, females are in effect choosing males that have been able to find more high-quality food than average.

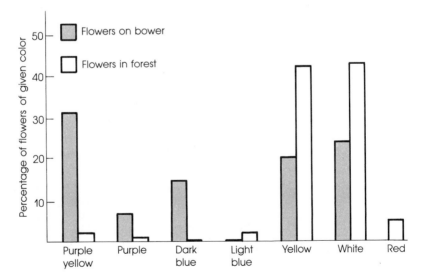

32 **Satin bowerbird males** prefer rare ornaments for their bower decorations. The flowers appearing at a sample of bowers were generally taken from species that were scarce in the woodlands in which the bowers occurred. (Flower abundances were measured by censusing a transect route through the woods.) *Source:* Borgia [86].

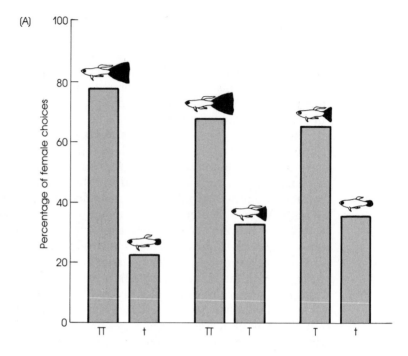

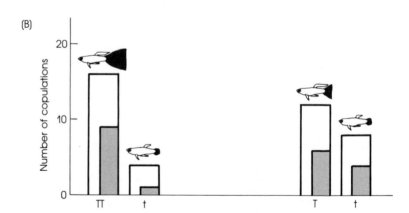

33 **Female preference for large-tailed males** in the guppy. (A) Females consistently approached males with larger tails in choice experiments, and (B) they mated with these males first, thus enabling large-tailed males to sire more offspring (narrow bar inside wider bar). TT, large tail; T, experimentally shortened tail; t, naturally small tail. *Source:* Bischoff et al. [68].

William D. Hamilton and Marlene Zuk tested the hypothesis that coloration might be an honest signal of male physiological condition in birds [334]. They proposed that only healthy individuals can succeed in maintaining showy, brightly colored plumage in top condition. Bird species vary in their susceptibility to attack by debilitating parasites, and this variation enabled Hamilton and Zuk to test the prediction that "male showiness" would be correlated with the risk of parasite infection for a species. The logic of the prediction is that when males of a species differ considerably in their parasite load, it is to a female's advantage to choose a partner who honestly signals that he is healthy, because his offspring may inherit his resistance. In species with little chance of parasite infection, this information is of little significance to selective females, and thus they should not favor males with bright, health-sensitive feathers.

To test their prediction, Hamilton and Zuk used data from the parasitology literature to determine the risk that a member of a particular species would be infected. They ranked species "showiness" by having a person unfamiliar with the parasite data rank each species from 1 (very dull plumage) to 6 (very brilliant plumage). They found a highly significant correlation between the showiness of a species and the risk that individuals of that species would be parasitized.

Subsequently, A. F. Read tested the parasite–plumage hypothesis with another sample of 109 bird species [661]. He found the same correlation between gaudy feathers and the presence of protozoan blood parasites. Further, he examined whether the correlation might be spurious. For example, it might be that parasites tend to infect large bird species for some reason related to body mass. If large birds were also brightly colored as a rule, then the correlation between parasites and bright plumage would be an incidental effect of the causal relationship between body size and parasite infestation. By controlling for body size and a host of other factors, Read showed that these were not responsible for the finding that birds susceptible to parasites are more likely to have colorful feathers. This discovery supports the argument that such plumage has evolved via functional mate choice.

All of the above examples are merely consistent with a good genes effect, for they can also be interpreted as arising from a Fisherian process. The really direct way to test the good genes hypothesis would be to demonstrate that female choice of certain types of purely sperm-donating males resulted in higher viability of the resulting offspring. These are difficult experiments to do. The few that exist have produced conflicting results. Christine Boake showed that some male flour beetles released more pheromone and mated more females than the average male does [72]. But the survival rate of the offspring of the preferred males was no greater than that of offspring from less favored males.

On the other hand, Bruce Woodward has found important paternal effects on offspring viability in two species of spade-foot toads in which females appear to be capable of choosing among males [867, 868]. Woodward took eggs from females in the laboratory and divided them into two batches; one half were fertilized by sperm taken from a large male and the others were fertilized by sperm from a smaller male. The eggs and tadpoles were then held under identical experimental conditions, and their growth rate and survival measured. In one species, eggs fertilized by a large male's sperm became faster growing tadpoles than eggs fertilized by a small male's sperm (Table 3). Spade-foot toad tadpoles develop in small temporary ponds in the southwestern United States, and in such an environment survival might

TABLE 3
Weight of juvenile toad offspring having the same mother but different (large or small) fathers

Offspring character	Large male's offspring	Small male's offspring	F value	P value
Initial weight (grams)	0.49 (44)	0.48 (44)	0.05	> 0.05
Days to transformation	37.6 (42)	38.1 (44)	1.99	0.16
Weight after 15 days	0.71 (42)	0.60 (42)	8.09	< 0.01
Change in weight from day 0 to day 30	0.40 (40)	0.25 (39)	5.66	< 0.05

Source: Woodward [867].

often depend on speedy development. In a second species of spade-foot toad, growth rate was the same for tadpoles sired by large and small males. However, significantly more toadlets survived if they had been fathered by a large male. These experiments, and a few others [724], suggest that females of some species could gain good genes by choosing certain kinds of males.

There are, however, so few strong tests of the good genes and the runaway selection hypotheses that most biologists feel the verdict is still out. Probably the two kinds of processes combine to drive the evolution of male characteristics and female preferences in many animal species [19]. What is needed now are analyses that tell us the relative importance of these intriguing possible contributors to sexual selection.

SUMMARY

1 Sexual reproduction creates a social environment of conflict and competition among individuals, as each tries to maximize its genetic contribution to subsequent generations. Males usually make many small gametes and exhibit little parental care, and try to fertilize as many eggs as possible. Because females make many fewer, larger gametes and often provide additional parental investment in their offspring, they are the object of male mating competition and may have the opportunity to choose among many potential partners.

2 Evolution by sexual selection occurs if genetically different individuals differ in their reproductive success (1) as a result of competition within one sex for mates or (2) because of their differential attractiveness to members of the other sex. The competition for mates component of sexual selection appears to have caused the evolution of many elements of male reproductive behavior, including a readiness to fight to monopolize mates, forced copulation as a male mating tactic, and alternative mating tactics within a single species. When more than one route to copulations exists, the multiple routes often represent behavioral options within one conditional strategy, which is an inherited program of behavior in which the tactics exercised by an individual reflect the social or environmental conditions that the individual happens to encounter.

3 A copulation may or may not yield egg fertilizations for a male. The risk that a male's sperm will be superseded by those of a rival has led to many postcopulatory traits, most commonly the guarding of a mate until she has laid her eggs.

4 Demonstrating the mate choice component of sexual selection requires that nonrandom mating by one sex occur because of preferences of the other, and not because of competitive interactions within one sex, usually the males. Females of some species prefer males that offer more material benefits, whether these be nuptial gifts, resources monopolized in territories, or parental care. Mate choice by males occurs when copulation is costly and females differ in their fecundity.

5 Mate choice by females occurs even in some species in which there is no male parental investment or other donation of material benefits by males to mates. The evolution of mate choice of this sort could arise as a result of selection by females of males whose genes will enhance the viability of their offspring (the good genes hypothesis). Alternatively, extravagant male features could spread through a population because of runaway sexual selection in which even arbitrary elements of male appearance or behavior become the basis for female preferences. Exaggerated variants of these elements can be selected strictly because females prefer to mate with individuals that have them. The relative importance of these mechanisms of sexual selection remains to be determined.

SUGGESTED READING

An entertaining article by James Lloyd is recommended as an introduction to the behavioral ecology of sexual selection [493]. General reviews of the effects of sexual selection are found in books by David Barash, Richard Dawkins, Michael Ghiselin, Michael Ryan, Robert Trivers, E. O. Wilson, and J. F. Wittenberger [41, 182, 278, 687, 787, 853, 865]. Stevan Arnold [31] and Richard Dawkins [185] have a go at explaining the genetical models of runaway selection developed by Russell Lande and Mark Kirkpatrick.

Richard Dawkins, William Cade, and Wallace Dominey have written understandable accounts of intraspecific variation in mating tactics [123, 183, 203]. Jonathan Waage's articles on the black-winged damselfly provide an unusually clear description of dramatic male competition for eggs to fertilize [805, 806]. Randy Thornhill's studies of hangingflies show how mate choice operates [763, 764, 767], and Gerald Borgia's work with satin bowerbirds helps discriminate among alternative hypotheses on the evolution of bizarre sexual behavior [84].

DISCUSSION QUESTIONS

1. Male rats, sheep, cattle, rhesus monkeys, and humans that have copulated to satiation with one female are speedily rejuvenated if they gain access to a new female. This is called the "Coolidge effect," supposedly because when Mrs. Coolidge learned that roosters copulate dozens of times each day, she said, "Please tell that to the President." When the President was told, he asked, "Same hen every time?" Upon learning that roosters select a new hen each time, he said, "Please tell that to Mrs. Coolidge." Why might sexual selection have favored males of some species that exhibit the Coolidge effect? What kinds of animals will lack the Coolidge effect?

2. There are some species in which males form a prolonged association with an immature female, rather than searching for or attempting to defend mature receptive females [400]! What are the costs of this behavior? What conditions are required for its evolution? In other words, what do you predict about the abundance of mature receptive females in these species, or the costs of locating them, or the egg fertilizations to be gained from mating with previously mated as opposed to virgin females? If males sometimes fight for possession of immature females, what prediction can you make about the time to maturity of a female and male investment in fighting?

3. In species with no male parental investment or other material dona-tion to their mates, female choice for "good genes" should in theory reduce the variation among males with respect to good genes. If there were no genetic variation among males with respect to genes that affect the surviv-ability of offspring, could selection in these cases favor female choice? What happens to the good genes hypothesis if males do not vary genetically? How might the coevolutionary race between hosts and their pathogens, parasites and predators maintain variance in viability genes possessed by males? See also [830].

4. To collect their nuptial gifts, male hangingflies incur time, energy and risk costs (many die in spider webs while hunting). What conditions were required for the spread of nuptial gift-giving in an ancestral popula-tion of hangingflies in which almost all males did *not* offer insect prey to their mates, and so did not incur these costs?

14

The Ecology of Mating Systems

Sexual selection shapes the often conflicting reproductive strategies of males and females. Together the tactics of the sexes define the mating system of a species, which have traditionally been categorized on the basis of three criteria: (1) the number of partners an individual copulates with in a breeding season, (2) whether males and females form pair bonds and cooperate in parental care, and (3) how long the pair bond is maintained. Because the many permutations of these factors creates an unwieldy list of mating systems, I favor focusing on the number of mates acquired by an individual in a breeding season. Even with this one variable, one can identify considerable diversity in mating systems, and explaining that diversity is the point of this chapter. The ecological classification of mating systems developed by Stephen Emlen and Lewis Oring [239] is heavily relied on here. These researchers argue that sexual selection is influenced not just by the disparity in parental investment made by males and females of a species but also by a number of key ecological factors, including predation and the distribution of food and other important resources, that determine where females can be found. In turn, female distribution typically defines what mating tactics will be productive for males, and understanding this point has helped greatly to explain why animals exhibit so many different kinds of mating systems.

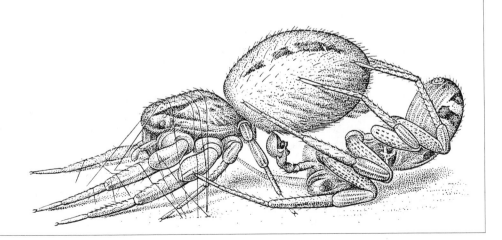

The Prevalence and Varieties of Polygyny   The vast majority of animal species practice POLYGYNY. A polygynous species is one in which some males are able to mate with more than one female in a breeding season. Because the sex ratio of most species is roughly 1:1, the fact that some males achieve polygyny means that other males will be MONOGAMOUS (mate with a single female) and others may not mate at all, given that females typically mate with just one male. The polygynous satin bowerbird is a case in point, with a few males copulating with a great many females, and many males failing to secure a single mate in any one nesting period [82].

Polygyny is common because male reproductive success is often a simple function of the number of females inseminated, a relation generating intense competition among males for mates. Those males that "win" this competition are, by definition, polygynists.

Stephen Emlen and Lewis Oring proposed that how males compete to be polygynists is dramatically affected by the spatial distribution of females, which in turn is a function of predation pressure, food dispersion, and other ecological factors operating on females. The essence of their argument is that various environmental pressures determine whether females will be clumped or scattered; clumped females can be monopolized economically by a male, whereas scattered females cannot be controlled without a far greater expenditure of time and energy. You will note that Emlen and Oring's argument rests on the cost–benefit approach to territoriality discussed in Chapter 9. The larger the defended area, the higher the costs for a would-be territory holder, and the less likely territoriality is to evolve.

Female Defense Polygyny

When females live together in a herd or group, the time and energy needed to patrol the area around them is modest, favoring the evolution of direct defense of clustered females, or FEMALE DEFENSE POLYGYNY. For example, gorilla females travel together in a band, perhaps for protection against leopards. Males fight for groups of females, and the winner prevents other males from mating with "his" females [342].

Female tree lizards of the Sonoran Desert do not form tight-knit herds or bands, but three or four commonly coexist in a single mesquite tree. Males try to defend a couple of neighboring mesquites, and the winner will generally have access to six, seven, or eight mates. If males are territorial in order to mate with local females, experimental removal of the "harem" should cause males to abandon the trees they have been guarding, and this is precisely what happens [548].

Yellow-bellied marmot females do form close social units. Females often remain near their birth site, as is typical for mammals (Chapter 9); consequently, clusters of mothers, daughters, and aunts arise, presumably for improved protection against coyotes and eagles [25]. Males search for locations with one or more resident females, and those able to find and defend a meadow with two or more mates practice female defense polygyny. As one might expect, the reproductive success of a male increases as the number of females living in his territory increases. Perhaps surprisingly, however, the number of young produced by a female each year falls as

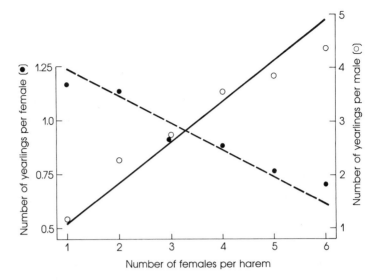

1 **Polygyny versus monogamy.** Annual reproductive success of male and female marmots. Polygynous males have more offspring that reach 1 year of age than monogamous males do. Females in large groups have fewer offspring that survive for at least 1 year than monogamous females do. *Source:* Armitage [25].

group size increases (Figure 1) [208]. Why then do females live in groups and permit some males to be polygynous? Females would seem to be better off seeking to live monogamously with a territorial male.

To analyze the fitness of two alternatives we must consider *lifetime* reproductive success, not just one year's output. Perhaps females form large groups of individuals willing to sacrifice a certain amount of annual reproductive output in order to thwart predators better. It could be that such females will live longer and reproduce for more years than monogamous females [231]. However, harem size and female survivorship are not positively correlated [25].

A second hypothesis reminds us that what is adaptive for an individual depends on what its other options are [865]. It could be that female marmots that lived alone with a mate would enjoy higher lifetime reproductive success than that achieved by living in a group. But in order to do this, they would have to exclude other females, and this may be costly for a variety of reasons. Incoming females may join a group and accept reduced reproductive success if the attempt to find better unoccupied habitat would be futile.

Predation pressure is not the only ecological variable that favors sociality among females and the direct female defense tactic of males. To the extent that lionesses form prides to hunt cooperatively and defend their kills against hyenas and the like, female foraging ecology creates "harems" that males then struggle to monopolize [700]. Likewise, females of a tropical bat form groups that forage together at night (as shown by radio-tracking studies), always returning to the same spot in their home cave to roost during the day. These females appear to be social in order to cooperate in finding and sharing food. The existence of the roosting groups is tailor-made for female defense polygyny. One male can easily keep intruders away from all the females in a cluster (Figure 2). Gary McCracken and Jack Bradbury have shown through electrophoretic analysis that successful territory holders father 60 to 90 percent of the offspring of the females in their roost and that one male may sire as many as 50 pups during his tenure as harem master [542].

In some species it may be possible for males to create and control clusters of mates by herding them together. Certain tiny siphonoecetine amphipods live in great numbers in the fine gravel of shallow marine bays. They typically construct elaborate cases composed of bits of pebble, fragments of mollusk shells, and the like. Males move about in their houses, searching for females in their abodes. Single females are "captured" and their houses glued to that of the male. A male may succeed in constructing an apartment complex containing his case cemented to the abodes of three females (Figure 3) [415]. One presumes that he monitors and monopolizes the sexual activity of his apartment-mates.

Resource Defense Polygyny
Females of some species do not live in clusters, and this changes the options available to males competing for mates. However, a male still may be able

2 **Female defense polygyny in a bat.** The male at the top left guards a roosting cluster of females. Photograph by Gary McCracken.

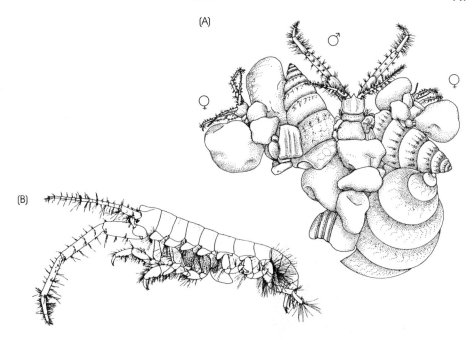

3 **Female defense polygyny in a marine amphipod.** (A) The male has glued the houses of several females to his case. (B) An individual without his house. Drawings courtesy of Jean Just.

to defend a territory that makes him polygynous, *if* the resources females need are clumped spatially. For example, sometimes the foods females need are distributed in discrete patches. By defending valuable food centers, a male may get to mate with any number of females in search of a meal. Males of an African antelope, the impala, compete to control rich grazing pastures that attract roving herds of females. The larger and richer the field, the longer females will remain to harvest the food; consequently, more females will have time to come into estrus and mate with the resource-defending male [401]. The unrelated vicuña of South America has evolved a convergent system; the males battle for the best foraging areas, which are in short supply, and mate with the females that join them on their feeding territories [267].

The orange-rumped honeyguide is a representative of polygynous, resource-defending birds [160]. This species has a very specialized diet, namely, beeswax, which is available in exposed hanging combs of the formidable giant honeybees of southeast Asia. The birds are immune to bee stings and feed safely on the comb. The wax produced by one large colony of giant honeybees can support several birds. Males compete for possession of a bee colony and permit several females, but no adult males, to feed on the wax in their domain (Figure 4). In return, females sometimes copulate with the territory owner.

In systems of this sort, the amount of resources controlled by a male

4 **Resource defense polygyny in the orange-rumped honey guide.** Males defend nests of the giant honeybee and permit females to feed on the beeswax resource under their control.

should be correlated with his opportunities to copulate. This prediction has been tested experimentally with black-winged damselflies. These agreeable insects are perfectly willing to defend strips of floating vegetation installed in a stream by an observer. By monitoring the number of females attracted to territories with different amounts of the same kind of oviposition resource, one can determine the effects of this variable on the number of matings that take place there (Figure 5) [7]. Although resource-rich sites do attract more females, they also attract more competing males, an effect causing a higher turnover in territory possession. Thus, there are trade-offs for males attempting to monopolize sites that have a great deal of oviposition resource needed by females.

In those species of birds whose males defend resource-based territories, there are also trade-offs for females attracted to males with unusually productive, valuable sites. Males of most birds, but not orange-rumped honeyguides, offer parental care to the offspring of their mates. A female that pairs with a male that already has a partner is forced to share the parental care of the male with his first mate. In order for a polygynous

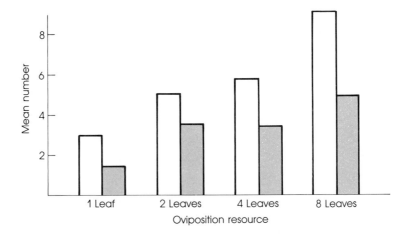

5 **Resource defense polygyny in the black-winged damselfly.** Over a 10-day experimental period, territories with more oviposition resource (strips of floating plant material) attracted a greater mean number of egg-laying females per day (open bars). Males that controlled these sites generally copulated with more females than males that defended sites with fewer leaves (solid bars). *Source:* Alcock [7].

arrangement to be advantageous to females in a species with parental males, the resources gained by occupying the territory must be exceptional. If a female were not compensated by an abundance of food (or an especially good nest site), selection would favor females that avoided already-mated males.

We can test this prediction by examining the reproductive success of females paired with males with different numbers of mates. Michael Monahan studied a population of red-winged blackbirds nesting in a hayfield and found that the number of fledglings produced by females in harems of different sizes did not vary significantly (Figure 6) [565]. This finding suggests that females can in some way assess their reproductive chances in different territories, weighing the number of nesting females already present against the foraging value or safety of the site.

Michael Carey and Val Nolan conducted a similar study with the indigo bunting, a small songbird whose parental males may attract either 0, 1, or 2 mates to their territories in old fields [133, 134]. Monogamous females whose relationship with a male lasted the whole breeding season had only slightly greater reproductive success than females that participated in a polygynous arrangement (Table 1). Because the second female in a "harem" usually arrives on the breeding ground after most males with acceptable territories have been claimed, her "expected" reproductive success should she pick an unmated male is probably very low. By joining a mated male on a superterritory, she does better.

But in the pied flycatcher, females that mate with a polygynous male have significantly lower reproductive success than those that have the

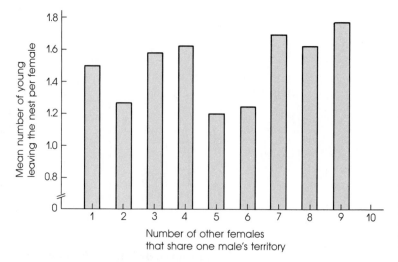

6 **Polygyny and reproductive success.** The number of offspring fledged per female red-winged blackbird in territories
with different numbers of females mated to the resident male shows
that polygyny imposes no reproductive penalty on females. *Source:*
Monahan [565].

undivided attention of the father of their offspring. The inability of males
with two mates to help both females as much as a monogamous male could
hurts each female's chances of producing fledglings. So why do females
choose to mate with polygynist flycatchers?

Rauno Alatalo and his co-workers have suggested that they do not choose
to do so but are deceived by their males, who manage to maintain two
separate territories, each containing a tree with a nest hole (a critical
resource for female flycatchers). By shuttling back and forth between the
two territories, males can attract two mates that do not know of each other's
existence. Each female lacks the information that would permit her to avoid
a bad decision, and so she settles down with her mate on one of his terri-

TABLE 1

Reproductive success of male and female indigo buntings engaged
in different mating systems

Pair Bond	Number of pairs in which age of male was		Mean number of fledglings per	
	≥ 2 Years	1 year	Male	Female
Polygyny	12	2	2.6	1.3
Monogamy				
Season-long	34	6	1.6	1.6
Short-term[a]	22	30	0.4	0.4

Source: Carey and Nolan [133].
[a]Pair bond lasted less than three-quarters of the breeding season.

tories only to receive relatively little help feeding her young later from her "polyterritorial" male. Even though each of his mate's average reproductive success may be lower than that of monogamous females, the *sum* of their output is greater than that of monogamous females, and this confers an advantage on polygynist males [4, 6].

Lek Polygyny

The direct defense of females or the resources they need makes intuitive sense as a male tactic for securing many mates. There are, however, species whose males are highly territorial despite the fact that their very small territories contain no harem of mates nor any resource of value to potential mates. Females visit these "symbolic" territories, but just to select a mate; after copulating, the female departs and may never see the male again. This odd mating system gets its title, LEK POLYGYNY, from the fact that often male territories are clustered in a traditional display site, or lek.

The South American white-bearded manakin is a typical lekking species [483]. Each male's territory contains nothing of value to a female, consisting only of a sapling or two in the forest and a bare patch of ground underneath the perch site, which the bird clears of leaves and debris to make a display court (Figure 7). There may be as many as 70 display courts in an area only 150–200 meters square. The male begins his display routine with a series of rapid jumps between perches, loudly snapping his clublike wing feathers together in flight. The male then pauses with body tensed before jumping to the ground with a snap and immediately back to the perch with a buzz, and then back and forth "so fast he seems to be bouncing and exploding like a firecracker" [736].

The arrival of a female at the lek encourages many males to display simultaneously, producing an uproar that can be heard far away in the forest. If the female is receptive and chooses a partner, she will fly to his perch for a series of mutual displays, followed by copulation. Afterward, she leaves to begin nesting, and the male manakin remains behind to court newcomers.

Great variation in the mating success of males characterizes lekking species. Alan Lill found that in a lek with 10 manakins, one male chalked up nearly 75 percent of the 438 recorded copulations; a second male mated 56 times (13 percent) while 6 males accumulated a minor total of only 10 matings [482].

There are obvious similarities here to the mating system of satin bowerbirds (Chapter 12), with females visiting male territories only to mate with males whose courtship display borders on the absurd and with some males enjoying great success while others hardly mate at all. Lill was unable to find any characteristic that set successful males apart from the failures at the lek, but we have seen that male satin bowerbirds with well-made and well-decorated bowers secure more matings. For some other lekking birds, copulations go disproportionately to males able to produce their strange displays at a high rate [282, 368].

Lek behavior is by no means restricted to birds. For example, during the breeding season, males of the bizarre West African hammer-headed bat

7 **Lek polygyny in the white-bearded manakin.** Males defend tiny display sites; females (shown here on far left and far right) visit the lek to select a mate from among the several males present.

(Figure 8) gather in the evening along river edges at traditional display areas [90]. From a perch high in a tree, each bat defends a territory that is 10 meters in diameter. While hanging upside down, males produce loud cries that sound like "a glass being rapped hard on a porcelain sink." Receptive females fly to the lek and visit several males, each of which responds with a paroxysm of wing-flapping displays and strange vocalizations (note the convergence with manakins and satin bowerbirds). Most males fail to attract mates; 6 percent of the males at one lek in 1974 were responsible for nearly 80 percent of all matings.

Why do male manakins, bowerbirds, and hammer-headed bats behave the way they do? Jack Bradbury has argued that were females of these species clustered, males would exhibit another mating system. Lekking is what males do as a tactic of last resort, because alternative tactics are not

practicable when females are widely and evenly dispersed. In the lek-forming sage grouse of western North America, females travel over an area of about 1000 hectares, whereas in nonlekking grouse, female home ranges are between 1 and 10 hectares [283]. Similarly, female manakins, bower-birds, and hammer-headed bats do not live in permanent groups but travel great distances in search of widely scattered sources of food, especially figs and other fruits of tropical trees. Defense of the food resource is made difficult by the unpredictable and irregular production of fruits by these trees. A male that tried to hold a territory around one tree might have a long wait before it began to bear attractive fruit; and when a tropical fig tree does begin to produce, it often yields so much that hordes of consumers descend upon the tree, making it all but impossible for one male to defend the fruit. Thus, the feeding ecology of females of these species makes it hard for males to monopolize females directly or indirectly. Instead, they must display their merits to choosy females that come to leks to inspect them.

8 A mammal that practices lek polygyny: the hammer-headed bat. Photograph by A. R. Devez.

The hypothesis that female diet controls male mating systems has been explored in the birds-of-paradise, a group with some resource-based territorial species and some lekking species. Bruce Beehler and Stephen Pruett-Jones discovered a strong relationship between the proportion of fruit in the diet and the occurrence of lek mating systems (Figure 9) [52]. Insect-eating birds-of-paradise defend territories, presumably in locations with superior supplies of insects. Fruit-eaters defend little display sites that females visit to pick a mate. The fruit trees that New Guinean birds-of-paradise rely on are also scattered and often heavily used by numerous species.

There are many unresolved or only partly resolved puzzles about lek polygyny. For example, why do satin bowerbirds display at dispersed sites, while male manakins and hammer-headed bats pack themselves into relatively small areas? Are the tightly clustered species those in which the females engage in active mate choice and refuse to mate with isolated males, preferring to select among a group of simultaneously displaying males? Or is the dispersion of males a function of the nature of female

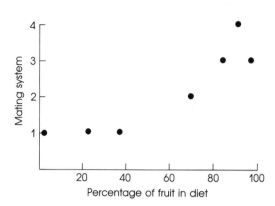

9 **Relation between diet and mating system** in birds-of-paradise. The greater the percentage of fruit in the diet, the more likely the species is to practice lek polygyny. Mating systems: 1, monogamy with birds in dispersed territories; 2, monogamy but pairs do not defend an all-purpose territory; 3, lek polygyny with male display sites somewhat scattered; 4, lek polygyny with display sites highly clumped. *Source:* Beehler and Pruett-Jones [52].

home ranges? Do some patterns of female distribution create "hot spots" where numbers of females will appear, a distribution pattern favoring males that aggregate at these locations [91]? Or do lekking males cluster only when they are especially vulnerable to predators (Chapter 11)?

For lekking *birds*, we can ask, Why don't males help their mates, given that monogamy with parental care is the preferred option for the vast majority of birds (see p. 458)? Perhaps in lekking species, even though it would be to a female's advantage to have the assistance of a male in rearing the young, the parental benefit *to the male* in terms of increased offspring survival is outweighed by the cost of reduced mating opportunities. In tropical rain forests, manakins and other lekking males can breed nearly year round, a condition creating multiple mating chances for males that are not tied up with one female's brood. However, many birds that form leks are not tropical species but breed in northern temperate regions.

Alternatively, females may actually gain by not having the male helping at the nest. Predation is apparently *the* major problem for nesting tropical birds, and it could be that having a second adult moving to and from the nest would only increase its conspicuousness to observant predators. This decreases the benefit of male parental care to the female (and the male). Moreover, because many fruit-eating female birds feed themselves and their offspring on abundant fruits that can be quickly gathered, the utility of male parental care is diminished. Discriminating among these and other competing hypotheses will require inventive tests.

Scramble Competition Polygyny

Polygyny is not always associated with territoriality. Males of many animals compete by trying to outrace rivals to receptive females; frequent winners will be highly polygynous. James Lloyd has tracked flashing males of a *Photinus* firefly that search nonaggressively for females through Florida woodland. Males search, and search, and search some more, for their scarce and widely distributed females. Lloyd walked 10.9 miles while tracking 199 signaling males and saw exactly two matings. When he did spot a female, she had to signal for an average of only 6 minutes before a male found her [495]. Males of this firefly ignore each other and do not try to defend territories. Instead, mating success goes to individuals that are the most persistent, durable, and perceptive searchers, to males able to outrace other males to the occasional receptive female who may appear almost anywhere over a broad area.

The thirteen-lined ground squirrel is a vertebrate equivalent of the searching firefly, and it, too, practices SCRAMBLE COMPETITION POLYGYNY, with reproductive rewards going primarily to individuals that are adept at finding females rather than at beating up other males [708]. Aggression occurs in this species, but not in defense of an area and not to monopolize a particular female. Instead, males wander widely trying to find estrous females during the brief period each year (less than 2 weeks) when females will mate.

The most dramatic form of scramble competition polygyny, the EXPLOSIVE BREEDING ASSEMBLAGE, is associated with breeding seasons that are even more highly compressed. In the wood frog, all the females in some populations are receptive for just one night each year. Naturally, this situation creates intense selection on males, favoring those that assemble at the pond on the mating night. The resulting high density of rivals raises the costs of repelling opponents from a defended area. Therefore, male wood frogs

10 **Scramble competition polygyny.** In horseshoe crabs large numbers of males compete by trying to find a receptive female before other males can mate with her. Photograph by James Lloyd.

eschew territorial behavior and instead hurry about trying to encounter as many highly fecund females in as short a time as possible [96].

Even if the breeding season is not truncated, other factors may cause large numbers of males to aggregate, raising the costs of territoriality to prohibitive levels (Figure 10). If high male density is one factor that promotes the evolution of scramble competition polygyny, then we can predict that in species that typically practice a territorial mating system, an increase in male numbers will change male tactics. This happens. When male dungflies are few and far between, some defend spots on dung pats that attract egg-laden females. But as density rises, males give up resource defense and instead simply try to outscramble rivals for incoming females [84].

Thus, under some circumstances, dispersed females and resources (or high male density) make female defense or resource defense uneconomical—although why this should lead in some cases to lek polygyny and in others to scramble competition polygyny remains puzzling.

The Rarity and Varieties of Monogamy Given that the quintessential male tactic is to fertilize as many eggs as possible, it will rarely be to a male's advantage to restrict himself voluntarily to one female. Species-wide monogamy is generally rare but not unknown. The evolution of monogamy may occur when the potential for polygyny is so reduced by the scarcity of females that a male, having found a potential mate, can gain more by guarding her than by trying to find additional mates.

Males of some insects achieve monogamy in a spectacular fashion, appearing to guard their partner posthumously. Consider honeybee drones, which leave their genitalia in the queen bee if they succeed in copulating with her in flight [554]. Afterward, the drone promptly dies, thus ensuring that he remain monogamous. His genitalia (Figure 11) remain in the queen unless removed by another drone (and drones can copulate despite the presence of a "mating sign" donated by a previous partner of the queen) or until workers take the genitalia from their queen on her return [438].

It remains to be demonstrated whether the suicidal donation of the mating sign in any way makes it more difficult for other males to mate

11 Monogamy. No male honeybee can mate more than once, because the male fires his genitalia (the pale structure) into the tip of the queen's abdomen. Photograph of a mated queen by N. Koeniger.

with a queen even though it clearly does not absolutely prevent multiple mating by her. But if—and note the "if"—the genitalic "plug" reduces sperm competition even slightly by reducing the total number of partners a queen accepts, it might still be advantageous for the drone to die, because the OPERATIONAL SEX RATIO in the traditional mating sites of honeybees is extraordinarily male biased. For every receptive female in the area, there are thousands of drones eager to mate. The enormously skewed sex ratio occurs because many more drones than queens are produced by each colony of bees. Given that there is almost no chance for a male to copulate twice in his lifetime, it pays a drone to die after one mating if his manner of dying improves his egg fertilization rate even a little bit [766].

For a less speculative case of monogamy via mate guarding, consider the shrimp *Hymenocera picta* (Figure 12), which also has a strongly male-biased operational sex ratio [838]. Females of this species are receptive only for a few hours after they have molted, which occurs every 3 weeks or so. Given this and the scattered distribution of females, it would be difficult, risky, and energetically expensive for a male to find one receptive female after another. Instead, when a male locates a female, he stays with her, repels other males if they appear, and mates with his partner just after she molts.

Monogamy in Birds

By far the most familiar examples of monogamy are found in birds, where about 90 percent of all species have males that pair bond with a female during a breeding season [463]. Although mate guarding plays a role in the male's sacrifice of opportunities to acquire more than one mate (Chapter 13), it may not be the primary factor that makes monogamy adaptive for male birds. The fact that female birds lay eggs means that their male

12 **A monogamous shrimp.** Males of this shrimp remain with a potential mate when they encounter one, because females are scarce, widely distributed, and rarely receptive. Photograph by Uta Seibt and Wolfgang Wickler.

partners have the potential to provide useful parental services—incubating eggs, and hatchlings, too, as well as feeding and defending the young. Thus, paternal parental care can result in a substantial increase in the number of descendants a male bird can produce with one partner. This increase can in theory outweigh any potential reproductive gains a *nonparental* male might secure through pursuit of many females [602].

The advantages of male parental care are particularly great for species whose females breed synchronously. By the time courtship is complete, the number of unpaired females available for a would-be polygynist is small. Under these circumstances, males can gain by raising their mate's reproductive success, and females gain by selecting unbonded males as partners, rather than sharing their mate's assistance with other females. The fact that in any number of birds a pair cares for a single offspring suggests that in these species male assistance is essential for a female to have any chance at producing young. If the male were to divide his efforts between two or more females, he might very well wind up with no offspring at all [865]. In other species with larger broods, the male's parental contribution to his reproductive success has been tested by the experimental removal of some males. Sure enough, the widowed females reared fewer young than the nonwidowed controls did (Figure 13) [511].

Thus, monogamy might be advantageous to both males and females under some circumstances, but it is also possible that it occurs only because monogamous females enjoy much greater fitness than those mated to polygynous males. Under these circumstances, females will not pair with an already-mated male, and they should attack other females that attempt to share their partner, his territory, and his resources [602].

In order to discriminate among these possibilities, Mats Bjorklund and Bjorn Westman looked at the effects of removing a male on the reproductive success of the great tit [69]. You may recall that this is a bird that nests in habitats of varying quality (Chapter 9). One reason why females might choose to mate with a male in a second-rate habitat is that they can gain

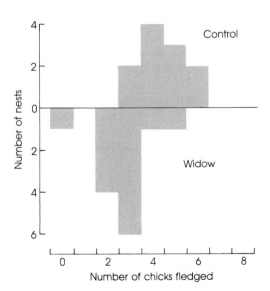

13 **The value of male assistance.** An experimental test of the hypothesis that male assistance raises female reproductive success in a monogamous bird, the snow bunting. Few females that have been experimentally widowed by removal of the male partner are able to fledge more than three chicks. Control females that have not been widowed usually rear four or more with the help of their partner. *Source:* Lyon et al. [511].

more by mating with a monogamous male there than by doubling up with a male in the preferred habitat. This prediction appears to be true for the great tits that Bjorklund and Westman studied. In the primary habitat (deciduous woods), a female paired with a male fledged nearly eight young with an average weight of 18.5 grams. In the secondary habitat (coniferous forest), monogamous pairs produced about seven young with an average weight of 17.5 grams. Even though titmice in conifers did not do as well as those in deciduous woods, they did better than experimental pairs from which the male partners were removed a few days after hatching of the eggs (less than six young fledged with a mean weight of 16.5 grams). The fact that a polygynous male with two females on his territory would have eleven fledglings to his credit, even if he helped neither female, suggests that polygyny would be advantageous to male great tits. But female mate choice prevents males from becoming polygynous. The nesting habitats of great tits vary, but not enough to compensate a female that picked a mated male and lost half (or all) of his assistance in feeding the young.

Susan Hannon looked at the question of who gains by monogamy in the willow ptarmigan. Like the great tit, this is a monogamous animal. But Hannon was able to induce polygyny in this species by removing many males from a population. As a result, some females wound up sharing a male, while others remained monogamous. In willow ptarmigan, males help their mates by keeping watch for predators while the female forages for food prior to and during incubation and by defending the brood after the chicks hatch. If male assistance influences success in producing and rearing young in the willow ptarmigan, then there should be differences in the numbers of chicks fledged by females with monogamous mates and by those with polygynous partners. Monogamous females averaged 5.1 fledglings, whereas polygynous females fledged 4.7—not a statistically significant difference. Moreover, the difference in the average weight of the chicks in the two groups was small and not significant, so their survival chances were probably the same.

Hannon then examined whether females that had mated monogamously were more likely to be present the next year than were females from polygynous pairs, and here she did find a significant difference. Nearly half of the monogamous females returned, whereas only a quarter of the polygynous females reproduced in the area again. If these differences reflect higher mortality for polygynous females, as seems likely, then the advantage of monogamy to females becomes apparent—they get to live longer and have more broods even though in any one season they can compensate for reduced male assistance. This finding also helps explain why female willow ptarmigan attack other females, thus enforcing monogamy on their partners.

Monogamy in Mammals

In contrast to birds, monogamy in mammals is very rare, probably because female mammals nurture their embryos internally and feed infants with their milk [602]. As long as female mammals can adequately feed and defend their offspring, males may gain little by remaining monogamous with a mate, and instead should attempt to fertilize as many females as possible. But when female mammals can be assisted in ways that greatly improve the care given offspring, males that provide the assistance can be favored over those that abandon their mates.

In this context, exceptions to the rule that mammals are polygynous are revealing. For example, monogamy does occur in many carnivorous mammals, including foxes and wolves. Males of these species can potentially raise the reproductive success of their mates by defending a feeding territory, chasing away infanticidal intruders, and by bringing in animal prey while the female remains at home. Because a male carnivore can potentially improve the nutrition and survival of his offspring and mate, a monogamous male can enjoy relatively high fitness [239].

An alternative hypothesis is that monogamous male carnivores would gain if they could become polygynous, but their mates enforce monogamy by repelling intruder females in solitary species or by preventing subordinate females from reproducing in social species. The female-enforced monogamy hypothesis has been tested with a solitary monogamous primate, the Bornean gibbon [560]. When John Mitani played tape-recorded calls of female gibbons within the range of a mated pair, it was the resident female that initiated the approach of the pair to the speaker. Just the reverse was true if the recorders played a tape of a calling male. Gibbon pairs regularly engage in aggressive interactions with neighboring pairs, which come together after bouts of calling. The fact that females discriminate between male and female calls suggests that they are particularly eager to repel a female intruder, who represents a threat to their sole use of the foraging resources in the pair's territory.

In some *social* monogamous primates, the dominant breeding female may physiologically suppress breeding in subordinate troop members [1]. Even when ovulatory cycling is not blocked in subordinate female mammals, they may not produce offspring. For example, a wolf pack usually contains just one dominant breeding female, often a mother, sister, or aunt to the nonbreeding females (Figure 14) [609]. Interestingly, the nonbreeders may

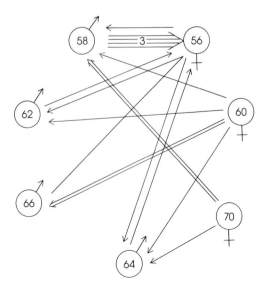

14 **Monogamy in wolves.** In this captive pack with three females and four males, only the dominant female (56) actually copulated with a male (58), on three occasions. Although the other two females came into estrus and solicited matings from the males in the pack, they did not copulate and so did not produce litters. Arrows show which individuals initiated sexual interactions and the targets of their solicitations. *Source:* Packard [609].

be ovulating and courting males, but they will only produce a litter if the dominant female is removed. Their failure to reproduce when the dominant female is present could be to their advantage, if their progeny would be killed by the dominant female, as occurs in some other monogamous canids [512], or if their pups would starve to death through food competition with the offspring of the dominant female. Whatever the ultimate reasons for the nonbreeding status of subordinate females in a wolf pack, the immediate effect is to permit the dominant female to be monogamous with a male in her group.

The Rarity and Varieties of Polyandry As a general rule, females rarely mate with a number of males in a breeding season. A single male can generally provide all the sperm a female needs to fertilize her limited number of eggs. After having selected a mate, females may only suffer from the time and energy costs of additional matings with no compensatory advantages, unlike males, which may gain more offspring from each female mated.

But in species in which parental males pair bond with a female, one might think (correctly) that females would benefit from having a bevy of male assistants. However, males will rarely gain by mating exclusively with and helping a female that already has another partner, who is likely to fertilize some or all of her eggs [602].

Because polyandry usually decreases male fitness, males often act in ways that make it difficult for females to be polyandrous. Nevertheless, females that lay eggs at intervals throughout a protracted breeding season, like anoles (Chapter 6) and fruit flies [652], may mate a number of times, often with different males, in order to fertilize different eggs or clutches of eggs. If the function of these "supplementary" copulations is to restore depleted sperm supplies, then the fertility of polyandrous females should exceed that of females that have mated just once [517, 652] (Figure 15).

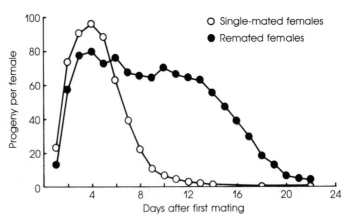

15 **Sperm-replenishment polyandry** is adaptive for female fruit flies. By copulating at intervals, females secure additional sperm and so are able to maintain high fecundity for more days than single-mated females. *Source:* Markow [517].

16 **"Prostitution" polyandry.** Multiple mating by female megachilid bees gives them access to flowers in the territories of their mates. Photograph by the author.

Another kind of polyandry may occur if, by copulating, females gain access to resources that males control. Females of some polygynous hummingbirds may mate with a male that owns a nectar-producing flower patch each time they come to feed there [866]. In exactly the same way, females of some bees visit clumps of flowers owned by territorial males, copulating with each male in turn in order to be permitted onto the territory to collect pollen and nectar (Figure 16) [8]. In species like hangingflies in which males offer food presents as inducements to mate, females may copulate with more than one male to receive nuptial gifts in return for accepting the male's sperm [323, 682, 766]. Likewise, females of some butterflies take the initiative in seeking matings [681, 682], perhaps because male butterflies transfer a substantial spermatophore during copulation. This kind of mating system has been labeled, uncharitably, "prostitution" polyandry [866].

Sex Role Reversal Polyandry
In the cases we have described, females for the most part are fairly selective in their choice of mates and more likely to engage in parental behavior than males. Moreover, multiple mating by females of these species does not

preclude polygyny by their males. Polyandry presents a much more intriguing challenge to evolutionary biologists in species in which there is a partial or complete sex role reversal, with males assuming all or most parental responsibilities, and females even competing for access to mates. We discussed this phenomenon earlier when analyzing the evolution of male parental care in fishes, which appears to occur because the costs of parenting are less for males than for females [319]. In many fishes of this sort, males behave more or less like typical polygynous males except that they provide parental care. Paternal behavior is not incompatible with acquiring multiple mates, and indeed may advance this goal.

Male polygyny coupled with female polyandry also occurs in a few birds whose social systems closely resemble those of paternal fishes. The partridge-like tropical tinamous are ground-nesting birds whose eggs appeal to many predators. Males incubate the eggs without aid from their mates. But males with one clutch often attract several females, who add their eggs to his nest. After producing one clutch, a female sets about making another one. She may donate this second clutch to her first mate if his first nest has been destroyed in the interim, or she may seek another male to incubate her eggs [602].

Whatever the basis for the role reversal in parental care, once it is established in a population we can expect females to take the initiative in courting male parents if they are in limited supply. In the male-brooding pipefish, females display to potential mates and attempt to induce them to mate and accept their eggs [62]. Likewise in giant water bugs, females approach and court males (although males do advertise their availability

17 **Sex role reversal.** Male waterbugs guard and brood eggs glued on their backs by their mates. Photograph by R. L. Smith.

with underwater "push-ups"). After copulating, a male permits his mate to glue her eggs on his back, eggs that he will tend until they hatch (Figure 17) [731].

Sex role reversal can be even more profound. For example, female spotted sandpipers, subject of Lewis Oring's long-term research, not only take the lead in courtship but also behave like males in many other ways [603, 604]. They are larger and more aggressive than males; they arrive on the breeding grounds first, whereas in most birds males precede females. Once on the breeding site, females compete for territories, which they defend against others of their sex (Figure 18). A female's territory may then attract two or three males that set up their own smaller territories within their mate's domain. Each male mates with the female and gets one clutch of eggs to incubate and rear on his own, while the female continues to defend her territory and produce new clutches for additional mates.

Lewis Oring believes that understanding polyandry in spotted sandpipers is advanced by recognizing a key "historical constraint" (Chapter 7) affecting this species. Females of all sandpipers lay a clutch of four eggs and cannot adjust the clutch size even when resources are abundant. Because spotted sandpipers are "locked into" an inflexible four-egg clutch—a trait inherited from a distant ancestor—they can capitalize on rich food resources only by laying more than one four-egg clutch, not by increasing the number of eggs laid in any one batch [603].

Polyandry may arise in spotted sandpipers because of a confluence of several unusual ecological features operating in the context of the set clutch size [466]. First, spotted sandpipers nest in areas with immense mayfly

18 **Resource defense polyandry.** Female spotted sandpipers fight for possession of territories that may attract several males to them. Photograph by Steven Maxson, courtesy of Lewis Oring.

hatches, which provide a superabundant food for females (and the young when they hatch). Second, a single parent can care for a clutch about as well as two parents (in part because the young are *precocial*, able to move about, feed themselves, and thermoregulate shortly after hatching). Third, the adult sex ratio is characterized by a slight excess of males. This combination of factors means that selection can favor females that desert a partner, leaving him in charge of one clutch of eggs, and freeing them to pursue additional males.

Female spotted sandpipers, by deserting their mates, can invest in the defense of a resource—good nesting habitat—that attracts several males to them. A similar case of resource defense polyandry is exhibited by the northern jacana, a tropical marshland bird. Like the spotted sandpiper, sex role reversal for this bird involves large, aggressive females that compete for possession of territories big enough to attract several males, each of whom cares for one clutch of eggs by himself [408, 409]. Females of this species may even practice ovicide, destroying the eggs of males on neighboring territories to free these males to move into their site and accept a clutch from them [744].

Another polyandrous bird, the red-necked phalarope (Figure 19), differs from spotted sandpipers and jacanas in that females directly defend mates rather than a territory that attracts males to them [664, 665]. These little shorebirds feature a reversed sexual dimorphism in coloration, with females usually more brightly colored than males. Like female spotted sandpipers, phalarope females arrive on their northern breeding grounds before males in order to increase their chances of securing a mate and his parental care. As individual males arrive, females pursue and court them, jabbing other females with their sharp beaks to keep rivals at a distance. Once paired, a female lays her clutch in her mate's nest and then leaves him to it, while she either goes away to forage prior to migration south or stays to find and guard another male to whom she donates a second clutch of eggs that he alone incubates and defends.

The Mating System of the Dunnock   Figure 20 summarizes the ecological classification of mating systems we have discussed in the chapter. Many of these mating systems occur in the dunnock, whose cloaca-pecking behavior we discussed in Chapter 13. This ordinary looking European song-

19 Mate defense polyandry. A pair of red-necked phalaropes, a species in which the female guards her mate from other females. Female phalaropes are as brightly or more brightly plumaged than their mates. Polyandrous females defend two males in sequence. Photograph by John D. Reynolds.

POLYGYNY

Individual males may mate with more than one female per breeding season.

A. Some males can economically monopolize access to several females.
1. FEMALE DEFENSE POLYGYNY :   Females live in permanent groups, which males defend directly.
2. RESOURCE DEFENSE POLYGYNY :   Females do not live in permanent groups but are spatially concentrated at food, nesting sites, or other resources, which some males can control.
B. Males cannot economically monopolize access to females directly or indirectly.
1. LEK POLYGYNY :   Males compete for high dominance ranking within a group, usually at a traditional display arena; the winner is often selected by many females.
2. SCRAMBLE COMPETITION POLYGYNY :   Females may be scattered or clustered spatially; but if clumped, they cannot be defended because of high male density; males race to contact as many receptive females as possible without engaging in territorial defense.

MONOGAMY

An individual male or female mates with only one partner per breeding season.

A. Males can economically monopolize only one female.
1. MATE-GUARDING MONOGAMY :   Females are dispersed; males defend one mate against other males.
2. MATE-ASSISTANCE MONOGAMY :   Males can provide such useful parental care that they raise the reproductive success of a partner enough to compensate for mating opportunities lost by staying with one female.

POLYANDRY

Individual females may mate with more than one male per breeding season.

A. Polyandry without sex role reversal.
1. SPERM REPLENISHMENT POLYANDRY :   Females may mate more than once to secure additional sperm with which to fertilize a new clutch of eggs.
2. "PROSTITUTION" POLYANDRY :   Females may mate with more than one male in order to gain access to the resources that males control and offer only to their mates.
B. Polyandry with sex role reversal.
1. RESOURCE DEFENSE POLYANDRY :   Females control resources attractive to more than one male; males may provide even greater parental care per offspring than females.
2. MALE DEFENSE POLYANDRY :   Females provide neither resources nor greater parental investment per offspring than male; they compete directly to monopolize male mates in female aggregations.

---

20 Ecological classification of mating systems. In each category males and females have a variety of mating strategies.

bird was recommended as an inspiration to humans by the Reverend F. O. Morris in 1856, because it was supposedly "unobtrusive, quiet, and retiring, without being shy, humble and homely in its deportment and habits, sober and unpretending in its dress, while still neat and graceful." Nicholas Davies notes that it is just as well the good reverend and his parishioners were unaware of the sexual behavior of this bird, which is anything but humble and homely [175, 178]. Davies and his colleague Arne Lundberg found within one population of the dunnock an almost complete spectrum of mating possibilities: one male paired with one female, one male paired with two females, one female paired with two males, and several males and females joined in a sexual liaison on the same territory.

Intraspecific diversity of this sort provides an ideal opportunity to look for the responsible ecological causes, because phylogenetic (historical) factors are held constant (dunnocks are all members of the same species). Remember that the essence of the Emlen and Oring argument is that female ecology determines the evolution of male tactics. Variation in the food available to female dunnocks affects how they distribute themselves spatially, and this in turn is a major contributor to variation in the mating systems of individual birds.

Dunnocks feed on tiny seeds and very small invertebrates, whose density varies from place to place. Some females range over a large area because prey density is low there; other females can find all they need in a small area because it is rich in tiny bugs. One male can economically monopolize a small territory and the female in it, and monogamy is the result. If, however, the female happens to have a large foraging area, a single male has difficulty keeping other males out (Figure 21). If other males skulk about in the territory, they try to mate with the female and they sometimes succeed despite the mate-guarding behavior of the primary male. Females appear to encourage cuckoldry by slipping away at times from their primary mate, which they can do because they forage in dense cover. Secondary males that mate with the female often help her feed her young at the nest. Such a female practices polyandry, which is advantageous to her and to

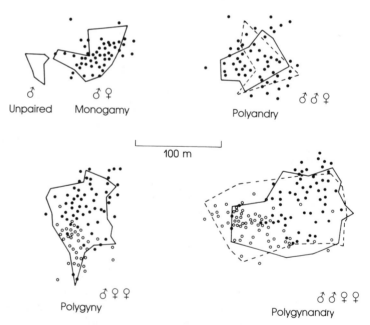

21 **Variation in mating system within a species.** The size of the home range occupied by a female dunnock determines whether her mate is able to be polygynous or monogamous or is forced to accept a polyandrous arrangement. Dots indicate sightings of known females. Lines show the area defended by singing territorial males. *Source:* Davies and Lundberg [178].

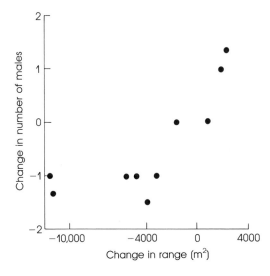

22 **Home range size and mating system.** Experimental test of the hypothesis that female home range size in the dunnock controls the mating system. Food supplements caused females to decrease their home range and increased the probability that their mates were monogamous. *Source:* Davies and Lundberg [178].

her secondary mate, who typically is a young bird unable to defend a female of his own elsewhere. To the extent that females use the sperm of secondary males to fertilize some of their eggs, polyandry reduces the fitness of the primary male.

Polygynous males are those exceptional birds able to defend a territory so large that it encompasses the foraging habitat of two females. Only very rarely is a male capable of this, and more often two monogamous pairs merge their adjacent and overlapping territories to form combinations of polygynous males and polyandrous females.

If it is correct that the size of a female's foraging area determines the ease with which a male can monopolize a mate, then compression of a female's territory should reduce the probability of polyandry [178]. Davies and Lundberg put out feeders with oats and mealworms in some female home areas and kept them stocked for months. Females with access to these supplemental foods decreased their foraging range by an average of 40 percent. These females were more likely to participate in monogamous relationships than were control females in areas that did not receive a feeding station (Figure 22).

These observations and experiments give us more confidence that female ecology establishes the rules of mating competition for males and shapes the diversity of mating systems of animal species.

SUMMARY

1 Mating systems can be defined purely on the number of mates an individual has during a breeding season. Sexual selection generally favors males that try to mate with more than one female, a selective pressure leading to polygyny if some males can succeed in copulating with more than one partner. Ecological factors that determine the spatial distribution of females affect the payoff for different tactics employed by males competing to be polygynous.

2 If females or resources used by females are clumped, female defense polygyny or resource defense polygyny results because some males can economically defend the productive locations for mating. If, however, these forms of territoriality are too costly, because of a dispersed distribution of females or a high density of rivals, males may either attempt to demonstrate social dominance to females through "symbolic" lek territoriality or they may engage in nonterritorial scramble competition for access to mates.

3 Monogamy by both males and females can occur through a combination of female preferences for unmated helpful partners and various ecological factors that reduce the ability of a male to monopolize more than one female.

4 Polyandry is the rarest mating system because a male rarely gains if his mate copulates with additional males. Nevertheless, in some species females mate with more than one male to secure additional sperm or added material benefits. In its most extreme form, polyandry is associated with sex role reversal in which males provide most or all of the parental care and females defend mates directly or guard territories whose resources attract several males to them.

SUGGESTED READING

Stephen Emlen and Lewis Oring's paper on the ecology of mating systems has changed the modern analysis of mating systems [239]. Randy Thornhill and I have applied their approach to insects [766]. As usual, the articles by Nicholas Davies (on the dunnock's mating system) are insightful and a pleasure to read [175, 178].

DISCUSSION QUESTIONS

1. The two hypotheses on why yellow-bellied marmots live in groups despite having reduced annual reproductive success generate different predictions on how females will react to a newcomer attempting to settle on their mate's territory. What are these predictions? How would you test them experimentally?

2. Honey possums are small, mouselike marsupials that live in Western Australia. Males have the largest testes relative to their body size of *any* mammal. What do you predict is the mating system of this animal? Produce a brief research proposal to study the honey possum's mating system.

3. Strictly speaking, the argument that female distribution determines the optimal mating tactic of males does not take into account the possibility that what other males in the population are doing will affect the payoffs associated with alternative tactics. Imagine a species with a scramble competition mating system in which most receptive females are in loose clusters but some are scattered between the areas of greatest female density. Use optimality theory and game theory to produce predictions on where we might expect to find males searching for mates. Are the predictions identical?

15

The Ecology of Social Behavior

The preceding chapters on behavioral ecology have analyzed different behavior patterns as adaptations to different environments. We have seen repeatedly that no one kind of behavioral tactic is superior for all ecological circumstances. This is a particularly important point to keep in mind when analyzing the evolution of social behavior. Unfortunately, people often believe that the highly complex societies of some vertebrates, most conspicuously our own human societies, represent a crowning achievement of evolution. To counteract the mistaken impression that complex social behavior is always superior to "primitive" sociality or solitary behavior, Chapter 15 begins with a discussion of the costs, as well as the benefits, of social living. The cost–benefit approach of behavioral ecology suggests that the advantages of sociality will exceed the disadvantages only under special ecological conditions. Other circumstances can favor the evolution of solitary living as the more adaptive mode of existence. After an exploration of this point, I shall examine one of the most intriguing problems in social behavior, the evolution of altruism. There are a variety of examples of this phenomenon, including the complete rejection of personal reproduction practiced by some animals. There are various hypotheses to account for the evolution of helping, and the chapter examines the different proposals and shows that there is more than one way in which extreme cooperation among social creatures can arise.

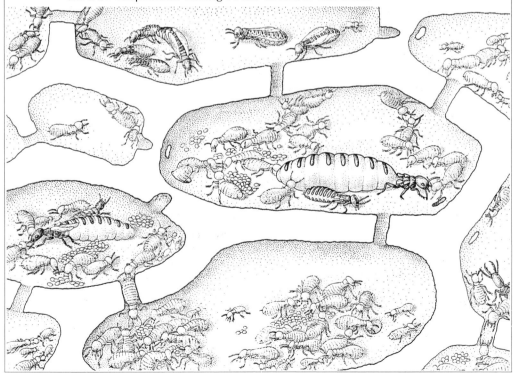

The Costs and Benefits of Sociality   Treating animal behavior as a form of economics has proved remarkably useful, as we have seen in the analysis of communication, territoriality, mating systems, and many other topics in behavioral ecology. We shall now apply the cost–benefit approach to living in groups (Table 1) [10]. Because humans are a social species, we like to flatter ourselves that sociality must be the "most advanced" way of life. Previous chapters have documented that living and cooperating with others has a variety of benefits, under some circumstances. Black-headed gulls capture food more easily when they hunt in flocks than when they forage by themselves [291]. A pride of lions can kill larger prey and better defend their kills and their hunting territory than can a lion on its own [612]. A pair of birds can sometimes care for their brood better than a single parent can [511]. And social prey can often spot danger quickly, or dilute the effects of predatory attacks, or cooperatively repel the enemy, or simply hide behind other members of the group, all things that single animals cannot do as well or at all (Chapter 11).

If there are so many advantages to social life, why aren't all species social? If an adaptationist perspective is correct, there must be ecological conditions for which the costs of living together exceed all possible benefits. We have seen in preceding chapters that social life can create problems for individuals. The bee-eater whose mate is forced to copulate with other males pays a price for nesting in a colony [240]. Langur females that associate with males run the risk of losing their offspring by infanticide [393]. The members of a mixed-species flock may gain protection from their enemies because they are social, but they also face steeper competition for food [5]. Here I shall focus on a cost of sociality not previously covered, namely, the increased chance of contracting a contagious disease or damaging parasite.

The cliff swallow is a colonial bird that builds mud nests shoulder to shoulder under overhanging cliffs, bridges, and culverts. Charles and Mary Brown examined the costs of sociality by exploiting natural variation in the size of cliff swallow colonies, which range from a handful of breeding pairs to hundreds of nests [108]. They knew that adult birds inadvertently transport the swallow bug, a blood-sucking relative of the notorious bedbug.

TABLE 1

Some major advantages and disadvantages of sociality

Advantages	Disadvantages
Reduction in predator pressure by improved detection or repulsion of enemies	Increased competition within the group for food, mates, nest sites, nest materials, or other limited resources
Improved foraging efficiency for large game or clumped ephemeral food resources	Increased risk of infection by contagious diseases and parasites
Improved defense of limited resources (space, food) against other groups of conspecific intruders	Increased risk of exploitation of parental care by conspecifics
Improved care of offspring through communal feeding and protection	Increased risk that conspecifics will kill one's progeny

Source: Alexander [10].

In large nesting groups there should be a greater chance for swallow bugs to be brought into the colony, and once there, the bugs might build up their population and spread among the many nests available to them. The Browns found that, sure enough, levels of infestation were consistently higher in large colonies than in small ones.

Swallow bugs drink the blood of swallow babies, but does this do lasting damage? When the Browns weighed nestlings and counted the number of bugs attacking them, they found that the more bugs per bird, the less a 10-day-old nestling weighed (Figure 1). A nestling in a bug-infested nest lost about 15 percent of its body weight to the socially transmitted parasites; and in heavily infested colonies, survivorship of the young declined as much as 50 percent.

If the bugs were responsible for these negative effects, then their removal should be advantageous to baby cliff swallows, a prediction the Browns tested by fumigating a sample of nests in an infested colony while leaving other control nests untreated with the insecticide. The weight of 10-day-old nestlings from fumigated nests was consistently much greater than the weight of youngsters of the same age from neighboring nests in which swallow bugs had not been killed (Figure 2).

The cliff swallow study shows that social living is far from an unmitigated blessing. Therefore, if sociality is to evolve, there must be special ecological conditions that so raise the benefits to individuals that these gains compensate for the costs the individuals endure by associating with

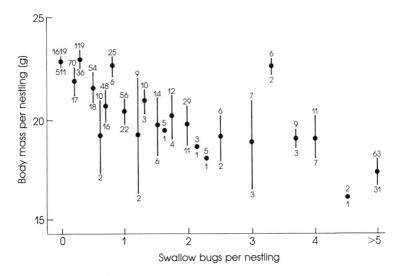

1 **Swallow bugs harm cliff swallow nestlings.** In large colonies of cliff swallows, there are more swallow bug parasites per nestling. The more parasites per nestling, the less the swallow weighs. The circles represent mean values. The numbers at the top of the error bars are the number of nestlings weighed; the numbers below are the number of nests sampled. *Source:* Brown and Brown [108].

2 **Effect of killing parasites** on cliff swallow body weight. The nestling on the right comes from an insecticide-treated nest; the one on the left occupied an parasite-infested nest. *Source:* Brown and Brown [108].

others. Mart Gross and Anne MacMillan have used a cost–benefit approach to explain why the bluegill sunfish is social, whereas its close relative, the pumpkinseed sunfish, is a solitary species [317].

Bluegills become social during the breeding season, when groups of 50 to 100 males may build their nests (depressions in a sandy lake bottom) side by side (Figure 3). The primary benefit to colonial males is almost certainly a reduction in predator pressure on the eggs deposited in the nests by spawning females (Chapter 11). For example, by defending overlapping territories, social males "cooperate" in expelling egg-eating catfish.

But social bluegills do not get their antipredator benefits for free. An individual that nests in a group must contend with the tendency of his neighbors (and other nonnesting bluegills attracted to the group) to consume the eggs in his nest, which he has fertilized. Sexual interference in courtship and spawning is another problem that occurs because of the cuckolder males (Chapter 13) that gather at large colonies. Moreover, fungi that destroy eggs may be transmitted from nest to nest in a dense colony. These costs reduce the net benefit enjoyed by social bluegills but do not eliminate it entirely.

Pumpkinseed sunfish do not breed in colonies. They are much less affected by predators than bluegills are, because they have mouthparts adapted for picking up, crushing, and consuming heavy-bodied, molluscan prey. Bluegills, on the other hand, have delicate, small mouths designed for "inhaling" small, soft-bodied, insect larvae. Thus, although a bluegill cannot pick up a snail and cart it away from the nest, pumpkinseeds are easily able to do this (and may consume their egg-loving enemy to boot).

In addition, a bluegill can be ignored by a nest-raiding bullhead catfish, but a pumpkinseed's attack has considerably more bite to it. Because pumpkinseeds are relatively free from nest predation, there are insufficient advantages to compensate for the costs of social breeding [317]. Pumpkinseed sunfish are in no way inferior or less well-adapted than bluegills because they are solitary; they simply face different ecological circumstances, for which colonial nesting would yield reduced individual fitness.

Degrees of Sociality in Prairie Dogs

John Hoogland has applied a cost–benefit analysis of sociality to two species of prairie dogs that live in "towns" containing different numbers and densities of individuals [380, 381]. Prairie dogs are similar to bluegills in that antipredator benefits are likely to have been the major impetus for the evolution of their colonial way of life. Hoogland tested the response of prairie dogs to potential predators by pulling a stuffed badger or weasel through a colony to his blind at a programmed rate of speed. He recorded the length of the interval between the moment at which the "predator" was activated and the first alarm call. For both the white-tailed prairie dog and the black-tailed prairie dog, the smaller the colony, the longer before the enemy was detected and an alarm sounded (Figure 4). Animals that hear an alarm call scamper back to their burrows, where they are relatively safe. Because black-tails tend to live in larger groups than white-tails, they generally react with alarm sooner than their relatives. Moreover, black-

3 **A fish society.** The colonial bluegill nests in groups, each male defending a territory bordered by the nest sites of other males and trying to protect its nests against bass (above), bullhead catfish (left), snails, and pumpkinseed sunfish (right foreground). Courtesy of Mart Gross.

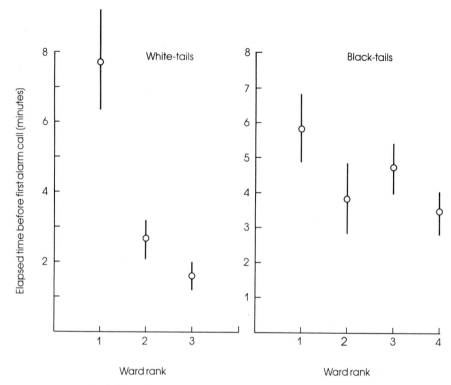

4 **A benefit of sociality.** In larger groups of prairie dogs, a model predator is detected sooner than in small groups. Wards (groups) are ranked by numbers living together with rank 1 the smallest group. *Source:* Hoogland [381].

tailed prairie dogs can afford to devote significantly less time (35 percent of their average daily time budget) to scanning for predators than can white-tails (43 percent).

Why don't white-tailed prairie dogs live in large groups, too, so that individuals can enjoy the multiple benefits that come from coping with predators efficiently? As with bluegills, there are costs to individuals that live in large groups. Although a prairie dog may spend less time looking about for danger in a large group, it will spend *more* time fighting with its neighbors (Figure 5). In addition, fleas that bear sylvatic plague, a devastating disease, are more abundant in large colonies. In dense black-tail groups Hoogland counted an average of 3.3 fleas per burrow entrance, whereas white-tailed prairie dogs are afflicted by an average of only 0.5 flea per burrow entrance. Black-tails pay a greater price for living in large groups in terms of disease risks and social interference. Do they also derive greater benefits?

Black-tailed prairie dogs live in an exposed environment where predator pressure may be especially severe. They are plains-dwellers, unlike the white-tailed prairie dogs, which live in shrubby habitat where they can

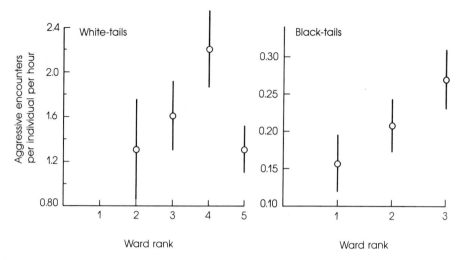

5 **A cost of sociality.** Within a species of prairie dog, individuals living in larger groups engage in more frequent aggression than those living in smaller groups (ward rank 1 is the smallest group). *Source:* Hoogland [380].

better hide from their enemies. (Hoogland documented that a significantly smaller proportion of feeding white-tails are visible at any moment to human observers than are black-tails.) This observation suggests that black-tails rely on one another to detect their predators, whereas white-tails can use naturally occurring shrubs as cover. Black-tails therefore have more to gain from group living, and this raises their benefit-to-cost ratio for forming large, dense colonies [380, 381].

The Evolution of Helpful Behavior   When one is tempted to rhapsodize on the benefits of social life, it is salutary to recall that these benefits are rarely distributed equally among the members of a group. Large bluegills get the safest central positions in nesting colonies; smaller males are forced to the outside. The clearest demonstration of the occurrence of competition within animal societies is the formation of dominance hierarchies. We have discussed this phenomenon before, and here we need only remind ourselves of the consequences of social status. The dominant baboon, or elephant seal, or wolf, or hyena reproduces; subordinates either fail altogether or have a more difficult time of it, thanks to interference by their social superiors. And when there is competition for foraging sites, dominants claim the safer or more productive locations (Figure 6). During a hard winter, subordinate adult acorn woodpeckers were about twice as likely to disappear and (presumably) die as the dominant males in the population. The still lower ranking juveniles disappeared at a rate nearly four times that of the dominant males [341].

On the other hand, animals that live in groups often appear to help one another, as when two or three bluegills jointly attack a bullhead, or when

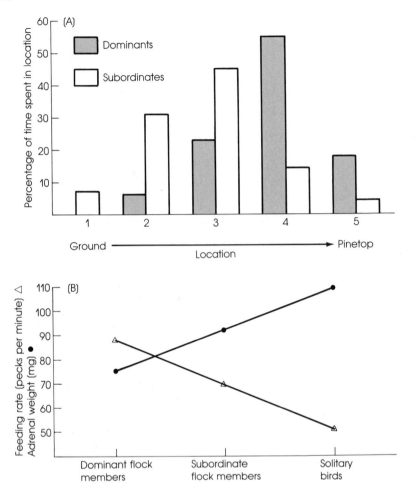

6 **Costs of sociality for subordinates.** (A) In willow tits, dominant birds forage in the safest parts of pines, forcing subordinates to forage in the sparser vegetation lower on the trees, where they are more exposed to hawks and owls. (Removal of dominants results in upward shifts in foraging by subordinates.) Locations: 1, the ground; 2–5, pine tree quarters ranging from the lowest fourth to the top fourth of the trees. *Source:* Ekman [230]. (B) In woodpigeons, subordinate flock members have lower feeding rates and larger adrenal glands (a sign of stress) than do dominant birds. Note, however, that subordinate flock members are still better off than solitary woodpigeons. *Source:* Murton et al. [580].

a prairie dog gives an alarm call after it spots a badger, or when a pride of lionesses overwhelms a dangerous Cape buffalo. Until recently, biologists were not especially surprised by this sort of thing, because they assumed that the members of a group should help each other for the benefit of the species as a whole. But with the recognition that group selection is less potent than individual selection (Chapter 1), helpful actions become considerably more interesting [845]. Here, as in the parallel analysis of communication (Chapter 8), the trick is to ask, Who benefits from a helpful action (Table 2)?

An individual need not lower its reproductive success by helping another. In many cases, helping immediately benefits both helper and recipient, as, for example, when one lioness drives a wildebeest toward her fellow pride members. If the antelope is killed by the ambushers, the driver will feed upon the prey as well. Likewise, if several male bluegills succeed in fending off a bullhead catfish that has entered the space they mutually claim and defend, the eggs guarded by all these males are at once safer as a result. If it can be demonstrated that both helper and recipient enjoy reproductive gains through their interaction, then there is no special evolutionary puzzle about the evolution of these helpful tendencies, which are labeled MUTUALISM or COOPERATION.

In addition, it is possible that animals can raise their reproductive success by participating in reciprocal arrangements in which individuals take turns helping one another. Robert Trivers invented the label RECIPROCAL ALTRUISM (also known as RECIPROCITY) for helpful actions that result in deferred reproductive gain for the helper [783]. He pointed out that if an individual could help another now at relatively little cost and later receive valuable "repayment" from the helped animal, the original helper would experience a net reproductive benefit from its initial helpful action. Critics of this notion, however, used the logic of game theory to raise questions about the evolutionary stability of reciprocal altruism. A population composed of reciprocal altruists would seem to be vulnerable to invasion by individuals that accepted help but later conveniently refused to pay back their helpers. "Cheaters" reduce the fitness of "noncheaters" in such a system, presumably making it unlikely that reciprocal altruism would evolve.

However, Robert Axelrod and W. D. Hamilton have shown that *if* there is a certain proportion of reciprocators in a population and *if* potential cooperators interact repeatedly, then individuals that use the simple behavioral rule "Do unto individual X as he did unto you the last time you met" can reap greater fitness gains than individuals that always try to accept assistance while never helping in return [33]. The strategy of taking

TABLE 2

Effects on reproductive success of interactions between a helpful donor and a recipient of assistance

Interaction type	Effect on individual reproductive success	
	Donor	Recipient
Mutualism = Cooperation	+	+
Altruism	−	+
Selfish behavior	+	−
Spiteful behavior	−	−

advantage of a cooperator, although possibly providing a large return for the cheater in that one interaction, in the long haul may yield less than a strategy of cooperating with those individuals that in the past cooperated with you.

Imagine a group of lionesses that excluded the female that drove the game toward them from a kill that they made as a result of the driver's help. The cheaters would enjoy the fruit of the driver's labor at her expense. But if this caused the driver to leave the group, the exploiters would have taken a short-term gain at the cost of losing long-term assistance. Cheating could be a short-sighted, maladaptive strategy when drivers that share a kill will cooperate in future hunts.

Altruism and Indirect Selection

Reciprocal altruism is really a special kind of mutualism in which the reciprocating individuals eventually earn a net increase in their reproductive success. In evolutionary biology, unadorned ALTRUISM is a term restricted to cases in which the donor really does lose reproductively over the long haul as a result of its help, while the recipient gains more surviving offspring. These actions, if they exist, are an exciting puzzle for an adaptationist, because they cannot have evolved via natural selection for increased reproductive success. Similarly, SPITEFUL BEHAVIOR (Table 2) poses a special challenge, because it is difficult to see how an animal could leave more copies of its genes by performing a reproductively costly action just because it harmed another individual's reproductive success. In fact, no completely documented cases of spite exist.

On the other hand, there are now dozens of examples of altruism in which the helpful individual is not "repaid" by receiving deferred assistance from the animal that it helped, nor is the helper actually "selfishly" manipulating the recipient with its "assistance" so as to increase its reproductive success. Cases of altruism need a special explanation, which W. D. Hamilton provided in 1964. He showed that if a helper directs its aid to its genetic relatives, it may be more than compensated for a reduction in its own reproductive success by an increase in the reproductive success of the related individuals [331]. Remember that from an evolutionary viewpoint the point of reproduction is to propagate one's alleles. Personal reproduction is an admirably direct way to achieve this ultimate goal. But helping relatives survive to reproduce provides an indirect route to the very same end.

To understand why this is so, we must discuss the concept of degrees of relatedness. There are ways to determine the probability that any two individuals will share a particular allele as a result of having inherited it from a common ancestor. For a parent with a genotype (amy^1, amy^2), the probability that its offspring will inherit, say, the amy^1 allele, is 0.5 (Chapter 3). There is one chance in two that the egg or sperm that the parent produced and that helped form the offspring will contain the amy^1 allele. The probability of possessing an allele by common descent is the COEFFICIENT OF RELATEDNESS or r. Between nonrelatives, r is essentially 0; between parents and offspring, r is 0.5.

TABLE 3
Coefficients of relatedness (r) between an individual and certain of its relatives[a]

Relationship	r	Relationship	r
Full siblings	1/2	Parent to offspring	1/2
Half-siblings	1/4	Uncle/aunt to niece/nephew	1/4
First cousins	1/8	Grandparent to grandchild	1/4

[a]The term r refers to the probability that a gene picked at random from the genome of an individual will be present in its relative because they had a common ancestor; or, r refers to the expected fraction of the total genome of the individual that the relatives will share as a result of having a common ancestor.

The coefficient of relatedness can be calculated for any pair of individuals that are descended from a common ancestor. The value of r reflects the number of meiotic events that separate the two individuals. Table 3 presents a sample of r values. With this information, we can determine the impact of a helpful, self-sacrificing action on the transmission of an allele that predisposes its bearer to behave in an altruistic manner. The key question is whether an individual can leave more copies of an altruism-promoting allele by helping relatives reproduce or by reproducing personally. Let us say that the animal could potentially have an average of two offspring of its own or give up reproduction, invest entirely in relatives, and help create an average of three siblings that would not have survived without its help. Offspring share half their genes with a parent; siblings also share half their genes with each other. Therefore, in this example, the cost of abandoning personal reproduction is ($r \times 2 = 0.5 \times 2 = 1$); the benefit in terms of genes passed on indirectly in the bodies of relatives is ($r \times 3 = 0.5 \times 3 = 1.5$). The altruistic route therefore offers greater genetic success than the "selfish" personal reproduction route.

If the cost of an altruistic act were the loss of one offspring ($C = 0.5 \times 1$), but if the altruistic act led to the survival of three nephews that would have otherwise perished ($B = 0.25 \times 3$), the action would also create a net genetic gain for the altruist. B must be greater than C if the action is to increase the frequency of the allele underlying the helpful act. When one thinks in these terms, it becomes clear that there can be a form of selection that occurs when genetically different individuals differ in their effects on the reproductive success of relatives. Jerram Brown calls this form of selection INDIRECT SELECTION, which he contrasts with DIRECT SELECTION for traits that raise individual reproductive success (Figure 7) [112].

A brief digression is needed at this point to deal with yet another term, KIN SELECTION. Although many people have treated this term as a synonym for indirect selection, Jerram Brown has pointed out that kin selection, as originally defined by John Maynard Smith, refers to the evolutionary effects of aid given to both descendant kin (offspring) and nondescendant kin (relatives other than offspring). Brown (but not everyone [714]) believes that it is useful to make a distinction between aid given to offspring (parental care) and aid given to nonoffspring (altruism).

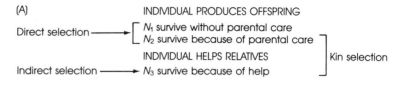

(A)   INDIVIDUAL PRODUCES OFFSPRING

Direct selection ──────────→ ⎡ N_1 survive without parental care
                              ⎣ N_2 survive because of parental care ⎤
                                             INDIVIDUAL HELPS RELATIVES   ⎥ Kin selection
Indirect selection ──────────→ N_3 survive because of help   ⎦

(B)

Direct fitness = $(N_1 \times r) + (N_2 \times r)$ ⎤
                                                    ⎥──────→ Inclusive fitness
Indirect fitness = $N_3 \times r$ ⎦

7 **Selection and fitness.** (A) The components of selection: direct, indirect and kin selection; and (B) the components of fitness: direct and indirect. After Brown [112].

Biologists have long recognized that parents can affect the evolutionary process by having surviving offspring, and this has made it easy for persons to understand that parental care might be adaptive. As we discussed in Chapter 12, parental care can evolve if the increase in the survival of aided offspring more than compensates a parent for the loss of opportunities to produce additional offspring in the future. In genetic terms, the reason that parents can theoretically gain via parental investment is that parents and offspring share 50 percent of their genes in common. By the same token, individuals can potentially promote the survival of certain of their genes by helping relatives other than offspring. Altruism can be favored by the component of kin selection that Brown has called indirect selection, provided that the gain in genes transmitted via relatives more than compensates for the loss of opportunities for personal reproduction in the future. The use of the term indirect selection, rather than kin selection, keeps the focus clearly on the distinction between parental effects on offspring and an aid-giver's effect on nondescendant kin, so we shall use it here.

W. D. Hamilton [331] was the first to recognize that an individual's total impact on evolution via transmission of genes to a subsequent gene pool required a combined measure of the animal's direct and indirect contributions of genes. Therefore, he invented the term INCLUSIVE FITNESS as a quantitative measure of total genetic success via personal reproduction (DIRECT FITNESS) and effects on reproduction by nondescendant kin (INDIRECT FITNESS).

Note that the concept of inclusive fitness is *not* a population genetics measure but a property of individuals. We do not quantify inclusive fitness by adding up an individual's genetic representation in his offspring plus that of all his other relatives; instead we add up only his effects on gene propagation directly in the bodies of surviving offspring and indirectly via *nondescendant kin that would not exist except for his actions.* One point of measuring inclusive fitness is to permit an observer of animal behavior to weigh the evolutionary consequences of alternative strategies, to identify what strategy is adaptive, and thereby to predict what course of action individuals will take, namely, the strategy that produces highest individual inclusive fitness.

Indirect Selection, Inclusive Fitness, and Helping in a Kingfisher

A concrete example of the utility of measuring inclusive fitness might be helpful at this stage. The pied kingfishers studied by Heinz-Ulrich Reyer offer the clearest demonstration that indirect selection *and* direct selection can both lead to the evolution of helping behavior [663]. These attractive African kingfishers nest colonially in tunnels in banks by large lakes. Breeding pairs may be assisted by one to four nonbreeding male helpers that bring food for the young (and the nesting female) and defend the nest against predators such as snakes and mongooses. These actions appear altruistic, because helpers do not reproduce personally while assisting other individuals.

Because Reyer had banded the kingfishers in his study area, he knew who was related to whom, and he therefore realized there were two categories of male helpers. In one group (the *primary* helpers) were the helpful sons of the breeding pair. These males were with their parents from the start of the breeding season. They brought fish first to their mother and later to their baby siblings. The other group of males (the *secondary* helpers) were unrelated to the breeding pair and were not tolerated at the nest until after the eggs had hatched. If accepted then, they, too, brought fish to the nest, but they tended to feed the breeding female in preference to the nestlings. Furthermore, secondary helpers contributed less total energy to the individuals they helped than the primary helpers did (Figure 8).

The fact that primary helpers worked very hard on behalf of their parents probably literally killed some of them (if we accept Reyer's conclusion that kingfishers failing to return to the breeding colony have died). Just 54 percent of the primary helpers returned for a second season, whereas 74

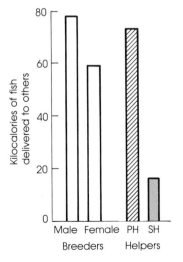

8 Altruism and relatedness in pied kingfishers. Primary helpers (PH) are sons that deliver more calories per day in fish to their mother and her offspring than do the unrelated secondary helpers (SH). *Source:* Reyer [663].

percent of the secondary helpers, who took a more relaxed approach to helping, returned the next season. Only two of every three surviving primary helpers found a mate in their second year and reproduced personally. The other 2-year-olds in this group either helped their parents again, or became secondary helpers elsewhere. In contrast, 91 percent of the males that were secondary helpers in year one succeeded in breeding if they survived to their second year. A substantial number of those that did breed reproduced with the female they had helped the preceding year (10 of 27 in Reyer's sample).

These data show that the two kinds of helpers are using two very different tactics to propagate their genes. Primary helpers throw themselves into helping their parents produce offspring *at the cost of having less chance of reproducing personally in the next year.* To the extent that they increased their parents' reproductive success, they created siblings that would not otherwise have existed and so indirectly propagated their genes in this fashion. Secondary helpers apparently help as the price of being permitted near a breeding female who may be available as a mate next year. Their assistance cannot yield an indirect fitness gain for them, because they are unrelated to the nestlings they feed as helpers.

The question is, What are the genetic returns for these two kinds of tactics, and how do they compare with two other options, which are to breed in the first year of life and to forgo breeding in this first year but not to help at all? Reyer's meticulous observations provided the numbers he needed to calculate the inclusive fitness of the four alternatives. We shall begin by comparing primary and secondary helpers. For simplicity's sake, we shall restrict our comparison to primary helpers that helped their parents rear siblings in year 1 (with no other helpers present), and then bred on their own the next year (if they survived) versus secondary helpers that helped nonrelatives (by themselves) in year 1 and then reproduced on their own (if they survived to the next year).

Table 4 presents the key data needed to determine the inclusive fitness of helpers in their first year. Primary helpers increased the reproductive success of their parents by an average of 1.8 siblings. Some primary helpers assisted their genetic mother and father, in which case the extra siblings were full brothers and sisters with a coefficient of relatedness of 0.5. But in other cases, one parent had died and the other had remated, so the

TABLE 4

Effect of assistance on number of fledglings produced by pairs of pied kingfishers

Type of assistance	Mean number of helpers per nest	Mean number of fledglings per nest	Mean increase in number fledged per helper
No helpers	0	1.8	—
Primary helpers	1.0	3.6	1.8
Secondary helpers	1.45	3.7	1.3

Source: Reyer [663].

offspring produced were only half-siblings ($r = 0.25$). The average coefficient of relatedness for sons helping a breeding pair was between one-quarter and one-half (0.32). Therefore, the average indirect fitness gained by helper sons was 1.8 sibs $\times$ 0.32 $= 0.58$. Secondary helpers have a coefficient of relatedness of 0 to the extra 1.3 offspring they helped rear to fledging, for an indirect fitness gain of $0 \times 1.3 = 0$. Thus, at the end of the first year, primary helpers are far ahead of secondary ones. But fitness is a lifetime function, and so we must consider what happens in the second year for pied kingfisher helpers.

Primary and secondary helpers that live to reproduce for the first time in their second year cannot have any primary helpers (because they had no sons in year 1), but they may attract some secondary helpers. The combined average for first-time breeders without helpers and those with secondary helpers is 2.5 offspring fledged ($r = 0.5$), for a gain of 1.25 units of direct fitness. A primary helper, however, had a probability of 0.54 of living to its second year, and then only a probability of 0.60 of finding a mate. Thus, the *average* contribution to direct fitness in the second year for a primary helper is $1.25 \times 0.54 \times 0.60 = 0.41$. The corresponding calculations for secondary helpers yield a figure of 0.84 (Table 5), because these birds are more likely to live to their second birthday and are more likely to secure a mate as well. The average inclusive fitness of the two kinds of birds for their first two years of life is shown in Table 5. Secondary helpers go a long ways toward catching up to primary helpers in their second year, but they are still behind.

Birds that sit out their first year altogether (delayers) have no better chance than secondary helpers of surviving to breed in their second year, and they have a *lower* probability of finding a mate than secondary helpers, which have established a special relationship with the female they helped. Thus, "delayers" generally do worse than secondary helpers.

Birds that breed in the first year have no primary helpers, and so produce an average of just 1.9 offspring, yielding $1.9 \times r = 1.9 \times 0.5 = 0.95$ direct

TABLE 5

Inclusive fitness estimates for male pied kingfishers exercising different behavioral options in their first year and then trying to breed in their second year

Behavioral option	First year[a]			Second year[a]					Inclusive fitness
	y	$\times r$	$= f_1$	y	$\times r$	$\times s$	$\times m$	$= f_2$	$f_1 + f_2$
Primary helper	1.8	0.32	0.58	2.5	0.50	0.54	0.60	0.41	0.99
Secondary helper	1.3	0	0	2.5	0.50	0.74	0.91	0.84	0.84
Delayer	0	0	0	2.5	0.50	0.70	0.33	0.29	0.29
Breeder	1.9	0.50	0.95	3.0	0.50	0.59	0.98	0.87	1.82

Source: Reyer [663].
[a]Symbols: y, extra young "produced" by helped parents or offspring of a breeding bird; r, the coefficient of relatedness between the adult and "y"; s, probability of survival from year 1 to year 2; m, probability of finding a mate in year 2; f_1, fitness in year 1; f_2 fitness in year 2.

fitness units then. A breeding male has only a 50 percent chance of surviving to reproduce again. But if he does, he may have sons to help him in his second year and secondary helpers as well; so survivors fledge an average of three offspring. Their second-year gain in the direct fitness column is 0.87. The 2-year total for breeding birds is therefore $0.95 + 0.87 = 1.82$ units of inclusive fitness.

For a variety of technical reasons, the actual inclusive fitness figures calculated by Reyer are slightly different than those given here, but his numbers also showed that first-year breeders had higher inclusive fitness than primary helpers, which outscored secondary helpers, with delayers doing poorest of all [663].

The calculations secured in this exercise are useful for two reasons. First, and very importantly, they confirm that some males, the primary helpers, sacrifice future personal reproduction in exchange for *increasing* the number of nondescendant kin. Their altruism is repaid by the heightened reproductive success of their parents, an outcome that gives primary helpers indirect fitness gains. Second, the results show clearly that the four options available to male pied kingfishers are not equally advantageous. The measures of inclusive fitness enable us to predict that first-year males will try to breed on their own. In the pied kingfisher, the sex ratio is skewed toward males, and no females are left unmated, supporting the first prediction. Males that cannot find a mate should return to help their parents. The fact that primary helping is common, even though males cannot be forced to help their mother and father, supports this prediction. Some males, however, will find their parents have disappeared or another male sibling will have claimed the primary helper role. These males should try to get on as secondary helpers elsewhere, and only if they cannot find an accepting pair, should they become delayers. These predictions have not been fully tested yet, and they depend heavily on the validity of Reyer's conclusion that disappearance of a male from a breeding colony was caused by the death of the bird, not its dispersal to a new breeding site. Nevertheless, the key point for our purposes is that calculations of inclusive fitness permit predictions that are decidedly different from those that we would make if we were to consider only the direct fitness consequences of different options.

Helpers at the Nest in Other Species

Even within pied kingfishers we see that some males help because of the indirect fitness gains they receive, whereas others help as a way to raise their direct fitness. Studies of helpers in other animals reinforce the point that helpful behavior may be altruism in some cases, mutualism in others. Each case needs to be analyzed in its own right.

Helpers-at-the-nest have been found in a surprisingly large number of bird species [112]. In order to understand why helpers exist in a particular species, the answers to the following questions are useful.

1. Does a helper raise the reproductive success of the birds it appears to be helping? If the helper does not, we can be sure that it gets no indirect fitness gains through its help, and the finding would strongly implicate selfish manipulation of the helped birds for the benefit of the helper.

2. What is the helper's degree of relatedness, if any, to the pair that it helps? If helpers are unrelated to those they help, we can again rule out the possibility of indirect fitness gains as the basis for the evolution of altruism in such a species.

3. Does a helper benefit by helping? If it can be shown that there are no indirect or direct benefits for the helper, then the hypothesis that the helpers are somehow coerced into giving aid becomes more likely. On the other hand, if helpers are related to the breeding birds they assist and if they raise the reproductive success of these individuals, they will gain some indirect fitness. Alternatively, as in the case of secondary helpers in pied kingfishers, the helpers may cooperate now to receive direct fitness gains later.

4. What are the other options open to helping birds, and what are the fitness payoffs associated with these tactics? Helpers delay or forego personal reproduction. In order for this to be the superior option for them, one must demonstrate that the gains from trying to reproduce personally during the season(s) they help would be less than the benefits, direct or indirect, that come from helping. For helper pied kingfishers, a skewed sex ratio makes it literally impossible for all males to find mates, leaving them with making the best of a bad job—which is to be a primary helper, if possible.

Let us ask these questions of the Florida scrub jay, a population of which has been studied in remarkable detail by Glen Woolfenden and John Fitzpatrick [869]. A breeding pair of jays may have as many as six other individuals living with them in their territory. These nonbreeding adults may be 2 or 3 years old, or even older; they are physiologically capable of producing offspring of their own but instead defend the breeding pair's territory, feed the nestlings, and repel predators (Figure 9). If one measures the number of offspring fledged by pairs with and without helpers, one finds an increase, albeit a modest one (Table 6), for pairs with helpers. Thus, helpers are really helping, answering the first question above.

By color-banding large numbers of scrub jays and observing their behavior in succeeding years, Woolfenden and Fitzpatrick showed that helper jays are almost without exception the offspring of the pair they help. There are almost no "secondary helpers" of the sort Uli Reyer found in pied kingfishers.

Because helpers keep alive more full and half-siblings than would exist without their assistance, they raise their inclusive fitness by adding at least $0.30 \times r = 0.30 \times 0.5 = 0.15$ units in the indirect fitness column (Table 6) each time they help their parents produce more siblings.

But is helping the option of first resort for a young scrub jay? The direct fitness gains for a jay that is breeding for the first time will be $1.03 \times r = 0.52$ units (the average gain for pairs breeding without helpers). This is higher than that secured by helper jays. Therefore, we can predict that given a chance to breed, helper birds will do so. In fact, helpers compete with one another for dominance; and top-ranking birds abandon helping to breed when a vacancy occurs through death of one of their parents or the demise of a breeder from a neighboring territory.

Helper jays do not sign a life-long pledge of celibacy, although many fail

9 **Cooperation among scrub jay relatives.** Helpers at the nest in the Florida scrub jay provide care for the young, defense for the territory, and protection against snakes. *Source:* Wilson [853].

to reproduce at all because opportunities never arise. The population of scrub jays studied by Woolfenden remained stable for years, and all the habitat suitable for breeding was occupied by pairs. If the probability of breeding independently is very low, then low-ranking jays forfeit the indirect genetic benefits they receive from helping if they leave their parents, brothers, and sisters in a futile attempt to reproduce personally. Further, helpers do gain experience while helping, so that when they breed for the first time on their own they are better parents than they would be otherwise.

If it is true that the habitat of scrub jays is saturated with territorial groups, then selection will tend to favor young birds that delay dispersal from their home range. A delay in breeding follows, if parents prevent their nondispersing young from breeding on the natal territory. If a nonbreeding individual stays at its birthplace, it is preadapted to become a helper. Its options under these conditions are to wait without helping or to help its parents while waiting. The helping option in scrub jays yields indirect fitness gains; the costs of helping have not been quantified, but if helping decreases survivorship only slightly, and if the probability of finding a

TABLE 6

Effect of scrub jay helpers at the nest on the reproductive success of their parents and their own inclusive fitness

	Inexperienced pairs[a]	Experienced pairs
Average number of fledglings produced with no helpers	1.03	1.62
Average number of fledglings produced with helpers	2.06	2.20
Average number of helpers	1.7	1.9
Increase due to help	1.03	0.58
Indirect contribution per helper	0.60	0.30

Source: Emlen [235].
[a]Inexperienced pairs are defined as those pairs in which one or both members are breeding for the first time.

territorial vacancy is very low, a helper's indirect fitness gains could outweigh the loss in future direct fitness gains, just as is true for primary helpers in pied kingfishers.

Indirect Selection and the Evolution of Alarm Calls

Helping at the nest is not the only type of potential altruism. In an earlier chapter, we discussed the advantages to social animals of giving alarm signals upon detection of a predator. There we noted that by uttering a special call when a falcon or hawk came hunting a Belding's ground squirrel improved its chances of survival, and so did its fellow squirrels that responded to the hawk alarm by dashing for cover. This then is a case of mutualism or cooperation with immediate direct fitness benefits for signal-giver and signal-receiver [716].

Belding's ground squirrels produce a special staccato whistle when they spot a *terrestrial* predator, like a coyote or badger (Figure 10). This alarm call causes other squirrels to rush for safety. Is this another case of cooperation, or is it altruism? We can ask the same questions about this call as we can about helping at the nest by scrub jays. The answers have been thoughtfully provided by Paul Sherman whose findings enable us to discriminate among the following alternative hypotheses [713]:

1. *Direct selection hypotheses.* The caller enhances its personal chances for reproductive success by giving an "alarm" signal.
a. *Predator confusion hypothesis.* The call's function is to alert the caller's neighbors so that together they create a chaotic mass dash for safety that confuses the terrestrial predator, using the same kind of tactic that works against aerial predators that helps everyone except the predator.
b. *Predator deterrence hypothesis.* Although one might think that the caller risks drawing attention to itself and should slip off quietly when it spots danger, actually predators that know they have been detected are likely to give up the hunt. The caller signals to communicate with the predator to save his own skin. Any benefits enjoyed by others are purely incidental and not the evolved function of the alarm call.

10 **Alarm call** of a Belding's group squirrel that has spotted a terrestrial predator such as a coyote. Photograph by George Lepp.

c. *Reciprocal altruism hypothesis.* A caller may give the signal at some immediate risk to itself, but later it will be more than repaid by others in the group when they return the favor. This presupposes a stable relationship between at least some group members who have the capacity to withhold help if they do not receive it from their fellows.

d. *Parental care hypothesis.* A caller gives the signal to warn its offspring, increasing their chances of survival and thus the caller's direct contribution to subsequent generations.

2. *Indirect selection hypothesis.* The caller reduces its lifetime chances for reproductive success by sounding the alarm, but the altruism nevertheless raises its inclusive fitness. Parents, aunts, uncles, brothers, sisters, or cousins are alerted by the signal, and the gain in indirect fitness via an increase in reproductive output by these nondescendant relatives outweighs the direct fitness loss paid by the altruist.

We can quickly show that there is a big difference between the alarm call given when an aerial predator approaches and that given when the enemy is a terrestrial predator (Figure 11). The reproductive status and the presence of relatives has little or no effect on the probability that a female will give the aerial alarm call, but they have a large effect on whether a female will warn others about a coyote or badger. This result appears to stem from differences in the costliness of the two alarm calls. Calling when a terrestrial predator is present puts the caller at considerable risk. Terrestrial predators are not confused or deterred from a hunt when they hear a squirrel call. Sherman and his gang of observers saw weasels, badgers, and coyotes stalk alarm callers and kill them at a rate higher than that experienced by noncalling, fleeing ground squirrels. Moreover, the probability that an individual will give an alarm call is not correlated with familiarity or length of association between the caller and the animals that benefit from its signal. This result decreases the probability that alarm calling is maintained through reciprocity, because stable, long-term associations are necessary for the kind of tit-for-tat arrangements that characterize evolutionarily stable reciprocity [715].

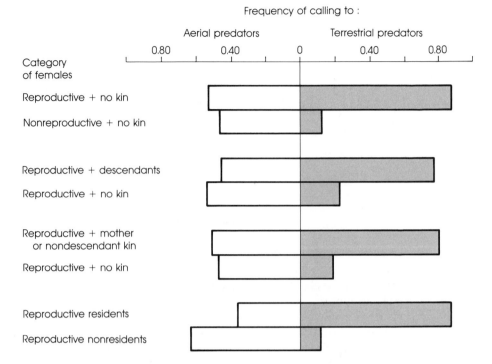

11 **Frequency of alarm calling** by female Belding's ground
squirrels to aerial and terrestrial predators. The presence
of offspring or other relatives has no effect on the probability that a
female will signal when she sees a hawk, but has a major effect on
the probability that a female will sound an alarm to a terrestrial
predator. The category of reproductive females includes those that
are pregnant, lactating, or living with weaned offspring. *Source:*
Sherman [716].

Both the parental care hypothesis and the altruism hypothesis predict
that females, rather than males, will be more likely to give risky alarm
calls. Female Belding's ground squirrels tend to be sedentary; therefore, a
female often lives with her daughters, sisters, aunts, and nieces. Males, on
the other hand, move away from their natal burrow (Chapter 9) and do not
live near offspring that they might help. According to the parental care
hypothesis, females should give more alarm calls than males because only
female callers might be compensated for their risky help by the improved
chances of survival of their offspring. The altruism hypothesis suggests
that females will give more calls because the loss in future reproduction
that they suffer from attracting predators to them is more than repaid
indirectly through the increased probability of survival of kin *other than
offspring*.

Sherman found that females are, in fact, much more likely to give alarm
calls when they spot a predator than are males, and females with living
relatives nearby called more frequently than females without living rela-

TABLE 7

Who gives alarm calls among Belding's ground squirrels?

Category of squirrel	Exposure to predator[a]	Number of squirrels observed to give alarm call (%)	Number of squirrels expected to call if alarms are given randomly (%)
Males, >1 year old	67	12 (18)	19 (28)
Females, >1 year old, with living relatives[b]	190	75 (39)	53 (28)
Females, >1 year old, without living relatives	168	31 (18)	46 (28)

[a]Number of times ground squirrels in each category were present when a terrestrial predator appeared, 1974–1979. Data courtesy of Paul Sherman.
[b]Relatives consist of daughters, granddaughters, mothers, or sisters.

tives as neighbors (Table 7) [713]. This finding suggests that the parental care and the altruism hypotheses may both apply to female Belding's ground squirrels, which gain direct fitness by helping their offspring escape from predators and indirect fitness to the extent than aunts, nieces, and sisters escape as well.

The fact that females without young, but with nondescendant kin nearby, will sound the alarm is strong evidence that indirect selection alone contributes to the maintenance of this behavior in the squirrel population, complementing the action of direct selection. Further evidence for the effects of indirect selection on ground squirrel behavior comes from studies by Warren Holmes and Paul Sherman, who showed that females help close female relatives in addition to their offspring in territorial conflicts with intruders (Figure 12).

This example shows that direct and indirect selection can contribute to the evolution of the same trait. In Belding's ground squirrels, alarm calling can serve a parental function even while simultaneously assisting individuals other than offspring, which also share genes in common with the alarm caller. One unit of inclusive fitness gained by keeping offspring alive is no different from one unit of inclusive fitness gained by saving the lives of assorted nondescendant kin.

Cooperation among Males

Although parental care and altruism are restricted to females in Belding's ground squirrels, some male animals cooperate to a high degree in acquiring mates. This is very surprising, given that males so often compete fiercely to copulate with as many females as possible. Yet male lions join forces to oust other males from a pride of females [64, 122]. After the new group has taken charge of a pride, males in the ruling coalition rarely fight openly over access to receptive females despite the fact that some males enjoy greatly disproportionate reproductive success. In one group of six lions that owned a pride, the top male copulated 3.5 times as frequently as the lowest male in the hierarchy. In this case, however, the males were a cluster of brothers, half-brothers, and cousins that were expelled from their natal pride at the same time and then remained together to compete with other male groups for possession of a harem of females. In Brian Bertram's terms, each male was to some degree "reproducing by proxy through his companions," because they had a common ancestor and shared a relatively high proportion of the same alleles. Therefore, to the extent that some males voluntarily gave up chances to mate in order to benefit a close relative, their reproductive loss may have been compensated in part by a gain in indirect fitness.

Considerable evidence exists, however, that the behavior of male lions in coalitions of relatives is regulated by selfish reproductive competition within the group. Subordinates may give way to dominants, not because the gains from indirect fitness outweigh their losses in direct fitness, but because they would be attacked and defeated by the other male if they behaved aggressively. They may have more to gain in direct fitness by being a low-ranking member of a coalition than by striking out on their own or by fighting for higher status within their band.

Indirect selection certainly cannot account for all helpful alliances among male lions, because some coalitions consist entirely of unrelated individuals [613]. If a male does not have brothers or half-brothers to assist him, his options are to try to control a cluster of females by himself or to join other unrelated males, using their combined force to repel opponents. Inasmuch as the probability that a solitary male will be able to subdue two or more cooperating rivals is vanishingly small, the second alternative is likely to be the more productive. Even if a male in a coalition of nonrelatives is low lion on the dominance totem pole and therefore gets to mate relatively

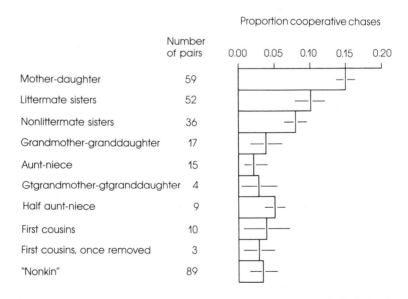

12 **Cooperative defense.** Only close relatives help defend a territory in Belding's ground squirrels. Mothers assist daughters, and sisters cooperate in the defense of space. *Source:* Sherman [715].

little, a few copulations are better than none at all in terms of direct fitness. A low-ranking male also gets to feed from kills made by the females of the pride, and so may live long enough to move higher in the dominance order of his coalition.

The helpfulness of male lions pales in comparison with that exhibited by long-tailed manakins [266, 512]. Although not much larger than English sparrows, long-tailed manakins are outfitted in stunning dark blue-black plumage, with a powder-blue cape and crimson cap. Two long trailing tail feathers complete the picture. But the behavior of the long-tailed manakin is even more exotic than its colorful plumage, for it is an animal in which males cooperate to attract and court females.

One manifestation of the helpfulness of the bird lies in its call, a whistled "toe-lay-doe" that is given over and over in the dry forests of northwestern Costa Rica. What sounds like one male's "toe-lay-doe" is actually produced by two individuals calling in perfect synchrony from a perch in a display area. The cooperating pair spend much of the prime mating months singing together, generating as many as 300 calls per hour.

The point of all the toe-lay-doeing is to attract females to a central display perch, often a horizontal section of liana vine that lies a foot or so above the ground. Females, which are decked out in drab olive, visit the area and, by alighting on the perch, invite an astonishing response from the duo that defends the site. The two males dart in and land within a few inches of each other. The female, a foot or two distant, watches as the males begin the cartwheel display. In this cooperative venture, the male farther from the female stretches his body out and leaps into the air to come down where his partner had been perching an instant before. To make room for the descending bird, the other male sidles quickly back to the jumping off point used by bird 1, and he then leaps up to exchange places with his partner (Figure 13). After a series of cartwheels, one male may leap over the head of the observant female, with the other member of the pair instantly following in order to perform a run of cartwheels on the other side of the female.

As if this were not sufficiently dramatic, long-tailed manakins interrupt bouts of cartwheeling with two other joint ventures. In one display, the birds fly out from the perch in what David McDonald calls the "butterfly flight." The males hold their wings above their heads so as to appear to be helicoptering back and forth in an exaggerated, fluttering flight. From this slow flight display, the males may switch to the other extreme in which they fire back and forth at top speed over the head of the perched female. When I watched the birds from within a little plastic blind with mosquitoes as my intimate companions, the males appeared to be mere black and blue blurs during this segment of the display program.

Male long-tailed manakins regularly perform 5 to 10 minutes of mutual display in which they run through their routines several times, all the while keeping in synchrony with their coperformer. Female visitors much more often than not respond to these impressive shows by flying away, leaving the males to resume their singing from various spots about the territory. But occasionally a female signals that she is ready to mate, by

13 **Cooperative display** of the long-tailed manakin. The two
    males are in the cartwheeling portion of their dual display
to a female, who is perched on the vine to the right.

beginning to jump about on the perch. At this cue, one member of the
display duo discretely leaves, while the remaining male begins a bout of
butterfly flight that brings him to the female. If she is truly receptive, he
mounts and copulates with her for a few seconds before she flies away. The
mated male then calls for his male partner, who flies back to resume his
cooperative calling and courtship duties.

By marking both members of several pairs, Mercedes Foster [266] and
David McDonald [512] found that one of the males was dominant to the
other and this individual secured almost all the copulations. (In McDonald's
study, the beta male accounted for only 2 of 121 copulations.) It is unlikely
that the helper male is the brother of the helpee, given the small clutch
size of manakins (one or two eggs). In order for two brothers to display
together, both would have to survive for at least 8 to 10 years to dominate
a display site. Indirect selection can hardly be responsible for the evolution

of mutual display in manakins. Therefore, we must explain the helper's actions in terms of their contribution to direct fitness, something that would seem to be hard to do given the near-total failure of the helper to copulate.

McDonald speculates, however, that the *lifetime* mating success of the subordinate manakin might be improved through his cooperation. As it turns out, any one display site has a resident alpha male, a resident beta male, and a number of other males that come to the area, and sometimes engage in mutual display with either male should one of the regulars be absent. There is in effect a dominance hierarchy at each display location, and only the alpha and beta get to perform regularly there. Only by moving up the dominance ladder can a bird become a beta male and have even a tiny chance of mating; and only by becoming a beta male can a long-tailed manakin inherit the alpha position upon the eventual demise of the top male. By displaying with the alpha male, a beta individual establishes his claim to be next, keeping other (younger?) birds at bay, while (unconsciously) counting the years until he can become the top male at the display site. Then his chances of copulating will improve sharply. (This hypothesis could be tested by predicting that removal of an alpha male will almost invariably result in the ascension of beta male to the top position at a display site.) Thus, perhaps cooperation between alpha and beta males has evolved through direct selection, although many questions remain to be answered about this case, not the least of which is, If joint display is an advantageous mutualism in long-tailed manakins, why is it not practiced by most other manakin species?

Reciprocity among Male Baboons
In lions and manakins, males do not take turns copulating; but in the olive baboon, cooperating males do reciprocate in this manner. Male baboons regularly form political alliances to dominate a high-ranking competitor who has formed a consort relationship with an estrous female [560] (Figure 14). Another closely matched male may threaten the consort male and may attempt to enlist the aid of another baboon in this task. If he is successful, the helper male will respond to his head-turning movements and the two monkeys will together drive the consort male from the female. If this happens, the male that enlisted the support of the helper always takes the female. The helper baboon runs a small risk of being attacked by the consort male, who is naturally not enthusiastic about relinquishing a fertilizable female. What does the helper gain for his risk?

The adult male baboons in a troop have emigrated from other bands and are not thought to be close relatives. This situation reduces the likelihood that indirect selection is responsible for the helpful behavior of one male to another. Instead, this appears to be a genuine case of reciprocity [610]. A helper is likely to receive assistance from the baboon he helped when he wishes to form an alliance to steal a female from another male. Males that participate in mutual threat displays often have favorite partners that take turns helping and being helped. Males that fail to give assistance when solicited are far less likely to receive aid than individuals that will help in threatening an opponent. Reciprocity can readily evolve in a baboon troop,

14 **Reciprocity in baboons.** The two males on the right have formed an alliance to defeat the male on the left in a dispute over possession of an estrous female. One male solicited assistance from the other; in return he will help his assistant later in a similar situation. Photograph by Leanne Nash.

because males are able to recognize each other as individuals and because male baboons have good memories and well-developed learning abilities. The cost of the altruistic act is small because it is rare that a single male dares withstand a threatening cooperating pair. The benefit of the act for the helpful male may be great if he eventually is repaid and inseminates a female that would otherwise have mated with another male.

These examples from social vertebrates demonstrate that extreme degrees of helpful behavior can evolve in a number of different ways, through indirect selection in favor of altruism toward relatives, through direct selection in favor of mutually beneficial forms of cooperation, and through direct selection in favor of reciprocal aid-giving.

The Evolution of Eusocial Insects   Although a Martian ethologist would surely admire the complex social lives of long-tailed manakins, scrub jays, and olive baboons, he or she would find being stung by a honeybee even more remarkable. Although the pain of a bee sting is attention-getting, a Martian aware of the theory of evolution by natural selection would be most impressed by the discovery that the worker bee dies when her stinger catches in its skin (Figure 15) [803].

One does not have to be a Martian to appreciate that the evolution of sterility and a readiness for suicidal self-sacrifice by a worker caste pose an intriguing evolutionary problem. Extreme altruism occurs all the time in the colonies of termites, ants, some bees, and some wasps, colonies that may contain hundreds or tens of thousands of workers that cannot reproduce but instead sacrifice themselves in many ways for the welfare of others in their group.

The contrast between a solitary wasp and a eusocial (caste-containing)

15 **Suicidal altruism.** When a honeybee stings a vertebrate, she leaves her stinger and associated poison sac in the body of the victim. She dies as a result. Photograph by Bernd Heinrich.

wasp helps illustrate the extent to which helpful behavior dominates the social systems of eusocial insects. A solitary wasp like *Ammophila novita* captures a single large prey, digs a nest, deposits the prey in the nest, lays an egg on the victim, and closes the nest as she leaves. The solitary female has no contact with her progeny and receives no assistance in her reproductive efforts from any other individual (Figure 16).

A radically different pattern occurs in the familiar *Polistes* paper wasps. Almost everybody living in the Americas has probably been stung by a paper wasp at least once (and a most unpleasant experience it is), because paper wasps regularly build their nests (Figure 17) in the shelter provided by the eaves on a house. Despite their intimidating stings, *Polistes* are well worth getting to know, because they are remarkable social beings [248]. Females capable of reproduction emerge from cells in the nest late in the breeding season (in a number of temperate-zone species). They mate with unrelated males, which are also being produced at this time, and then spend the winter hibernating in a sheltered spot. In the spring, they rouse themselves and start a nest, which is constructed of chewed plant fibers. The nest contains a series of cells, each of which receives a single egg from the new queen. A foundress female may be joined by other overwintering females; this generates dominance contests and the formation of a hierarchy, with the dominant female reproducing and the subordinate(s) helping her rear the larvae and protect the nest against predators and parasites.

The eggs that are laid early in the season are destined to become daugh-

16 **A solitary wasp, *Ammophila novita*.** The female provisions her nests with moth larvae, like the one she holds in her jaws, without help from any other member of her species. Photograph by the author.

ter wasps, because they are fertilized by the "queen" wasp, which uses sperm stored from last fall's copulation. When it is adaptive to produce sons, the female simply lays unfertilized eggs.

As the eggs hatch, the newborn larvae are fed water, nectar, and fragments of insect prey. When the first females emerge, they assist their mother in raising still more daughters (their full- and half-sisters) rather than flying off to start new colonies in which to rear their own offspring.

17 **A social wasp of the genus *Polistes*.** Females of this paper wasp cooperate in the construction and defense of a nest and in the feeding and care of the young. One of the females of the four adults here is the queen; the others are workers that assist in the care of the eggs (see central cell) and larvae (the translucent grubs in several of the cells). Photograph by the author.

Several broods are produced in this fashion during the colony's life, with more and more females joining the work force. They increase the size of the nest, add cells, feed the larvae, detect and drive off parasites, and sting predators into retreat. However, as the summer progresses in temperate regions, the queen produces increasing numbers of females that do not join the workers but instead lounge about on the nest, appropriating food from their working sisters. Later in the summer, males emerge for the first time and these, too, do little to aid the welfare of the colony. Activity at the nest dramatically decreases at this time, the "lazy" females and males fly off, mating occurs with members of other colonies, the males die, and the mated females—future colony foundresses—hibernate through the winter months to resume the cycle the succeeding spring [248].

Indirect Selection and the Evolution of Eusociality

The by-now familiar question is, Why should a daughter forgo reproduction in order to help rear additional sisters? In *Polistes* and many other highly social insects, daughter workers may not have the option to reproduce personally should a suitable opportunity arise. They are chained to their helper roles for life, unlike the helper scrub jay. This is altogether more extreme than the reproductive cooperation exhibited by social vertebrates. The fact that obligatorily sterile workers are known only from certain Hymenoptera (ants, bees, and wasps) and the termites reinforces the unusual nature of the eusocial phenomenon [852].

The link between the Hymenoptera and eusociality led W. D. Hamilton to consider the genetic consequences of the peculiar mode of sex determination practiced by this group [331]. As indicated already, males are haploid, having been produced from unfertilized eggs, whereas females are diploid, with two sets of chromosomes, one from the egg and the other from the sperm. If a female hymenopteran mates with just one male, all the sperm she has received will be identical (the male having only one set of chromosomes, which he copies when making gametes). Therefore, all the daughters a mother produces will carry the same set of male genes. Any one daughter will share these genes (50 percent of her total genotype) with all her sisters. The other set of chromosomes comes from her mother. The foundress' eggs are not identical genetically because the mother is diploid; gamete formation in animals with two sets of chromosomes involves producing a cell with just one set. The statistically average egg made by a female wasp, bee, or ant will have 50 percent of the same alleles carried in her other eggs. Thus, when eggs unite with identical sperm, genotypes are created that are identical for an average of 75 percent of their alleles—50 percent of those alleles carried in common are from the father and 0 to 50 percent are from the mother (Figure 18). The mean probability that an allele present in one female will also be carried by her sister is 0.75: $r = 0.75$.

Because of the haplodiploid nature of sex determination in the Hymenoptera, sisters may be genetically very similar to one another, more so than a mother to her daughters and sons, with which she has a coefficient of relatedness of 0.50. Because sisters are so closely related in the Hymenop-

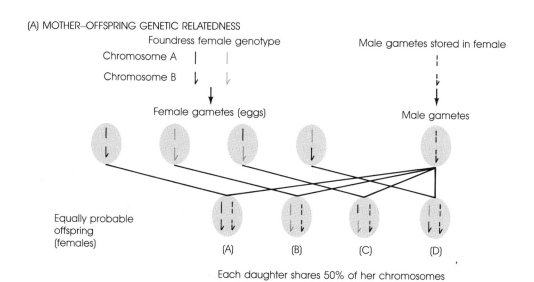

(A) MOTHER–OFFSPRING GENETIC RELATEDNESS

Foundress female genotype

Chromosome A

Chromosome B

Male gametes stored in female

Female gametes (eggs)

Male gametes

Equally probable offspring (females)

(A) (B) (C) (D)

Each daughter shares 50% of her chromosomes (and thus her genes) with her mother

(B) SISTER–SISTER GENETIC RELATEDNESS

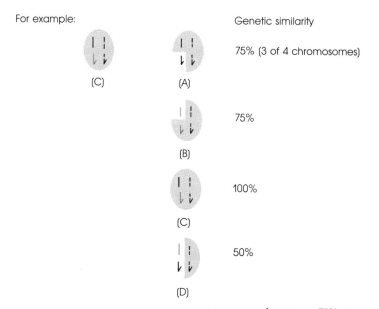

Pick any daughter genotype and compare it with the possible genotypes of her sisters

For example:

Genetic similarity

(C)   (A)   75% (3 of 4 chromosomes)

(B)   75%

(C)   100%

(D)   50%

Average = 75% of genes shared between sisters

18 **The significance of haplodiploidy** on the evolution of sociality in the Hymenoptera. The degree of genetic relatedness of a female wasp to her offspring (A) and among sisters (B). For the sake of simplicity only two chromosomes are considered. Sisters may be more closely related to one another than they would be to their own progeny (see text).

tera, Hamilton suggested, and many others have agreed, that indirect selection is the driving force behind the evolution of the social insects [331, 852]. The high proportion of identical alleles increases the probability that a helpful act between sisters will elevate the inclusive fitness of the altruist. In support of the indirect selection hypothesis, observers have noted that the workers of social Hymenoptera are sisters. Males are not as highly related to each other and have an average of only 25 percent of their alleles in common with their sisters. They show no special cooperative interactions with siblings of either sex. Moreover, in one case in which larvae of both sexes contribute silk for nest construction, the males have smaller silk glands and provide less material than their sisters [857].

The essence of the Hamiltonian hypothesis is that female wasps are really helping their reproductively competent sisters (future queens) and only incidentally assisting their mother. That is, the genetic goal of their behavior is to increase the chances of survival of their own genes by aiding very closely related siblings. The advantages of altruism are so great in this case that indirect selection can favor females that, so to speak, put all their eggs (alleles) in a sister's basket and give up personal reproduction entirely. Of course, this depends on there being sisters to help and the positive effect of a helper's assistance in increasing the number of surviving siblings. Under the appropriate conditions, however, it is theoretically possible for a worker to gain more inclusive fitness by being a nonreproducing helper than by being a reproducing queen.

That sterility can arise through indirect selection is supported by the occurrence of nonreproducing soldier aphids that defend the reproducing members of their clone, which are genetically identical to them [22]. A sterile soldier caste has also evolved in a parasitic wasp whose females lay numerous eggs within a single host. Some members of a female's brood develop into large-jawed larvae (Figure 19) which attack and dispatch other parasitic competitors within the host that endanger the siblings of the defender. The large-jawed types never metamorphose into reproducing adults, sacrificing their reproductive chances to help their close relatives survive [161].

These cases do not prove, however, that workers in the social Hymenoptera have evolved via indirect selection. To test this hypothesis, Robert Trivers and Hope Hare looked at the sex ratio of colonies of eusocial ants [788]. The queen of a colony has 50 percent of her alleles in common with both her male and female progeny; she presumably gains no genetic advantage by investing more resources in the production of females than males or vice versa. Her daughters, however, should benefit if more sisters are produced than brothers, because $r = 0.75$ for sisters whereas $r = 0.25$ between sister and brother.

Thus, there is a potential conflict between the queen and the workers. The queen's offspring in large colonies are cared for and fed by their sisters. Workers could materially influence the weight and even survival of the male larvae in a colony by withholding food from their brothers and channeling it instead to their sisters. If female Hymenoptera behave in ways that benefit their alleles rather than their mother's alleles, we should expect that the *weight* of all the female reproductives (a measure of total resources

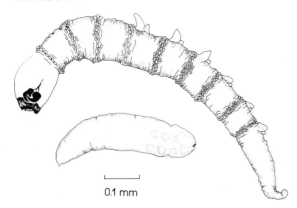

19 **Altruism in a parasitic wasp.** The larval soldier forms of this wasp (top) have jaws but will not develop into an adult. Instead they protect their brothers and sisters (bottom), which live in the same host, by dispatching other species of parasites in their host. *Source:* Cruz [161].

0.1 mm

devoted to female production) made by a colony should exceed the combined weight of male reproductives by 3 to 1 (sisters having 75 percent of their alleles in common with other sisters and only 25 percent with their brothers). Acquiring the data needed to test this prediction is difficult, but Trivers and Hare's review indicated a strong skewing of investment toward females, although there are alternative hypotheses for this result [14].

Individual Selection and the Evolution of Eusociality

Even skeptics of the Hamiltonian hypothesis and its tests to date admire Hamilton's genius for having perceived the possible causal link between the haplodiploid system of sex determination and the evolution of sterile female castes in the Hymenoptera. A brilliant idea, however, may still be incorrect or at least only part of the story. Critics point to the following problems with the hypothesis, problems that were for the most part acknowledged by Hamilton himself [18]:

1. Termites are every bit as highly social as honeybees and paper wasps, despite the fact that males and females are diploid (Figure 20). Thus, haplodiploidy is not essential for the evolution of eusociality. Moreover, there are many (haplodiploid) bees and wasps that have remained solitary, and the same is true for some other insect groups that use the haplodiploid system of sex determination. Something other than or in addition to haplodiploidy favors the evolution of sterile castes.

2. Many species of eusocial insects have colonies in which the queen has mated more than once, as in the honeybee. When queens mate with several males, they receive more than one kind of male gamete for the fertilization of their eggs. The greater the number of male mating partners, the greater the genetic diversity among sperm and the lower the average degree of genetic relatedness among sisters. For example, Kenneth Ross demonstrated that queens of two species of highly social wasps usually mate with more than one male and that the *maximum* average coefficient of relatedness among workers and their reproducing sisters cannot exceed 0.4 [675].

3. Furthermore, many colonies of social insects have several queens ruling conjointly, each producing eggs cared for by the worker force at

20 **Soldier termites.** Although termites are diploid organisms, they have evolved eusocial behavior similar to that of the eusocial Hymenoptera. Here some large sterile soldiers guard a group of foraging sterile workers. Photograph by E. S. Ross.

large. This decreases sharply the probability that members of a worker caste will be able to direct their altruism toward sisters that have a coefficient of relatedness of 0.75.

These factors suffice to cast doubt on Hamilton's indirect selection hypothesis with its emphasis on the close genetic relationships between workers and the reproductive females they help rear. But what are the alternative explanations for the evolution of sterile, altruistic castes in eusocial insects?

The alternatives have their own problems. A mutualism hypothesis notes that females that nest together may better defeat parasites, predators, and conspecific rivals [18]. In areas with a high density of *Polistes* females, a nest with a single foundress has little chance of producing offspring, because other females will raid the nest and take it over when she is off collecting food or nest material (Figure 21). Under these conditions, a female that joins an established nest and protects it against usurpers, parasites, and other enemies may inherit an established productive nest later in the season, should the foundress die [271, 748]. Under various conditions, a solitary female may have very low reproductive potential and therefore joining may be the superior option even though a joiner female's chances of reproducing personally are small [829].

However, although these explanations are persuasive with respect to the formation of alliances among unrelated *reproductive* females, they do not apply to the evolution of alliances among a reproductive female and *sterile* workers who have no chance to produce offspring. In order to account for sterility, some authors have, like Trivers, envisioned a conflict of interest between parents and offspring; but instead of the offspring winning, they see the worker caste as evidence that the queen has won [10]. If a queen mother carries genes that predispose her to generate some sterile progeny that do not reproduce but instead help her make reproductively competent individuals, the queen's fitness may be elevated. There are two general strategies a queen could follow to achieve this end. One would be to produce offspring that, regardless of their genes, were coerced to behave in ways that benefited the queen's genes [10]. For example, the queens of some social species lay eggs for the express purpose of feeding them to other progeny (Figure 22). The trophic eggs contain genotypes whose chances for personal reproduction are sacrificed entirely when they are eaten by a fellow colony member, but this might be a way for the queen to raise her own genetic success. For adherents of the parental manipulation hypothesis, workers are merely grown-up trophic eggs that are forced to do things that lower their inclusive fitness but increase that of their mother [829].

Parental manipulation of offspring need not take so dramatic a form. Instead, a mother hymenopteran might produce some small progeny with low reproductive chances who, in order to maximize their inclusive fitness, would have to do just those things that also raised their mother's fitness [829]. The parent would be taking advantage of her offspring's conditional capacity to make the best of a bad situation by helping other sisters with superior reproductive potential. In this way a female with zero prospects of reproducing personally might gain some indirect fitness.

Thus, a combination of direct selection on queens and indirect selection

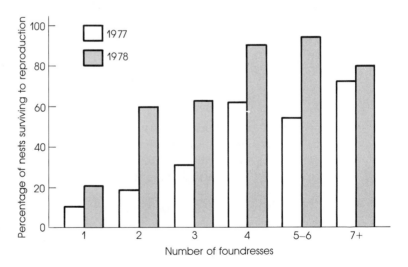

2 1 **Nest survival** is a function of the number of foundress females in many social Hymenoptera. In 1977 and 1978 the more cooperating female paper wasps founding a nest, the more likely it was to survive to produce future nest foundresses and drones. *Source:* Queller and Strassman [653].

22 **Trophic eggs.** Female social insects, like these Australian bulldog ants, may produce eggs (above left) that are fed to the larval offspring of the queen (above right). A queen accepts a trophic egg from a worker (below). She may consume it or feed it to a larva. Photographs by Jenny Barnett.

on potential workers who have close relatives to assist in their mother's colony may work together to produce sterility in workers. Whether or not this has actually happened, and the relative strength of the two processes, remain to be examined empirically for particular cases of eusociality.

The Behavioral Ecology of Eusociality

Malte Andersson has approached the problem of why complex altruism has evolved in some species and not others by considering how ecological pressures shared by the social species differ from those operating on solitary species [18]. Almost all social insects have (1) an elaborate nest in which offspring are reared (Figure 23), and (2) unusually effective devices to protect their young (such as the stings of female Hymenoptera). To start from scratch to reconstruct a monumental nest is a major task for foundress reproductives, and one destined to fail in a very large proportion of cases. This factor raises the cost of dispersal for potential reproductives, and therefore acts as a predisposing condition for the formation of cooperative alliances of relatives within the natal nest. Nondispersers might wait in the nest without helping or they might help relatives while waiting. Given that they may be able to defend their relatives effectively, the benefits of helping are relatively high for social insects, while the costs of personal reproduction are high, skewing the cost–benefit ratio in favor of helping. Insects, unlike most vertebrates, have adult life spans so short that an individual has little chance of devoting its early life to increasing its indirect fitness and a later portion to personal reproduction. If the chances of doing both are tiny, indirect selection may favor those individuals that allocate all their developmental energy to produce a body that facilitates the helper role. A soldier does not need functional ovaries.

Although sterility is an insect specialty, the ecological similarities between social insects and social vertebrates with helpers are strong. This is especially true for the naked mole rat, a mammal whose societies resemble those of eusocial insects (Figure 24). These burrowing animals live in colonies with a central nest in a complex network of underground tunnels. In the nest a big "queen" and several "kings" live with a retinue of smaller nonreproducing helpers that play specialized roles in the life of the colony (Figure 25) [460, 402]. The value of the nest and the difficulty of starting a new one may have favored delayed dispersal, which sets the stage for helping reproductives in the colony.

In general, as we discussed earlier, helper vertebrates are ones in which some ecological factors, such a biased sex-ratio or limited open habitat for

23 **The nests of social insects** are often exceptionally elaborate and difficult to construct, as seen in these monumental Australian termite mounds. Photographs by the author.

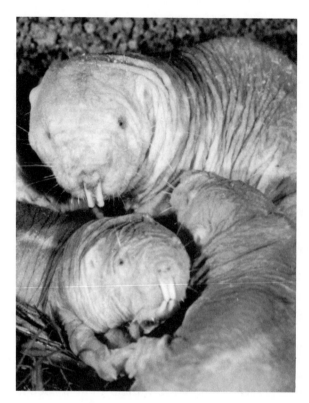

24 **Naked mole rats** are eusocial mammals with a worker caste. Here a large female is attended by two smaller workers below her. Photograph by Raymond Mendez, courtesy of Animals, Animals.

new breeders to colonize, reduce the success of would-be breeders and permit helping to be an adaptive alternative. But helping is only advantageous if the would-be helper can perform useful services for its relatives. Helpers at the nest occur almost without exception in altricial birds, species that produce helpless young that require care and feeding for some time, and not in precocial species whose young are feathered, mobile, and partly independent within a few days or even hours of hatching. Helpers are especially common in carnivorous mammals that rely on scarce and difficult-to-capture game to feed the young and that often defend a valuable den in which the offspring are raised. These ecological conditions increase the value of altruism by nonparents and decrease the chances that dispersing first-time breeders will succeed in rearing their offspring on their own. Thus, although social life is characterized by different degrees of altruism and has evolved independently many times, there may be a limited set of ecological conditions that promotes the evolution of social systems.

SUMMARY

1 In animal societies, individuals tolerate the close presence of other conspecifics despite the increased competition for limited resources and the heightened risk of disease and intraspecific exploitation that this entails. Under some ecological circumstances, the advantages of sociality (usually improved defense against predators and sometimes improved foraging effi-

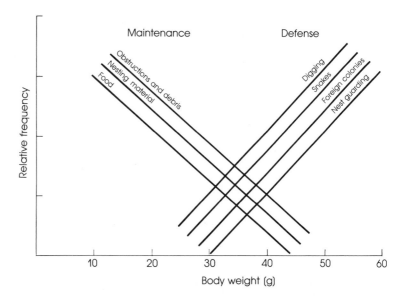

25 **Division of labor** in naked mole rat colonies. Small individuals engage primarily in maintenance activities; larger individuals undertake digging and defense duties. Only the very largest members of the colony breed. *Source:* Lacey and Sherman [460].

ciency) are great enough to outweigh the costs of social living. By no means is social life always superior to solitary life.

2 A central problem for evolutionary biology is to explain the evolution of helpful behavior in which an individual sacrifices a portion of its direct fitness to elevate the reproductive success of another. This behavior is far from rare in animal societies and includes the alarm calls of some birds and mammals, help given to rear offspring that do not belong to the helper, and, most amazing of all, the complete rejection of reproduction in favor of helping other individuals reproduce.

3 Several alternative hypotheses apply to the evolution of helpful behaviors. Some cooperative acts may immediately elevate the personal reproductive success of the cooperator. Other actions may require that costly help be provided now in order to receive reciprocal assistance later that will advance the reproductive chances of the helper. Both these categories of cooperation can, therefore, be the product of direct (natural) selection. If a helper really does permanently reduce its personal reproduction by helping, it may nevertheless raise its inclusive fitness (total genetic propagation) if its altruism is directed to relatives other than offspring. Costly helpful acts that increase the fitness of nondescendant kin can be the product of indirect selection.

4 To discriminate between the alternative hypotheses for a given case, some kinds of information are especially useful—in particular, the degree of genetic relatedness between the helper and the individual(s) it helps and the costs and benefits of the act for both the donor and recipient(s). Even

if this information is available, however, the relative importance of indirect selection and direct selection in the evolution of a trait may not be entirely clear, as illustrated by the continuing debate on the evolutionary basis for the sterile castes of some social insects.

SUGGESTED READING

Edward O. Wilson's *Sociobiology* [853] is a masterly compendium of material on all aspects of the evolution of social behavior and societies. I also like David Barash's *Sociobiology and Behavior* [41], Richard Dawkins's *The Selfish Gene* [182], and J. F. Wittenberger's *Animal Social Behavior* [865].

Important theoretical articles on the evolution of sociality include W. D. Hamilton's work [331], which helped initiate the revolution in animal behavior, and more recent reviews by Richard Alexander [10], Mary Jane West Eberhard [829], and Stephen Emlen [237, 238]. I have relied heavily on Jerram Brown's *Helping and Communal Breeding in Birds* as a guide for understanding the kinds of selection that affect the evolution of social behavior [112]. Heinz-Ulrich Reyer's study of the social pied kingfisher is superb [663].

Other good descriptions and analyses of the social behavior of different groups of animals are provided by E. O. Wilson [852] (on Hymenoptera and termites), by George Schaller [700] (on lions), by Paul Sherman [713, 715] (on ground squirrels), by John Hoogland [380, 381] (on prairie dogs), and by Mart Gross and Anne MacMillan [317] (on sunfish).

DISCUSSION QUESTIONS

1. If a helper pied kingfisher increases by 20 percent the survival chances of the pair it helps, how much does this add to the direct fitness, indirect fitness, and inclusive fitness of (a) the parents, (b) a primary helper, and (c) a secondary helper? Assume that if the breeding pair does survive, it will have 3.0 offspring.

2. Some vampire bats regurgitate blood meals to other hungry individuals [844]. What is the minimum information that you would need to determine whether this helpful behavior evolved via indirect selection? By direct selection? By kin selection?

3. Let's say that in calculating the inclusive fitness of an individual in a coalition of lions, you calculated direct fitness by multiplying the number of offspring produced by the male by r and then calculated the indirect fitness by multiplying the total number of all offspring produced by the other members of the coalition times the mean r of the male to these offspring. Your calculation would be challenged on what grounds?

4. If in a social species a female could help produce more females with an r of 3/4, why would there be any females that reproduced personally, because reproducers are related to their offspring by 1/2?

16

An Evolutionary
Approach to
Human Behavior

Humans are members of an animal species with an evolutionary history. It is true that we are an unusual species; kittiwakes and hangingflies are unique and wonderful creatures, too. If we can apply evolutionary thinking to kittiwakes and hangingflies, perhaps we can do the same for humans. Not everyone agrees this is desirable or useful, and I shall review why so many of us are reluctant to accept the possibility that our behavior has been shaped by natural selection in ways that promote the inclusive fitness of individuals. Evolutionary theory suggests that our genes influence the development of our brains in ways that "encourage" us to engage in adaptive (fitness-maximizing) behavior. The complexity and variety of human behavior in cultures around the world has convinced some students of human behavior that an evolutionary approach cannot apply to our species. The challenge for both proponents and opponents of an evolutionary approach is to test whether underlying the great diversity of human cultural pursuits there are nevertheless basic elements of the human psyche and behavior that make sense from an evolutionary perspective. The first step in this process is to develop hypotheses about human psychological or behavioral attributes that are consistent with direct (or indirect) selection. Here I shall present a sample of these hypotheses, some more speculative than others, some better tested than others. None of these ideas has been exhaustively tested against a full range of alternative hypotheses. They are offered as a conceptual exercise to show that it is possible to formulate and test plausible evolutionary hypotheses about the ultimate significance of human behavior, including some elements that at first glance seem maladaptive. I do not present these examples as the Final Word, but I believe that an evolutionary approach has much to say that is interesting and useful if we wish to understand ourselves.

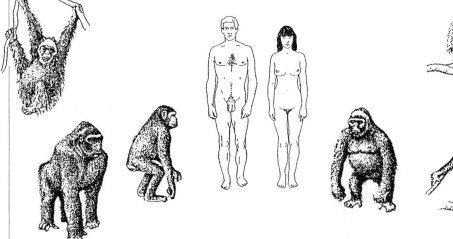

The Sociobiology Controversy   When E. O. Wilson published *Sociobiology: The New Synthesis* [853] in 1975, he was praised warmly by some readers but vilified in the strongest terms by others, including his own colleagues at Harvard University, Stephen J. Gould and Richard Lewontin [15, 297]. To read the comments of the most vehement critics, one would suppose that Wilson's book is a polemic on the biological basis of human behavior. Actually, only one of 27 chapters in *Sociobiology* discusses this topic in detail, but this was more than enough to upset a great many people.

There are still many persons who are disturbed by the sociobiological approach to human behavior, although happily Wilson is no longer called a closet racist or genocidal fascist. Sociobiology still elicits strong emotional reactions, however, from those who believe that human behavior is almost infinitely malleable and can be understood strictly in terms of environmental causes. I am not sure that it is possible to convince these persons that their objections are based largely on misunderstandings, but I do know that misconceptions about the sociobiological approach are widespread. My first goal, therefore, is to explain what sociobiology is *not*, before illustrating what it is, and here I examine, one by one, five major misunderstandings about this field of study [168, 446].

"Sociobiology is E. O. Wilson's new theory of human behavior."

Wilson did not invent sociobiology, nor is the discipline exclusively or even primarily concerned with human behavior. What Wilson did was to give a simple and easily remembered label to what otherwise would be ponderously called a Darwinian evolutionary approach to social behavior, or the study of the ultimate (adaptive) significance of social behavior. *Sociobiology* is Wilson's synthesis of information gathered by many evolutionary biologists about the social behavior of the entire animal kingdom, from slime molds to human beings.

As stressed in earlier chapters, Darwinian evolutionary theory provides a way to identify interesting problems in animal behavior, social behavior included. Application of this approach has enormously improved our understanding of why nonhuman social species behave as they do, a point that is almost never disputed by even the most dedicated critics of human sociobiology (but see [303, 432]).

In his last chapter, Wilson used an evolutionary approach to generate a number of hypotheses about the possible adaptive significance of elements of our own behavior. Sociobiology is part and parcel of evolutionary biology, and it would be surprising if Darwinian theory had nothing helpful to say about *human* biology [168].

"No one has ever identified the genes responsible for any human behavior."

Many persons are under the mistaken impression that sociobiology's goal is to discover genes for various human behaviors like altruism and aggression. In order to employ an evolutionary approach to the behavior of any species, one has to assume that the behavior of interest has evolved. In order for behavior to evolve, it must have a genetic foundation and in the past individuals with different alleles must have exhibited different behaviors that affected their inclusive fitness. Critics of human sociobiology have claimed, therefore, that it is not acceptable to talk about the possible adaptive value of human behavior, because no one has ever demonstrated precisely which genes underlie particular behavioral traits in humans.

But sociobiology offers a way to analyze behavior at the ultimate level. Studies of the proximate role that genes play in the development of human behavior are interesting and valuable (see Chapter 3). But one does not have to know everything there is to know about the proximate causes of a behavior in order to test hypotheses about its adaptive consequences. We know even less about the genetics of Florida scrub jay behavior and ground squirrel alarm calling than we do about human behavior genetics. Fortunately, this has not prevented evolutionary biologists from asking questions and solving problems about the evolution of scrub jay and ground squirrel behavior. One does not have to know which gene(s) contribute to the development of a behavior pattern in order to test ideas about the reproductive significance of the attribute. As we shall see, it is entirely possible to test sociobiological hypotheses about the adaptive nature of certain human characteristics without knowing any details about the genetic or developmental basis for these characteristics.

"But humans don't do things just because they want to raise their inclusive fitness."

Some opponents of sociobiology have pointed out that humans rarely, if ever, seem to be motivated to do the things they do because of a desire for reproductive success [691]. If you were to ask an artist why he wished to produce attractive paintings, or Bill why he wanted to marry Jane, the answer would probably not refer to inclusive fitness. But if a baby cuckoo could talk, it, too, surely would not tell you that it rolled its host's eggs out of the nest "because I want to propagate as many copies of my genes as possible." No animal, cuckoo or human, need possess a *conscious* awareness of the ultimate reasons for its activities. It is enough that proximate mechanisms predispose individuals to behave in ways that lead to direct or indirect fitness gains. On the proximate level we enjoy sweet foods, we fall in love, we desire approval from others, and we learn a language because we possess physiological mechanisms that make us want to do these things. We do not have to be aware of the caloric content of honey and its contribution to cell growth and maintenance in order to eat it because it tastes good [41].

"But not all human behavior is biologically adaptive!"

Critics of sociobiology often claim that certain cultural practices, like circumcision, prohibitions against eating some perfectly edible foods, and a celibate priesthood seem most unlikely to advance individual fitness. If some human practices reduce fitness, they would have us believe that sociobiology cannot be correct. Implicit in this declaration is the argument that evolutionary biologists accept on faith the belief that every aspect of every organism is adaptive. But as we noted in Chapters 1 and 8, biologists use the adaptationist method to identify interesting puzzles and to create hypotheses from which predictions can be derived and tests made. As outlined before, the assumption of adaptation is an unavoidable and necessary basis for the formulation of testable hypotheses about the adaptive function of a trait. If a sociobiologist were to examine the practice of celibacy by priests of some religions, he might choose to assume that the trait was adaptive. But he would not be asserting that the trait is certain to be adaptive, because it might well be a maladaptive by-product of psychic mechanisms that produced very different behavior in the past but which in novel present environments induce a few individuals to engage in lifelong celibacy. If a sociobiologist were to assume that celibacy was adaptive in the present or recent past, it would be to produce a testable hypothesis on how celibacy might paradoxically enable priests to leave more copies of their genes than they could leave if they were not celibate. Needless to say, this would be a challenge, but perhaps not an insuperable one for someone aware of indirect selection.

Even if one or more adaptationist hypotheses on celibacy were developed, there is no guarantee that any would withstand testing. This is as it should be. T. H. Huxley, the great defender of Darwinian theory, wrote, "There is a wonderful truth in [the] saying [that] next to being right in this world, the best of all things is to be clearly and definitely wrong, because you will come out somewhere" [396]. If a sociobiological hypothesis about the celibate priesthood is incorrect, effective tests will reveal this. One wrong idea would then be discarded, narrowing the search for an appropriate explanation.

"Sociobiology provides aid and comfort for those who would maintain social injustice and inequality."

A fifth argument of the critics, and one that appears to have strongly motivated their attempt to rebut sociobiology, is that sociobiology provides "scientific" justification for immoral social policies [15]. They claim that to say a trait is adaptive is to imply that it is both genetically determined and good and, therefore, *cannot* and *should not* be changed. Noting that racist and fascist demagogues have misused biological theories in the past to promote evil political programs, they have argued that human sociobiology could easily be used again in this fashion. After all, if one claims that male dominance is adaptive, isn't this saying that the status quo in our society is desirable and that feminist claims fly in the face of what is biologically fixed and necessary?

There is no doubt that scientific investigation can be employed in ways that often surprise, and even horrify, the investigator. Einstein's basic research on the relation between energy and matter contributed to the development of atomic weapons, much to his dismay; Darwin's theory of evolution has been used to defend the principle that the rich are evolutionarily superior beings [308], as well as to promote unabashedly racist plans for the "improvement of the species" by selective breeding of humans [268].

We can hope that political perversions of evolutionary theory have been so discredited that they will not happen again. The critical point, however, is that sociobiology is a discipline that attempts to explain why social behavior exists, *not to justify the behavior*. This distinction is easily understood in cases not involving humans [41]. Biologists who study infanticide by male langurs or the AIDS virus are not accused of approving of infanticide or the AIDS disease. To say that something is biologically or evolutionarily adaptive means only that the something tends to elevate the inclusive fitness of individuals with the trait—nothing more.

Moreover, a hypothesis that a behavioral ability is adaptive does not demand that the characteristic be "determined" by genes, only that there be a correlation between genotype and behavior [181]. There are no alleles for altruism or aggression or any other trait, in the sense of a gene that encodes the characteristic and produces it no matter what (Chapter 3). Behavioral development depends on the interaction between the genetic recipe and the environment of the developing individual. Change the environment and you may well change the behavioral outcome.

A classic example in human biology involves the disease phenylketonuria. Possession of two copies of the recessive form of one gene in the human genotype means that these individuals will under some environmental conditions exhibit mental retardation and die young. But the allele does not make the disease, it codes information related to the production of a particular enzyme. Individuals with two recessive forms of the gene do not produce an enzyme that is used by our cells to alter a particular substance. In the absence of the enzyme, the substance tends to build up and this causes developmental changes in brain cells that lead to mental retardation. *But* if individuals with the alleles "for" phenylketonuria are not fed certain foods that contribute to the build up of the damaging substance, they will not develop the disease [181].

The proposition that natural selection has endowed us with genes "for" aggression or sexual jealousy or the urge to reproduce does not condemn us to exhibit these characteristics for all time. As an evolutionary biologist, I recognize that in evolutionary terms we exist solely to propagate the genes within us. But thanks to an environment in which effective means of birth control exist, my wife and I have chosen to have only two children.

A Strategy for Developing Sociobiological Hypotheses

Sociobiology offers an evolutionary framework for the development and testing of ultimate hypotheses about social behavior. Using this approach for human social activities has proved to be difficult, however, with much debate on the proper procedures [137, 611, 759]. One reason why human behavior poses a special challenge for evolutionary analysis becomes apparent when we realize just how differently humans behave in cultures around the world. How can it be said that humans tend to behave in ways that elevate their fitness when there are polyandrous, polygynous, and monogamous societies, cultures in which females make important political decisions and others in which males dominate females, human groups for which warfare is a constant fact of life and other groups that never fight. The list of cultural peculiarities is nearly endless; depending on where you were born, you might have to memorize passages from the Koran or the Bible, you might speak Jivaro or Japanese, you might be allowed to marry your first cousin, you might be forbidden to look at your mother-in-law, and you might be forced to have your penis cut from stem to stern at adolescence or you might not (Figure 1).

For some students of human behavior, the diversity of cultural traditions is evidence enough that we have in some sense escaped from our evolutionary history, that we are unique among animals in creating our own rules, our own completely changeable societies. Advocates of this view argue that to search for adaptation in the fondness of some humans for a Bach cantata or for the Boston Red Sox or for Buddhist religious beliefs is misguided and futile because such human traits are simply arbitrary artifacts of our intellect [297, 691].

Faced with the bewildering diversity of cultural practices, sociobiologists could be forgiven for giving up and returning to the study of paper wasps and ground squirrels. But some have stayed on to apply the adaptationist approach to various elements of human behavior. The hypotheses that sociobiologists have produced fall into three general categories. First, the human behavior in question may currently have an average positive effect on the reproductive success of the individuals that engage in the activity. Second, it may not currently tend to raise reproductive success, but it did so in the past under different environmental conditions. Third, the characteristic is not now nor ever was adaptive per se but instead is a reflection of psychic mechanisms that have an overall positive effect on fitness (currently or in the past) despite sometimes causing incidental outcomes that are nonadaptive or even maladaptive.

Take celibacy, for example. It is conceivable, but I think most unlikely, that in modern societies men who enter the priesthood are those who for a variety of reasons would be unlikely to secure mates and produce offspring directly. Using the game theory approach, we might argue that if some men are likely to monopolize available mates, then other men can, by becoming a member of the celibate priesthood, engage in a profession that may improve the fitness of their close relatives (as, for example, when resources given to the church are channeled to relatives of the priests). This could result in a sufficiently high indirect fitness benefit for the celibate individual to compensate for the absence of direct reproductive success.

I suspect that if someone chooses to test this hypothesis rigorously, he will find it is incorrect. Discarding this hypothesis would not rule out the second possibility, which is that lifelong celibacy *in the past* raised some individuals' indirect fitness sufficiently to favor the maintenance of a hypothetical conditional celibacy mechanism in the human population until the present, when it causes a few individuals to behave maladaptively. Such an argument would parallel that used to explain the craving for salt, which causes so many modern humans to injure themselves through overconsumption of table salt. In the past, a readiness and desire to eat quantities of salt may well have been adaptive, because salt provides valuable ions and was so scarce in the foods available to past generations of humans that people could almost never overindulge. Only modern technologies have

1 **Cultural diversity in body ornamentation.** A North American Indian (left) with considerable body painting, photographed in western Arizona in the early twentieth century. Photograph courtesy of the Arizona State Museum. A New Guinean tribesman (right) with an elaborate wig and facial painting. Photograph by Lyle Steadman.

provided most people with salt in great abundance, thereby permitting an evolved desire to now reduce survival and reproductive success in some people.

The third possibility is that a characteristic may have appeared in past and present as a nonadaptive or maladaptive by-product of an otherwise adaptive mechanism. This is the pleiotropic or incidental effect argument discussed earlier in the context of explaining why not every action is necessarily adaptive. Such effects may be common throughout the animal kingdom, because "rules of thumb" are so common. The nervous system of a small warbler "tells" it to feed whichever of its nestlings begs the most actively. The usual consequence of this rule is that a parent adaptively feeds the hungriest and healthiest of its nestlings. Sometimes, however, the warbler maladaptively feeds a baby cowbird that parasitizes its parental behavior, which is organized by this simple and usually effective rule of thumb.

A great deal of human behavior also appears to depend upon species-wide rules of thumb: "Eat foods that taste salty and sweet; avoid those that taste bitter;" "Enjoy plump, happy babies;" "Seek to acquire the esteem of others." Satisfying these rules of thumb may yield superficially different behavior from culture to culture, and yet might produce generally adaptive outcomes. But a rule of thumb may sometimes fail. A fondness for sweets may ensure the ingestion of high-calorie foods, but this can be overdone in some circumstances. The desire to be around babies may cause adults to adopt a genetic stranger. A desire for social approval may motivate individuals to acquire useful allies, but an eagerness for acceptance may also cause some persons to be exploited and deceived by those whose esteem they seek.

Lifelong celibacy might arise as a maladaptive consequence of some rules of thumb little understood or investigated to date. Although no one has developed an incidental effect hypothesis for celibacy, it could be done and tested in the same way that current adaptation and past adaptation hypotheses on human behavior can also be tested. Ideally, of course, all three categories should be considered, unless there is compelling evidence to the contrary. Moreover, alternative hypotheses should be taken from each category and discriminating predictions produced. Unfortunately, this is relatively rarely done. Instead, investigators tend to follow a hunch and explore one possibility at a time. In the pages that follow, we shall explore what these researchers have revealed about the evolutionary basis of our behavior, but please keep in mind the limitations of the one hypothesis–one test approach.

Warfare   Warfare is a human activity that has attracted the attention of some sociobiologists. In order to analyze a behavior that (1) takes on many different forms even in the cultures that do go to war (and not all societies do) and (2) superficially appears to be biologically irrational, evolutionary biologists have adopted a strategic approach whose elements have been clearly outlined by John Tooby and Irven DeVore [781]. Although some persons believe that war was and is a largely capricious invention of some cultures [566], the starting point for evolutionary biologists is the adaptationist assumption that humans have over evolutionary time usually engaged in war under conditions that tended to raise the fitness of the instigators of group violence.

The hypothesis that warfare is (or was) an adaptive conditional strategy emphatically does not imply that warfare is inevitable or that it is good in a moral or ethical sense, for reasons outlined earlier. What it does imply is that humans have evolved brain mechanisms and rules of thumb that promote cooperative aggression against others when conditions are such that the promoters of war stand to improve their fitness. Given the obvious risks involved in warfare, there must be substantial benefits if the proximate mechanisms that facilitate this activity are to spread through populations. One suggestion has been that warring individuals may raise their inclusive fitness by acquiring or defending valuable resources [213].

Many researchers have in effect tested this same hypothesis in studies of nonhuman animals that employ cooperative aggression against other groups. When two hyena clans clash, or when two lion prides fight for a territory, they do so for a valuable resource, a rich hunting area [797]. Individuals can get killed in a battle with another band. Natural selection will favor taking this risk only if the reward for winning is great.

We can employ the cost–benefit approach to group aggression in humans, too. Cultural variation in the proclivity for warfare could be more or less random with respect to ecological factors that effect the costs and benefits of this activity, or humans could have the same sort of patterns that occur in other species.

The hypothesis that a clumped distribution of valuable resources promoted the cultural practice of warfare in the past appears to be supported by anthropological accounts of tribal aggression. For example, the western Shoshone and Paiute Indians lived in the Great Basin area of western North America, but they differed strongly in social organization and in a readiness to fight over food resources. Although both tribes relied heavily on grass seeds as a food staple during summer, the Paiute lived along watercourses where seed-producing grasses grew predictably and in abundance. These Indians formed small villages and aggressively defended a stretch of river. The Shoshone, in contrast, lived in dry, short-grass prairie habitat in regions without permanent streams and in an environment with scarce and patchy rainfall. As a result, the seed-producing plants that they searched for were scattered, ephemeral, and unpredictable in location. These Indians lived in small family bands that wandered widely and did not fight with other family groups living in the prairie [236].

Just as divergent cultural evolution in neighboring tribes provides a test of the prediction that warfare has an economic function, so, too, convergent cultural practices in groups living apart can be used for this purpose. W. H. Durham notes that warring cultures evolved independently in tropical South America, Borneo, and New Guinea (Figure 2) [213]. These excep-

2 **An aggressive culture.** A New Guinean tribe prepares for war with a neighboring tribe. Photograph courtesy of the Harvard Film Study Center.

tionally aggressive cultures engaged in intensive agricultural production of a carbohydrate-rich, but protein-poor, food staples such as manioc or sweet potato. Protein was either secured as wild game from forests adjacent to villages or grown in the form of pigs. Warfare seems to have been correlated with population growth and the associated pressure placed on the protein-producing sector of the environment.

In the Mundurucu of Brazil, for example, villages apparently began to carry out head-hunting raids when the preferred animal prey, wild peccary, became scarce in their traditional hunting areas because of increased hunting pressure from neighboring villages. The warfare that broke out had as its proximate goal the collection of severed heads of enemy tribesmen. But interestingly, the Mundurucu called a warrior that came home, head in hand, a "mother of the peccary," as if they saw a correlation between successful warfare and improved hunting.

These examples constitute a weak test of the cost–benefit approach to human warfare. Strong tests of these and other sociobiological predictions about warfare are needed, but the evidence reviewed above suggests that the approach is not lunatic and that additional investigations of sociobiological hypotheses on the subject are justified.

Warfare and the Evolution of Human Intelligence

If human groups fought for limited resources in the distant past, as well as more recently, we may have created selection pressures on ourselves that resulted in the remarkable features of human intelligence. This is the thesis of Richard Alexander [12], and I present a sketch of it here because the topic is fascinating, although the hypothesis may not be easy to test.

Our brain is our most extraordinary feature, in terms of both size and capabilities. Brain weight is linked to body weight in animals generally, with larger species possessing brains larger than those of smaller species. By plotting brain weight against body weight for a large sample of mammals, one can easily see this relationship and plot the average brain weight for a mammal of given size. When this is done, we find that the human brain is about *seven times larger* than would be predicted for a mammal of our size [411].

What selective factors could possibly have been so strong as to favor such inflated human brains, despite the well-documented developmental and metabolic costs of brain tissue (see Chapter 2)? The problems of finding food and shelter, of avoiding predators, and of finding mates are problems basic to animals in general. If larger brains were reproductively advantageous in solving these problems, then we would expect that most animals would devote a higher proportion of body mass to brain tissue than they currently do. In order to account for the unique evolutionary inflation of human brains, we must find a unique ecological problem for our species.

Alexander believes that what is special about human evolution is the potential for hostility between groups. If bands can readily kill or displace members of competing groups, they create a kind of runaway selection (Chapter 13) on human *social intelligence*. The capacity for lethal violence is particularly great in humans, thanks to our technological skills that have been applied to weapons throughout human history. Any advance in intelligence, dependent on increased brain size, in the members of one group creates selection favoring unusually intelligent, big-brained members of rival groups. In other words, an evolutionary "game" is set in motion in which what some individuals are doing determines the fitness of alternatives used by other members of the population. If it is easier to take resources from neighboring groups by force than by collecting them on your own, then the capacity for violent aggression will spread through the population. In the end, all groups will be raiding just to stay even; and although everyone might be better off if they stopped going to war, no one can break out of an evolutionarily stable system (until the next technological or intellectual or social advance that gives one group an edge in the fighting).

Warfare among neighboring bands not only favors individuals with a special kind of intelligence but also highly social people. Individuals able to live with and cooperate with many others can better defend themselves or overwhelm the defensive capacity of smaller groups. Life in large groups where an individual has to interact with distant relatives and genetic strangers is itself a powerful impetus for the evolution of social intelligence, according to Alexander's game theory approach. The calculus of social life becomes immensely more complicated when one is simultaneously dependent on others for assistance but also in reproductive competition with potential helpers (as well as with enemies in other bands).

The exceptional features of our intellect, our self-awareness, consciousness, sense of conscience, sense of morality, contribute to a life of great social complexity. In bands of any size, individual reproductive success and that of one's kin will depend upon what other members of the group remember about one's conduct in the past. A person's treatment may even depend upon what people have heard about the individual secondhand from other members in a social network. Given this factor, planning a fitness-maximizing course of action for the days, months, even years ahead favors a very special and elaborate sort of intelligence.

The kinds of things humans are good at include being aware of one's self and how others view us, being able to keep track of one's (moral) obligations and the reciprocal obligations of others, being able to calculate how much you can get away with and yet not jeopardize friendships that may be handy in the future. Our large brains act in some sense like a computer with complicated social programs that enable us to run through mental simulations of alternative scenarios. "If I delay paying Bill back, he is likely to react in the following way; if this happens, I can expect Fred to do this, and this in turn will have consequences A, B, and C."

This broad sketch of the Alexanderian argument sweeps across a whole battery of hypotheses, each one of which ideally should be dissected and subjected to test against competing alternatives. Limited space prevents this here, but I can at least take one small element of the scenario and show how it might be applied to analyze an aspect of human behavior in a way that is usually not considered by persons unaware of evolutionary perspectives.

A number of opponents of sociobiology have claimed that donating blood for free is a kind of "pure" altruism practiced for no possible reproductive advantage. The donors are supposedly behaving simply out of a moral concern for others whom they will never meet, from whom they can receive no repayment. According to Peter Singer, "Common sense tells us that people who give blood do it to help others, not for a disguised benefit to themselves" [725].

Alexander, however, suggests that "giving blood is a fine way of suggesting that one is so altruistic that he is willing to give up a most dear possession for a perfect stranger." This could yield payoffs in the future, not from the stranger but from those who live and interact with the donor and who view him as potentially more cooperative and helpful than others who are not known to be so "altruistic." Alexander is obviously *not* proposing that humans have a gene for blood donating; blood banks are after all a very recent phenomenon. But giving blood, although purely a modern phenomenon, might tap a part of the human psyche that has evolved and is adaptive. Martin Daly and Margo Wilson suggest that our brains tell us that it feels good to engage in low-cost "do-goodery." At the proximate level, this feeling currently motivates us to give blood occasionally or exhibit other inexpensive forms of culturally approved charity. Such a mechanism could have the generally adaptive consequences of maintaining a good reputation and facilitating reciprocal relations as Alexander claims.

No one has yet formally presented a set of adaptationist alternatives for blood donating, but in principle this could be done, and Alexander's argument is testable on its own. We predict that unpaid blood donors almost always let some other people know that they have given blood. We predict that if people who give blood are given the chance to receive priority access to blood for themselves or their kin in the future, there will be more willing blood donors. We predict that given a hypothetical choice of choosing between a known blood donor and one known to refuse to give blood, most people would opt for the blood donor as a friend or co-worker. Maybe these predictions will eventually be shown to be incorrect, but if we listened just to "common sense," rather than being willing to test an evolutionary hypothesis, we would not have bothered to put an idea on the line.

Questions about Human Sexual Behavior   We have barely begun to apply evolutionary theory to questions about human warfare, morality, and intelligence, and the same can be said for analyses of human sexual behavior. A number of researchers, however, have begun the analysis of human sexuality, using sexual selection theory (Chapter 13) as the foundation for speculation and hypothesis testing [165, 758]. They begin with the well-known fact that human males, like males of most other animals, produce such vast quantities of sperm than one man potentially could, at little physiological expense, father numerous offspring. (The record for a man is 888 [165].) To achieve extensive fatherhood, a male would have to inseminate more than one female, because each woman's reproductive potential is severely limited by the relatively small number of fertilizable eggs she produces during a lifetime and by the demands of pregnancy and nursing. (The record for a woman is 69 children, many of which were triplets [165].)

Given the realities of human reproductive physiology, we would expect males to compete for females (if sexual selection has shaped the evolution of our behavior), because sexual access to females (and their eggs) limits male reproductive success. All other things being equal, the more females inseminated, the greater the probability of fertilizing eggs. On the other hand, there is no obvious reason why a female's fitness should increase as the number of her copulatory partners increases. We would expect females to exercise choice among potential partners, because there should be many willing candidates and any variation among men that affected female fitness would favor female selectivity. A female's success in leaving descendants should depend far more on the quality of a husband than on the quantity of her mates.

If this simple argument makes sense, we are in position to try to answer a number of questions about the evolution of human sexual behavior. In the pages that follow I have selected a sample of these questions, and I review some tentative answers offered by human sociobiologists. This is done not to make the claim that we now understand all there is to know about the ultimate aspects of human sexuality. Sociobiological ideas on these issues have rarely been subjected to a thorough analysis of a set of alternative hypotheses. But as long as we keep this in mind, we will not be tempted to make too much of the arguments presented below, but will accept them as illustrations of the ideas and approach of human sociobiology.

Before beginning, I will remind my readers again that if one proposes that a human behavior is adaptive, this carries no moral or ethical message nor any implication that the behavior develops independently of the environment and so cannot be changed.

Why are males more likely to engage in adultery than females?

Our first question arises because there is considerable information, much of it reviewed in Donald Symons's *The Evolution of Human Sexuality*, that married men are more likely to be unfaithful than are their wives [758]. In human societies around the world, most men and women get married, making a primary parental commitment to the offspring that arise from the marriage. The parental investment route is not, however, the only fitness-raising option available to males. Imagine a married man who helps rear two children with his wife, and who fathers another child by another woman whom he does not assist. If this child survives, the male has increased his fitness 50 percent without making any additional parental investment.

Unlike human males, females that have an offspring as a result of an adulterous liaison could not (over 99.9 percent of human history) escape a major parental investment by their action. They will nurture the fetus and they will also care for the baby in all likelihood. This removes a major fitness benefit for copulation outside marriage by married women, and so makes it less likely to be practiced by females, so the argument goes.

Symons suggests that we can test this hypothesis by making a prediction about the *proximate* causes of male and female sexuality. He argues that male brains should have evolved in ways that provide men with psychological drives and desires that increase the likelihood that they will seek out multiple copulatory partners. The value that males place on sexual variety for its own sake will, he predicts, be absent or much reduced in women because of sex-linked differences in the operation of female brains.

The evidence in support of this prediction is substantial. Female prostitution is common, whereas women almost never pay men to copulate with them. Males, not females, support a huge pornography industry in Western societies because men, not women, are willing to pay even to look at nude individuals of the opposite sex. Note that in modern societies these aspects of male behavior probably are *maladaptive*; prostitutes almost universally employ effective birth control or undergo abortions when pregnant, and payment for pornographic materials is unlikely to do much for a man's reproductive success. But maladaptive responses may reveal what people really care about, giving us insight into the nature of the human psyche, which evolved when birth control and *Playboy* were not available [168]. The prostitution and pornography industries take advantage of the male psyche. They exploit those neuronal mechanisms that attach pleasure to copulation and motivate males to seek many sexual partners. These mechanisms may be adaptive in most situations, by stimulating males to engage in adultery under some circumstances or by encouraging males to acquire multiple wives in cultures in which polygyny is permitted. But they can also lead to costly incidental effects that are not adaptive.

The proposed differences in psychic functioning can be tested in yet another way. Donald Symons notes that in Western society the behavior of male homosexuals is very different from that of female lesbians. Not only is male homosexuality much more common than the female variety, but males until recently were likely to have a progression of partners, whereas their female counterparts far more often had long-lasting, stable relationships. Symons's argument is that male homosexuals were free to express the proximate mechanisms that motivate males in general to achieve sexual variety [758].

Why has homosexual behavior persisted over evolutionary time?

One might think that homosexuality is evidence itself that evolutionary arguments do not apply to human sexual behavior. How can it be adaptive to engage in sex of this sort? Exclusive homosexuality may be purely a twentieth-century phenomenon, and if so, how can it be useful to analyze its evolutionary basis? As just illustrated, however, one way to account for maladaptive behavior in humans is to propose that it is an incidental by-product of an underlying proximate mechanism that generally induces humans to do things that raise rather than lower their fitness. The desire for sexual activity, the low threshold for sexual arousal, and the search for sexual variety may usually contribute to an active heterosexuality by males but may also lead some men as an incidental effect of these mechanisms to engage in sexual behavior that cannot produce offspring. (Please note that this hypothesis is *not* passing moral or ethical judgment on homosexuals.

What is biologically appropriate, that is, fitness-elevating, may or may not be socially desirable. I personally believe, for what it is worth, that a person's sexual preferences are his or her business, but my views on sexual morality are irrelevant to evolutionary analyses.)

There are other sociobiological hypotheses on the ultimate significance of homosexuality. E. O. Wilson has suggested that the trait was adaptive per se in the past, evolving via indirect selection [854] if males under some circumstances made a greater contribution to the propagation of their genes by refraining from marrying and instead working on behalf of their nephews and nieces. Such males might have been more likely to forgo marriage if they possessed the nonreproductive sexual outlet provided by homosexuality. You may not feel that this hypothesis is terribly plausible, but the important point is that it can be tested.

A straightforward prediction from the indirect selection hypothesis is that homosexuals should show some special predilection for helping relatives. The evidence on this point is lacking. But if it were shown that male homosexual activity commonly occurred in other animal species in which males lacked the option of assisting their relatives, this would weaken the indirect selection hypothesis, which argues that it is the opportunity for indirect fitness gains that has promoted homosexuality. In fact, males of solitary species throughout the animal kingdom have been observed courting, attempting to copulate, and even copulating with other males [766]. Observations of this sort and the almost complete absence of female homosexuality in nonhuman animals support the hypothesis that male homosexual behavior arises incidentally out of the typically high sexual drive of males and is not the product of indirect selection.

Predictions from the incidental effect hypothesis include (1) that there be no genetic difference between exclusively homosexual and heterosexual males that is correlated with their sexual differences, and (2) that *exclusive* homosexuality will be very rare in males. These are weak predictions, because they can also be derived from other hypotheses on homosexuality not presented here. But for what it is worth, on prediction (1), there is evidence from twin studies that genetic factors do contribute to the differences between sexual preferences of homosexual and heterosexual men [362]. If one member of a pair of identical twins reared apart is homosexual, his brother is almost always homosexual as well [220]. Although the sample size is small, these findings are damaging to the incidental effect hypothesis.

The second prediction, however, is amply supported by evidence that most homosexuals are actually bisexual, and many have families with children. Moreover, homosexuality occurs more commonly in populations of males unable to secure heterosexual partners (i.e., men in prisons and adolescent, unmarried men) [429]. It is entirely possible, but not yet demonstrated, that the average reproductive success (direct fitness) of males that engage in some homosexual behavior during their lifetime is the same, or even higher, than that of exclusive heterosexuals.

This difficult case illustrates the general point that not only do predictions vary in their power to discriminate among competing hypotheses, but also some follow more clearly from a given hypothesis than others. Perhaps it is not surprising that debate continues on the evolutionary basis of homosexuality and most other aspects of human behavior.

Why do some men commit rape?

Although the sociobiological analysis of homosexuality is far from complete, current evidence points toward the behavior arising as a nonadaptive by-product of brain mechanisms that *usually* promote adaptive sexual behavior. Rape may also be an outcome of these same mechanisms if the elements of the male psyche that motivate men to copulate sometimes lead men to inseminate females against their will. According to this hypothesis, the motivating systems regulating male sexuality have a net positive effect on fitness, even if coercive copulation actually reduces the fitness of the rapist in a given society.

The argument that there is a sexual aspect to rape is often disputed by those who claim that the act is purely one of violence and aggression. Note that this is one of those cases in which proximate and ultimate factors are confounded. At a proximate level, rapists might be motivated by a desire to attack women violently, but if this behavior sometimes resulted in the fertilization of raped women one could legitimately discuss its ultimate sexual function.

As is true for homosexuality, there are alternative hypotheses to account for rape, one of which has been developed by Randy and Nancy Thornhill, who propose that rape is one of several optional tactics in a conditional sexual strategy [768]. They argue that rather than being a nonadaptive or maladaptive incidental effect of the male sexual psyche, sexual selection in the past favored males with the capacity for rape *under some conditions* as a means of fertilizing eggs and leaving descendants. According to this view, rape in humans is analogous to forced copulation in *Panorpa* scorpionflies (Chapter 13) in which males excluded from more productive avenues of reproductive competition engage in a low-gain, high-risk alternative. Male *Panorpa* that are able to offer material benefits to females do so in return for copulations; males that cannot offer nuptial gifts attempt to force females to copulate with them. Human males unable to attract willing sexual partners might also rape as a reproductive option of last resort.

Some feminists angrily interpreted the Thornhills' adaptationist hypothesis to be a statement justifying and excusing rape [e.g., 258]. But my readers, I hope, will recognize that a hypothesis that rape is an adaptive component of a conditional strategy is not a declaration of approval for rapists, but is a claim about the possible fitness consequences of the action.

There are still other hypotheses, including one presented by Susan Brownmiller in her favorably reviewed and influential book *Against Our Will* [114]. In her view, rapists act on behalf of all men to instill fear into women, the better to intimidate and control women, keeping them "in their place."

We can test these hypotheses by looking at the predictions they produce (Table 1). Let's begin with the intimidation hypothesis, the only hypothesis that relies on a form of group selection as its theoretical base. Brownmiller

TABLE 1

Alternative ultimate hypotheses on why some human males
commit rape and some predictions taken from these hypotheses

Hypothesis 1: Rake is an *incidental effect* of adaptive neural mechanisms controlling male sexual behavior	**Hypothesis 2:** Rape is an *adaptive effect* of adaptive neural mechanisms controlling a conditional male sexual strategy	**Hypothesis 3:** Rape is *adaptive (for all men)* as a tactic designed to subjugate all women
Prediction: Victims of rape are likely to be in their peak reproductive years	Prediction: Victims of rape are likely to be in their peak reproductive years	Prediction: Victims of rape should tend to be older women in positions of power in society, or young women who pursue nontraditional careers
Prediction: Rapists should come from all strata of society	Prediction: Victims of rape will sometimes become pregnant following insemination by the rapist	Prediction: Rapists should come from all strata of society
	Prediction: Unmarried, poor men will be heavily overrepresented in the rapist category	

implies that some males are willing to take the risks associated with rape, which is a capital crime in many cultures, in order to provide a benefit for the rest of male society. This argument suffers from all the logical problems inherent in a "for-the-good-of-the-group" hypothesis (with the added difficulty that groups composed of only one sex cannot be the focus of any realistic sort of group selection), but we can test it anyway. If the evolved function of the trait is to subjugate all women, then the rapist element in male society can be predicted to target older, dominant women (or young women who aspire to positions of power) to demonstrate the penalty that comes from overstepping the traditional subordinate role of women. This prediction is not met. Most rape victims are young, *poor* women [768].

Moreover, this hypothesis gives no reason for thinking that rapists will come predominantly from any socioeconomic sector, and yet the vast majority of apprehended rapists are young, poor men. This is exactly the class of individuals expected to engage in rape if rape is a last-resort tactic of males unlikely to attract females to them [768]. This result is consistent with the conditional sexual strategy hypothesis.

Rape could not be (or have been) an adaptive tactic in a conditional sexual strategy if raped women always failed to produce an offspring of the rapist. But raped women do sometimes become pregnant [768], even in modern societies where many women cannot be fertilized because they employ chemical birth control technology. In the past, when access to reliable birth control pills and abortion procedures were not available, it seems likely that rapists would have experienced a higher rate of reproductive success than currently.

If adaptive brain mechanisms regulating sexual behavior are involved in the production of rape (hypotheses 1 and 2), then we can predict that rapists will tend to assault women that are most likely to become pregnant, whatever the immediate feelings of the rapist toward his victim. This prediction was tested by the Thornhills, who used crime statistics to compare the age distribution of females as a whole against the population of raped women (Figure 3). Women who report rape do not constitute a random sample; they are much more likely to fall into the years of peak fertility. Moreover, if the function of rape were merely to inflict violence on women, then we might expect the age distribution of murder victims to parallel that of rape victims, but it does not. But cautions are in order. It could well be that women of peak fertility are more likely to report rape than older women, and if this were true it would bias crime statistics and make them an unreliable source of data with which to test hypotheses on rape.

Nevertheless we can say with some confidence that hypothesis 3 is less likely than 1 or 2. Rapists in the past, and perhaps even in the present, may occasionally father a child through their action, although whether this benefit can outweigh the costly effects of the behavior for the male is unknown. Many other questions remain to be answered, including the emotionally charged issue of whether rape represents a truly distinctive

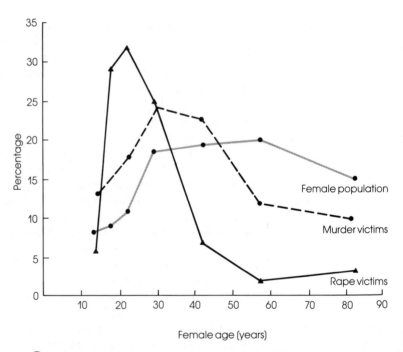

3 **The proximate basis of rape.** A test of the popular notion that rape is motivated purely by the intent to inflict violence on women. If this were true, then we would expect that the distribution of rape victims would match that of female murder victims. Instead, rape victims are especially likely to be young (fertile) women. After Thornhill and Thornhill [768].

"tactic," or whether criminal rape is at the end of a continuum, a spectrum of male sexual efforts that ranges from the violently coercive into aggressive and psychologically coercive, but noncriminal, activity.

Questions about Parental Care Rapists and some adulterous men fail to provide parental care for certain of their offspring. Other men, however, make a considerable parental investment in each of their children, assisting their wives in rearing offspring to reproductive age (and beyond). As someone with a son in college and another a few years away, I know what I am talking about. Parental males are something of an anomaly among mammals, and the question arises, Why are human males so likely to provide costly parental services to the children of their mates?

Here I shall present one scenario for the evolution of male parental care in humans. As in preceding sections, my point is not to claim that this hypothesis is established beyond question but to illustrate the sociobiological approach.

Some sociobiologists have proposed that paternal behavior arose in the following manner. Originally human females may have regularly mated with more than one male over short periods [732]. Multiple mating by females creates the possibility of sperm competition (Chapter 13) in which the sperm of different males compete to fertilize an egg. One way to deal with sperm competition is by guarding a mate, preventing her from copulating with other individuals. This tactic requires considerable time and energy focused on one female and reduces the guarding male's chances of finding other mates. However, the association between a male and a female may permit a male to provide useful services for his mate, including the care of her offspring, if the association lasts long enough. If a male's assistance raises his partner's reproductive success, he raises his own fitness, assuming that he fathers many or most of the offspring his partner produces.

The evidence that sperm competition has been a factor in the evolution of human characteristics has been reviewed by Robert L. Smith. He suggests that several distinctive features of human male anatomy and physiology are devices that give a male an advantage in competing with rival ejaculates within a female. For example, the penis of human males is proportionally much longer than that of any of our closest primate relatives (Figure 4). Smith notes that a long penis permits delivery of the ejaculate farther up the female reproductive tract and closer to the egg than does a small penis. This anatomical attribute might advance the success of sperm contained in the ejaculate.

Moreover, human ejaculates contain many more sperm than, for example, a gorilla's ejaculate. Gorillas have a mating system in which one male absolutely monopolizes access to several females. In such a system the risk that a fertilizable female will receive ejaculates from more than one male is slight. Therefore, delivery of large quantities of sperm deep within the female carries no special advantage for the male.

Given our differences with gorillas, Smith argues that human males have over evolutionary time faced a moderate probability that a mate may

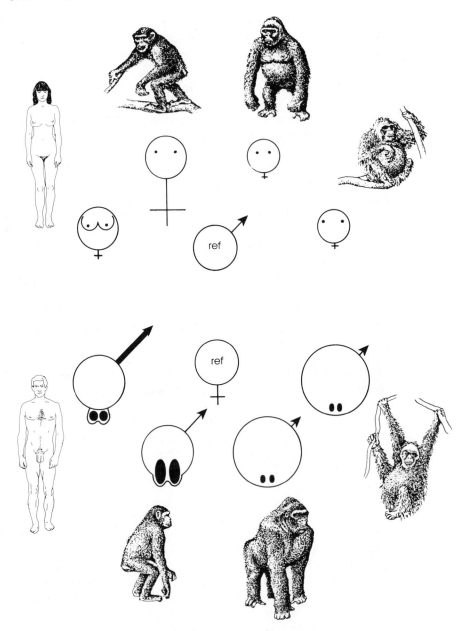

4 **Sexual dimorphism and anatomical characteristics** of humans (left), chimpanzees (left center), gorillas (right center), and orangutans (right). Females are on the upper half of the figure, males on the bottom. Symbols labeled "ref" are reference figures to show the degree of sexual dimorphism in body size exhibited by a species. The relative size of the external genitalia and breasts of females and the penis and testes of males can also be deduced from the symbols for each species. *Source:* Smith [732].

have recently copulated or will copulate again soon with a rival male. This possibility has selected for the ability to deliver large quantities of sperm close to the ovum, stacking the sperm competition odds in favor of the individual able to deliver more, closer. (For a review of alternative hypotheses on this issue, see [732].)

Mate Guarding, Marriage, and Male Parental Care

The key counter to sperm competition, however, is to prevent the introduction of rival ejaculates into one's mates. Mate-guarding by males is practiced widely throughout the animal kingdom in species with multiple-mating females, including many primates. Humans employ exceptionally prolonged mate-guarding, culturally institutionalized in the form of marriage.

One of several alternative scenarios [118, 165] for the origin of prolonged mate-guarding in humans rests on two points, a sexual division of labor in food-gathering and the loss of estrus by females at some point in human evolution. On the first point, there is ample evidence that hunting has long been a key source of food for human beings [128]. Women differ from men in pelvic structure, body size, the proportion of body mass devoted to muscle, and muscle physiology. Average running speeds of men are faster than those of women at all distances [355], a fact suggesting that males may early on have specialized in the capture of animal prey, while females devoted themselves primarily to the collection of plant foods. Essentially all modern hunter–gatherer societies exhibited such a division of economic labor; this finding supports the notion that a similar division was likely to apply in the more distant past.

The second assumption about our ancestors is that female protohominids at one time had a well-defined estrous phase in which they were sexually receptive, as is true for most living primates [414]. During estrus, an early woman might have attracted a consort male who guarded her during the time she was most likely to ovulate, just as happens in some multimale bands of living social primates.

During the consort phase, a male might share prey he had recently captured with the female. High-protein gifts might enhance the reproductive chances of his mate, and so yield a genetic gain for the male. If "nuptial presents" became an established component of courtship and consort guarding, then a female might gain if she could extend the period when she was guarded and given gifts. Instead of blatantly advertising the time of ovulation by becoming receptive only just prior to the event, female protohominids might have partly hidden the time of ovulation by lengthening the period of receptivity.

Robert Smith proposes that the conspicuous breasts of human females may also have evolved in the context of removing information about the reproductive state of the female, the better to encourage prolonged mate-guarding [732]. In our closest relatives, mammaries only become enlarged during the period of pregnancy in preparation for lactation and infant-feeding (Figure 4). In humans this correlation no longer applies; conspicuous breasts appear during puberty and remain enlarged thereafter. Pregnant female protohominids may have at one stage been less likely to attract a helpful consort male; such a female cannot ovulate and therefore cannot produce an additional offspring at that time for a male. But if females were to maintain enlarged breasts even after they had completed breast-feeding an infant and after resuming ovulation, males would be placed under selection pressure to consort with (and provide food for) them over longer periods, given that these females could potentially be fertilized even though they had removed a signal of their immediate reproductive condition to mate-seeking males.

The more difficult it is for a male to determine when a female can be fertilized successfully, the more days he must remain with her (copulating at intervals) in order to increase his chances of fertilizing her eggs when they do become available. Widespread marriage reduces the number of unguarded females and reduces the time a married man has to find these females. These conditions may make it adaptive to be a devoted husband and father and to expend resources on one's presumptive offspring, which are likely to bear one's genes and which can derive major benefits from the scarce, high-protein foods that males can secure for mates and children. This expenditure may increase the offspring's chances of surviving to reproduce, thereby providing a fitness gain that may under some circumstances exceed a Casanova strategy of attempts at multiple sexual conquests.

Mate Choice and Parental Care

In a species such as our own, in which some males do provide resources, including parental care, to their offspring, we can make two fundamental predictions about female mate choice. First, females will make marriage judgments, to the extent that they have a say in the matter, primarily on the basis of the material benefits offered by potential partners, rather than on relative genetic quality. Second, females that gain access to more resources and paternal care for their young will experience superior reproductive success.

In many modern societies females appear to have little choice in a marriage partner, with their fathers or male kin making this decision for them. In the past, however, women may have had more autonomy; and even in modern male-controlled cultures, women may influence their kinsmen's selection of a husband or they may thwart the will of these authorities by refusing to cooperate. Although conflict between a woman and her male kin may be expected occasionally, often their interests will coincide, with all parties benefiting if the woman marries a relatively wealthy or paternal partner.

The foundation for the claim that female choice will center on male economic status is based on the argument that the resources a male controls will have much more impact on the survival chances of a woman's offspring than the genes he provides to fertilize her eggs. In the Ache tribesmen of eastern Paraguay, the children of men who are good hunters are more likely to survive than are the children of men who bring less game to their group [418]. Good hunters are also more likely to have extramarital affairs and produce illegitimate children than are poor hunters, an outcome suggesting that Ache females find the capacity to acquire food resources attractive in men. In many other cultures, courting males are expected to provide gifts to either the female or her family. These demands constitute an economic test of a partner and offer a way to assess the capacity of a male to provide material benefits.

David Buss points out that studies of mate preferences in our society show that men and women consistently agree on the relative importance of most factors—except that men rank "physical attractiveness" substantially higher than women, whereas women especially value "good earning power" in a potential mate [121]. Men are predicted to find "physically attractive" those features in women that are correlated with youthfulness, because ultimately youthfulness in women is highly correlated with reproductive value. Monique Borgerhoff-Mulder tested the prediction that reproductive value determines the value of a woman to a man by examining data on how much men of the Kipsigi tribe of Kenya paid to the families of their brides. Bridewealth payments were higher for women that had reached puberty sooner (Figure 5). Early maturing women did in fact have more children than women that matured later [78].

A woman's reproductive success, however, is not likely to be increased by marriage to a youthful man, but should depend more on the resources her husband can offer her. If females prefer wealthy males, then in general

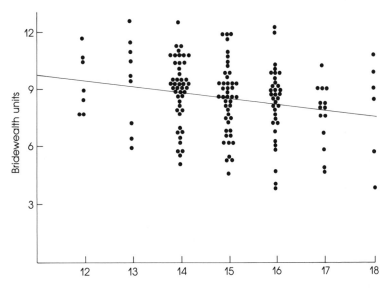

5 **Test of a sociobiological hypothesis.** Bridewealth payments in the Kipsigis reveal that men pay more for brides who have undergone clitorectomy earlier in life. Women are subjected to this operation at menarche. *Source:* Borgerhoff-Mulder [78].

women with such husbands ought to have more surviving offspring than women with poorer partners [399]. In the Kipsigis, the amount of land owned by a husband is positively correlated with the number of surviving offspring a woman has. Husband's wealth also confers a reproductive advantage on females in the different herding culture of the Yomut of Iran (Figure 6). Note, however, that the reproductive edge enjoyed by Yomut women in the wealthier half of the population is less than that of their husbands [399]. The Yomut permit polygyny, and richer men attract women prepared to divide their husband's resources with a co-wife. The same is true for the Kipsigis [77, 79]. Women married to polygynists need not pay a fitness penalty; one-half or one-third of a great deal can be more than 100 percent of a little.

Although the correlation between wealth and reproductive success may apply to "traditional" cultures, and thus by implication to early humans, isn't it true that the advent of civilized society has erased this relationship? A revealing study of genealogical records of Portuguese nobility by James Boone demonstrates that during the 15th- and 16th-century, wealth in this Western civilization almost certainly had a major effect on an individual's reproductive chances, particularly for men [75]. The higher the social status attained by a male, the greater the number of offspring he produced, the more illegitimate children he sired, and the more likely he was to marry more than once (Table 2). Women married to the men of highest nobility

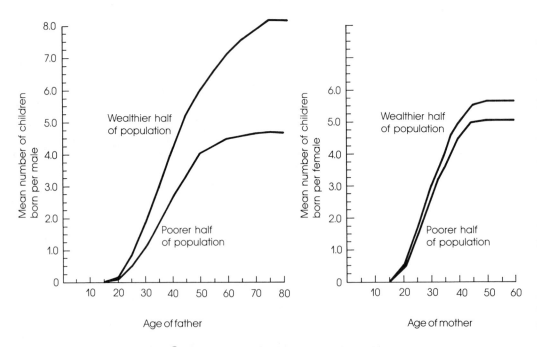

6 **Reproductive success and wealth** in a polygynous society, the Yomuts of Iran. As predicted, both males and females in the wealthier half of the population have higher reproductive success than poorer Yomut men and women. *Source:* Irons [399].

TABLE 2

Reproductive performance of married males in relation to
their social rank: The Portuguese nobility in the Middle Ages

Rank	All offspring		Illegitimate offspring only		Percentage married more than once
	Sample size	Mean per male	Sample size	Mean per male	
Royalty	96	4.75	50	0.52	22.6
Royal bureaucracy	168	4.62	43	0.36	18.5
Landed aristocrats	216	4.54	80	0.37	13.9
Untitled/Military	553	2.33	127	0.23	10.3

Source: Boone [75].

(and presumed greatest wealth) had more children than women whose husbands were untitled or in the military, although the variance in reproductive success of women was much less than for men, just as is true for the Yomut.

This may apply for the recent past, but we have all heard that in our welfare state poor women actually outreproduce wealthy ones. There is much debate on this point [800], but Martin Daly and Margo Wilson show through an analysis of 1973 census data from the United States that for all age classes of married women, there is a positive correlation between the income of husbands and the proportion of the women who have two or more offspring [165].

The Allocation of Paternal Care

We have presented the hypothesis that marriage decisions by women are influenced by the resources under a male's control and that these resources influence the number of children a man's wife or wives has. These children are guaranteed to be the offspring of the mother that bore them; if she chooses to care for them, she will with 100 percent certainty be assisting offspring that carry her genes into the next generation.

The same cannot be said for her husband, who may have been cuckolded. Here again is a simple fact of biology that should have had evolutionary consequences on the operation of human brains, with selection favoring motivating mechanisms that encourage males to do those things that will raise, rather than lower, individual reproductive success.

The double standard of sexual morality characteristic of our own and many other cultures reflects one such element of male psychology. On the one hand married men are much more likely than their wives to engage in extramarital sexual activity. But far from promoting a generalized tolerance of this behavior, men typically view adultery by wives as a grave offense against them. One commonly hears about so-called sexually permissive societies in which complete sexual freedom is supposed to be the norm. There are certainly differences in cultural attitudes about sexual

7 **A Yanomamö uncle and his nephew.** In many societies, uncles treat their sister's children in a highly paternal fashion. Photograph by Napoleon Chagnon.

matters, but the notion that there are cultures that have a completely relaxed view of adultery appears to have been a (wistful?) misinterpretation on the part of outside observers. In *all* cultures studied to date, ample evidence exists to indicate that men regard the sexual favors of their wives as their "property." Adultery (or even suspicion of it) often precipitates a violent response by a husband against his wife or her lover. Disputes over women are a major cause of murders by men in many societies [164], and violence by a cuckolded individual is often considered legally justified [169].

In our society, men are far more likely than women to state that adultery by a spouse caused them to seek a divorce [758]. Concern about paternity is so obsessive that husbands of rape victims may divorce their unfortunate wives, an action accepted or even encouraged by a diversity of religious groups and legal codes [114].

There are societies more sexually permissive than our own, such as the South Sea Island cultures of the Pacific. Even though the extent of sexual freedom may have been greatly overestimated by some Western observers, still it seems likely that the average certainty of paternity for a male in these cultures may be relatively low. The standard practice in cultures of this sort is for the male to withhold some or all parental care from the children of his wife and instead help the children of his sister (Figure 7). Richard Alexander has shown that if a male can expect to father only one of four children produced by his wife, he would share 5/32 of his genes with his nephews and nieces on his sister's side, but only 4/32 of his genes with his wife's children (Figure 8). Although the uncle-father role is adaptive only when the certainty of paternity is very low, nevertheless the correlation between sexual permissiveness and this cultural tradition supports the hypothesis that males take this factor into consideration when allocating their childcare [11].

Why Does Child Abuse Occur?

A likely ultimate function of male sexual jealousy is to avoid caring for another male's child. We predict that when a male knows or suspects that a child of his wife is not his biological offspring, the child will receive less

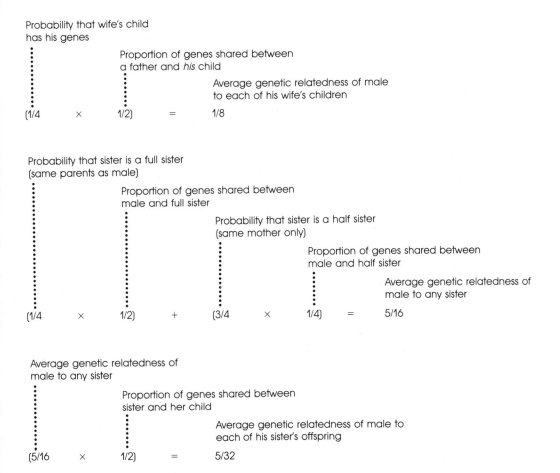

8 **Reliability of paternity** and the coefficient of relationship between a male and the offspring of his wife. In a society in which husbands father only one in four of their wives' offspring, a male's average coefficient of relatedness to the offspring of his wife will be lower than his relatedness to his sister's offspring.

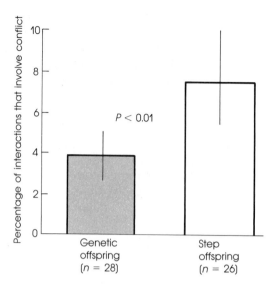

9 **Conflict in families.** The percentage of all interactions that are agonistic (involving conflict) between men and their genetic offspring and step-offspring in households in which men live with both categories of offspring. *Source:* Flinn [263].

paternal care. In a Trinidad village studied by Mark Flinn, there were a number of families in which a stepfather lived with children of his own as well as those his wife had by another man. In these families, there was significantly greater frequency of conflict between stepfathers and their stepchildren than between these men and the children acknowledged to be their genetic offspring (Figure 9) [263]. Furthermore, stepchildren were far more likely to leave a household with a stepfather and move in with relatives (their grandparents, for example) than were children growing up in a household without a stepfather present.

Let me emphasize that the Trinidadian stepchildren were *not* abused. Martin Daly and Margo Wilson have, however, suggested that child abuse may be a side effect of the psychological mechanisms that guide human decisions on the allocation of parental care. The tendency of individuals to exhibit less solicitude for children in whom they have no genetic stake may reach an extreme, and almost certainly maladaptive, expression in criminal child abuse [166, 168].

Note that the argument employed here is fundamentally the same as that employed by Donald Symons in his analysis of male sexuality. The idea is that selection has shaped the evolution of the human brain in a way that promotes certain kinds of behavior that generally lead to increased fitness but that may, as a by-product of their operation, lead some individuals to behave in ways that reduce their fitness (for example, when an adulterous male is wounded or killed by an enraged husband, or when a couple goes to jail for the criminal abuse of children under their control).

Daly and Wilson tested the hypothesis that adults will tend to bias their care toward their biological offspring by predicting that stepparents will contribute a disproportionately large fraction of the cases of criminal child abuse in Western societies. They found that in the Canadian city of Hamilton, children 4 years old or younger were *40 times more likely* to suffer child abuse in families with a stepparent than in families in which both

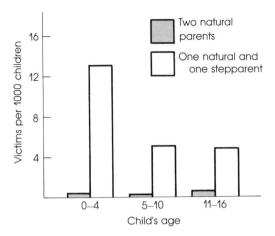

10 **Child abuse** is far more likely to be reported in a household with a stepparent than in households with two biological parents. *Source:* Daly and Wilson [166].

biological parents were present. Note that for both categories the absolute likelihood that a child would be abused was small (Figure 10), but the *relative* risk was far, far greater for children in households with a stepparent [166, 167]. Daly and Wilson's point is that discriminating parental care is an adaptive aspect of human behavior because it prevents individuals from providing for the children of others. But the psychological mechanisms that promote selective parentalism cause emotional difficulties for adults that embark on the stepparent role. The vast majority cope, but a few do not.

Inheritance Rules

Children may receive parental benefits even after their parents have died, thanks to procedures for distributing the wealth of individuals after their death. There is great cultural variation in the rules and regulations surrounding matters of inheritance. The great diversity of traditions on inheritance appears to support the hypothesis that culturally conditioned behavior is biologically arbitrary, with people following whatever tradition they happen to invent.

John Hartung, however, proposed that inheritance be treated as a final parental allocation, with fitness consequences that can vary depending on (1) how the inheritance is distributed among a family's children and (2) the mating system of the culture [346]. He pointed out that whether or not there is a bride price depends on the extent to which the society is polygynous (Table 3). Over 90 percent of all cultures classified as generally polygynous have rules demanding payment, usually to the bride's family, before marriage can take place. This practice reflects the fact that when some males can monopolize several women, brides become a scarce and an especially valuable commodity, a fact that is then exploited by families with marriageable females. In contrast, payments for wives are not required in a majority of monogamous societies.

When polygyny is commonly practiced, male reproductive success becomes highly dependent upon the ability to pay bride prices, with exceptional fitness available for the unusually wealthy person. Under these conditions, parents can potentially secure many descendants posthumously

TABLE 3

The relations between the mating system of a culture, the occurrence of a bride price to be paid by a husband-to-be, and an inheritance system that favors sons

	Price for wife		Inheritance system	
Mating system	No cost (%)	Bride price (%)	Even distribution (%)	Sons favored[a] (%)
Monogamy	62	38	42	58
Limited polygyny	47	53	20	80
General polygyny	9	91	3	97

Source: Hartung [346].

[a]Inheritance rules are such that sons either get the bulk of the inheritance or all of it, whereas under "even" systems daughters are only slightly disfavored or receive an equal portion. Data are derived from Murdock's *Ethnographic Atlas* [579], which codes for the various cultural components appearing in this table. The percentages are based on 112 monogamous cultures, 290 that practice limited polygyny (with less than 20 percent of males actually polygynous), and 448 in which general polygyny occurs (with more than 20 percent of married men with more than one wife).

if they direct their wealth into the hands of a male, who can then use it to sequester a number of wives. The successful polygynist can father many more grandchildren for the deceased parent than a daughter, whose reproductive success is limited by the number of embryos she personally can produce and nurture. On the basis of this analysis, Hartung predicted that there will be a significant correlation between the practice of polygyny in a society and the occurrence of inheritance rules that favor sons, concentrating wealth in individuals who then may have the economic capacity to become polygynous (Table 3).

Monique Borgerhoff Mulder has recently demonstrated that inherited wealth actually does have greater reproductive benefits for sons than for daughters in the traditional (polygynous) society of the Kipsigis. The number of cows owned by a man's parents strongly correlates with his reproductive success, but there is no correlation between parental cow wealth and a daughter's reproductive success [80].

Even in supposedly monogamous Western society, very rich men may have opportunities for unusual reproductive success because their wealth makes them attractive to women and able to support many children. If so, we can predict that very wealthy parents will be more inclined than poorer people to bias their inheritance in favor of sons rather than daughters. A group of Canadian researchers tested this prediction by examining 1000 probated wills. The data from these records support the hypothesis [730] (Figure 11).

Adoption

The findings of sociobiology appear to paint a grim picture of human beings that are driven in ultimate terms by unceasing reproductive competition and self-interest. But many couples adopt the children of complete strangers, youngsters that they treat with the same love and affection that parents

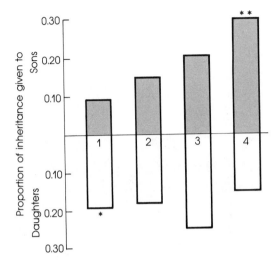

11 Inheritance decisions. A test of the hypothesis that wealthy parents will bias their inheritance distribution to their sons, who are more likely than females to convert exceptional wealth into exceptional reproductive success. Categories 1–4 represent the sample divided into fourths with wills involving less than $20,000, $20,000 to $53,000, $53,000 to $111,000, and more than $111,000 respectively. *, In this category, daughters receive significantly more; **, in this group, males receive significantly more. *Source:* Smith et al. [730].

typically supply to their own biological children. Yet these adoptive parents subsidize competitor genes. Surely this means that at least some aspects of human behavior cannot be analyzed successfully from a sociobiological perspective.

Or does it? As we have shown throughout this chapter, a sociobiological (evolutionary) approach does not require that all aspects of behavior be adaptive. Here, as in other aspects of human behavior, we can consider the hypothesis that maladaptive responses arise as the by-product of underlying proximate mechanisms that generate adaptive reactions, or did so in the past under conditions that existed then but not now. For example, adopting a nonrelative may be one outcome of the motivational system that causes adult humans to want to have children and raise a family. Although it is true that adults who adopt infant strangers reduce their fitness, the urge to have a family and the love of children that cause them to behave this way are generally beneficial. Because these traits tend to elevate fitness, they are maintained in human populations even though they *sometimes* motivate people to behave maladaptively.

The point is that all adaptations have costs and benefits, and one cost is that a desire for something, be it sugary or salty foods, sexual variety or fidelity of a mate, or even the desire for children, can be misdirected when viewed from a reproductive perspective. This is true especially in rare or unusual circumstances. In many species of animals, adoption occurs when adults have lost their offspring and are presented fortuitously with a substitute (Figure 12). Cardinals have been known to feed goldfish [823], and a white whale has been seen trying to lift a floating log part way out of the water, as if the log were a distressed infant that needed help to reach the surface to breathe [94]. Adopting a goldfish or a log did nothing to promote the cardinal's or the white whale's genes. But these individuals had almost certainly lost their young recently and were employing generally adaptive parental behaviors in unusual circumstances.

The misdirected parental care hypothesis for adoption generates a test-

12 **Adoption occurs in nonhuman animals,** often when adults have just lost their brood and encounter a substitute offspring. Here two emperor penguins compete for "possession" of a youngster. Photograph by Yvon Le Maho.

able prediction, which is that husbands and wives who have lost an only child or who fail to produce children themselves should be especially prone to adopt strangers. Their frustrated desire to be parents could act as a proximate basis for adopting a substitute for genetic offspring.

One can also hypothesize that the adoption of complete strangers is a novel, recent phenomenon that occurs primarily because of changes in the environments of humans over the past several hundred years. Throughout the vast majority of human evolution, during which people lived in small bands or villages, the probability of adopting a complete stranger would have been vanishingly small. We can predict that acceptance of the children of relatives was the most common kind of adoption in the past [829]. Males that practice the uncle-father tactic are adopting their sister's children as their own. Grandparents in Trinidad that take in grandchildren who would otherwise live with their mother and a stepfather are adopting relatives. To the extent that this increases the reproductive success of these relatives, the adopters are indirectly increasing the representation of their alleles in subsequent generations.

In large industrial societies, however, children are made available to people who do not know the parents of the adoptee. Under these novel conditions, our psychic mechanisms, which evolved in the past, can cause us to behave maladaptively in the present and in so doing, reveal something about the design features of our psyches that exist because of natural selection.

I believe human sociobiology involves a straightforward application of evolutionary theory to human affairs. *If* one's goal is the satisfaction of curiosity about ourselves, it seems foolish to foreclose any possible avenue of understanding, particularly the Darwinian approach that has proved so helpful with respect to nonhuman animals. I am not convinced, however, that even when an evolutionary analysis of behavior is reasonably complete (and it is not likely to be soon) we will put this information to uses that most people would consider good. The brain mechanisms that push us to try to solve mysteries about ourselves and our world may lead us to discoveries that will be used to worsen rather than alleviate the social and military crises that seem to dominate modern society. But whatever our wishes, the fact that we are an evolved animal species is not going to change, so we might as well understand the significance of this fact, if only to give us insight into why Pogo was probably right when he declared, "We have met the enemy and it is us."

SUMMARY

1 Human beings are an evolved animal species. Human sociobiology employs an evolutionary approach to generate testable hypotheses about our behavior based on the assumption that what individuals do tends to propagate their genes.

2 The study of human sociobiology has been marked with severe controversy in part because of confusion surrounding the goals and methods of the approach. Sociobiology does *not* demand acceptance of the notion that if a human behavior is adaptive, it is developmentally unchangeable and socially desirable.

3 Sociobiologists employ a set of general hypotheses that they use to analyze specific human behaviors. A behavior may currently raise the inclusive fitness of the individuals that exhibit the trait. Or it may be currently maladaptive, but did promote fitness in the past under different environments. Or it may never have been adaptive, but is a sometimes revealing by-product of psychic mechanisms that may be or have been adaptive. These general arguments can be used as a foundation for the production of working hypotheses on the evolutionary significance of a behavior of interest, like child abuse, adoption, homosexuality, and so on.

4 Not all societies practice war. For those that do, there appears to be a correlation between group aggression and competition for limited resources of reproductive value to humans, a relation suggesting that at least in the past, humans engaged in war over materials related to fitness.

5 Cultural influences permeate human reproductive behavior, but it is possible to develop testable hypotheses about the fitness effects of a broad range of human sexual activities. Men and women are predicted to differ in their sexual tactics, given that male fitness is often a function of the number of different females the male inseminates, whereas a woman can only produce about one child a year at most, regardless of the number of mates she takes.

6 Human males often sacrifice opportunities to fertilize many females in order to invest parentally in the children of their wives. If male behavior is the product of natural selection, we can predict that men will behave in ways that increase the probability that the children they help carry their genes, and not those of some other male. Mate guarding by males is a proximate mechanism that helps achieve this ultimate goal.

7 To the extent that we are curious about the evolution of human behavior, a sociobiological approach offers avenues of exploration that other approaches do not. Although there are risks in employing this approach, the same can be said about any scientific endeavor that touches on human concerns.

SUGGESTED READING

The conflict over *Sociobiology: The New Synthesis* [853] can be reviewed by reading the last chapter in the book, and then the exchange between E. O. Wilson and his opponents that appears in *BioScience* [15, 855]. For more recent and general critiques of sociobiology, see articles by Stephen J. Gould [e.g. 303, 304] and the book by Philip Kitcher [432], which can be contrasted with the arguments of Wilson [856, 858], Richard Alexander [11, 12], Michael Ruse [680], and Donald Symons, whose book on human sexual behavior is especially enjoyable reading [758]. John Tooby and Irven DeVore offer a clear description of the way in which persons who wish to analyze the evolution of human behavior can go about their work [781].

For examples of human sociobiological research that sets up hypotheses and then tests them, I recommend the controversial paper by Randy and Nancy Thornhill [768] and the papers of Martin Daly, Margo Wilson, and their associates [e.g., 166, 168].

DISCUSSION QUESTIONS

1. Marshall Sahlins has argued that sociobiology is contradicted by the fact that most cultures do not even have words to express fractions. He points out that without fractions, a person cannot possibly calculate coefficients of relationship, and without this information (he claims) people could not be expected to determine how to behave in order to maximize indirect fitness [691]. Is Sahlins right to believe that this evidence is a serious blow to sociobiological theory?

2. Philip Kitcher states that "socially relevant science," like sociobiology, demands "higher standards of evidence," because if a mistake is made

(a hypothesis presented as confirmed when it is false) then the societal consequences may be especially severe. For example, the hypothesis that men are more disposed to seek political power and high status in business and science than women is dangerous because it "threaten[s] to stifle the aspirations of millions" [432]. Do you agree that scientific standards should be raised for human sociobiology? How would you go about doing this? Do you agree that sociobiological hypotheses carry with them the special likelihood of social damage? If so, how do you envision a specific sociobiological hypothesis would exercise its damaging effects, and how would you prevent these effects?

3. How might sociobiologists defend themselves from the accusation of Stephen J. Gould [298] that sociobiology presents us with "unproved and unprovable speculations about the adaptive and genetic basis of specific human behaviors?"

GLOSSARY

Action potential The neural signal; a self-regenerating change in membrane electrical charge that travels the length of the axon of a nerve cell.

Adaptation (1) The tendency of a nerve cell to cease responding to unvarying stimulation. (2) A characteristic that confers higher inclusive fitness to individuals than any other existing alternative exhibited by other individuals within the population; a trait that has spread or will spread through the population as a result of natural selection or indirect selection.

Adaptationist A behavioral biologist who uses the assumption that traits are adaptive to generate testable hypotheses on the fitness consequences of a particular trait.

Allele A form of a gene; different alleles typically code for distinctive variants of the same enzyme.

Altruism Helpful behavior that raises the recipient's direct fitness while lowering the donor's direct fitness.

Artificial selection *See* Selection.

Axon That part of the nerve cell that carries a message to other cells in a neural network.

Behavioral ecology The study of the adaptive value of behavioral attributes in solving the environmental obstacles to reproduction by individuals.

Brood parasite An animal that exploits the parental care of individuals other than its parents.

Central pattern generator Nerve cells in the central nervous system capable of producing an organized set of messages that control some element of an animal's locomotion.

Circadian rhythm A roughly 24-hour cycle of behavior that expresses itself independent of environmental changes.

Circannual rhythm A annual cycle of behavior that expresses itself independent of environmental changes.

Coefficient of relatedness A measure of genes shared as a result of having a common ancestor; the probability that an allele present in one individual will be present in a relative; the proportion of the total genotype of one individual present in the other, as a result of shared ancestry.

Communication The cooperative transfer of information from a signaler to a receiver.

Comparative method A procedure for testing evolutionary hypotheses that is based on disciplined comparisons among species.

Competition for mates *See* Selection (Sexual Selection).

Competitive release The expansion of the range of resources utilized by a species in locations where a competitor species is absent.

Concorde fallacy The argument that what an individual has already invested (generally in an existing offspring) should determine what its future investment should be.

Conditional strategy *See* Strategy.

Conflict behavior The actions that result when two different motivational systems are activated simultaneously within an individual.

Convergent evolution   The independent acquisition over time through natural selection of similar characteristics in two or more unrelated species.

Cooperation   A mutually helpful action.

Critical period   A phase in an animal's life when certain experiences are particularly likely to have a potent developmental effect.

Developmental homeostasis   The capacity of developmental mechanisms within individuals to keep the course of development guided toward an adaptive end point despite potentially disruptive effects of mutant genes and suboptimal environmental conditions.

Dilution effect   Safety in numbers; the decreased probability of being taken by a predator that stems from being a member of a large, as opposed to a small, group of prey.

Divergent evolution   The naturally selected changes in related species that once shared a characteristic in common (as a result of having inherited it from a common ancestor) but have come to be different.

Dominance hierarchy   A social ranking system within a group, in which some individuals give way to others, often conceding useful resources to others without a fight.

Diploid   Having two copies of each gene in one's genotype.

Direct fitness   *See* Fitness.

Direct selection   *See* Selection.

Display   A stereotyped action used as a communication signal by individuals.

Ethology   The study of the proximate mechanisms and adaptive value of animal behavior.

Evolutionarily stable strategy   *See* Strategy.

Evolutionary benefit   That aspect of a trait that tends to raise the inclusive fitness of individuals.

Evolutionary cost   That aspect of a trait that tends to decrease the inclusive fitness of individuals.

Explosive breeding assemblage   The formation of large groups of reproducing individuals in species that breed on only a few days each year.

Fitness   A measure of the genes contributed to the next and succeeding generations by an individual, often stated in terms of the number of surviving offspring produced by the individual.

   Direct fitness   The genes contributed by an individual via personal reproduction in the bodies of surviving offspring.

   Indirect fitness   The genes contributed by an individual indirectly by helping nondescendant kin, in effect creating relatives that would not have existed without the help of the individual.

   Inclusive fitness   The sum of an individual's direct and indirect fitness.

Fixed action pattern   An innate, highly stereotyped response that is triggered by a well-defined simple stimulus; once the pattern is activated, the response is performed in its entirety.

Forced copulation   Rape, in which a male inseminates a female against her will and with the consequence that her fitness is reduced.

Frequency-dependent selection   *See* Selection.

Game theory   The quantitative study of the effects of competition among individuals in which the payoffs to individuals of different actions are dependent on the tactics employed by rivals in the species or group.

Genetic mosaic   An individual whose tissues are a mix of different genotypes.

Genotype   The genetic constitution of an individual; may refer to the alleles of one gene possessed by the individual or to its complete set of genes.

"Good genes" hypothesis   The argument that mate choice advances individual fitness because it provides the offspring of choosy individuals with genes that promote the development of traits useful in solving ecological problems.

Group selection   *See* Selection.

Habituation A form of learning in which individuals stop responding to stimuli that in the past experience of the individuals have had no reinforcing consequences.

Haploid Having only one copy of each gene in one's genotype, as is true of the sperm and eggs of diploid organisms.

Home range An area that an animal occupies but does not defend, in contrast to a "territory," which is defended.

Imprinting A form of learning in which individuals exposed to certain key stimuli, usually during an early stage of behavioral development, form an association with the object and may later show sexual behavior toward similar objects.

Inbreeding depression The tendency of inbred organisms to have lower fitness than non-inbred members of their species.

Indirect selection *See* Selection.

Individual selection *See* Selection.

Instinct A behavior pattern that reliably develops in individuals that receive adequate nutrition and that is given in functional form on its first performance.

Interneuron A nerve cell that relays messages either from receptor neurons to the central nervous system (a sensory interneuron) or from the central nervous system to neurons commanding muscle cells (a motor interneuron).

Kin recognition The capacity of an individual to use various cues to identify individuals that are likely to be its genetic relatives.

Kin selection *See* Selection.

Lateral inhibition A property of groups of neurons in which activity in one cell reduces the probability that its neighbors will fire.

Learning A durable and usually adaptive change in an animal's behavior traceable to a specific experience in the individual's life.

Lek A traditional display site where males gather to defend small territories that lack resources useful to females; nevertheless females visit the site to mate.

Lek polygyny *See* Polygyny.

Mass recruitment The attraction of large numbers of individuals to a resource using signals given by the discoverers of the resource.

Mate choice *See* Sexual selection.

Mobbing behavior Activity occurring when (usually) several prey individuals closely approach a predator, often attempting to harass it.

Monogamy A mating system in which one male mates with just one female in a breeding season.

Mutualism A mutually beneficial relationship or action.

Natural selection *See* Selection.

Neuron A nerve cell.

Neurotransmitter A chemical that is released by one neuron and affects the probability that the next cell in the network will fire.

Nondescendant kin Relatives other than offspring or grandchildren.

Nuptial gift A food item transferred by a male to a female just prior to or during copulation.

Operant conditioning A kind of learning, allied with trial-and-error learning, in which an action (or operant) that is rewarded will become more frequently performed.

Operational sex ratio (OSR) The ratio of receptive males to receptive females in a population at any one period.

Optimality theory The development of quantitative models, and thus quantitative predictions, about a behavior based on the assumption that an animal's actions are optimal (yield higher inclusive fitness than other traits).

Organizing substance A chemical produced by cells in the developing organism that plays a key role in triggering a particular pattern of cell development.

Parental investment The risks taken by a parent and the time and energy it invests in an existing offspring that reduce its chances of producing additional offspring in the future.

Phenotype matching A proximate mechanism of kin recognition in which an individual's behavior toward another is based on how similar they are in appearance, odor, or other phenotype.

Pheromone A volatile chemical released by one individual to communicate with another.

Photoperiod The number of hours of light in 24 hours.

Pleiotropy The capacity of one gene to have multiple effects on the development of an individual.

Polyandry A mating system in which one female mates with several males during a breeding season.

Polygeny The capacity of several genes to influence the development of a trait.

Polygyny A mating system in which one male mates with several females during a breeding season.

Female defense polygyny The polygynous males directly defend their mates.

Lek polygyny The polygynous males use displays to attract several females to them at small, resourceless display sites.

Resource defense polygyny The polygynous males monopolize useful resources that receptive females use.

Scramble-competition polygyny The polygynous males are nonterritorial and are better able to find receptive females than their rivals.

Receptive field That portion of the receptor surface that is monitored by a higher order nerve cell.

Reciprocity A synonym for reciprocal altruism; a helpful action that will be repaid in the future by the recipient.

Releaser A sign stimulus given by one individual as a social signal to another.

Reproductive success The number of surviving offspring produced by an individual.

Runaway selection *See* Selection.

Satellite male A male that waits by a dominant rival to intercept females attracted to the advertisements of the rival.

Search image A hypothetical mental picture of a prey item used by a predator to search specifically for a cryptic, common, edible prey.

Selection The differential capacity of individuals to transmit copies of their genes to the next generation.

Artificial selection A process that is identical with natural selection except that humans control the reproductive success of alternative types within the selected population.

Frequency-dependent selection A form of natural selection in which those individuals that happen to belong to the less common of two types in the population are the ones that are more fit because of their lower frequency in the population.

Group selection The process that occurs when groups differ in their attributes and the differences are correlated with differences in group survival.

Indirect selection The process that occurs when individuals differ in their effects on the survival of nondescendant kin, creating differences in the indirect fitness of the individuals interacting with this category of kin.

Individual selection *See* Natural selection.

Kin selection The process that occurs when individuals differ in ways that affect their parental care or helping behavior, and thus the survival of their own offspring, or the survival of nondescendant kin.

Natural selection (Also called "direct selection" and "individual selection.") The process that occurs when individuals differ in their traits and the differences are correlated with differences in reproductive success. Natural selection can produce evolutionary change when these differences are inherited.

Runaway selection A form of sexual selection that occurs when female mating preferences for certain male attributes create a positive feedback loop favoring males with these attributes and females that prefer them.

Sexual selection A form of natural selection that occurs when individuals vary in their ability to compete with others for mates or to attract members of the opposite sex. As with natural selection, when the variation among individuals is correlated with genetic differences, sexual selection leads to genetic changes in the population.

Selfish herd A group of individuals, each one of which uses the others as living shields against predators.

Sexual dimorphism A difference between males and females in a species.

Sexual refractory period A time when animals cannot be stimulated to mate, even when internal and external conditions are manipulated to match those occurring when mating normally takes place.

Sexual selection *See* Selection.

Sign stimulus The effective component of an action or object that triggers a fixed action pattern in an animal.

Sociobiology A discipline that uses evolutionary theory as the foundation for the study of social behavior.

Sperm competition When the sperm of two or more males are received by a female with fertilizable egg(s).

Spite Behavior that reduces the fitness of both the spiteful individual and the individual that is the object of spite.

Startle response An animal's alarmed withdrawal from an unexpected and threatening change in a potential prey.

Stimulus filtering The capacity of nerve cells and neural networks *not* to respond to some stimuli.

Strategy A genetically distinctive set of rules for behavior exhibited by individuals.

Conditional strategy A set of rules that provides for different tactics under different environmental conditions; the behavioral capacity to be flexible in response.

Evolutionarily stable strategy (ESS) That set of rules of behavior that when adopted by a certain proportion of the population cannot be replaced by any alternative strategy.

Supernormal stimulus A sign stimulus that is more effective in eliciting a response than naturally occurring actions or objects.

Synapse The point of near contact between one nerve cell and another.

Tactic One of the options that may be exercised by individuals whose behavior is guided by a conditional strategy.

Working hypothesis A speculative explanation that is presented to be tested.

BIBLIOGRAPHY

1 Abbot, D. H. 1987. Behaviourally mediated suppression of reproduction in female primates. *Journal of Zoology* 213:455–470.

2 Ackerman, S. L., and R. W. Siegel. 1986. Chemically reinforced conditioned courtship in *Drosophila*: responses of wild-type and the *dunce, amnesiac*, and *don giovanni* mutants. *Journal of Neurogenetics* 3:111–123.

3 Afton, A. D. 1985. Forced copulation as a reproductive strategy of male lesser scaup: a field test of some predictions. *Behaviour* 92:146–167.

4 Alatalo, R. V., A. Carlson, A. Lundberg, and S. Ulfstrand. 1981. The conflict between male polygamy and female monogamy: the case of the pied flycatcher *Ficedula hypoleuca. American Naturalist* 117:738–753.

5 Alatalo, R. V., D. Eriksson, L. Gustafsson, and K. Larsson. 1987. Exploitation competition influences the use of foraging sites by tits: experimental evidence. *Ecology* 68:284–290.

6 Alatalo, R. V., A. Lundberg, and K. Ståhlbrandt. 1984. Female mate choice in the pied flycatcher *Ficedula hypoleuca. Behavioral Ecology and Sociobiology* 14:253–262.

7 Alcock, J. 1987. The effects of experimental manipulation of resources on the behavior of two calopterygid damselflies that exhibit resource-defense polygyny. *Canadian Journal of Zoology* 65:2475–2482.

8 Alcock, J., G. C. Eickwort, and K. R. Eickwort. 1977. The reproductive behavior of *Anthidium maculosum* and the evolutionary significance of multiple copulations by females. *Behavioral Ecology and Sociobiology* 2:385–396.

9 Alexander, R. D. 1961. Aggressiveness, territoriality and sexual behavior in field crickets (Orthoptera: Gryllidae). *Behaviour* 17:130–223.

10 Alexander, R. D. 1974. The evolution of social behavior. *Annual Review of Ecology and Systematics* 5:325–383.

11 Alexander, R. D. 1979. *Darwinism and Human Affairs*. University of Washington Press, Seattle, WA.

12 Alexander, R. D. 1987. *The Biology of Moral Systems*. Aldine de Gruyter, New York.

13 Alexander, R. D., J. L. Hoogland, R. D. Howard, K. M. Noonan, and P. W. Sherman. 1979. Sexual dimorphism and breeding systems in pinnipeds, ungulates, primates, and humans. In *Evolutionary Biology and Human Social Behavior: An Anthropological Perspective*. N. A. Chagnon and W. Irons (eds.). Duxbury Press, North Scituate, MA.

14 Alexander, R. D., and P. W. Sherman. 1977. Local mate competition and parental investment in social insects. *Science* 196:494–500.

15 Allen, G. E. et al. 1976. Sociobiology—another biological determinism. *BioScience* 26:183–186.

16 Andersson, M. 1981. Central place foraging in the whinchat, *Saxicola rubetra. Ecology* 62:538–544.

17 Andersson, M. 1982. Female choice selects for extreme tail length in a widowbird. *Nature* 299:818–820.

18 Andersson, M. 1984. The evolution of eusociality. *Annual Reviews of Ecology and Systematics* 15:165–189.

19 Andersson, M. 1986. Sexual selection and the importance of viability differences: a reply. *Journal of Theoretical Biology* 120:251–254.

20 Andersson, M. 1987. Genetic models of sexual selection: some aims, assumptions and tests. In *Sexual Selection: Testing the Alternatives*. J. W. Bradbury and M. Andersson (eds.). Springer-Verlag, Heidelberg.

21 Andersson, M., F. Gotmark, and C. G. Wiklund. 1981. Food information in the black-headed gull, *Larus ridibundus. Behavioral Ecology and Sociobiology* 9:199–202.

22 Aoki, S. 1977. *Colophina clematis* (Homoptera: Pemphigidae), an aphid species with "soldiers." *Kontyu* 45:276–282.

23 Appleby, M. C. 1982. The consequences and causes of high social rank in red deer stags. *Behaviour* 80:259–282.

24 Arak, A. 1983. Sexual selection by male–male competition in natterjack toad choruses. *Nature* 306:261–262.

25 Armitage, K. B. 1986. Marmot polygyny revisited: the determinants of male and female reproductive strategies. In *Ecological Aspects of Social Evolution: Birds and Mammals.* D. I. Rubenstein and R. W. Wrangham (eds.). Princeton University Press, Princeton, NJ.

26 Arms, K. P., P. Feeny, and R. C. Lederhouse. 1974. Sodium: stimulus for puddling behavior by tiger swallowtail butterflies, *Papilio glaucus. Science* 185:372–374.

27 Armstrong, E. 1983. Relative brain size and metabolism in mammals. *Science* 220:1302–1304.

28 Arnold, S. J. 1976. Sexual behavior, sexual interference, and sexual defense in the salamanders *Ambystoma maculatum, Ambystoma tigrinum,* and *Plethodon jordani. Zeitschrift für Tierpsychologie* 42:247–300.

29 Arnold, S. J. 1980. The microevolution of feeding behavior. In *Foraging Behavior: Ecology, Ethological, and Psychological Approaches.* A. Kamil and T. Sargent (eds.). Garland STPM Press, New York.

30 Arnold, S. J. 1982. A quantitative approach to antipredator performance: salamander defense against snake attack. *Copeia* 1982:247–253.

31 Arnold, S. J. 1983. Sexual selection: the interface of theory and empiricism. In *Mate Choice.* P. P. G. Bateson (ed.). Cambridge University Press, Cambridge.

32 Arora, K., V. Rodrigues, S. Joshi, S. Shanbhag, and O. Siddiqi. 1987. A gene affecting the specificity of the chemosensory neurons of *Drosophila. Nature* 330:62–63.

33 Axelrod, R., and W. D. Hamilton. 1981. The evolution of cooperation. *Science* 211:1390–1396.

34 Baker, R. R. 1978. *The Evolutionary Ecology of Animal Migration.* Hodder & Stoughton, London.

35 Balda, R. P. 1980. Recovery of cached seeds by a captive *Nucifraga caryocatactes. Zeitschrift für Tierpsychologie* 52:331–346.

36 Baldaccini, N. E., S. Benvenuti, V. Fiaschi, and F. Papi. 1975. Pigeon navigation: effects of wind deflection at home cage and homing behavior. *Journal of Comparative Physiology* 99:177–186.

37 Bänziger, H. 1986. Skin-piercing blood-sucking moths. IV: Biological studies on adults of 4 *Calyptra* species and 2 subspecies (Lep., Noctuidae). *Bulletin de la Société Entomologique Suisse* 59:111–138.

38 Bänziger, H. 1971. Bloodsucking moths of Malaya. *Fauna* 1:4–16.

39 Baptista, L. F., and L. Petrinovich. 1984. Social interaction, sensitive phases, and the song template hypothesis in the white-crowned sparrow. *Animal Behaviour* 32:172–181.

40 Baptista, L. F., and L. Petrinovich. 1986. Song development in the white-crowned sparrow: social factors and sex differences. *Animal Behaviour* 34:1359–1371.

41 Barash, D. P. 1982. *Sociobiology and Behavior,* Second Edition. Elsevier, New York.

42 Barnosky, A. D. 1985. Taphonomy and herd structure of the extinct Irish elk, *Megaloceros giganteus. Science* 228:340–344.

43 Bateman, A. J. 1948. Intra-sexual selection in *Drosophila. Heredity* 2:349–368.

44 Bateson, P. P. G. 1976. Rules and reciprocity in behavioural development. In *Growing Points in Ethology.* P. P. G. Bateson and R. A. Hinde (eds.). Cambridge University Press, Cambridge.

45 Bateson, P. P. G. 1982. Preferences for cousins in Japanese quail. *Nature* 295:236–237.

46 Baylies, M. K., T. A. Bargiello, F. R. Jackson, and M. W. Young. 1987. Changes in abundance or structure of the *per* gene product can alter periodicity of the *Drosophila* clock. *Nature* 326:390–392.

47 Beach, F. 1976. Sexual attractivity, proceptivity, and receptivity in female mammals. *Hormones and Behavior* 7:105–138.

48 Beecher, M. D. 1982. Signature systems and kin recognition. *American Zoologist* 22:477–490.

49 Beecher, M. D. 1988. Some comments on the adaptationist approach to learning. In *Evolution and Learning.* R. C. Bolles and M. D. Beecher (eds.). Lawrence Erlbaum Associates, Hillsdale, NJ.

50 Beecher, M. D., and I. M. Beecher. 1979. Sociobiology of bank swallows: reproductive strategy of the male. *Science* 205:1282–1285.

51 Beecher, M. D., M. B. Medvin, P. K. Stoddard, and P. Loesche. 1986. Acoustic adaptations for parent-offspring recognition in swallows. *Experimental Biology* 45:179–183.

52 Beehler, B., and S. G. Pruett-Jones. 1983. Display dispersion and diet of birds of paradise: a comparison of nine species. *Behavioral Ecology and Sociobiology* 13:229–238.

53 Beletsky, L. D., and G. H. Orians. 1987. Territoriality among male red-winged blackbirds. I. Site fidelity and movement patterns. *Behavioral Ecology and Sociobiology* 20:21–34.

54 Bell, G. 1982. *The Masterpiece of Nature: The Evolution and Genetics of Sexuality.* University of California Press, Berkeley, CA.

55 Belovsky, G. E. 1981. Food plant selection by a generalist herbivore: the moose. *Ecology* 62:1020–1030.

56 Bentley, D., and R. R. Hoy. 1974. The neurobiology of cricket song. *Scientific American* 231 (Aug.):34–44.

57 Bentley, D., and H. Keshishian. 1982. Pathfinding by peripheral pioneer neurons in grasshoppers. *Science* 218:1082–1088.

58 Benzer, S. 1973. Genetic dissection of behavior. *Scientific American* 229 (Dec.):24–37.

59 Bercovitch, F. B. 1986. Male rank and reproductive activity in savanna baboons. *International Journal of Primatology* 7:533–550.

60 Berens von Rautenfeld, D. 1978. Bemerkungen zur Austauschbarkeil von Küken der Silbermöve (*Larus argentatus*) nach der ersten Lebenswoche. *Zeitschrift für Tierpsychologie* 47:180–181.

61 Berger, J. 1983. Induced abortion and social factors in wild horses. *Nature* 303:59–61.

62 Berglund, A., G. Rosenqvist, and I. Svensson. 1986. Mate choice, fecundity, and sexual dimorphism in two pipefish species (Syngnathidae). *Behavioral Ecology and Sociobiology* 19:301–307.

63 Bernstein, I. L. 1978. Learned taste aversion in children receiving chemotherapy. *Science* 200:1302–1303.

64 Bertram, B. C. R. 1978. Kin selection in lions and evolution. In *Growing Points in Ethology.* P. P. G. Bateson and R. A. Hinde (eds.). Cambridge University Press, New York.

65 Bildstein, K. L. 1983. Why white-tailed deer flag their tails. *American Naturalist* 121:709–715.

66 Birch, H. C. 1978. Chemical communication in pine bark beetles. *American Scientist* 66:409–41.

67 Birkhead, T. R. 1978. Behavioural adaptations to high density nesting in the common guillemot *Uria aalge*. *Animal Behaviour* 26:321–331.

68 Bischoff, R. J., J. L. Gould, and D. I. Rubenstein. 1985. Tail size and female choice in the guppy (*Poecilia reticulata*). *Behavioral Ecology and Sociobiology* 17:253–256.

69 Björklund, M., and B. Westman. 1986. Adaptive advantages of monogamy in the great tit (*Parus major*): an experimental test of the polygyny threshold model. *Animal Behaviour* 34:1436–1440.

70 Black, A. H. 1971. The direct control of neural processes by reward and punishment. *American Scientist* 59:236–245.

71 Blest, A. D. 1957. The function of eye-spot patterns in the Lepidoptera. *Behaviour* 11:209–256.

72 Boake, R. B. 1986. A method for testing adaptive hypotheses of mate choice. *American Naturalist* 127:654–666.

73 Boggess, J. 1984. Infant killing and male reproductive strategies in langurs (*Presbytis entellus*). In *Infanticide, Comparative and Evolutionary Perspectives.* G. Hausfater and S. B. Hrdy (eds.). Aldine, Chicago.

74 Bolles, R. C. 1973. The comparative psychology of learning: the selection association principle and some problems with "general" laws of learning. In *Perspectives in Animal Behavior.* G. Bermant (ed.). Scott, Foresman & Company, Glenview, IL.

75 Boone, J. L., III. 1986. Parental investment and elite family structure in preindustrial states: a case study of late medieval-early modern Portuguese genealogies. *American Anthropologist* 88:859–878.

76 Boppré, M. 1986. Insects pharmacophagously utilizing defensive plant chemicals (pyrrolizidine alkaloids). *Naturwissenschaften* 73:17–26.

77 Borgerhoff Mulder, M. 1987. On cultural and reproductive success: Kipsigis evidence. *American Anthropologist* 89:617–634.

78 Borgerhoff Mulder, M. 1987. Kipsigis bridewealth payments. In *Human Reproductive Behavior: A Darwinian Perspective.* L. L. Betzig, M. Borgerhoff Mulder, and P. Turke (eds.). Cambridge University Press, Cambridge.

79 Borgerhoff Mulder, M. 1987. Resources and reproductive success in women with an example from the Kipsigis of Kenya. *Journal of Zoology* 213:489–505.

80 Borgerhoff Mulder, M. 1988. Reproductive consequences of sex-biased inheritance. In *Comparative Socioecology of Mammals and Man.* V. Standen and R. Foley (eds.). Blackwell, London.

81 Borgia, G. 1980. Sexual competition in *Scatophaga stercoraria*: size- and density-related changes in male ability to capture females. *Behaviour* 75:155–206.

82 Borgia, G. 1985. Bower quality, number of decorations and mating success of male satin bowerbirds (*Ptilonorhynchus violaceus*). *Animal Behaviour* 33:266–271.

83 Borgia, G. 1985. Bower destruction and sexual competition in the satin bowerbird (*Ptilonorhynchus violaceus*). *Behavioral Ecology and Sociobiology* 18:91–100.

84 Borgia, G. 1986. Sexual selection in bowerbirds. *Scientific American* 254 (June):92–100.

85 Borgia, G., and M. A. Gore. 1986. Feather stealing in the satin bowerbird (*Ptilonorhynchus violaceus*): male competition and the quality of display. *Animal Behaviour* 34:727–738.

86 Borgia, G., I. M. Kaatz, and R. Condit. 1987. Flower choice and bower decoration in the satin bowerbird (*Ptilonorhynchus violaceus*): a test of hypotheses for the evolution of male display. *Animal Behaviour* 35:1129–1139.

87 Bouchard, T. J., Jr., and M. McGue. 1981. Familial studies of intelligence: a review. *Science* 212:1055–1059.

88 Bowman, R. I. 1961. Morphological differentiation and adaptation in the Galapagos finches. *Occasional Papers of the California Academy of Sciences* 58:1–302.

89 Brachmachary, R. L., and J. Dutta. 1981. On the pheromones of tigers: experiments and theory. *American Naturalist* 118:561–567.

90 Bradbury, J. W. 1977. Lek mating behavior in the hammer-headed bat. *Zeitschrift für Tierpsychologie* 45:225–255.

91 Bradbury, J. W. 1982. The evolution of leks. In *Natural Selection and Social Behavior*. R. D. Alexander and D. W. Tinkle (eds.). Chiron Press, New York.

92 Brady, R. 0. 1976. Inherited metabolic diseases of the nervous system. *Science* 193:733–739.

93 Breed, M. D., J. H. Fewell, A. J. Moore, and K. R. Williams. 1987. Graded recruitment in a ponerine ant. *Behavioral Ecology and Sociobiology* 20:407–411.

94 Bremmer, F. 1986. White whales on holiday. *Natural History* 95 (Jan.):40–49.

95 Brenowitz, E. A., and A. P. Arnold. 1986. Interspecific comparisons of the size of neural song control regions and song complexity in duetting birds: evolutionary implications. *Journal of Neuroscience* 6:2875–2879.

96 Breven, K. A. 1981. Mate choice in the wood frog, *Rana sylvatica*. *Evolution* 35:707–722.

97 Brockway, B. F. 1964. Social influences on reproductive physiology and ethology of budgerigars (*Melopsittacus undulatus*). *Animal Behaviour* 12:493–501.

98 Brockway, B. F. 1965. Stimulation of ovarian development and egglaying by male courtship vocalizations in budgerigars (*Melopsittacus undulatus*). *Animal Behaviour* 13:575–578.

99 Brodie, E. D., Jr. 1977. Hedgehogs use toad venom in their own defense. *Nature* 268:627–628.

100 Brodie, E. D., Jr., J. L. Hensel, and J. A. Johnson. 1974. Toxicity of urodele amphibians *Taricha*, *Notophthalmus*, *Cynops*, and *Paramesotriton* (Salamandridae). *Copeia* 1974:506–511.

101 Brower, J. V. 1958. Experimental studies of mimicry in some North American butterflies. 1. The monarch, *Danaus plexippus*, and viceroy, *Limenitis archippus*. *Evolution* 12:3–47.

102 Brower, J. V., and L. P. Brower. 1962. Experimental studies of mimicry. 6. The reaction of toads (*Bufo terrestris*) to honeybees (*Apis mellifera*) and their dronefly mimics (*Eristalis vinetorum*). *American Naturalist* 96:297–307.

103 Brower, L. P. 1969. Ecological chemistry. *Scientific American* 220 (Feb.):22–29.

104 Brower, L. P. 1985. Foraging dynamics of bird predators on overwintering monarch butterflies in Mexico. *Evolution* 39:852–868.

105 Brower, L. P., and W. H. Calvert. 1984. Chemical defence in butterflies. In *The Biology of Butterflies*. R. I. Vane-Wright and P. R. Ackery (eds.). Academic Press, London.

106 Brower, L. P., W. N. Ryerson, J. L. Coppinger, and S. C. Glazier. 1968. Ecological chemistry and the palatability spectrum. *Science* 161:1349–1351.

107 Brown, C. H., and P. M. Waser. 1984. Hearing and communication in blue monkeys (*Cercopithecus mitis*). *Animal Behaviour* 32:66–75.

108 Brown, C. R., and M. B. Brown. 1986. Ectoparasitism as a cost of coloniality in cliff swallows (*Hirundo pyrrhonota*). *Ecology* 67:1206–1218.

109 Brown, J. H., and G. A. Lieberman. 1973. Resource utilization and coexistence of seed eating desert rodents in sand dune habitats. *Ecology* 54:788–797.

110 Brown, J. L. 1975. *The Evolution of Behavior.* Norton, New York.

111 Brown, J. L. 1982. The adaptationist program. *Science* 217:884–886.

112 Brown, J. L. 1987. *Helping and Communal Breeding in Birds: Ecology and Evolution.* Princeton University Press, Princeton, NJ.

113 Brown, J. L., and G. H. Orians. 1970. Spacing patterns in mobile animals. *Annual Review of Ecology and Systematics* 1:239–262.

114 Brownmiller, S. 1975. *Against Our Will. Men, Women, and Rape.* Simon & Schuster, New York.

115 Buchsbaum, M. S., R. D. Coursey, and D. L. Murphy. 1976. The biochemical high-risk paradigm: behavioral and familial correlates of low platelet monoamine oxidase activity. *Science* 194:339–341.

116 Burger, J. 1980. The transition to independence and postfledging parental care in seabirds. In *Behavior of Marine Animals.* J. Burger, B. L. Olla, and H. E. Winn (eds.). Plenum, New York.

117 Burkhardt, D., and I. de la Motte. 1983. How stalk-eyed flies eye stalk-eyed flies: observations and measurements of the eyes of *Cyrtodiopsis whitei* (Diopsidae, Diptera). *Journal of Comparative Physiology* A 151:407–421.

118 Burley, N. 1979. The evolution of concealed ovulation. *American Naturalist* 114:835–858.

119 Burtt, E. H., Jr. 1984. Colour of the upper mandible: an adaptation to reduce reflectance. *Animal Behaviour* 32:652–658.

120 Bush, G. L., R. W. Neck, and G. B. Kitto. 1976. Screwworm eradication: inadvertent selection for noncompetitive ecotypes during mass rearing. *Science* 193:491–493.

121 Buss, D. M. 1987. Sex differences in human mate selection criteria: an evolutionary perspective. In *Sociobiology and Psychology: Ideas, Issues, and Applications.* C. Crawford, M. Smith, and D. Krebs (eds.). Lawrence Erlbaum Associates, Hillsdale, NJ.

122 Bygott, J. D., B. C. R. Bertram, and J. P. Hanby. 1979. Male lions in large coalitions gain reproductive advantage. *Nature* 282:839–841.

123 Cade, W. 1980. Alternative male reproductive strategies. *Florida Entomologist* 63:30–45.

124 Cade, W. 1981. Alternative male strategies: genetic differences in crickets. *Science* 212:563–564.

125 Calvert, W. H., and L. P. Brower. 1986. The location of monarch butterfly (*Danaus plexippus* L.) overwintering colonies in Mexico in relation to topography and climate. *Journal of the Lepidopterists' Society* 40:164–187.

126 Calvert, W. H., L. E. Hedrick, and L. P. Brower. 1979. Mortality of the monarch butterfly (*Danaus plexippus* L.): avian predation at five overwintering sites in Mexico. *Science* 204:847–851.

127 Camhi, J. M. 1984. *Neuroethology.* Sinauer Associates, Sunderland, MA.

128 Campbell, B. 1966. *Human Evolution.* Aldine, Chicago.

129 Caple, G., R. P. Balda, and W. R. Willis. 1983. The physics of leaping animals and the evolution of preflight. *American Naturalist* 121:455–476.

130 Caraco, T. 1979. Time budgeting and group size: a test of theory. *Ecology* 60:618–627.

131 Caraco, T., and L. L. Wolf. 1975. Ecological determinants of group sizes of foraging lions. *American Naturalist* 109:343–352.

132 Carew, T. J., and C. L. Sahley. 1986. Invertebrate learning and memory: from behavior to molecules. *Annual Review of Neuroscience* 9:435–487.

133 Carey, M., and V. Nolan, Jr. 1975. Polygyny in indigo buntings: a hypothesis tested. *Science* 190:1296–1297.

134 Carey, M., and V. Nolan, Jr. 1979. Population dynamics of indigo buntings and the evolution of avian polygyny. *Evolution* 33:1180–1192.

135 Caro, T. M. 1986. The functions of stotting: a review of the hypotheses. *Animal Behaviour* 34:649–662.

136 Caro, T. M. 1986. The functions of stotting in Thomson's gazelles: some tests of the predictions. *Animal Behaviour* 34:663–684.

137 Caro, T. M., and M. Borgerhoff Mulder. 1987. The problem of adaptation in the study of human behavior. *Ethology and Sociobiology* 8:61–72.

138 Carpenter, F. L., D. C. Paton, and M. H. Hixon. 1983. Weight gain and adjustment of feeding territory size in migrant hummingbirds. *Proceedings of the National Academy of Sciences* 80:7259–7263.

139 Carr, A. 1967. *So Excellent a Fish.* Anchor Books, Garden City, New York.

140 Carr, A. 1967. Adaptive aspects of the scheduled travel of Chelonia. In *Animal Orientation and Navigation.* R. M. Storm (ed.). Oregon State University Press, Corvallis.

141 Carter-Saltzman, L. 1980. Biological and sociocultural effects on handedness: comparison between biological and adoptive families. *Science* 209:1263–1265.

142 Catchpole, C. K. 1980. Sexual selection and the evolution of complex songs among European warblers of the genus *Acrocephalus*. *Behaviour* 74:149–166.

143 Chen, J.-S., and A. Amsel. 1980. Recall (versus recognition) of taste and immunization against aversive taste anticipations based on illness. *Science* 209:831–833.

144 Clarke, B. 1976. The ecological genetics of host–parasite relationships. In *Genetic Aspects of Host–Parasite Relationships*, Symposia of the British Society for Parasitology 14. A. E. R. Taylor and R. Muller (eds.). Blackwell, Oxford.

145 Clutton-Brock, T. H., and S. D. Albon. 1979. The roaring of red deer and the evolution of honest advertisement. *Behaviour* 69:145–170.

146 Clutton-Brock, T. H., S. D. Albon, R. M. Gibson, and F. E. Guinness. 1979. The logical stag: adaptive aspects of fighting in red deer. *Animal Behaviour* 27:211–225.

147 Clutton-Brock, T. H., and P. H. Harvey. 1984. Comparative approaches to investigating adaptation. In *Behavioural Ecology: An Evolutionary Approach*. J. R. Krebs and N. B. Davies (eds.). Blackwell, Oxford.

148 Conner, W. E., T. Eisner, R. K. vander Meer, A. Guerrero, D. Ghiringelli, and J. Meinwald. 1980. Sex attractant of an arctiid moth (*Utetheisa ornatrix*): a pulsed chemical signal. *Behavioral Ecology and Sociobiology* 7:55–63.

149 Cooper, W. E., Jr., and L. J. Vitt. 1985. Bluetails and autotomy: enhancement of predation avoidance in juvenile skinks. *Zeitschrift für Tierpsychologie* 70:265–276.

150 Cosmides, L., and J. Tooby. 1987. From evolution to behavior: evolutionary psychology as the missing link. In *The Latest on the Best*. J. Dupré (ed.). MIT Press, Cambridge, MA.

151 Cowan, T. M., and R. W. Siegel. 1984. Mutational and pharmacological alterations of neuronal membrane function disrupt conditioning in *Drosophila*. *Journal of Neurogenetics* 1:333–344.

152 Cox, G. W. 1985. The evolution of avian migration systems between temperate and tropical regions of the New World. *American Naturalist* 126:451–474.

153 Craig, J. L., and M. E. Douglas. 1986. Resource distribution, aggressive asymmetries, and variable access to resources in the nectar feeding bellbird. *Behavioral Ecology and Sociobiology* 18:231–240.

154 Crespi, B. J. 1986. Size assessment and alternative fighting tactics in *Elaphrothrips tuberculatus* (Insecta: Thysanoptera). *Animal Behaviour* 34:1324–1335.

155 Crews, D. 1975. Psychobiology of reptilian reproduction. *Science* 189:1059–1065.

156 Crews, D. (ed.). 1987. *Psychobiology of Reproductive Behavior: An Evolutionary Perspective*. Prentice-Hall, Englewood Cliffs, NJ.

157 Crews, D., and N. Greenberg. 1981. Function and causation of social signals in lizards. *American Zoologist* 21:273–294.

158 Crews, D., and M. C. Moore. 1986. Evolution of mechanisms controlling mating behavior. *Science* 231:121–125.

159 Crews, D., and M. C. Moore. In press. Reproductive biology of parthenogenetic whiptail lizards. In *Biology of the* Cnemidophorus. J. Wright (ed.). Allen Press, Lawrence, KS.

160 Cronin, E. W., Jr., and P. W. Sherman. 1977. A resource-based mating system: the orange-rumped honey guide. *Living Bird* 15:5–32.

161 Cruz, Y. P. 1981. A sterile defender morph in a polymorphic hymenopterous parasite. *Nature* 294:446–447.

162 Cullen, E. 1957. Adaptations in the kittiwake to cliff nesting. *Ibis* 99:275–302.

163 Curio, E. 1976. *The Ethology of Predation*. Springer-Verlag, New York.

164 Daly, M., and M. Wilson. 1982. Homicide and kinship. *American Anthropologist* 84:372–378.

165 Daly, M., and M. Wilson. 1983. *Sex, Evolution, and Behavior,* Second Edition. Willard Grant Press, Boston.

166 Daly, M., and M. Wilson. 1985. Child abuse and other risks of not living with both parents. *Ethology and Sociobiology* 6:197–210.

167 Daly, M., and M. Wilson. 1987. Evolutionary psychology and family violence. In *Sociobiology and Psychology*. C. Crawford, M. Smith, and D. Krebs (eds.). Lawrence Erlbaum Associates, Hillsdale, NJ.

168 Daly, M., and M. Wilson. 1987. Children as homicide victims. In *Biosocial Perspectives on Child Abuse*. R. Gelles and J. Lancaster (eds.). Aldine, New York.

169 Daly, M., M. Wilson, and S. J. Weghorst. 1982. Male sexual jealousy. *Ethology and Sociobiology* 3:11–27.

170 Darwin, C. 1859. *On the Origin of Species*. Murray, London.

171 Darwin, C. 1871. *The Descent of Man and Selection in Relation to Sex*. Murray, London.

172 Darwin, C. 1872. *The Expression of Emotions in Man and Animals*. Murray, London.

173 Davies, N. B. 1977. Prey selection and social behaviour in wagtails (Aves: Motacillidae). *Journal of Animal Ecology* 46:37–57.

174 Davies, N. B. 1983. Polyandry, cloaca-pecking and sperm competition in dunnocks. *Nature* 302:334–336.

175 Davies, N. B. 1985. Cooperation and conflict among dunnocks, *Prunella modularis*, in a variable mating system. *Animal Behaviour* 33:628–648.

176 Davies, N. B., and T. R. Halliday. 1978. Deep croaks and fighting assessment in toads *Bufo bufo*. *Nature* 275:683–685.

177 Davies, N. B., and A. I. Houston. 1981. Owners and satellites: the economics of territory defense in the pied wagtail, *Motacilla alba*. *Journal of Animal Ecology* 50:157–180.

178 Davies, N. B., and A. Lundberg. 1984. Food distribution and a variable mating system in the dunnock, *Prunella modularis*. *Journal of Animal Ecology* 53:895–912.

179 Davis, W. J., G. J. Mpitsos, and J. M. Pinneo. 1974. The behavioral hierarchy of the mollusk *Pleurobranchaea*. II. Hormonal suppression of feeding associated with egg-laying. *Journal of Comparative Physiology* 90:225–243.

180 Davison, J. 1976. *Hydra hymanae*: regulation of the life cycle by time and temperature. *Science* 194:618–620.

181 Dawkins, M. S. 1986. *Unravelling Animal Behaviour*. Longman, Harlow, Essex.

182 Dawkins, R. 1977. *The Selfish Gene*. Oxford University Press, New York.

183 Dawkins, R. 1980. Good strategy or evolutionarily stable strategy? In *Sociobiology: Beyond Nature/Nurture?* G. W. Barlow and J. Silverberg (eds.). Westview Press, Boulder, CO.

184 Dawkins, R. 1982. *The Extended Phenotype*. W. H. Freeman, San Francisco.

185 Dawkins, R. 1986. *The Blind Watchmaker*. W. W. Norton, New York.

186 Dawkins, R., and J. Krebs. 1978. Animal signals: information or manipulation? In *Behavioural Ecology: An Evolutionary Approach*. J. R. Krebs and N. B. Davies (eds.). Blackwell, Oxford.

187 de la Motte, I., and D. Burkhardt. 1983. Portrait of an Asian stalk-eyed fly. *Naturwissenschaften* 70:451–461.

188 den Boer, P. J. 1986. The present status of the competitive exclusion principle. *Trends in Ecology and Evolution* 1:25–28.

189 Denton, E. J., P. J. Herring, E. A. Widder, M. F. Latz, and J. F. Case. 1985. The role of filters in the photophores of oceanic animals and their relation to vision in the oceanic environment. *Proceedings of the Royal Society of London* B 225:63–97.

190 Dethier, V. G. 1962. *To Know a Fly*. Holden-Day, San Francisco.

191 Dethier, V. G. 1976. *The Hungry Fly: A Physiological Study of the Behavior Associated with Feeding*. Harvard University Press, Cambridge, MA.

192 Dethier, V. G., and D. Bodenstein. 1958. Hunger in the blowfly. *Zeitschrift für Tierpsychologie* 15:129–140.

193 Dewsbury, D. A. 1982. Ejaculate cost and male choice. *American Naturalist* 119:601–610.

194 Dhondt, A. A., and J. Schillemans. 1983. Reproductive success of the great tit in relation to its territorial status. *Animal Behaviour* 31:902–912.

195 Dial, B. E., and L. C. Fitzpatrick. 1983. Lizard tail autotomy: function and energetics of postautotomy tail movement in *Scincella lateralis*. *Science* 219:391–393.

196 Diamond, J. M. 1983. The biology of the wheel. *Nature* 302:572–573.

197 Diamond, J. M., E. Cooper, C. Turner, and L. Macintyre. 1976. Trophic regulation of nerve sprouting. *Science* 193:371–377.

198 Dilger, W. C. 1962. The behavior of lovebirds. *Scientific American* 206 (Jan.):88–98.

199 Dill, L. M., and A. H. G. Fraser. 1984. Risk of predation and the feeding behavior of juvenile coho salmon (*Oncorhynchus kisutch*). *Behavioral Ecology and Sociobiology* 16:65–71.

200 Dobson, F. S. 1985. The use of phylogeny in behavior and ecology. *Evolution* 39:1384–1388.

201 Doherty, J. A., and H. C. Gerhardt. 1983. Hybrid tree frogs: vocalizations of males and selective phonotaxis of females. *Science* 220:1078–1080.

202 Dominey, W. J. 1980. Female mimicry in male bluegill sunfish—a genetic polymorphism? *Nature* 284:546–548.

203 Dominey, W. J. 1984. Alternative mating tactics and evolutionarily stable strategies. *American Zoologist* 24:385–396.

204 Domjan, M., and B. Burkhard. 1986. *The Principles of Learning and Behavior*, Second Edition. Brooks/Cole Publishing, Monterey, CA.

205 Dow, H., and S. Fredga. 1983. Breeding and natal dispersal of the goldeneye, *Bucephala clangula*. *Journal of Animal Ecology* 52:681–695.

206 Downes, J. A. 1970. The feeding and mating behaviour of the specialized Empididinae (Diptera): observations on four species of *Rhamphomyia* in the arctic and a general discussion. *Canadian Entomologist* 102:769–791.

207 Downes, J. A. 1973. Lepidoptera feeding at puddle-margins, dung, and carrion. *Journal of the Lepidopterists' Society* 27:89–99.

208 Downhower, J. F., and K. B. Armitage. 1971. The yellow-bellied marmot and the evolution of polygamy. *American Naturalist* 105:355–370.

209 Duffey, S. S. 1970. Cardiac glycosides and distastefulness: some observations on the palatability spectrum of butterflies. *Science* 169:78–79.

210 Dumont, J. P. C., and R. M. Robertson. 1986. Neuronal circuits: an evolutionary perspective. *Science* 233:849–853.

211 Dunn, J. 1976. How far do early differences in mother–child relations affect later development? In *Growing Points in Ethology*. P. P. G. Bateson and R. A. Hinde (eds.). Cambridge University Press, Cambridge.

212 Durant, J. 1986. From amateur naturalist to professional scientist. *New Scientist* 111 (July)24:41–44.

213 Durham, W. H. 1976. Resource competition and human aggression. *Quarterly Review of Biology* 51:385–415.

214 Duvall, D., M. B. King, and K. J. Gutzwiller. 1985. Behavioral ecology and ethology of the prairie rattlesnake. *National Geographic Research* 1:80–111.

215 Dyer, F. C., and J. L. Gould. 1981. Honey bee orientation: a backup system for cloudy days. *Science* 214:1041–1042.

216 Dyer, F. C., and J. L. Gould. 1983. Honey bee navigation. *American Scientist* 71:587–597.

217 Earle, M. 1987. A flexible body mass in social carnivores. *American Naturalist* 129:755–760.

218 Eberhard, W. G. 1980. The natural history and behavior of the bolas spider *Mastophora dizzydeani* sp. n. (Araneidae). *Psyche* 87:143–169.

219 Eberhard, W. G. 1985. *Sexual Selection and Animal Genitalia*. Harvard University Press, Cambridge, MA.

220 Eckert, E. D., T. J. Bouchard, J. Bohlen, and L. L. Heston. 1986. Homosexuality in monozygotic twins reared apart. *British Journal of Psychiatry* 148:421–425.

221 Eckert, R., and Y. Naitoh. 1972. Bioelectric control of locomotion in the ciliates. *Journal of Protozoology* 19:237–243.

222 Edmunds, M. 1974. *Defence in Animals*. Longman, Harlow, Essex.

223 Edwards, J. S. 1966. Observations on the life history and predatory behaviour of *Zelus exsanguis* (Stål) (Heteroptera; Reduviidae). *Proceedings of the Royal Entomological Society of London* A 41:21–24.

224 Ehrman, L., and P. A. Parsons. 1976. *The Genetics of Behavior*. Sinauer Associates, Sunderland, MA.

225 Eibl-Eibesfeldt, I. 1975. *Ethology: The Biology of Behavior*, Second Edition. Holt, Rinehart & Winston, New York.

226 Eimas, P. D. 1975. Speech perception in early infancy. In *Infant Perceptions from Sensation to Cognition*. L. B. Cohen and P. Salapatek (eds.). Academic Press, New York.

227 Eisner, T. E. 1966. Beetle spray discourages predators. *Natural History* 75 (Feb.):42–47.

228 Eisner, T. E. 1970. Chemical defense against predation in arthropods. In *Chemical Ecology*. E. Sondheimer and J. B. Simeone (eds.). Academic Press, New York.

229 Eisner, T. E., D. F. Weimer, L. W. Haynes, and J. Meinwald. 1978. Lucibufagins: defense steroids from the fireflies *Photinus ignitus* and *Photinus arginellus* (Coleoptera, Lampyridae). *Proceedings of the National Academy of Sciences* 75:905–908.

230 Ekman, J. 1987. Exposure and time use in willow tit flocks: the cost of subordination. *Animal Behaviour* 35:445–452.

231 Elliott, P. F. 1975. Longevity and the evolution of polygamy. *American Naturalist* 109:281–287.

232 Emlen, J. T., and R. L. Penney. 1966. The navigation of penguins. *Scientific American* 218 (Oct.):104–113.

233 Emlen, S. T. 1975. Migration: orientation and navigation. In *Avian Biology*. D. S. Farner and J. R. King (eds.). Academic Press, New York.

234 Emlen, S. T. 1975. The stellar-orientation system of a migratory bird. *Scientific American* 223 (Aug.):102–111.

235 Emlen, S. T. 1978. Cooperative breeding. In *Behavioural Ecology: An Evolutionary Approach*. J. R. Krebs and N. B. Davies (eds.). Blackwell, Oxford.

236 Emlen, S. T. 1980. Ecological determinism and sociobiology. In *Sociobiology: Beyond Nature/Nurture?* G. W. Barlow and J. Silverberg (eds.). Westview Press, Boulder, CO.

237 Emlen, S. T. 1982. The evolution of helping. I. An ecological constraints model. *American Naturalist* 119:29–39.

238 Emlen, S. T. 1982. The evolution of helping.

II. The role of behavioral conflict. *American Naturalist* 119:40–53.

239 Emlen, S. T., and L. W. Oring. 1977. Ecology, sexual selection and the evolution of mating systems. *Science* 197:215–223.

240 Emlen, S. T., and P. H. Wrege. 1986. Forced copulations and intra-specific parasitism: two costs of social living in the white-fronted bee-eater. *Ethology* 71:2–29.

241 Emlen, S. T., W. Wiltschko, N. J. Demong, R. Wiltschko, and S. Berian. 1976. Magnetic direction finding: evidence for its use in migratory indigo buntings. *Science* 193:505–508.

242 Endler, J. A. 1986. *Natural Selection in the Wild*. Princeton University Press, Princeton, NJ.

243 Epstein, R., C. E. Kirshnit, R. P. Lanza, and L. C. Rubin. 1984. "Insight" in the pigeon: antecedents and determinants of an intelligent performance. *Nature* 308:61–62.

244 Epstein, R., R. P. Lanza, and B. F. Skinner. 1980. Symbolic communication between two pigeons (*Columba livia domestica*). *Science* 207:543–545.

245 Esch, H. 1967. The evolution of bee language. *Scientific American* 216 (Apr.):96–104.

246 Evans, H. E. 1966. *Life on a Little Known Planet*. Dell, New York.

247 Evans, H. E. 1973. *Wasp Farm*. Anchor Press, Garden City, NY.

248 Evans, H. E., and M. J. W. Eberhard. 1970. *The Wasps*. University of Michigan Press, Ann Arbor.

249 Ewald, P. W. 1985. Influence of asymmetries in resource quality and age on aggression and dominance in black-chinned hummingbirds. *Animal Behaviour* 33:705–719.

250 Ewald, P. W., and R. J. Bransfield. 1987. Territory quality and territorial behavior in two sympatric species of hummingbirds. *Behavioral Ecology and Sociobiology* 20:285–293.

251 Ewer, R. F. 1973. *The Carnivores*. Cornell University Press, Ithaca, NY.

252 Ewert, J.-P. 1974. The neural basis of visually guided behavior. *Scientific American* 230 (Mar.):34–42.

253 Ewert, J.-P. 1980. *Neuro-Ethology*. Springer-Verlag, New York.

254 Falk, D. 1984. The petrified brain. *Natural History* 93:36–39.

255 Farentinos, R. C., P. J. Capretta, R. E. Kepner, and V. M. Littlefield. 1981. Selective herbivory in tassel-eared squirrels: role of monoterpenes in ponderosa pines chosen as feeding trees. *Science* 213:1273–1275.

256 Farner, D. S. 1964. Time measurement in vertebrate photoperiodism. *American Naturalist* 95:375–386.

257 Farner, D. S., and R. A. Lewis. 1971. Photoperiodism and reproductive cycles in birds. *Photophysiology* 6:325–370.

258 Fausto-Sterling, A. 1985. *Myths of Gender*. Basic Books, New York.

259 Feduccia, A., and H. B. Tordoff. 1970. Feathers of *Archeopteryx*: asymmetric vanes indicate aerodynamic function. *Science* 203:1021–1022.

260 Fink, L. S., and L. P. Brower. 1981. Birds can overcome the cardenolide defence of monarch butterflies in Mexico. *Nature* 291:67–70.

261 Fisher, R. A. 1930. *The Genetical Theory of Natural Selection*. Clarendon Press, Oxford.

262 Fjeldså, J. 1983. Ecological character displacement and character release in grebes Podicepididae. *Ibis* 125:463–481.

263 Flinn, M. V. In press. Step-parent/step-offspring interactions in a Caribbean village. *Ethology and Sociobiology*.

264 Ford, E. B. 1955. *Moths*. Collins, London.

265 Forsyth, A. 1986. *The Natural History of Sex*. Scribners, New York.

266 Foster, M. S. 1977. Odd couples in manakins: a study of social organization and cooperative breeding in *Chiroxiphia linearis*. *American Naturalist* 111:845–853.

267 Franklin, W. L. 1974. The social behaviour of the vicuña. In *The Behaviour of Ungulates and Its Relation to Management*. V. Giest and F. Walther (eds.). IUCN Publications, Morges, Switzerland.

268 Freedman, D. 1983. *Margaret Mead and Samoa*. Harvard University Press, Cambridge, MA.

269 Gadgil, M. 1972. Male dimorphism as a consequence of sexual selection. *American Naturalist* 106:574–580.

270 Galef, B. G., Jr., and H. C. Kaner. 1980. Establishment and maintenance of preferences for maternal and artificial olfactory stimuli in juvenile rats. *Journal of Comparative and Physiological Psychology* 94:588–595.

271 Gamboa, G. J. 1978. Intraspecific defense: advantage of social cooperation among paper wasp foundresses. *Science* 199:1463–1465.

272 Ganetzky, B., and C.-F. Wu. 1982. Indirect suppression involving behavioral mutant with altered nerve excitability in *Drosophila melanogaster*. *Genetics* 100:597–614.

273 Garcia, J., and F. R. Ervin. 1968. Gustatory-visceral and telereceptor-cutaneous conditioning: adaptation in internal and external milieus. *Communications in Behavioral Biology* (A) 1:389–415.

274 Garcia, J., W. G. Hankins, and K. W. Rusiniak. 1974. Behavioral regulation of the milieu interne in man and rat. *Science* 185:824–831.

275 Gaulin, S. J. C., and R. W. FitzGerald. 1986. Sex differences in spatial ability: an evolutionary hypothesis and test. *American Naturalist* 127:74–88.

276 Gaulin, S. J. C., and R. W. FitzGerald. In press. Darwinian psychology: sexual selection for spatial ability. *Animal Behaviour.*

277 Geist, V. 1986. The paradox of the great Irish stag. *Natural History* 95 (Mar.):54–65.

278 Ghiselin, M. T. 1974. *The Economy of Nature and the Evolution of Sex*. University of California Press, Berkeley.

279 Ghiselin, M. T. 1985. A moveable feaster. *Natural History* 94 (Sept.):54–60.

280 Ghysen, A. 1978. Sensory neurones recognize defined pathways in *Drosophila* central nervous system. *Nature* 274:869–872.

281 Gibbons, J. A., and H. B. Lillywhite. 1981. Ecological segregation, color matching, and speciation in lizards of the *Amphibolurus decresii* species complex (Lacertilia: Agamidae). *Ecology* 62:1573–1584.

282 Gibson, R. M., and J. W. Bradbury. 1985. Sexual selection in lekking sage grouse: phenotypic correlates of male mating success. *Behavioral Ecology and Sociobiology* 18:117–123.

283 Gibson, R. M., and J. W. Bradbury. 1986. Male and female mating strategies on sage grouse leks. In *Ecological Aspects of Social Evolution: Birds and Mammals*. D. I. Rubenstein and R. W. Wrangham (eds.). Princeton University Press, Princeton, NJ.

284 Gilbert, L. E. 1976. Postmating female odor in *Heliconius* butterflies: a male-contributed antiaphrodisiac? *Science* 193:419–420.

285 Gill, F. B., and L. L. Wolf. 1975. Foraging strategies and energetics of East African sunbirds at mistletoe flowers. *American Naturalist* 109:491–510.

286 Gill, F. B., and L. L. Wolf. 1975. Economics of feeding territoriality in the golden-winged sunbird. *Ecology* 56:333–345.

287 Gill, F. B., and L. L. Wolf. 1978. Comparative foraging efficiencies of some montane sunbirds in Kenya. *Condor* 80:391–400.

288 Gittleman, J. L., and P. H. Harvey. 1980. Why are distasteful prey not cryptic? *Nature* 286:149–150.

289 Glander, K. E. 1981. Feeding patterns in mantled howling monkeys. In *Foraging Behavior: Ecological, Ethological, and Psychological Approaches*. A. C. Kamil and T. D. Sargent (eds.). Garland Press, New York.

290 Goodman, C. S., and N. C. Spitzer. 1979. Embryonic development of identified neurones: differentiation from neuroblast to neurons. *Nature* 280:208–214.

291 Gotmark, F., D. W. Winkler, and M. Andersson. 1986. Flock-feeding on fish schools increases individual success in gulls. *Nature* 319:589–591.

292 Gould, J. L. 1975. Honey bee recruitment. *Science* 189:685–693.

293 Gould, J. L. 1980. The case for magnetic field sensitivity in birds and bees. *American Scientist* 68:256–267.

294 Gould, J. L. 1982. *Ethology: The Mechanisms and Evolution of Behavior*. Norton, New York.

295 Gould, J. L. 1982. Why do honeybees have dialects? *Behavioral Ecology and Sociobiology* 10:53–56.

296 Gould, J. L., and W. F. Towne. 1987. Evolution of the dance language. *American Naturalist* 130:317–338.

297 Gould, S. J. 1974. This view of life: the non-science of human nature. *Natural History* 83 (Apr.):21–25.

298 Gould, S. J. 1980. Vision with a vengeance. *Natural History* 89 (Sept.):16–20.

299 Gould, S. J. 1981. Hyena myths and realities. *Natural History* 90 (Feb.):16–24.

300 Gould, S. J. 1981. Kingdoms without wheels. *Natural History* 90 (Mar.):42–48.

301 Gould, S. J. 1984. Only his wings remained. *Natural History* 93 (Sept.):10–18.

302 Gould, S. J. 1986. Evolution and the triumph of homology, or why history matters. *American Scientist* 74:60–69.

303 Gould, S. J. 1986. Review of *Vaulting Ambition*. *New York Review of Books* 33 (Sept. 25):47ff.

304 Gould, S. J., and R. C. Lewontin. 1981. The spandrels of San Marco and the Panglossian paradigm: a critique of the adaptationist programme. *Proceedings of the Royal Society of London* B 205:581–598.

305 Grant, P. R. 1986. *Ecology and Evolution of Darwin's Finches*. Princeton University Press, Princeton, NJ.

306 Graves, J. A., and A. Whiten. 1980. Adoption of strange chicks by herring gulls, *Larus argentatus*. *Zeitschrift für Tierpsychologie* 54:267–278.

307 Greene, H. W., and J. A. Campbell. 1972. Note on the use of caudal lures by arboreal green pit vipers. *Herpetologica* 28:32–34.

308 Greene, J. C. 1981. *Science, Ideology, and World View*. University of California Press, Berkeley.

309 Greenwood, P. J. 1980. Mating systems, philopatry, and dispersal in birds and mammals. *Animal Behaviour* 28:1140–1162.

310 Griffin, D. R. 1958. *Listening in the Dark.* Yale University Press, New Haven, CT.

311 Griffin, D. R., F. A. Webster, and C. R. Michael. 1960. The echolocation of flying insects by bats. *Animal Behaviour* 8:141–154.

312 Griffiths, M. 1978. *The Biology of Monotremes.* Academic Press, New York.

313 Grillner, S., and P. Wallén. 1985. Central pattern generators for locomotion, with special reference to vertebrates. *Annual Review of Neuroscience* 8:233–261.

314 Gross, M. R. 1982. Sneakers, satellites, and parentals: polymorphic mating strategies in North American sunfishes. *Zeitschrift für Tierpsychologie* 60:1–26.

315 Gross, M. R. 1987. Evolution of diadromy in fishes. *American Fisheries Society Symposium* 1:14-25.

316 Gross, M. R., and E. L. Charnov. 1980. Alternative life histories in bluegill sunfish. *Proceedings of the National Academy of Sciences* 77:6937–6940.

317 Gross, M. R., and A. M. MacMillan. 1981. Predation and the evolution of colonial nesting in bluegill sunfish (*Lepomis macrochirus*). *Behavioral Ecology and Sociobiology* 8:163–174.

318 Gross, M. R., and D. P. Philipp. In press. Inheritance of alternative male reproductive patterns in fish.

319 Gross, M. R., and R. C. Sargent. 1985. The evolution of male and female parental care in fishes. *American Zoologist* 25:807–822.

320 Gross, M. R., and R. Shine. 1981. Parental care and mode of fertilization in ectothermic vertebrates. *Evolution* 35:775–793.

321 Gurney, M. E., and M. Konishi. 1980. Hormone-induced sexual differentiation of brain and behavior in zebra finches. *Science* 208:1380–1383.

322 Gwinner, E., and W. Wiltschko. 1980. Circannual changes in migratory orientation of the garden warbler, *Sylvia borin. Behavioral Ecology and Sociobiology* 7:73–78.

323 Gwynne, D. T. 1981. Sexual difference theory: Mormon crickets show role reversal in mate choice. *Science* 213:779–780.

324 Gwynne, D. T. 1984. Courtship feeding increases female reproductive success in bushcrickets. *Nature* 307:361–363.

325 Hailman, J. P. 1967. The ontogeny of an instinct. *Behaviour Supplements* 15:1–159.

326 Hall, J. C. 1977. Portions of the central nervous system controlling reproductive behavior in *Drosophila melanogaster. Behavior Genetics* 7:291–312.

327 Hall, J. C. 1985. Genetic analysis of behavior in insects. In *Comprehensive Insect Physiology, Biochemistry, and Pharmacology.* G. A. Kerkut and L. I. Gilbert (eds.). Pergamon, Oxford.

328 Hall, J. C., R. J. Greenspan, and W. A. Harris. 1982. *Genetic Neurobiology.* MIT Press, Cambridge, MA.

329 Hall, K. R. L., and G. B. Schaller. 1964. Tool-using behavior of the California sea otter. *Journal of Mammalogy* 45:287–298.

330 Halliday, T. 1983. Do frogs and toads choose their mates? *Nature* 306:226–227.

331 Hamilton, W. D. 1964. The evolution of social behavior. *Journal of Theoretical Biology* 7:1–52.

332 Hamilton, W. D. 1971. Geometry for the selfish herd. *Journal of Theoretical Biology* 31:295–311.

333 Hamilton, W. D. 1975. Gamblers since life began: barnacles, aphids, elms. *Quarterly Review of Biology* 50:175–180.

334 Hamilton, W. D., and M. Zuk. 1982. Heritable true fitness and bright birds: a role for parasites. *Science* 218:384–387.

335 Hamilton, W. D., P. A. Henderson, and N. A. Moran. 1981. Fluctuation of environment and coevolved antagonist polymorphism as factors in the maintenance of sex. In *Natural Selection and Social Behavior.* R. D. Alexander and D. W. Tinkle (eds.). Chiron Press, New York.

336 Hamilton, W. J., and G. H. Orians. 1965. Evolution of brood parasitism in altricial birds. *Condor* 67:361–382.

337 Hamilton, W. J., R. L. Tilson, and L. G. Frank. 1986. Sexual monomorphism in spotted hyenas, *Crocuta crocuta. Ethology* 71:63-73.

338 Hamner, W. M. 1964. Circadian control of photoperiodism in the house finch demonstrated by interrupted-night experiments. *Nature* 203:1400–1401.

339 Hamner, W. M., P. P. Hamner, S. W. Strand, and R. W. Gilmer. 1983. Behavior of antarctic krill, *Euphausia superba*: chemoreception, feeding, schooling, and molting. *Science* 220:433–435.

340 Hanby, J. P., and J. D. Bygott. 1987. Emigration of subadult lions. *Animal Behaviour* 35:161–169.

341 Hannon, S. J., R. L. Mumme, W. D. Koenig, S. Spon, and F. A. Pitelka. 1987. Poor acorn crop, dominance, and decline in numbers of acorn woodpeckers. *Journal of Animal Ecology* 56:197–207.

342 Harcourt, A. H., P. H. Harvey, S. G. Larson, and R. V. Short. 1981. Testis weight, body weight and breeding system in primates. *Nature* 293:55–57.

343 Harlow, H. F., and M. K. Harlow. 1962. Social deprivation in monkeys. *Scientific American* 207 (Nov.):136–146.

344 Harlow, H. F., M. K. Harlow, and S. J. Suomi. 1971. From thought to therapy: lessons from a primate laboratory. *American Scientist* 59:538–549.

345 Harris, J. H. 1984. An experimental analysis of desert rodent foraging ecology. *Ecology* 65:1579–1584.

346 Hartung, J. 1982. Polygyny and inheritance of wealth. *Current Anthropology* 23:1–12.

347 Harvey, P., and S. J. Arnold. 1982. Female mate choice and runaway sexual selection. *Nature* 297:533–534.

348 Haseltine, F. P., and S. Ohno. 1981. Mechanisms of gonadal differentiation. *Science* 211:1272–1277.

349 Hausfater, G. 1975. Dominance and reproduction in baboons (*Papio cynocephalus*): a quantitative analysis. *Contributions in Primatology* 7:1–150.

350 Hausfater, G., and S. B. Hrdy (eds.). 1984. *Infanticide, Comparative, and Evolutionary Perspectives*. Aldine, Chicago.

351 Hay, R. L., and M. D. Leakey. 1982. The fossil footprints of Laetoli. *Scientific American* 246 (Feb.):50–57.

352 Hedrick, A. V. 1986. Female preferences based on male calling in a field cricket. *Behavioral Ecology and Sociobiology* 19:73–77.

353 Heinrich, B. 1979. *Bumblebee Economics*. Harvard University Press, Cambridge, MA.

354 Heinrich, B. 1979. Foraging strategies of caterpillars: leaf damage and possible predator avoidance strategies. *Oecologia* 42:325–337.

355 Heinrich, B. 1985. Men vs. women, marathoners vs. ultramarathoners. *Ultrarunning* 1985 (Jan.):16–18.

356 Heinrich, B., and G. A. Bartholomew. 1979. The ecology of the African dung beetle. *Scientific American* 241 (Nov.):146–156.

357 Heinrich, B., and S. L. Collins. 1983. Caterpillar leaf damage, and the game of hide-and-seek with birds. *Ecology* 64:592–602.

358 Hennessy, D. F., and D. H. Owings. 1978. Snake species discrimination and the role of olfactory cues in the snake-directed behavior of the California ground squirrel. *Behaviour* 65:115–124.

359 Henry, C. S. 1972. Eggs and rapagula of *Ululodea* and *Ascaloptynx* (Neuroptera: Ascalaphidae): a comparative study. *Psyche* 79:1–22.

360 Henson, O. W., Jr. 1970. The ear and audition. In *Biology of Bats*. W. A. Wimsatt (ed.). Academic Press, New York.

361 Heston, L. L. 1970. The genetics of schizophrenic and schizoid disease. *Science* 167:248–255.

362 Heston, L. L., and J. Shields. 1968. Homosexuality in twins: a family study and a registry study. *Archives of General Psychiatry* 18:149–160.

363 Heth, G., E. Frankenberg, A. Raz, and E. Nevo. 1987. Vibrational communication in subterranean mole rats (*Spalax ehrenbergi*). *Behavioral Ecology and Sociobiology* 21:31–33.

364 Hilborn, A., and S. C. Stearns. 1982. On inference in ecology and evolutionary biology: the problem of multiple causes. *Acta Biotheoretica* 31:145–164.

365 Hirth, D. H., and D. R. McCullough. 1977. Evolution of alarm signals in ungulates with special reference to white-tailed deer. *American Naturalist* 111:31–42.

366 Hogg, J. T. 1984. Mating in bighorn sheep: multiple creative male strategies. *Science* 225:526–529.

367 Hogg, J. T. 1987. Intrasexual competition and mate choice in Rocky Mountain bighorn sheep. *Ethology* 75:119–144.

368 Höglund, J., and A. Lundberg. 1987. Sexual selection in a monomorphic lek-breeding bird: correlates of male mating success in the great snipe *Gallinago media*. *Behavioral Ecology and Sociobiology* 21:211–216.

369 Högstedt, G. 1983. Adaptation unto death: function of fear screams. *American Naturalist* 121:562–570.

370 Holden, C. 1980. Identical twins reared apart. *Science* 207:1323–1328.

371 Holekamp, K. E. 1984. Natal dispersal in Belding's ground squirrels (*Spermophilus beldingi*). *Behavioral Ecology and Sociobiology* 16:21–30.

372 Hölldobler, B. 1971. Communication between ants and their guests. *Scientific American* 224 (Mar.):86–95.

373 Hölldobler, B. 1974. Communication by tandem running in the ant *Camponotus sericeus*. *Journal of Comparative Physiology* 90:105–127.

374 Hölldobler, B. 1980. Canopy orientation: a new kind of orientation in ants. *Science* 210:86–88.

375 Holley, A. J. F. 1984. Adoption, parent–chick recognition, and maladaptation in the herring gull *Larus argentatus*. *Zeitschrift für Tierpsychologie* 64:9–14.

376 Holmes, W. G. 1984. Predation risk and for-

aging behavior of the hoary marmot in Alaska. *Behavioral Ecology and Sociobiology* 15:293–302.

377 Holmes, W. G. 1986. Identification of paternal half-siblings by captive Belding's ground squirrels. *Animal Behaviour* 34:321–327.

378 Holmes, W. G., and P. W. Sherman. 1982. The ontogeny of kin recognition in two species of ground squirrels. *American Zoologist* 22:491–517.

379 Holmes, W. G., and P. W. Sherman. 1983. Kin recognition in animals. *American Scientist* 71:46–55.

380 Hoogland, J. L. 1979. Aggression, ectoparasitism and other possible costs of prairie dog (Sciuridae, *Cynomys* spp.) coloniality. *Behaviour* 69:1–35.

381 Hoogland, J. L. 1981. The evolution of coloniality in white-tailed and black-tailed prairie dogs (Sciuridae: *Cynomys leucurus* and *C. ludovicianus*). *Ecology* 62:252–272.

382 Hoogland, J. L. 1982. Prairie dogs avoid extreme inbreeding. *Science* 215:1639–1641.

383 Hoogland, J. L., and P. W. Sherman. 1976. Advantages and disadvantages of bank swallow (*Riparia riparia*) coloniality. *Ecological Monographs* 46:33–58.

384 Hopkins, C. D. 1974. Electric communication in fish. *American Scientist* 62:426–437.

385 Hotta, Y., and S. Benzer. 1979. Courtship in *Drosophila* mosaics: sex-specific foci for sequential action patterns. *Proceedings of the National Academy of Science* 73:4154–4158.

386 Houston, A. I., R. H. McCleery, and N. B. Davies. 1985. Territory size, prey renewal, and feeding rates: interpretation of observations on the pied wagtail (*Motacilla alba*) by simulation. *Journal of Animal Ecology* 54:227–239.

387 Howard, J. J. 1987. Diet selection by the leafcutting ant *Atta cephalotes*—The role of nutrients, water, and second chemistry. *Ecology* 68:503–515.

388 Howard, R. D. 1978. The evolution of mating strategies in bullfrogs, *Rana catesbiana*. *Evolution* 32:850–871.

389 Howard, R. D. 1980. Mating behaviour and mating success in woodfrogs, *Rana sylvatica*. *Animal Behaviour* 28:705–716.

390 Howard, R. R., and E. D. Brodie, Jr. 1973. A Batesian mimetic complex in salamanders: response of avian predators. *Herpetologica* 29:33–41.

391 Howell, T. R. 1984. Buried treasure of the Egyptian plover. *Natural History* 93 (Sept.):60–67.

392 Howlett, R. J., and M. E. N. Majerus. 1987. The understanding of industrial melanism in the peppered moth (*Biston betularia*) (Lepidoptera: Geometridae). *Biological Journal of the Linnean Society* 30:31–44.

393 Hrdy, S. B. 1977. *The Langurs of Abu*. Harvard University Press, Cambridge, MA.

394 Hrdy, S. B. 1977. Infanticide as a primate reproductive strategy. *American Scientist* 65:40–49.

395 Hubbell, S. P., and L. K. Johnson. 1987. Environmental variance in lifetime mating success, mate choice, and sexual selection. *American Naturalist* 130:91–112.

396 Huxley, T. H. 1910. *Lectures and Lay Sermons*. E. P. Dutton, New York.

397 Immelmann, K. 1969. Song development in the zebra finch and other estrildid finches. In *Bird Vocalizations*. R. A. Hinde (ed.). Cambridge University Press, Cambridge.

398 Inglis, I. R., and J. Lasarus. 1981. Vigilance and flock size in brent geese: the edge effect. *Zeitschrift für Tierpsychologie* 57:193–200.

399 Irons, W. 1979. Cultural and biological success. In *Evolutionary Biology and Human Social Behavior*. N. Chagnon and W. Irons (eds.). Duxbury, North Scituate, MA.

400 Jackson, R. R. 1986. Cohabitation of male and juvenile females: a prevalent mating tactic of spiders. *Journal of Natural History* 20:1193–1210.

401 Jarman, M. V. 1979. Impala social behaviour. Territory, hierarchy, mating and use of space. *Fortschritte Verhaltensforschung* 21:1–92.

402 Jarvis, J. U. M. 1981. Eusociality in a mammal: cooperative breeding in naked mole-rat colonies. *Science* 212:571–573.

403 Jaycox, E. R., and S. G. Parise. 1980. Homesite selection by Italian honeybee swarms, *Apis mellifera ligustica* (Hymenoptera: Apidae). *Journal of the Kansas Entomological Society* 53:171–178.

404 Jaycox, E. R., and S. G. Parise. 1981. Homesite selection by swarms of black-bodied honeybees, *Apis mellifera caucasia* and *A. m. carnica*. *Journal of the Kansas Entomological Society* 54:697–703.

405 Jeanne, R. L. 1970. Chemical defense of brood by a social wasp. *Science* 168:1465–1466.

406 Jeanne, R. L., H. A. Downing, and D. C. Post. 1983. Morphology and function of sternal glands in polistine wasps (Hymenoptera: Vespidae). *Zoomorphology* 103:149–184.

407 Jen, P. H.-S., and N. Suga. 1976. Coordinated activities of middle-ear and laryngeal muscles in echolocating bats. *Science* 191:950–952.

408 Jenni, D. A. 1974. Evolution of polyandry in birds. *American Zoologist* 14:129–144.

409 Jenni, D. A., and G. Collier. 1972. Polyandry in the American jacana. *Auk* 89:743–765.

410 Jeppsson, B. 1986. Mating by pregnant water voles (*Arvicola terrestris*): a strategy to counter infanticide by males? *Behavioral Ecology and Sociobiology* 19:293–296.

411 Jerison, H. J. 1973. *Evolution of the Brain and Intelligence*. Academic Press, New York.

412 Johnson, C. H., and J. W. Hasting. 1986. The elusive mechanisms of the circadian clock. *American Scientist* 74:29–37.

413 Johnston, T. D. 1982. Selective costs and benefits in the evolution of learning. *Advances in the Study of Behavior* 12:65–106.

414 Jolly, A. 1972. *The Evolution of Primate Behavior*. Macmillan, New York.

415 Just, J. In press. Siphonoecetinae (Crustacea, Amphipoda, Corophiidae) 5: a survey of phylogeny, distribution, and biology. *Crustaceana*, Supplement 13 (Studies on Amphipoda, Proceedings of the VIth International Colloquium on Amphipod Crustaceans).

416 Kacelnik, A., and J. R. Krebs. 1983. The dawn chorus in the great tit (*Parus major*): proximate and ultimate causes. *Behaviour* 83:287–309.

417 Kagan, J., and R. E. Klein. 1973. Cross-cultural perspectives on early development. *American Psychologist* 28:947–961.

418 Kaplan, H., and K. Hill. 1985. Hunting ability and reproductive success among male Ache foragers: preliminary results. *Current Anthropology* 26:131–133.

419 Keeton, W. T. 1974. The orientational and navigational basis of homing in birds. *Advances in the Study of Behavior* 5:47–132.

420 Keeton, W. T. 1974. The mystery of pigeon homing. *Scientific American* 231 (Dec.):96–107.

421 Kennedy, C. E. J., J. A. Endler, S. L. Poynton, and H. McMinn. 1987. Parasite load predicts mate choice in guppies. *Behavioral Ecology and Sociobiology* 21:291–296.

422 Kenward, R. E. 1978. Hawks and doves: factors affecting success and selection in goshawk attacks on wild pigeons. *Journal of Animal Ecology* 47:449–460.

423 Kessel, E. L. 1955. Mating activities of balloon flies. *Systematic Zoology* 4:97–104.

424 Kettlewell, H. B. D. 1955. Selection experiments on industrial melanism in the Lepidoptera. *Heredity* 9:323–343.

425 Kiepenheuer, J. 1985. Can pigeons be fooled about the actual release site position by presenting them information from another site? *Behavioral Ecology and Sociobiology* 18:75–82.

426 King, A. P., and M. J. West. 1983. Female perception of cowbird song: a closed developmental program. *Developmental Psychobiology* 16:335–342.

427 King, A. P., and M. J. West. 1983. Epigenesis of cowbird song—a joint endeavour of males and females. *Nature* 305:704–706.

428 King, M. C., and A. C. Wilson. 1975. Evolution at two levels in humans and chimpanzees. *Science* 188:107–116.

429 Kinsey, A. C., W. B. Pomeroy, and C. E. Martin. 1948. *Sexual Behavior of the Human Male*. W. B. Saunders, New York.

430 Kirkpatrick, M. 1982. Sexual selection and the evolution of female choice. *Evolution* 36:1–12.

431 Kitchener, A. 1987. Fighting behaviour of the extinct Irish elk. *Modern Geology* 11:1–28.

432 Kitcher, P. 1985. *Vaulting Ambition*. MIT Press, Cambridge, MA.

433 Klump, G. M., E. Kretzschmar, and E. Curio. 1986. The hearing of an avian predator and its avian prey. *Behavioral Ecology and Sociobiology* 18:317–324.

434 Knudsen, E. I., S. du Lac, and S. D. Esterly. 1987. Computational maps in the brain. *Annual Review of Neuroscience* 10:41–65.

435 Kodric-Brown, A. 1985. Female preference and sexual selection for male coloration in the guppy (*Poecilia reticulata*). *Behavioral Ecology and Sociobiology* 17:199–205.

436 Kodric-Brown, A. 1986. Satellites and sneakers: opportunistic male breeding tactics in pupfish (*Cyprinodon pecoensis*). *Behavioral Ecology and Sociobiology* 19:425–432.

437 Kodric-Brown, A., and J. H. Brown. 1984. Truth in advertising: the kinds of traits favored by sexual selection. *American Naturalist* 124:309–323.

438 Koeniger, G. 1986. Mating sign and multiple mating in the honeybee. *Bee World* 67:141–150.

439 Köhler, W. 1925. *The Mentality of Apes*. Harcourt Brace, New York.

440 Kojima, J. 1983. Defense of the pre-emergence colony against ants by means of a chemical barrier in *Ropalidia fasciata* (Hymenoptera, Vespidae). *Japanese Journal of Ecology* 33:213–223.

441 Konishi, M. 1965. The role of auditory feedback in the control of vocalization in the white-crowned sparrow. *Zeitschrift für Tierpsychologie* 22:770–783.

442 Konishi, M. 1985. Birdsong: from behavior to neurons. *Annual Review of Neuroscience* 8:125–170.

443 Kovac, M. P., and W. J. Davis. 1977. Behavioral choice: neural mechanisms in *Pleurobranchaea*. *Science* 198:632–634.

444 Krebs, J. R. 1971. Territory and breeding density in the great tit, *Parus major* L. *Ecology* 52:2–22.

445 Krebs, J. R. 1977. The significance of song repertoires: the Beau Geste hypothesis. *Animal Behaviour* 25:475–478.

446 Krebs, J. R. 1985. Sociobiology ten years on. *New Scientist* 108 (3 Oct.):40–43.

447 Krebs, J. R., and M. I. Avery. 1984. Chick growth and prey quality in the European bee-eater. *Oecologia* 64:363–368.

448 Krebs, J. R., and N. B. Davies. 1981. *An Introduction to Behavioural Ecology*. Blackwell Scientific, Oxford.

449 Krebs, J. R., R. Ashcroft, and M. Webber. 1978. Song repertoires and territory defense in the great tit. *Nature* 271:539–542.

450 Kroodsma, D. E., and R. A. Canady. 1985. Differences in repertoire size, singing behavior, and associated neuroanatomy among marsh wren populations have a genetic basis. *Auk* 102:439–446.

451 Kroodsma, D. E., and R. Pickert. 1984. Repertoire size, auditory templates, and selective vocal learning in songbirds. *Animal Behaviour* 32:395–399.

452 Kruuk, H. 1964. Predators and anti-predator behaviour of the black-headed gull *Larus ridibundus*. *Behaviour Supplements* 11:1–129.

453 Kruuk, H. 1972. *The Spotted Hyena*. University of Chicago Press, Chicago.

454 Kruuk, H. 1976. Feeding and social behavior of the striped hyaena (*Hyaena vulgaris* Desmaret). *East African Wildlife Journal* 14:91–111.

455 Kung, C., S.-Y. Chang, Y. Satow, J. van Houten, and H. Hansma. 1975. Genetic dissection of behavior in *Paramecium*. *Science* 188:898–904.

456 Kyriacou, C. P., and J. C. Hall. 1984. Learning and memory mutations impair acoustic priming of mating behaviour in *Drosophila*. *Nature* 308:62–65.

457 LaBarbera, M. 1983. Why the wheels won't go. *American Naturalist* 121:395–408.

458 Labov, J. B. 1981. Pregnancy blocking in rodents: adaptive advantages for females. *American Naturalist* 118:361–371.

459 Labov, J. B., V. W. Huck, R. W. Elwood, and R. J. Brooks. 1985. Current problems in the study of infanticidal behavior of rodents. *Quarterly Review of Biology* 60:1–20.

460 Lacey, E. A., and P. W. Sherman. In press. Social organization of naked mole rat colonies: evidence for divisions of labor. In *The Biology of the Naked Mole-Rat*. P. W. Sherman, J. U. M. Jarvis, and R. D. Alexander (eds.).

461 Lack, D. 1947. *Darwin's Finches*. Cambridge University Press, New York.

462 Lack, D. 1966. *Population Studies of Birds*. Clarendon Press, Oxford.

463 Lack, D. 1968. *Ecological Adaptations for Breeding in Birds*. Methuen, London.

464 Laitman, J. T. 1984. The anatomy of human speech. *Natural History* 93:20–27.

465 Lande, R. 1981. Models of speciation by sexual selection of polygenic traits. *Proceedings of the National Academy of Sciences* 78:3721–3725.

466 Lank, D. B., L. W. Oring, and S. J. Maxson. 1985. Mate and nutrient limitation of egg-laying in a polyandrous shorebird. *Ecology* 66:1513–1524.

467 Lawrence, E. S. 1985. Evidence for search image in blackbirds *Turdus merula* L. : long-term learning. *Animal Behaviour* 33:1301–1309.

468 LeBoeuf, B. J. 1974. Male–male competition and reproductive success in elephant seals. *American Zoologist* 14:163–176.

469 Leffelaar, D., and R. J. Robertson. 1984. Do male tree swallows guard their mates? *Behavioral Ecology and Sociobiology* 16:73–80.

470 Lehrman, D. S. 1953. A critique of Konrad Lorenz's theory of instinctive behavior. *Quarterly Review of Biology* 28:337–363.

471 Lehrman, D. S. 1970. Semantic and conceptual issues in the nature–nurture problem. In *Development and Evolution of Behavior*. L. R. Aronson, E. Tobach, D. S. Lehrman, and J. S. Rosenblatt (eds.). W. H. Freeman, San Francisco.

472 Lendrem, D. W. 1985. Kinship affects puberty acceleration in mice (*Mus musculus*). *Behavioral Ecology and Sociobiology* 17:397–400.

473 Lenington, S. 1983. Social preferences for partners carrying "good genes" in wild house mice. *Animal Behaviour* 31:325–333.

474 Lenneberg, E. H. 1968. *The Biological Foundations of Language*. Wiley, New York.

475 Leopold, A. S. 1977. *The California Quail*. University of California Press, Berkeley.

476 Levick, M. G. 1914. *Antarctic Penguins: A Study of Their Social Habits*. William Heineman, London.

477 Levine, S. 1966. Sex differences in the brain. *Scientific American* 214 (Apr.):84–90.

478 Lewin, R. 1983. Were Lucy's feet made for walking? *Science* 220:700–702.

479 Licht, P. 1973. Influence of temperature and photoperiod on the annual ovarian cycle of the lizard *Anolis carolinensis*. *Copeia* 1973:465–472.

480 Lifjeld, J. T., and T. Slagsvold. 1986. The function of courtship feeding during incubation in the pied flycatcher *Ficedula hypoleuca*. *Animal Behaviour* 34:1441–1453.

481 Ligon, J. D. 1978. Reproductive interdependence of pinyon jays and pinyon pines. *Ecological Monographs* 48:111–126.

482 Lill, A. 1974. Sexual behavior of the lek-forming white-bearded manakin (*Manacus manacus trinitatis* Hartert). *Zeitschrift für Tierpsychologie* 36:1–36.

483 Lill, A. 1974. Social organization and space utilization in the lek-forming white-bearded manakin, *M. manacus trinitatis* Hartert. *Zeitschrift für Tierpsychologie* 36:513–530.

484 Lima, S. L. 1985. Maximizing efficiency and minimizing time exposed to predators: a trade-off in the black-capped chickadee. *Oecologia* 66:60–67.

485 Lindauer, M. 1961. *Communication among Social Bees*. Harvard University Press, Cambridge, MA.

486 Linsenmair, K. E. 1972. Die Bedeutung familienspezifischer "Abseichen" für den Familienzusammenhalt bei der sozialen Wüstenassel *Hemilepistus reaumuri* Audouin u. Savigny (Crustacea, Isopoda, Oniscoidea). *Zeitschrift für Tierpsychologie* 31:131–162.

487 Liske, E., and W. J. Davis. 1986. Courtship and mating behaviour of the Chinese praying mantis, *Tenodera aridifolia sinensis*. *Animal Behaviour* 35:1524–1537.

488 Lissmann, H. W. 1963. Electric location by fishes. *Scientific American* 208 (Mar.):50–59.

489 Lively, C. M. 1987. Evidence from a New Zealand snail for the maintenance of sex by parasitism. *Nature* 328:519–521.

490 Lloyd, J. E. 1965. Aggressive mimicry in *Photuris*: firefly *femmes fatales*. *Science* 149:653–654.

491 Lloyd, J. E. 1966. Studies on the flash communication system in *Photinus* fireflies. *Miscellaneous Publications of the Museum of Zoology, University of Michigan* 130:1–95.

492 Lloyd, J. E. 1975. Aggressive mimicry in *Photuris* fireflies: signal repertoires by *femmes fatales*. *Science* 197:452–453.

493 Lloyd, J. E. 1979. Mating behavior and natural selection. *Florida Entomologist* 62:17–34.

494 Lloyd, J. E. 1984. On deception, a way of all flesh, and firefly signalling and systematics. In *Oxford Surveys of Evolutionary Biology*. R. Dawkins and M. Ridley (eds.). Oxford University Press, Oxford.

495 Lloyd, J. E. 1986. Firefly communication and deception: "Oh, what a tangled web". In *Deception: Perspectives on Human and Non-human Deceit*. R. W. Mitchell and N. S. Thompson (eds.). SUNY Press, Albany, NY.

496 Lockard, R. B. 1978. Seasonal change in the activity pattern of *Dipodomys spectabilis*. *Journal of Mammalogy* 59:563–568.

497 Lockard, R. B., and D. H. Owings. 1974. Seasonal variation in moonlight avoidance by bannertail kangaroo rats. *Journal of Mammalogy* 55:189–193.

498 Loher, W. 1972. Circadian control of stridulation in the cricket *Teleogryllus commodus* Walker. *Journal of Comparative Physiology* 79:173–190.

499 Loher, W. 1979. Circadian rhythmicity of locomotor behavior and oviposition in female *Teleogryllus commodus*. *Behavioral Ecology and Sociobiology* 5:383–390.

500 Loher, W. 1984. Behavioral and physiological changes in cricket-females after mating. *Advances in Invertebrate Reproduction* 3:189–201.

501 Loher, W., and L. J. Orsak. 1985. Circadian patterns of premating behavior in *Teleogryllus oceanicus* Le Guillou under laboratory and field conditions. *Behavioral Ecology and Sociobiology* 16:223–231.

502 Lore, R., and K. Flannelly. 1977. Rat societies. *Scientific American* 236 (May):106–116.

503 Lorenz, K. Z. 1952. *King Solomon's Ring*. Crowell, New York.

504 Lorenz, K. Z. 1965. *Evolution and Modification of Behavior*. University of Chicago Press, Chicago.

505 Lorenz, K. Z. 1969. Innate bases of learning. In *On the Biology of Learning*. K. H. Pribram (ed.). Harcourt Brace Jovanovich, New York.

506 Lorenz, K. Z. 1970. *Studies on Animal and Human Behavior*, Vols. 1 and 2. Harvard University Press, Cambridge, MA.

507 Lorenz, K. Z. 1970. Companions as factors in the bird's environment. In *Studies on Animal and Human Behavior*. Vol. 1. K. Z. Lorenz (ed.). Harvard University Press, Cambridge, MA.

508 Lott, D. F. 1979. Dominance relations and breeding rate in mature male American bison. *Zeitschrift für Tierpsychologie* 49:418–432.

509 Loyn, R. H., R. G. Runnalls, G. Y. Forward, and J. Tyers. 1983. Territorial bell miners and other birds affecting populations of insect prey. *Science* 221:1411–1413.

510 Luling, K. H. 1963. The archer fish. *Scientific American* 209 (July):100–109.

511 Lyon, B. E., R. D. Montgomerie, and L. D. Hamilton. 1987. Male parental care and monogamy in snow buntings. *Behavioral Ecology and Sociobiology* 20:377–382.

512 MacDonald, D. W. 1982. Social factors affecting reproduction amongst red foxes. *Biogeographica* 18:123–175.

513 Mace, R. 1987. The dawn chorus in the great tit *Parus major* is directly related to female fertility. *Nature* 330:745-746.

514 MacLusky, N. J., and F. Naftolin. 1981. Sexual differentiation of the central nervous system. *Science* 211:1294–1302.

515 Malcolm, S. B. 1987. Monarch butterfly migration in North America: controversy and conservation. *Trends in Ecology and Evolution* 2:135–139.

516 Markl, H., and J. Tautz. 1975. The sensitivity of hair receptors in caterpillars of *Barathra brassicae* L. (Lepidoptera, Noctuidae) to particle movement in a sound field. *Journal of Comparative Physiology* 99:79–87.

517 Markow, T. A. 1982. Mating systems of cactophilic *Drosophila*. In *Ecological Energetics and Evolution: The Cactus–Yeast–Drosophila Model System*. J. S. F. Barker and W. T. Starmer (eds.). Academic Press, New York.

518 Marler, P. 1955. Characteristics of some animal calls. *Nature* 176:6–8.

519 Marler, P. 1959. Developments in the study of animal communication. In *Darwin's Biological Work*. P. R. Bell (ed.). Cambridge University Press, New York.

520 Marler, P. 1970. Birdsong and speech development: could there be parallels? *American Scientist* 58:669–673.

521 Marler, P., and W. J. Hamilton. 1966. *Mechanisms of Animal Behavior*. Wiley, New York.

522 Marler, P., and V. Sherman. 1985. Innate differences in singing behaviour of sparrows reared in isolation from adult conspecific song. *Animal Behaviour* 33:57–71.

523 Marler, P., and M. Tamura. 1964. Culturally transmitted patterns of vocal behavior in sparrows. *Science* 146:1483–1486.

524 Marr, D. 1982. *Vision*. W. H. Freeman, San Francisco.

525 Marten, K., and P. Marler. 1977. Sound transmission and its significance for animal vocalization. I. Temperate habitats. *Behavioral Ecology and Sociobiology* 2:271–290.

526 Marten, K., D. Quine, and P. Marler. 1977. Sound transmission and its significance for animal vocalization. II. Tropical forest habitats. *Behavioral Ecology and Sociobiology* 2:291–302.

527 Maschwitz, U., and E. Maschwitz. 1974. Platzende Arbeiterinnen: Eine neue Art der Feindabwehr bei sozialen Hautflüglern. *Oecologia* 14:289–294.

528 Mason, W. A. 1968. Early social deprivation in non-human primates: implications for human behavior. In *Biology and Behavior: Environmental Influences*. D. C. Glass (ed.). Rockefeller University Press, New York.

529 Mason, W. A. 1978. Social experience and primate cognitive development. In *The Development of Behavior: Comparative and Evolutionary Aspects*. G. M. Burghardt and M. Bekoff (eds.). Garland STPM Press, New York.

530 Matsubara, J. A. 1981. Neural correlates of a nonjammable electrolocation system. *Science* 211:722–724.

531 Maynard Smith, J. 1974. The theory of games and the evolution of animal conflict. *Journal of Theoretical Biology* 47:209–221.

532 Maynard Smith, J. 1978. *The Evolution of Sex*. Cambridge University Press, Cambridge.

533 Mayr, E. 1961. Cause and effect in biology. *Science* 134:1501–1506.

534 Mayr, E. 1963. *Animal Species and Evolution*. Harvard University Press, Cambridge, MA.

535 Mayr, E. 1977. Darwin and natural selection. *American Scientist* 65:321–327.

536 Mayr, E. 1982. *The Growth of Biological Thought*. Harvard University Press, Cambridge, MA.

537 Mayr, E. 1983. How to carry out the adaptationist program? *American Naturalist* 121:324–334.

538 McAllister, L. B., R. H. Scheller, E. R. Kandel, and R. Axel. 1983. *In situ* hybridization to study the origin and fate of identified neurons. *Science* 222:800–808.

539 McCann, T. S. 1981. Aggression and sexual activity of male southern elephant seals, *Mirounga leonina*. *Journal of Zoology* 195:295–310.

540 McCosker, J. E. 1977. Flashlight fishes. *Scientific American* 236 (Mar.):106–115.

541 McCracken, G. F. 1984. Communal nursing in Mexican free-tailed bat maternity colonies. *Science* 223:1090–1091.

542 McCracken, G. F., and J. W. Bradbury. 1981. Social organization and kinship in the polygynous bat *Phyllostomus hastatus*. *Behavioral Ecology and Sociobiology* 8:11–34.

543 McDonald, D. B. 1987. Male–male Cooperation in a Neotropical Lekking Bird. Ph.D. thesis, University of Arizona.

544 McEwen, B. S. 1981. Neural gonadal steroid actions. *Science* 211:1303–1311.

545 McFarland, D. 1985. *Animal Behavior.* Benjamin/Cummings Publishing, Menlo Park, CA.

546 McGregor, P. K., and J. R. Krebs. 1982. Mating and song types in the great tit. *Nature* 297:60–61.

547 McGue, M., and T. J. Bouchard, Jr. In press. Genetic and environmental determinants of information processing and special mental abilities: a twin analysis. *Advances in the Psychology of Human Intelligence.*

548 M'Closkey, R. T., K. A. Baia, and R. W. Russell. 1987. Defense of mates: a territory departure rule for male tree lizards following sex-ratio manipulation. *Oecologia* 73:28–31.

549 McNicol, D., Jr., and D. Crews. 1979. Estrogen/progesterone synergy in the control of female sexual receptivity in the lizard, *Anolis carolinensis. General and Comparative Endocrinology* 38:68–74.

550 Mech, L. D. 1970. *The Wolf: The Ecology and Behavior of an Endangered Species.* Doubleday (Natural History Press), Garden City, New York.

551 Medvin, M. B., and M. D. Beecher. 1986. Parent–offspring recognition in the barn swallow (*Hirundo rustica*). *Animal Behaviour* 34:1627–1639.

552 Menaker, M., and S. Wisner. 1983. Temperature-compensated circadian clock in the pineal of *Anolis carolinensis? Proceedings of the National Academy of Sciences* 80:6119–6121.

553 Mendlewicz, J., and J. D. Ranier. 1977. Adoption study supporting genetic transmission of manic–depressive illness. *Nature* 268:327–329.

554 Michener, C. D. 1974. *The Social Behavior of the Bees.* Harvard University Press, Cambridge, MA.

555 Milinski, M. 1984. Parasites determine a predator's optimal feeding strategy. *Behavioral Ecology and Sociobiology* 15:35–38.

556 Miller, D. E., and J. T. Emlen, Jr. 1975. Individual chick recognition and family integrity in the ring-billed gull. *Behaviour* 52:124–144.

557 Miller, L. A. 1983. How insects detect and avoid bats. In *Neuroethology and Behavioral Physiology.* F. Huber and H. Markl (eds.). Springer-Verlag, Berlin.

558 Miller, L. A., and J. Olesen. 1979. Avoidance behavior in green lacewings. I. Behavior of free flying green lacewings to hunting bats and ultrasound. *Journal of Comparative Physiology* 131:113–120.

559 Millington, S. J., and T. D. Price. 1985. Song inheritance and mating patterns in Darwin's finches. *Auk* 102:342–346.

560 Mitani, J. C. 1984. The behavioral regulation of monogamy in gibbons (*Hylobates muelleri*). *Behavioral Ecology and Sociobiology* 15:225–229.

561 Mock, D. W. 1983. On the study of avian mating systems. In *Perspectives in Ornithology.* G. A. Clark, Jr. and A. Brush (eds.). Cambridge University Press, Cambridge.

562 Mock, D. W. 1984. Siblicidal aggression and resource monopolization in birds. *Science* 225:731–733.

563 Mock, D. W., and B. J. Ploger. 1987. Parental manipulation of optimal hatch asynchrony in cattle egrets: an experimental study. *Animal Behaviour* 35:150–160.

564 Moiseff, A., G. S. Pollack, and R. R. Hoy. 1978. Steering responses of flying crickets to sound and ultrasound: mate attraction and predator avoidance. *Proceedings of the National Academy of Sciences* 75:4052–4056.

565 Monahan, M. W. 1977. Determinants of male pairing success in the red-winged blackbird (*Agelaius phoeniceus*): a multivariate, experimental analysis. Ph. D. thesis, Indiana University.

566 Montagu, A. 1976. *The Nature of Human Aggression.* Oxford University Press, New York.

567 Moore, F. R. 1978. Interspecific aggression: toward whom should a mockingbird be aggressive. *Behavioral Ecology and Sociobiology* 3:173–176.

568 Moore, J., and R. Ali. 1984. Are dispersal and inbreeding avoidance related? *Animal Behaviour* 32:94–112.

569 Moore, M. C. 1986. Elevated testosterone levels during nonbreeding-season territoriality in fall-breeding lizard, *Sceloporus jarrovi. Journal of Comparative Physiology* A 158:159–163.

570 Moore, M. C., and B. Kranz. 1983. Evidence for androgen independence of male mounting behavior in white-crowned sparrows (*Zonotrichia leucophrys gambelii*). *Hormones and Behavior* 17:414–423.

571 Morse, D. H. 1977. Resource partitioning in bumble bees: the role of behavioral factors. *Science* 197:678–680.

572 Morton, E. S. 1975. Ecological sources of selection on avian sounds. *American Naturalist* 109:17–34.

573 Morton, E. S. 1977. On the occurrence and significance of motivation-structural rules in some bird and mammal sounds. *American Naturalist* 111:855–869.

574 Morton, M. L., M. E. Pereyra, and L. F. Baptista. 1985. Photoperiodically induced ovarian growth in the white-crowned sparrow (*Zonotrichia leucophrys gambelii*) and its augmentation by song. *Comparative Biochemistry and Physiology* A 80:93–97.

575 Moskowitz, B. A. 1978. The acquisition of language. *Scientific American* 239 (Nov.):91–108.

576 Mower, G. D., W. G. Christen, and C. J. Caplan. 1983. Very brief visual experience eliminates plasticity in the cat visual cortex. *Science* 221:178–180.

577 Munger, J. C., and J. H. Brown. 1981. Competition in desert rodents: an experiment with semipermeable enclosures. *Science* 211:510–512.

578 Munn, C. A. 1986. Birds that "cry wolf". *Nature* 319:143–145.

579 Murdock, G. P. 1967. *Ethnographic Atlas*. Pittsburgh University Press, Pittsburgh, PA.

580 Murton, R. K., A. J. Isaacson, and N. J. Westwood. 1971. The significance of gregarious feeding behaviour and adrenal stress in a population of wood-pigeons (*Columba palumbus*). *Journal of Zoology* 165:53–84.

581 Naftolin, F., and E. Butz. 1981. Sexual dimorphism. *Science* 211:1263–1324.

582 Narins, P. M. 1982. Effects of masking noise on evoked calling in the Puerto Rican coqui (Anura: Leptodactylidae). *Journal of Comparative Physiology* 147:439–446.

583 Narins, P. M. 1982. Behavioral refractory period in Neotropical treefrogs. *Journal of Comparative Physiology* 148:337–344.

584 Narins, P. M., and D. D. Hurley. 1982. The relationship between call intensity and function in the Puerto Rican coqui (Anura: Leptodactylidae). *Herpetologica* 38:287–295.

585 Narins, P. M., and E. R. Lewis. 1984. The vertebrate ear as an exquisite seismic sensor. *Journal of the Acoustic Society of America* 76:1384–1387.

586 Nemeroff, C. B., W. W. Youngblood, P. J. Manberg, A. J. Prange, Jr., and J. S. Kizer. 1983. Regional brain concentrations of neuropeptides in Huntington's chorea and schizophrenia. *Science* 221:972–975.

587 Newton, P. N. 1986. Infanticide in an undisturbed forest population of hanuman langurs, *Presbytis entellus*. *Animal Behaviour* 34:785–789.

588 Nicol, S. E., and I. I. Gottesman. 1983. Clues to the genetics and neurobiology of schizophrenia. *American Scientist* 71:398–404.

589 Nisbet, I. C. T. 1973. Courtship-feeding, egg size and breeding success in common terns. *Nature* 241:141–142.

590 Noonan, K. C. 1983. Female choice in the cichlid fish *Cichlasoma nigrofasciatum*. *Animal Behaviour* 31:1005–1010.

591 Norberg, U. M. 1985. Evolution of vertebrate flight: an aerodynamic model for the transition from gliding to active flight. *American Naturalist* 126:303–327.

592 Norris, K. S. 1967. Some observations on the migration and orientation of marine mammals. In *Animal Orientation and Navigation*. R. M. Storm (ed.). Oregon State University Press, Corvallis.

593 Nottebohm, F. 1981. A brain for all seasons: cyclical anatomical changes in song control nuclei of the canary brain. *Science* 214:1368–1370.

594 Nottebohm, F. 1984. Birdsong as a model in which to study brain processes related to learning. *Condor* 86:227–236.

595 Nottebohm, F. 1987. Plasticity in adult avian central nervous system: possible relation between hormones, learning, and brain repair. In *Higher Functions of the Nervous System*, Section 1 in *Handbook of Physiology*. F. Plum (ed.). American Physiological Society, Baltimore.

596 Nottebohm, F., and M. E. Nottebohm. 1971. Vocalization and breeding behaviour of surgically deafened ring doves (*Streptopelia risoria*). *Animal Behaviour* 19:313–327.

597 Novacek, M. J. 1985. Evidence for echolocation in oldest known bats. *Nature* 315:140–141.

598 Nusbaum, M. D., W. O. Friesen, W. B. Kristan, Jr., and R. A. Pearce. 1987. Neural mechanisms generating the leech swimming rhythm: swim-initiator neurons excite the network of swim oscillator neurons. *Journal of Comparative Physiology* A 161:355–366.

599 Oldroyd, H. 1964. *The Natural History of Flies*. Norton, New York.

600 O'Neill, W. E., and N. Suga. 1979. Target range-sensitive neurons in the auditory cortex of the mustache bat. *Science* 203:69–72.

601 Orians, G. H. 1962. Natural selection and ecological theory. *American Naturalist* 96:257–264.

602 Orians, G. H. 1969. On the evolution of mating systems in birds and mammal. *American Naturalist* 103:589–603.

603 Oring, L. W. 1985. Avian polyandry. *Current Ornithology* 3:309–351.

604 Oring, L. W., and M. L. Knudson. 1973. Monogamy and polyandry in the spotted sandpiper. *The Living Bird* 11:59–73.

605 Ostrom, J. H. 1974. *Archeopteryx* and the origin of flight. *Quarterly Review of Biology* 49:27–47.

606 Otte, D. 1974. Effects and functions in the evolution of signaling systems. *Annual Review of Ecology and Systematics* 5:385–417.

607 Owens, D. D., and M. J. Owens. 1979. Communal denning and clan associations in brown hyenas (*Hyaena brunnea*, Thunberg) in the central Kalahari Desert. *African Journal of Ecology* 17:35–44.

608 Owings, D. H., and R. G. Coss. 1977. Snake mobbing by California ground squirrels: adaptive variation and ontogeny. *Behaviour* 62:50–69.

609 Packard, J. M., U. S. Seal, L. D. Mech, and E. D. Plotka. 1985. Causes of reproductive failure in two family groups of wolves (*Canis lupus*). *Zeitschrift für Tierpsychologie* 69:24–40.

610 Packer, C. 1977. Reciprocal altruism in *Papio anubis*. *Nature* 265:441–443.

611 Packer, C. 1979. Inter-troop transfer and inbreeding avoidance in *Papio anubis*. *Animal Behaviour* 27:1–36.

612 Packer, C. 1986. The ecology of sociality in felids. In *Ecological Aspects of Social Evolution*. D. I. Rubenstein and R. W. Wrangham (eds.). Princeton University Press, Princeton, NJ.

613 Packer, C., and A. E. Pusey. 1982. Cooperation and competition within coalitions of lions: kin selection or game theory? *Nature* 296:740–742.

614 Packer, C., L. Herbst, A. E. Pusey, J. D. Bygott, S. J. Cairns, and M. Borgerhoff-Mulder. 1987. Reproductive success of lions. In *Reproductive Success*. T. H. Clutton-Brock (ed.). University of Chicago Press, Chicago.

615 Page, D. C., R. Mosher, E. M. Simpson, E. M. C. Fisher, G. Mardon, J. Pollack, B. McGillivray, A. de la Chapelle, and L. G. Brown. 1987. The sex-determining region of the human Y chromosome encodes a finger protein. *Cell* 51:1091–1104.

616 Page, T. L. 1985. Clocks and circadian rhythms. In *Comprehensive Insect Physiology, Biochemistry, and Pharmacology*. G. A. Kerkut and L. I. Gilbert (eds.). Pergamon Press, New York.

617 Palmer, J. D. 1974. *Biological Clocks in Marine Organisms: The Control of Physiological and Behavioral Tidal Rhythms*. Wiley, New York.

618 Palmer, J. D. 1976. *An Introduction to Biological Rhythms*. Academic Press, New York.

619 Papi, F. 1975. La navigazione dei colombi viaggiatori. *Le Scienze* 78:66–75.

620 Papi, F. 1986. Pigeon navigation: solved problems and open questions. *Monitore Zoologici Italiana* 20:471–517.

621 Papi, F., and H. G. Wallraff (eds.). 1982. *Avian Navigation*. Springer-Verlag, Berlin.

622 Parker, G. A. 1970. Sperm competition and its evolutionary consequences in the insects. *Biological Reviews* 45:525–567.

623 Parker, G. A. 1974. Assessment strategy and the evolution of fighting behaviour. *Journal of Theoretical Biology* 47:223–243.

624 Partridge, L. 1974. Habitat selection in titmice. *Nature* 247:573–574.

625 Partridge, L. 1976. Field and laboratory observations on the foraging and feeding techniques of blue tits (*Parus caeruleus*) and coal tits (*Parus ater*) in relation to their habitats. *Animal Behaviour* 24:534–544.

626 Paton, D. C., and H. A. Ford. 1983. The influence of plant characteristics and honeyeater size on levels of pollination in Australian plants. In *Handbook of Experimental Pollination Biology*. C. E. Jones and R. J. Little (eds.). Van Nostrand Reinhold, New York.

627 Pengelly, E. T., and S. J. Asmundson. 1974. Circannual rhythmicity in hibernating animals. In *Circannual Clocks*. E. T. Pengelley (ed.). Academic Press, New York.

628 Perrett, D. I., and E. T. Rolls. 1983. Neural mechanisms underlying the visual analysis of faces. In *Advances in Vertebrate Neuroethology*. J.-P. Ewert, R. R. Capranica, and D. J. Ingle (eds.). Plenum Press, New York.

629 Perrill, S. A., H. C. Gerhardt, and R. Daniel. 1978. Sexual parasitism in the green tree frog (*Hyla cinerea*). *Science* 200:1179–1180.

630 Petrie, M. 1983. Female moorhens compete for small fat males. *Science* 220:413–415.

631 Pfaff, D. W., and B. S. McEwen. 1983. Actions of estrogens and progestins on nerve cells. *Science* 219:808–814.

632 Pfennig, D. W., G. J. Gamboa, H. K. Reeve, J. S. Reeve, and I. D. Ferguson. 1983. The mechanism of nestmate discrimination in social wasps (*Polistes*, Hymenoptera: Vespidae). *Behavioral Ecology and Sociobiology* 13:299–305.

633 Pierce, N. E., and P. S. Mead. 1981. Parasitoids as selective agents in the symbiosis between lycaenid butterfly larvae and ants. *Science* 211:1185–1187.

634 Pietrewicz, A. T., and A. C. Kamil. 1977. Visual detection of cryptic prey by blue jays (*Cyanocitta cristata*). *Science* 195:580–582.

635 Pietrewicz, A. T., and A. C. Kamil. 1979. Search image formation in the blue jay *Cyanocitta cristata*. *Science* 204:1332–1333.

636 Pietsch, T. W., and D. B. Grobecker. 1978. The compleat angler: aggressive mimicry in the antennariid anglerfish. *Science* 201:369–370.

637 Pitcher, T. 1979. He who hesitates lives. Is stotting antiambush behavior? *American Naturalist* 113:453–456.

638 Plomin, R., and D. C. Rowe. 1978. Genes, environment, and development of temperament in young human twins. In *The Development of Behavior*. G. M. Burghardt and M. Bekoff (eds.). Garland STPM Press, New York.

639 Pohl-Apel, G., and R. Sossinka. 1984. Hormonal determination of song capacity in females of the zebra finch: critical phase of treatment. *Zeitschrift für Tierpsychologie* 64:330–336.

640 Pollak, G., D. Marsh, R. Bodenhamer, and A. Souther. 1977. Echo-detecting characteristics of neurons in inferior colliculus of unanesthetized bats. *Science* 196:675–677.

641 Porter, R. H., V. J. Tepper, and D. M. White. 1981. Experiential influences on the development of huddling preferences and "sibling" recognition in spiny mice. *Developmental Psychobiology* 14:375–382.

642 Porter, S. D., and D. A. Eastmond. 1982. *Euryopis coki* (Theridiidae), a spider that preys on *Pogonomyrmex* ants. *Journal of Arachnology* 10:275–277.

643 Pough, F. H. 1972. Newts, leeches, and agriculture. *New York's Food and Life Sciences Quarterly* 5:4–7.

644 Powell, G. V. N. 1974. Experimental analysis of the social value of flocking by starlings (*Sturnus vulgaris*) in relation to predation and foraging. *Animal Behaviour* 22:501–505.

645 Pritchard, P. C. H. 1976. Post-nesting movements of marine turtle (Cheloniidae and Dermochelyidae) tagged in the Guianas. *Copeia* 1976:749–754.

646 Provine, R. R. 1986. Yawning as a stereotyped action pattern and releasing stimulus. *Ethology* 72:109–122.

647 Purcell, J. E. 1980. Influence of siphonophore behavior upon their natural diets: evidence for aggressive mimicry. *Science* 209:1045–1047.

648 Pusey, A. E., and C. Packer. 1987. The evolution of sex-biased dispersal in lions. *Behaviour* 101:275–310.

649 Putman, R. J. 1988. The *Natural History of Deer*. Christopher Helm, Bromley, Kent.

650 Pyburn, W. F. 1980. The function of eggless capsules in leaf nests of the frog *Phyllomedusa hypochondrialis* (Anura: Hylidae). *Proceedings of the Biological Society of Washington* 93:153–167.

651 Pyke, G. 1979. The economics of territory size and time budget in the golden-winged sunbird. *American Naturalist* 114:131–145.

652 Pyle, D. W., and M. H. Gromko. 1978. Repeated mating by female *Drosophila melanogaster*: the adaptive importance. *Experientia* 34:449–450.

653 Queller, D. C., and J. E. Strassmann. 1988. Reproductive success and group nesting in the paper wasp, *Polistes annularis*. In *Reproductive Success*. T. Clutton-Brock (ed.). University of Chicago Press, Chicago.

654 Racey, P. A., and J. D. Skinner. 1979. Endocrine aspects of sexual mimicry in spotted hyenas, *Crocuta crocuta*. *Journal of Zoology* 187:315–326.

655 Ralls, K., K. Brugger, and J. Ballou. 1979. Inbreeding and juvenile mortality in small populations of ungulates. *Science* 206:1101–1103.

656 Ram, J. L., S. R. Salpeter, and W. J. Davis. 1977. *Pleurobranchaea* egg-laying hormone: localization and partial purification. *Journal of Comparative Physiology* 199:171–194.

657 Rand, A. S. 1964. Inverse relationship between temperature and shyness in the lizard *Anolis lineatopus*. *Ecology* 45:863–864.

658 Randall, J. A. 1984. Territorial defense and advertisement by footdrumming in bannertail kangaroo rats (*Dipodomys spectabilis*) at high and low population densities. *Behavioral Ecology and Sociobiology* 16:11–20.

659 Rasa, O. A. E. 1984. Dwarf mongoose and hornbill mutualism in the Taru Desert, Kenya. *Behavioral Ecology and Sociobiology* 12:181–190.

660 Ray, T. S., and C. C. Andrews. 1980. Antbutterflies: butterflies that follow army ants to feed on antbird droppings. *Science* 210:1147–1148.

661 Read, A. F. 1987. Comparative evidence supports the Hamilton and Zuk hypothesis on parasites and sexual selection. *Nature* 328:68–70.

662 Reichman, J. 1979. Subtly suited to a seedy existence. *New Scientist* 81 (Mar.):658–660.

663 Reyer, H.-U. 1984. Investment and relatedness: a cost/benefit analysis of breeding and helping in the pied kingfisher. *Animal Behaviour* 32:1163–1178.

664 Reynolds, J. D. 1987. Mating system and nesting biology of the red-necked phalarope *Phalaropus lobatus*. *Ibis* 129:225–242.

665 Reynolds, J. D., M. A. Colwell, and F. Cooke. 1986. Sexual selection and spring arrival times of red-necked and Wilson's phalaropes. *Behavioral Ecology and Sociobiology* 18:303–310.

666 Richardson, H., and N. A. M. Verbeek. 1986. Diet selection and optimization by northwestern crows feeding on Japanese littleneck clams. *Ecology* 67:1219–1226.

667 Ridley, M. 1983. *The Explanation of Organic Diversity*. Oxford University Press, Oxford.

668 Robbins, R. K. 1981. The "false head" hypothesis: predation and wing pattern variation of lycaenid butterflies. *American Naturalist* 118:770–775.

669 Roeder, K. D. 1963. *Nerve Cells and Insect Behavior*. Harvard University Press, Cambridge, MA.

670 Roeder, K. D. 1965. Moths and ultrasound. *Scientific American* 212 (Apr.):94–102.

671 Roeder, K. D. 1970. Episodes in insect brains. *American Scientist* 58:378–389.

672 Roeder, K. D., and A. E. Treat. 1961. The detection and evasion of bats by moths. *American Scientist* 49:135–148.

673 Rosenblatt, J. S., H. I. Siegel, and A. D. Mayer. 1979. Progress in the study of maternal behavior in the rat: hormonal, nonhormonal, sensory, and developmental aspects. *Advances in the Study of Behavior* 10:225–331.

674 Rosenzweig, M. L. 1973. Habitat selection experiments with a pair of coexisting heteromyid rodent species. *Ecology* 54:111–117.

675 Ross, K. G. 1986. Kin selection and the problem of sperm utilization in social insects. *Nature* 323:798–800.

676 Routtenberg, A. 1978. The reward system of the brain. *Scientific American* 239 (Nov.):154–165.

677 Rowe, M. P., R. G. Coss, and D. H. Owings. 1986. Rattlesnake rattles and burrowing owl hisses: a case of acoustic Batesian mimicry. *Ethology* 72:53–71.

678 Rowley, I., and G. Chapman. 1986. Crossfostering, imprinting, and learning in two sympatric species of cockatoos. *Behaviour* 96:1–16.

679 Rusak, B., and G. Groos. 1982. Suprachiasmatic stimulation phase shifts rodent circadian rhythms. *Science* 215:1407–1409.

680 Ruse, M. 1979. *Sociobiology: Sense or Nonsense?* D. Reidel Publishing, Dordrecht, Holland.

681 Rutowski, R. L. 1980. Courtship solicitation by females of the checkered white butterfly, *Pieris protodice*. *Behavioral Ecology and Sociobiology* 7:113–117.

682 Rutowski, R. L., and G. W. Gilchrist. 1986. Copulation in *Colias eurytheme* (Lepidoptera: Pieridae): patterns and frequency. *Journal of Zoology* 209:115–124.

683 Rutowski, R. L., C. E. Long, L. D. Marshall, and R. S. Vetter. 1981. Courtship solicitation by *Colias* females (Lepidoptera: Pieridae). *American Midland Naturalist* 105:334–340.

684 Ryan, M. J. 1980. Female mate choice in a Neotropical frog. *Science* 209:523–525.

685 Ryan, M. J. 1983. Sexual selection and communication in a Neotropical frog, *Physalaemus pustulosus*. *Evolution* 37:261–272.

686 Ryan, M. J. 1983. Frequency modulated calls and species recognition in a Neotropical frog, *Physalaemus pustulosus*. *Journal of Comparative Physiology* 150:217–221.

687 Ryan, M. J. 1985. *The Tungara Frog*. University of Chicago Press, Chicago.

688 Ryan, M. J., and M. D. Tuttle. 1983. The ability of the frog-eating bat to discriminate among novel and potentially poisonous frog species using acoustic cues. *Animal Behaviour* 31:827–833.

689 Ryan, M. J., M. D. Tuttle, and L. K. Taft. 1981. The costs and benefits of frog chorusing behavior. *Behavioral Ecology and Sociobiology* 8:273–278.

690 Ryker, L. C. 1984. Acoustic and chemical signals in the life cycle of a beetle. *Scientific American* 250 (June):112–123.

691 Sahlins, M. 1976. *The Use and Abuse of Biology*. University of Michigan Press, Ann Arbor.

692 Sakaluk, S. K. 1984. Male crickets feed females to ensure complete sperm transfer. *Science* 223:609–610.

693 Sakaluk, S. K. 1986. Is courtship feeding by male insects parental investment? *Ethology* 73:161–166.

694 Sargent, R. C., and M. R. Gross. 1985. Parental investment decision rules and the Concorde fallacy. *Behavioral Ecology and Sociobiology* 17:43–45.

695 Sargent, R. C., M. R. Gross, and E. P. van den Berghe. 1986. Male mate choice in fishes. *Animal Behaviour* 34:545–550.

696 Sargent, T. D. 1969. Behavioural adaptations of cryptic moths. III. Resting attitudes of two bark-like species, *Melanolophis canadaria* and *Catocala ultronia*. *Animal Behaviour* 17:670–672.

697 Sargent, T. D. 1976. *Legion of Night—The Underwing Moths*. University of Massachusetts Press, Amherst.

698 Savage-Rumbaugh, E. S. 1986. *Ape Lan-

guage: From Conditioned Response to Symbol. Columbia University Press, New York.

699 Schaller, G. B. 1964. *The Year of the Gorilla.* University of Chicago Press, Chicago.

700 Schaller, G. B. 1972. *The Serengeti Lion.* University of Chicago Press, Chicago.

701 Scheich, H., G. Langer, C. Tidemann, R. B. Coles, and A. Guppy. 1986. Electroreception and electrolocation in platypus. *Nature* 319:401–402.

702 Schlenoff, D. H. 1985. The startle response of blue jays to *Catocala* (Lepidoptera: Noctuidae) prey models. *Animal Behaviour* 33:1057–1067.

703 Schmidt-Koenig, K. 1987. Bird navigation: has olfactory orientation solved the problem? *Quarterly Review of Biology* 62:31–47.

704 Schneider, D. 1969. Insect olfaction: deciphering system for chemical messages. *Science* 163:1031–1036.

705 Schoener, T. W. 1982. The controversy over interspecific competition. *American Scientist* 70:586–595.

706 Schuckit, M. A., and V. Rayses. 1979. Ethanol ingestion: differences in blood acetaldehyde concentrations in relatives of alcoholics and controls. *Science* 203:54–55.

707 Schwagmeyer, P. L. 1979. The Bruce effect: an evaluation of male/female advantages. *American Naturalist* 114:932–938.

708 Schwagmeyer, P. L., and C. H. Brown. 1983. Factors affecting male-male competition in thirteen-lined ground squirrels. *Behavioral Ecology and Sociobiology* 13:1–6.

709 Searle, L. V. 1949. The organization of hereditary maze-brightness and maze-dullness. *Genetic Psychology Monographs* 39:279–335.

710 Seeley, T. D. 1977. Measurement of nest cavity volume by the honey bee (*Apis mellifera*). *Behavioral Ecology and Sociobiology* 2:201–227.

711 Seeley, T. D. 1985. *Honeybee Ecology: A Study of Adaptation in Social Life.* Princeton University Press, Princeton, NJ.

712 Seeley, T. D., R. H. Seeley, and P. Akratanakul. 1982. Colony defense strategies of the honeybees in Thailand. *Ecological Monographs* 52:43–63.

713 Sherman, P. W. 1977. Nepotism and the evolution of alarm calls. *Science* 197:1246–1253.

714 Sherman, P. W. 1980. The meaning of nepotism. *American Naturalist* 116:604–606.

715 Sherman, P. W. 1981. Kinship, demography, and Belding's ground squirrel nepotism. *Behavioral Ecology and Sociobiology* 8:251–259.

716 Sherman, P. W. 1985. Alarm calls of Belding's ground squirrels to aerial predators: nepotism or self-preservation? *Behavioral Ecology and Sociobiology* 17:313–323.

717 Sherman, P. W., and W. G. Holmes. 1985. Kin recognition: issues and evidence. In *Experimental Behavioral Ecology and Sociobiology.* B. Hölldobler and M. Lindauer (eds.). G. Fischer Verlag, Stuttgart.

718 Sherry, D. 1984. Food storage by blackcapped chickadees: memory of the location and contents of caches. *Animal Behaviour* 32:451–464.

719 Shields, W. M. 1983. Optimal inbreeding and the evolution of philopatry. In *The Ecology of Animal Movement.* I. R. Swingland and P. J. Greenwood (eds.). Clarendon Press, Oxford.

720 Shields, W. M. 1984. Barn swallow mobbing: self-defence, collateral kin defence, group defence, or parental care? *Animal Behaviour* 32:132–148.

721 Siegel, R. W., J. C. Hall, D. A. Gailey, and C. P. Kyriacou. 1984. Genetic elements of courtship in *Drosophila*: mosaics and learning mutants. *Behavior Genetics* 14:383–410.

722 Sillén-Tullberg, B. 1985. Higher survival of an aposematic form than of a cryptic form of a distasteful bug. *Oecologia* 67:411–415.

723 Silverin, B. 1980. Effects of long-acting testosterone treatment on free-living pied flycatchers, *Ficedula hypoleuca*, during the breeding period. *Animal Behaviour* 28:906–912.

724 Simmons, L. W. 1987. Female choice contributes to offspring fitness in the field cricket, *Gryllus bimaculatus* (De Geer). *Behavioral Ecology and Sociobiology* 21:313–322.

725 Singer, P. 1981. *The Expanding Circle: Ethics and Sociobiology.* Farrar, Straus, and Giroux, New York.

726 Siniff, D. B., I. Stirling, J. L. Bengston, and R. A. Reichle. 1979. Social and reproductive behavior of crabeater seals (*Lobodon carcinophagus*) during the austral spring. *Canadian Journal of Zoology* 57:2243–2255.

727 Skinner, B. F. 1966. Operant behavior. In *Operant Behavior.* W. K. Honig (ed.). Appleton-Century-Crofts, New York.

728 Slobodchikoff, C. N. 1978. Experimental studies of tenebrionid beetle predation by skunks. *Behaviour* 66:313–322.

729 Smigel, B. W., and M. L. Rosenzweig. 1974. Seed selection in *Dipodomys merriami* and *Perognatus penicillatus. Ecology* 55:328–339.

730 Smith, M. S., B. J. Kish, and C. B. Crawford. 1987. Inheritance of wealth as human
kin investment. *Ethology and Sociobiology*
8:171–182.

731 Smith, R. L. 1979. Paternity assurance and
altered roles in the mating behavior of a
giant water bug *Abedus herberti* (Heteroptera: Belostomatidae). *Animal Behaviour*
27:716–728.

732 Smith, R. L. 1984. Human sperm competition. In *Sperm Competition and the Evolution of Animal Mating Systems*. R. L. Smith
(ed.). Academic Press, New York.

733 Smith, S. M. 1977. Coral-snake pattern recognition and stimulus generalization by naive great kiskadees (Aves: Tyrannidae). *Nature* 265:535–536.

734 Smith, S. M. 1978. The "underworld" in a
territorial species: adaptive strategy for floaters. *American Naturalist* 112:571–582.

735 Snow, B. K. 1977. Territorial behavior and
courtship in the male three-wattled bellbird.
Auk 94:623–645.

736 Snow, D. W. 1956. Courtship ritual: the
dance of the manakins. *Animal Kingdom*
59:86–91.

737 Sommer, V. 1987. Infanticide among free-
ranging langurs (*Presbytis entellus*) at Jodh-
pur (Rajasthan/India): recent observations
and a reconsideration of hypotheses. *Primates* 28:163–197.

738 Sommer, V., and S. M. Mohnot. 1985. New
observations on infanticides among hanuman langurs (*Presbytis entellus*) near Jodh-
pur (Rajasthan/India). *Behavioral Ecology
and Sociobiology* 16:245–248.

739 Southwick, C. H., M. A. Beg, and M. R. Sid-
diqi. 1965. Rhesus monkeys in North India.
In *Primate Behavior*. I. DeVore (ed.). Holt,
Rinehart & Winston, New York.

740 Spangler, H. G. 1984. Silence as a defense
against predatory bats in two species of calling insects. *Southwestern Naturalist* 29:481–
488.

741 Stamps, J. A., and K. Tollestrup. 1984. Pro-
spective resource defense in a territorial
species. *American Naturalist* 123:99–114.

742 Stein, Z., M. Susser, G. Saenger, and F. Mar-
olla. 1972. Nutrition and mental performance. *Science* 178:708–713.

743 Stephens, D. W., and J. R. Krebs. 1986. *Foraging Theory*. Princeton University Press,
Princeton, NJ.

744 Stephens, M. 1982. Mate takeover and possible infanticide by a female northern jacana
(*Jacana spinosa*). *Animal Behaviour*
30:1253–1254.

745 Sternberg, D. E., D. P. van Kammen, P. Ler-

ner, and W. E. Bunney. 1982. Schizophrenia:
dopamine β-hydroxylase activity and treatment response. *Science* 216:1423–1425.

746 Stewart, K. J. 1987. Spotted hyaenas: the
importance of being dominant. *Trends in
Ecology and Evolution* 2:88–89.

747 Stoutamire, W. P. 1974. Australian terrestrial orchids, thynnid wasps, and pseudo-
copulation. *American Orchid Society Bulletin* 43:13–18.

748 Strassmann, J. E. 1981. Parasitoids, predators, and group size in the paper wasp, *Polistes exclamans*. *Ecology* 62:1225–1233.

749 Suga, N., and W. E. O'Neill. 1979. Neural
axis representing target range in the auditory cortex of the mustache bat. *Science*
206:351–353.

750 Suga, N., and T. Shimozawa. 1974. Site of
neural attenuation of responses to self-vocal-
ized sounds in echolocating bats. *Science*
183:1211–1213.

751 Sugiyama, Y. 1984. Proximate factors of infanticide among langurs at Dharwar: a reply to Boggess. In *Infanticide: Comparative
and Evolutionary Perspectives*. G. Hausfater
and S. B. Hrdy (eds.). Aldine, Chicago.

752 Sullivan, B. K. 1983. Sexual selection in
Woodhouse's toad (*Bufo woodhousei*). II. Female choice. *Animal Behaviour* 31:1011–
1017.

753 Sullivan, K. A. In press. Age-specific profit-
ability and prey choice. *Animal Behaviour*.

754 Susman, R. L., and J. T. Stern. 1982. Functional morphology of *Homo habilis*. *Science*
217:931–934.

755 Sutherland, W. J. 1985. Chance can produce
a sex difference in variance in mating success and explain Bateman's data. *Animal
Behaviour* 33:1349–1352.

756 Swan, L. W. 1970. Goose of the Himalayas.
Natural History 79 (Dec.):68–75.

757 Sweeney, B. W., and R. L. Vannote. 1982.
Population synchrony in mayflies: a predator satiation hypothesis. *Evolution* 36:810–
821.

758 Symons, D. 1979. *The Evolution of Human
Sexuality*. Oxford University Press, New
York.

759 Symons, D. 1987. If we're all Darwinians,
what's the fuss about? In *Sociobiology and
Psychology: Ideas, Issues, and Applications*.
C. Crawford, M. Smith, and D. Krebs (eds.).
Lawrence Erlbaum Associates, Hillsdale,
NJ.

760 Tautz, J., and H. Markl. 1978. Caterpillars
detect flying wasps by hairs sensitive to airborne vibration. *Behavioral Ecology and Sociobiology* 4:101–110.

761 Tellegen, A., D. T. Lykken, T. J. Bouchard, Jr., K. J. Wilcox, N. L. Segal, and S. Rich. In press. Personality similarity in twins reared apart and together. *Journal of Personality and Social Psychology.*

762 Tetsu, S. 1986. A brood parasitic catfish of mouthbrooding cichlid fishes in Lake Tanganyika. *Nature* 323:58–59.

763 Thompson, D. B. A. 1986. The economics of kleptoparasitism: optimal foraging, host and prey selection by gulls. *Animal Behaviour* 34:1189–1205.

764 Thornhill, R. 1976. Sexual selection and nuptial feeding behavior in *Bittacus apicalis* (Insecta: Mecoptera). *American Naturalist* 110:529–548.

765 Thornhill, R. 1981. *Panorpa* (Mecoptera: Panorpidae) scorpionflies: systems for understanding resource-defense polygyny and alternative male reproductive efforts. *Annual Review of Ecology and Systematics* 12:355–386.

766 Thornhill, R., and J. Alcock. 1983. *The Evolution of Insect Mating Systems.* Harvard University Press, Cambridge, MA.

767 Thornhill, R., and D. T. Gwynne. 1986. The evolution of sexual differences in insects. *American Scientist* 74:382–389.

768 Thornhill, R., and N. W. Thornhill. 1983. Human rape: an evolutionary analysis. *Ethology and Sociobiology* 4:137–173.

769 Tierney, A. J. 1986. The evolution of learned and innate behavior: contributions from genetics and neurobiology to a theory of behavioral evolution. *Animal Learning and Behavior* 14:339–348.

770 Tinbergen, L. 1960. The natural control of insects in pinewoods. 1. Factors influencing the intensity of predation by songbirds. *Archives Néerlandaises de Zoologie* 13:265–343.

771 Tinbergen, N. 1951. *The Study of Instinct.* Oxford University Press, New York.

772 Tinbergen, N. 1958. *Curious Naturalists.* Doubleday, Garden City, New York.

773 Tinbergen, N. 1959. Comparative studies of the behavior of gulls (Laridae): a progress report. *Behaviour* 15:1–70.

774 Tinbergen, N. 1960. *The Herring Gull's World.* Doubleday, Garden City, New York.

775 Tinbergen, N. 1963. The shell menace. *Natural History* 72 (Aug.):28–35.

776 Tinbergen, N. 1973. *The Animal in Its World: Explorations of an Ethologist.* Vols. 1 and 2. Harvard University Press, Cambridge, MA.

777 Tinbergen, N., and A. C. Perdeck. 1950. On the stimulus situations releasing the begging response in the newly hatched herring gull (*Larus argentatus* Pont.). *Behaviour* 3:1–39.

778 Tokarz, R. R., and D. Crews. 1980. Induction of sexual receptivity in the female lizard, *Anolis carolinensis*: effects of estrogen and the antiestrogen CI-628. *Hormones and Behavior* 14:33–45.

779 Tokarz, R. R., and D. Crews. 1981. Effects of prostaglandins on sexual receptivity in the female lizard, *Anolis carolinensis. Endocrinology* 109:451–457.

780 Tompkins, L., and J. C. Hall. 1983. Identification of brain sites controlling female receptivity in mosaics of *Drosophila melanogaster. Genetics* 103:179–195.

781 Tooby, J., and I. DeVore. 1987. The reconstruction of hominid behavioral evolution through strategic modeling. In *The Evolution of Human Behavior: Primate Models.* W. G. Kinzey (ed.). State University of New York Press, Albany, NY.

782 Topoff, H. 1977. The pit and the antlion. *Natural History* 86 (Apr.):64–71.

783 Trivers, R. L. 1971. The evolution of reciprocal altruism. *Quarterly Review of Biology* 46:35–57.

784 Trivers, R. L. 1972. Parental investment and sexual selection. In *Sexual Selection and the Descent of Man.* B. Campbell (ed.). Aldine, Chicago.

785 Trivers, R. L. 1974. Parent–offspring conflict. *American Zoologist* 14:249–264.

786 Trivers, R. L. 1976. Sexual selection and resource-accruing abilities in *Anolis garmani. Evolution* 30:253–269.

787 Trivers, R. L. 1985. *Social Evolution.* Benjamin/Cummings Publishing, Menlo Park, CA.

788 Trivers, R. L., and H. Hare. 1976. Haplodiploidy and the evolution of the social insects. *Science* 191:249–263.

789 Truman, J., and L. Riddiford. 1970. Neuroendocrine control of ecdysis in silkmoths. *Science* 167:1624–1626.

790 Truman, J. W., and S. E. Reiss. 1976. Dendritic reorganization of an identified motoneuron during metamorphosis of the tobacco hornworm moth. *Science* 192:477–479.

791 Tryon, R. C. 1940. Genetic differences in maze-learning ability in rats. *Yearbook of the National Society for the Study of Education* 39:111–119.

792 Tully, T., and W. G. Quinn. 1985. Classical conditioning and retention in normal and mutant *Drosophila melanogaster. Journal of Comparative Physiology* 157:263–277.

793 Turner, J. R. G. 1986. Drinking crocodile tears: the only use for a butterfly. *Antenna* 10:119–120.

794 Tuttle, M. D., L. J. Taft, and M. J. Ryan. 1982. Evasive behaviour of a frog in response to bat predation. *Animal Behaviour* 30:393–397.

795 Tuttle, R. H. 1969. Knuckle-walking and the problem of human origins. *Science* 166:953–961.

796 Urquhart, F. A. 1960. *The Monarch Butterfly*. University of Toronto Press, Toronto.

797 van Lawick, H., and J. van Lawick-Goodall. 1971. *Innocent Killers*. Houghton Mifflin, Boston.

798 van Lawick-Goodall, J. 1970. Tool-using in primates and other vertebrates. *Advances in the Study of Behavior* 3:195–249.

799 Vetter, R. S. 1980. Defensive behavior of the black widow spider *Latrodectus hesperus* (Araneae: Theridiidae). *Behavioral Ecology and Sociobiology* 7:187–193.

800 Vining, D. R., Jr. 1986. Social vs. reproductive success: the central theoretical problem of human sociobiology. *Behavioral and Brain Science* 9:167–186.

801 Visscher, P. K., and T. D. Seeley. 1982. Foraging strategy of honeybee colonies in a temperate deciduous forest. *Ecology* 63:1790–1801.

802 vom Saal, F. S., W. M. Grant, C. W. McMullen, and K. S. Laves. 1983. High fetal estrogen concentrations: correlation with increased adult sexual activity and decreased aggression in male mice. *Science* 220:1306–1309.

803 von Frisch, K. 1953. *The Dancing Bees*. Harcourt Brace Jovanovich, New York.

804 von Frisch, K. 1967. *The Dance Language and Orientation of Bees*. Harvard University Press, Cambridge, MA.

805 Waage, J. K. 1973. Reproductive behavior and its relation to territoriality in *Calopteryx maculata* (Beauvois) (Odonata: Calopterygidae). *Behaviour* 47:240–256.

806 Waage, J. K. 1979. Dual function of the damselfly penis: sperm removal and transfer. *Science* 203:916–918.

807 Waddington, C. H. 1957. *The Strategy of Genes*. Allen & Unwin, London.

808 Walcott, C. 1972. Bird navigation. *Natural History* 81 (June):32–43.

809 Walker, A., and R. E. F. Leakey. 1978. The hominids of East Turkana. *Scientific American* 239 (Aug.):44–56.

810 Walker, M. M., J. L. Kirschvink, S.-B. R. Chang, and A. E. Dizon. 1984. A candidate magnetic sense organ in the yellowfin tuna, *Thunnus albacares*. *Science* 224:751–753.

811 Wallraff, H. G. 1983. Relevance of atmospheric odours and geomagnetic field to pigeon navigation: what is the *map* basis? *Comparative and Biochemical Physiology* 76A:643–663.

812 Walter, H. 1979. *Eleonora's Falcon: Adaptations to Prey and Habitat in a Social Raptor*. University of Chicago Press, Chicago.

813 Ward, P., and A. Zahavi. 1973. The importance of certain assemblages of birds as "information-centres" for food finding. *Ibis* 115:517–534.

814 Warner, R. R., D. R. Robertson, and E. G. Leigh, Jr. 1975. Sex change and sexual selection. *Science* 190:633–638.

815 Waser, P. M. 1985. Does competition drive dispersal? *Ecology* 66:1170–1175.

816 Waser, P. M., and C. H. Brown. 1984. Is there a "sound window" for primate communication? *Behavioral Ecology and Sociobiology* 15:73–76.

817 Waser, P. M., and W. T. Jones. 1983. Natal philopatry among solitary mammals. *Quarterly Review of Biology* 58:355–390.

818 Wehner, R. 1983. Celestial and terrestrial navigation: human strategies—insect strategies. In *Neuroethology and Behavioral Physiology*. F. Huber and H. Markl (eds.). Springer-Verlag, Berlin.

819 Wehner, R. 1987. "Matched filters:" neural models of the external world. *Journal of Comparative Physiology* A 161:511-531.

820 Wells, K. D. 1977. Territoriality and male mating success in the green frog (*Rana clamitans*). *Ecology* 58:750–762.

821 Wells, K. D., and J. J. Schwartz. 1984. Vocal communication in a Neotropical treefrog, *Hyla ebraccata*: advertisement calls. *Animal Behaviour* 32:405–420.

822 Wells, K. D., and T. L. Taigen. 1986. The effect of social interactions on calling energetics in the gray treefrog (*Hyla versicolor*). *Behavioral Ecology and Sociobiology* 19:9–18.

823 Welty, J. 1975. *The Life of Birds*, Second Edition. Saunders, Philadelphia.

824 Wenner, A. M. 1971. *The Bee Language Controversy*. Educational Programs Improvement Company, Boulder, CO.

825 Werner, E. C., J. F. Gilliam, D. J. Hall, and G. G. Mittelbach. 1983. An experimental test of the effects of predation risk on habitat use in fish. *Ecology* 64:1540–1548.

826 Werren, J. H., M. R. Gross, and R. Shine. 1980. Paternity and the evolution of male

parental care. *Journal of Theoretical Biology* 82:619–631.

827 West, M. J., and A. P. King. 1985. Social guidance of vocal learning by female cowbirds: validating its functional significance. *Zeitschrift für Tierpsychologie* 70:225–235.

828 West, M. J., A. P. King, and D. H. Eastzer. 1981. The cowbird: reflections on development from an unlikely source. *American Scientist* 69:56–66.

829 West Eberhard, M. J. 1975. The evolution of social behavior by kin selection. *Quarterly Review of Biology* 50:1–33.

830 West Eberhard, M. J. 1979. Sexual selection, social competition, and evolution. *Proceedings of the American Philosophical Society* 123:222–234.

831 White, T. D. 1980. Evolutionary implications of pliocene hominid footprints. *Science* 208:175–176.

832 Whitehead, H. 1985. Why whales leap. *Scientific American* 252 (Mar.):84–93.

833 Whitham, T. G. 1979. Habitat selection by *Pemphigus* aphids in response to resource limitation and competition. *Ecology* 59:1164–1176.

834 Whitham, T. G. 1979. Territorial defense in a gall aphid. *Nature* 279:324–325.

835 Whitham, T. G. 1980. The theory of habitat selection: examined and extended using *Pemphigus* aphids. *American Naturalist* 115:449–466.

836 Whitham, T. G. 1986. Costs and benefits of territoriality: behavioral and reproductive release by competing aphids. *Ecology* 67:139–147.

837 Wickler, W. 1968. *Mimicry in Plants and Animals*. World University Library, London.

838 Wickler, W., and U. Seibt. 1981. Monogamy in Crustacea and man. *Zeitschrift für Tierpsychologie* 57:215–234.

839 Wiens, J. A. 1977. On competition and variable environments. *American Scientist* 65:590–597.

840 Wiggins, D. A., and R. D. Morris. 1986. Criteria for female choice of mates: courtship feeding and parental care in the common tern. *American Naturalist* 128:126–129.

841 Wiklund, C., and T. Jarvi. 1982. Survival of distasteful insects after being attacked by naive birds: a reappraisal of the theory of aposematic coloration evolving through individual selection. *Evolution* 36:998–1002.

842 Wiklund, C., and B. Sillén-Tullberg. 1985. Why distasteful butterflies have aposematic larvae and adults, but cryptic pupae: evi-

dence from predation experiments on the monarch and the European swallowtail. *Evolution* 39:1155–1158.

843 Wilcox, S. 1979. Sex discrimination in *Gerris remigis*: role of a surface wave signal. *Science* 206:1325–1327.

844 Wilkinson, G. S. 1984. Reciprocal food sharing in the vampire bat. *Nature* 308:181–184.

845 Williams, G. C. 1966. *Adaptation and Natural Selection*. Princeton University Press, Princeton, NJ.

846 Williams, G. C. 1975. *Sex and Evolution*. Princeton University Press, Princeton, NJ.

847 Williams, G. C. 1980. Kin selection and the paradox of sexuality. In *Sociobiology: Beyond Nature/Nurture?* G. W. Barlow and J. Silverberg (eds.). Westview Press, Boulder, CO.

848 Williams, G. C. 1985. A defense of reductionism in evolutionary biology. *Oxford Surveys in Evolutionary Biology* 2:1–27.

849 Williams, T. C., and J. M. Williams. 1978. An oceanic mass migration of land birds. *Scientific American* 239 (Oct.):166–176.

850 Wilson, E. O. 1962. Chemical communication among workers of the fire ant, *Solenopsis saevissima* (Fr. Smith). 1. The organization of mass-foraging. *Animal Behaviour* 10:134–147.

851 Wilson, E. O. 1963. Pheromones. *Scientific American* 208 (May):100–114.

852 Wilson, E. O. 1971. *The Insect Societies*. Harvard University Press, Cambridge, MA.

853 Wilson, E. O. 1975. *Sociobiology: The New Synthesis*. Harvard University Press, Cambridge, MA.

854 Wilson, E. O. 1975. Human decency is animal. *The New York Times Magazine* 1975 (12 Oct.):38–50.

855 Wilson, E. O. 1976. Academic vigilantism and the political significance of sociobiology. *BioScience* 26:187–190.

856 Wilson, E. O. 1978. *On Human Nature*. Harvard University Press, Cambridge, MA.

857 Wilson, E. O., and B. Hölldobler. 1980. Sex differences in cooperative silk-spinning by weaver ant larvae. *Proceedings of the National Academy of Science* 77:2343–2347.

858 Wilson, E. O., and C. Lumsden. 1981. *Genes, Mind, and Culture*. Harvard University Press, Cambridge, MA.

859 Wilson, R. S. 1972. Twins: early mental development. *Science* 176:914–917.

860 Wiltschko, R., D. Nohr, and W. Wiltschko. 1981. Pigeons with a deficient sun compass use the magnetic compass. *Science* 214:343–345.

861 Wiltschko, W., R. Wiltschko, and C. Walcott. 1987. Pigeon homing: different effects of olfactory deprivation in different countries. *Behavioral Ecology and Sociobiology* 21:333–342.

862 Wingfield, J. C. 1980. Fine temporal adjustment of reproductive functions. In *Avian Endocrinology*. A. Epple and M. H. Stetson (eds.). Academic Press, New York.

863 Wingfield, J. C., and D. S. Farner. 1980. Control of seasonal reproduction in temperate-zone birds. *Progress in Reproductive Biology* 5:62–101.

864 Wingfield, J. C., and M. C. Moore. 1987. Hormonal, social and environmental factors in the reproductive biology of free-living male birds. In *Psychobiology of Reproductive Behavior: An Evolutionary Perspective*. D. Crews (ed.). Prentice-Hall, Englewood Cliffs, NJ.

865 Wittenberger, J. F. 1981. *Animal Social Behavior*. Duxbury Press, Boston.

866 Wolf, L. L. 1975. "Prostitution" behavior in a tropical hummingbird. *Condor* 77:140–144.

867 Woodward, B. D. 1986. Paternal effects on juvenile growth in *Scaphiopus multiplicatus* (the New Mexico spadefoot toad). *American Naturalist* 128:58–65.

868 Woodward, B. D. 1987. Paternal effects on offspring traits in *Scaphiopus couchi* (Anura: Pelobatidae). *Oecologia* 73:626–629.

869 Woolfenden, G. E., and J. W. Fitzpatrick. 1984. *The Florida Scrub Jay: Demography of a Cooperative-Breeding Bird*. Princeton University Press, Princeton, NJ.

870 Wourms, M. K., and F. E. Wasserman. 1985. Butterfly wing markings are more advantageous during handling than during the initial strike of an avian predator. *Evolution* 39:845–851.

871 Wynne-Edwards, V. C. 1962. *Animal Dispersion in Relation to Social Behaviour*. Oliver & Boyd, Edinburgh.

872 Yalden, D. W. 1985. Forelimb function in *Archeopteryx*. In *The Beginnings of Birds*. M. K. Hecht, J. H. Ostrom, G. Viohl, and P. Wellnhofer (eds.). Freunde des Jura-Museums Eichstatt, Willibaldsburg, West Germany.

873 Yasukawa, K. 1981. Song repertoires in the red-winged blackbird (*Agelaius phoeniceus*): a test of the Beau Geste hypothesis. *Animal Behaviour* 29:114–125.

874 Yasukawa, K., and W. A. Searcy. 1985. Song repertoires and density assessment in red-winged blackbirds: further tests of the Beau Geste hypothesis. *Behavioral Ecology and Sociobiology* 16:171–176.

875 Ydenberg, R. C., and L. M. Dill. 1986. The economics of fleeing from predators. *Advances in the Study of Behavior* 16:229–249.

876 Yokoyama, K., and D. S. Farner. 1978. Induction of *Zugunruhe* by photostimulation of encephalic receptors in white-crowned sparrows. *Science* 201:767–779.

877 Young, V. R., and N. S. Scrimshaw. 1971. The physiology of starvation. *Scientific American* 225 (Oct.):14–21.

878 Yu, Q., A. C. Jacquier, Y. Citri, M. Hamblen, J. C. Hall, and M. Rosbash. 1987. Molecular mapping of point mutations in the *period* gene that stop or speed up biological clocks in *Drosophila melanogaster*. *Proceedings of the National Academy of Sciences* 84:784–788.

879 Yunis, J. J., and O. Prakash. 1982. The origin of man: a chromosomal pictorial legacy. *Science* 215:1525–1530.

880 Zach, R. 1979. Shell-dropping: decision-making and optimal foraging in northwestern crows. *Behaviour* 68:106–117.

881 Zahavi, A. 1975. Mate selection—a selection for a handicap. *Journal of Theoretical Biology* 53:205–214.

882 Zeiller, W. 1971. Naked gills and recycled stings. *Natural History* 80 (Dec.):36–41.

883 Zenone, P. G., M. E. Sims, and C. J. Erickson. 1979. Male ring dove behavior and the defense of genetic paternity. *American Naturalist* 114:615–626.

884 Zippelius, H. 1972. Die Karawanenbildung bie Feld- und Hausspitzmaus. *Zeitschrift für Tierpsychologie* 30:305–320.

885 Zucker, I. 1983. Motivation, biological clocks, and temporal organization of behavior. In *Handbook of Behavioral Neurobiology: Motivation*. E. Satinoff and P. Teitelbaum (eds.). Plenum Press, New York.

ILLUSTRATION CREDITS

Chapter 1

4 From F. J. Ayala, 1978. *Scientific American* 239(Sept.):57. Copyright © 1978 by Scientific American Inc. All rights reserved.

9 From V. Sommer, 1987. *Primates* 28:163–197.

Chapter 2

1 From I. Rowley and G. Chapman, 1986. *Behaviour* 96:1–16. E. J. Brill, Leiden.

2(A) From N. Tinbergen, 1951. *The Study of Instinct*, Oxford University Press, New York.

7 From H. H. Kendler, 1968. *Basic Psychology*, 2nd ed., W. A. Benjamin, Menlo Park, CA.

9 From N. Tinbergen, 1951. *The Study of Instinct*, Oxford University Press, New York.

10 From D. Sherry, 1984. *Animal Behaviour* 32:451–464.

14 From H. Zippelius, 1972. *Zeitschrift für Tierpsychologie* 30:305–320.

16 From R. C. Bolles, 1969. *Journal of Comparative and Physiological Psychology* 69:355–358. Copyright © 1969 by the American Psychological Association.

18 Data from F. Nottebohm, 1981. *Science* 214:1368–1370. Copyright © 1981 by the American Association for the Advancement of Science.

20 After S. J. C. Gaulin and R. W. FitzGerald, 1986. *American Naturalist* 127:74–88.

Table 1 From S. M. Smith, 1977. *Nature* 265:535–536. Copyright © 1977 by Macmillan Magazines Ltd.

Chapter 3

3 From R. S. Wilson, 1972. *Science* 175:914–917. Copyright © 1972 by the American Association for the Advancement of Science.

5 From J. A. Doherty and H. C. Gerhardt, 1983. *Science* 220:1078–1080. Copyright © 1983 by the American Association for the Advancement of Science.

6 After M. K. Baylies et al., 1987. *Nature* 326:390–392. Copyright © 1987 by Macmillan Magazines Ltd.

7 From S. L. Ackerman and R. W. Siegel, 1986. *Journal of Neurogenetics* 3:111–123.

9 From J. J. Yunis and O. Prakash, 1982. *Science* 215:1525–1530. Copyright © 1982 by the American Association for the Advancement of Science.

14–17 From S. J. Arnold, 1980. In *Foraging Behavior*, A. Kamil and T. Sargent (eds.), Garland Press, New York.

Table 1 From T. J. Bouchard, Jr. and M. McGue, 1981. *Science* 212:1055–1059. Copyright © 1981 by the American Association for the Advancement of Science.

Chapter 4

3 From F. vom Saal et al., 1983. *Science* 220:1306–1309. Copyright © 1983 by the American Association for the Advancement of Science.

5, 7 From P. Marler and M. Tamura, 1964. *Science* 146:1483–1486. Copyright © 1964 by the American Association for the Advancement of Science.

8 From L. F. Baptista and L. Petrinovich, 1984. *Animal Behaviour* 32:172–181.

9 Data from P. Marler and V. Sherman, 1985. *Animal Behaviour* 33:57–71.

10 From M. J. West et al., 1981. *American Scientist* 69:56–66.

12 After G. D. Mower et al., 1983. *Science* 221:178–180. Copyright © 1983 by the American Association for the Advancement of Science.

13, 14 From H. F. Harlow, 1962. In *Roots of Behavior*, E. L. Bliss (ed.), Harper & Row, New York.

15, 16 From W. A. Mason, 1978. In *The Development of Behavior: Comparative and Evolutionary Aspects*, G. M. Burghardt and M. Bekoff (eds.), Garland Press, New York.

15 From N. B. Davies and T. R. Halliday, 1978. *Nature* 274:683–685. Copyright © 1978 by Macmillan Magazines Ltd.

18 From I. de la Motte and D. Burkhardt, 1983. *Naturwissenschaften* 70:451–461.

20 From H. Spangler, 1984. *Southwestern Naturalist* 29:481–488.

21(A) From M. Ryan, 1985. *The Tungara Frog*, University of Chicago Press, Chicago.

21(B) From M. Ryan, 1983. *Journal of Comparative Physiology* 150:217–221.

23 From G. M. Klump et al., 1986. *Behavioral Ecology and Sociobiology* 18:317–323.

24 From P. Marler, 1959. In *Darwin's Biological Work*, P. R. Bell (ed.), Cambridge University Press, New York.

26 From R. D. Alexander, 1962. *Evolution* 16:443–467.

27 From J. E. Lloyd, 1966. *Miscellaneous Publications of the Museum of Zoology, University of Michigan* 130:1–95.

28 From E. S. Morton, 1977. *American Naturalist* 109:17–34.

29 After C. H. Brown and P. M. Waser, 1984. *Animal Behaviour* 32:66–75.

30 From A. Kacelnik and J. Krebs, 1983. *Behaviour* 83:287–309. E. J. Brill, Leiden.

31 After K. D. Wells and J. J. Schwartz, 1984. *Animal Behaviour* 32:405–420.

32 From E. H. Burtt, 1984. *Animal Behaviour* 32:652–658.

Table 1 From W. M. Shields, 1984. *Animal Behaviour* 32:132–148.

Table 4 From N. Tinbergen, 1963. *Natural History* 72:28–35.

Chapter 9

3 From K. E. Holekamp, 1984. *Behavioral Ecology and Sociobiology* 16:21–30.

5 From A. E. Pusey and C. Packer, 1987. *Behaviour* 101:275–310. E. J. Brill, Leiden.

6 From P. M. Waser, 1985. *Ecology* 66:1170–1175. Copyright © 1985 by the Ecological Society of America. Reprinted by permission.

11 From S. B. Malcolm, 1987. *Trends in Ecology and Evolution* 2:135–138.

13 From T. G. Whitham, 1979. *Nature* 279:324–325. Copyright © 1979 by Macmillan Magazines Ltd.

14 From T. G. Whitham, 1986. *Ecology* 67:139–147. Copyright © 1986 by the Ecological Society of America. Reprinted by permission.

15 From J. R. Krebs, 1971. *Ecology* 52:2–22. Copyright © 1971 by the Ecological Society of America. Reprinted by permission.

17 From N. B. Davies and A. I. Houston, 1981. *Journal of Animal Ecology* 50:157–180.

18 After L. Carpenter et al., 1983. *Proceedings of the National Academy of Sciences* 80:7259–7263.

19 After D. C. Paton and H. C. Ford, 1983. In *Handbook of Experimental Pollination Biology*, C. E. Jones and R. J. Little (eds.), van Nostrand Reinhold, New York.

20 Data from F. R. Moore, 1978. *Behavioral Ecology and Sociobiology* 3:172–176.

Table 1 From L. Partridge, 1976. *Animal Behaviour* 24:534–544.

Table 2 From H. Dow and S. Fredga, 1983. *Journal of Animal Ecology* 52:681–695.

Table 3 From T. G. Whitham, 1980. *American Naturalist* 115:449–466.

Table 4 From F. B. Gill and L. L. Wolf, 1975. *Ecology* 56:333–345. Copyright © 1975 by the Ecological Society of America. Reprinted by permission.

Chapter 10

2 From M. J. Ryan and M. D. Tuttle, 1983. *Animal Behaviour* 31:827–833.

3 From A. T. Pietrewicz and A. C. Kamil, 1977. *Science* 195:580–582. Copyright © 1977 by the American Association for the Advancement of Science.

4 From T. W. Pietsch and D. B. Grobecker, 1978. *Science* 201:369–370. Copyright © 1978 by the American Association for the Advancement of Science.

5 From W. G. Eberhard, 1978. *Science* 198:510–512. Copyright © 1978 by American Association for the Advancement of Science.

9 From C. Packer, 1986. In *Ecological Aspects of Social Evolution*, D. I. Rubenstein and R. W. Wrangham (eds.), Princeton University Press, Princeton, NJ.

11 From F. Gotmark et al., 1986. *Nature* 319:589–591. Copyright © by Macmillan Magazines Ltd.

14 From E. O. Wilson, 1963. *Animal Behaviour* 10:134–147.

15 From R. I. Bowman, 1961. *University of California Publications in Zoology* 58:1–302.

16 From J. Fjeldsa, 1983. *Ibis* 125:463–481.

17 From R. Alatalo et al., 1987. *Ecology* 68:284–290. Copyright © by the Ecological Society of America. Reprinted by permission.

19 From J. C. Munger and J. H. Brown, 1981. *Science* 211:510–512. Copyright © 1981 by the American Association for the Advancement of Science.

20 From R. Zach, 1979. *Behaviour* 68:106–117.
E. J. Brill, Leiden.
21 From H. Richardson and N. A. M. Verbeek,
1986. *Ecology* 67:1219–1226. Copyright ©
1986 by the Ecological Society of America.
Reprinted by permission.
22 From K. A. Sullivan, in press. *Animal Behaviour*.
23 From S. L. Lima, 1985. *Oecologia* 66:60–67.
24 From L. M. Dill and A. H. G. Fraser, 1984.
Behavioral Ecology and Sociobiology 16:65–
71.
Table 1 From G. Pyke, 1979. *American Naturalist* 114:131–145.

Chapter 11

3 Data from R. J. Howlett and M. E. N. Majerus,
1987. *Biological Journal of the Linnean Society* 30:31–44.
4 From T. D. Sargent, 1976. *Legion of the Night:
The Underwing Moths*, University of Massachusetts Press, Amherst.
5 From B. Heinrich and S. L. Collins, 1983.
Ecology 64:592–602. Copyright © 1983 by the
Ecological Society of America. Reprinted by
permission.
6 Data from A. S. Rand, 1964. *Ecology* 45:863–
864. Copyright © 1964 by the Ecological Society of America. Reprinted by permission.
8 Data from T. Caro, 1986. *Animal Behaviour*
34:663–684.
11 Data from M. K. Wourms and F. E. Wasserman, 1985. *Evolution* 39:845–851.
13 From D. H. Schlenoff, 1985. *Animal Behaviour* 33:1057–1067.
14 Data from G. Hogstedt, 1983. *American Naturalist* 121:562–570.
15 From R. S. Vetter, 1980. *Behavioral Ecology
and Sociobiology* 7:187–193.
17(C) Adapted from C. S. Henry, 1972. *Psyche*
79:1–22.
18 From S. J. Arnold, 1982. *Copeia* 1982:247–
253.
20 Data from B. Sillén-Tullberg, 1985. *Oecologia*
67:411–415.
23 From M. P. Rowe et al., 1986. *Ethology*
72:53–71.
24 From E. D. Brodie, 1977. *Nature* 268:627–
628. Copyright © 1977 Macmillan Magazines
Ltd.
27 From R. E. Kenward, 1978. *Journal of Animal Ecology* 47:449–460.
28 Data from O. A. E. Rasa, 1984. *Behavioral
Ecology and Sociobiology* 12:181–190.
29 From B. W. Sweeney and R. L. Vannote,
1982. *Evolution* 36:810–821.

30 From W. H. Calvert et al., 1979. *Science*
204:847–851. Copyright © 1979 by the American Association for the Advancement of
Science.
33 From Fred A. Urquhart, 1987. *Monarch Butterfly: International Traveler*, Nelson-Hall,
Chicago. Copyright © 1987 by Fred A.
Urquhart.
Table 2 From N. E. Pierce and P. S. Mead,
1981. *Science* 211:1185–1187. Copyright ©
1981 by the American Association for the Advancement of Science.
Table 3 From P. W. Sherman, 1985. *Behavioral
Ecology and Sociobiology* 17:313–323.
Table 4 From G. V. N. Powell, 1974. *Animal Behaviour* 22:501–505.
Table 5 From M. R. Gross and A. M.
MacMillan, 1981. *Behavioral Ecology and
Sociobiology* 8:163–174.

Chapter 12

4 From B. Clarke, 1976. In *Genetic Aspects of
Host–Parasite Relationships*, A. E. R. Taylor
and R. Muller (eds.), Blackwell, Oxford.
6 Data from C. Lively, 1987. *Nature* 328:519–
521. Copyright © 1987 by Macmillan Magazines Ltd.
10 Data from D. E. Miller and J. T. Emlen,
1975. *Behaviour* 52:124–144. E. J. Brill,
Leiden.
12 From M. D. Beecher, 1982. *American Zoologist* 22:477–490.
Table 2 From M. R. Gross and R. Shine, 1981.
Evolution 35:775–793.
Table 3 Data from D. W. Mock and B. J. Ploger,
1987. *Animal Behaviour* 35:150–160.

Chapter 13

2(A) From A. J. Bateman, 1942. *Heredity* 2:349–
368.
2(B) From C. Packer et al., 1987. In *Reproductive Success*, T. H. Clutton-Brock (ed.), University of Chicago Press, Chicago.
3 From T. S. McCann, 1981. *Journal of Zoology*
195:295–310.
4 From G. Hausfater, 1975. *Contributions to Primatology* 7:1–150.
6 From R. D. Alexander et al., 1979. In *Evolutionary Biology and Human Social Behavior:
An Anthropological Perspective*, N. A. Chagnon and W. Irons (eds.), Wadsworth Publishing Co., Belmont, CA.
7 From S. J. Arnold, 1976. *Zeitschrift für Tierpsychologie* 42:247–300.

10 Data from A. Kodric-Brown, 1986. *Behavioral Ecology and Sociobiology* 19:425–432.

11 From M. R. Gross, 1982. *Zeitschrift für Tierpsychologie* 60:1–26.

16(A) Data from A. D. Afton, 1985. *Behaviour* 92:146–167. E. J. Brill, Leiden.

16(B) From S. T. Emlen and P. H. Wrege, 1986. *Ethology* 71:2–29.

20 From N. B. Davies, 1983. *Nature* 302:334–335. Copyright © 1983 by Macmillan Magazines Ltd.

23 From R. Thornhill, 1976. *American Naturalist* 110:529–548.

25 From R. D. Howard, 1978. *Evolution* 32:850–871.

26 From K. Noonan, 1983. *Animal Behaviour* 31:1005–1010.

27 From C. K. Catchpool, 1980. *Behaviour* 74:148–155. E. J. Brill, Leiden.

30 From M. Andersson, 1982. *Nature* 299:818–820. Copyright © 1982 by Macmillan Magazines Ltd.

31 From M. J. Ryan, 1983. *Evolution* 37:261–272.

32 From G. Borgia et al., 1987. *Animal Behaviour* 35:1129–1139.

33 From R. J. Bischoff et al., 1985. *Behavioral Ecology and Sociobiology* 17:253–256.

Table 1 From D. T. Gwynne, 1984. *Nature* 307:361–363. Copyright © 1984 by Macmillan Magazines Ltd.

Table 2 From R. L. Rutowski and G. W. Gilchrist, 1986. *Journal of Zoology* 209:115–124.

Table 3 From B. D. Woodward, 1986. *American Naturalist* 128:58–65.

Chapter 14

1 From K. B. Armitage, 1986. In *Ecological Aspects of Social Evolution*, D. I. Rubenstein and R. W. Wrangham (eds.), Princeton University Press, Princeton, NJ.

2 From J. Alcock, 1987. *Canadian Journal of Zoology* 65:2475–2482.

9 From B. Beehler and S. G. Pruett-Jones, 1983. *Behavioral Ecology and Sociobiology* 13:229–238.

13 From B. Lyon et al., 1987. *Behavioral Ecology and Sociobiology* 20:377–382.

14 From J. M. Packard et al., 1985. *Zeitschrift für Tierpsychologie* 68:24–40.

15 From D. W. Pyle and M. H. Gromko, 1978. *Experientia* 34:449–450.

18 From L. W. Oring, 1982. In *Avian Biology*, Volume VI, D. S. Farner and A. P. King (eds.), Academic Press, New York.

20, 21 From N. B. Davies and A. Lundberg, 1984. *Journal of Animal Ecology* 53:895–912.

Table 1 From M. Carey and V. Nolan, Jr., 1979. *Evolution* 33:1180–1192.

Chapter 15

1, 2 From C. R. Brown and M. B. Brown, 1986. *Ecology* 67:1206–1218. Copyright © by the Ecological Society of America. Reprinted by permission.

4 From J. L. Hoogland, 1981. *Ecology* 62:252–272. Copyright © 1981 by the Ecological Society of America. Reprinted by permission.

5 From J. L. Hoogland, 1979. *Behaviour* 69:1–35. E. J. Brill, Leiden.

6(A) Data from J. Ekman, 1987. *Animal Behaviour* 35:445–452.

6(B) Data from R. K. Murton et al., 1971. *Journal of Zoology* 165:53–84.

7 After J. L. Brown, 1987. *Helping and Communal Breeding in Birds*, Princeton University Press, Princeton, NJ.

8 From H.-U. Reyer, 1984. *Animal Behaviour* 32:1163–1178.

9 From E. O. Wilson, 1975. *Sociobiology: The New Synthesis*, Harvard University Press, Cambridge, MA.

11 From P. W. Sherman, 1985. *Behavioral Ecology and Sociobiology* 17:313–323.

12 From P. W. Sherman, 1981. *Behavioral Ecology and Sociobiology* 8:251–259.

19 From Y. P. Cruz, 1981. *Nature* 294:446–447. Copyright © 1981 Macmillan Magazines Ltd.

21 From D. C. Queller and J. E. Strassman, 1988. In *Reproductive Success*, T. Clutton-Brock (ed.), University of Chicago Press, Chicago.

23 From E. A. Lacey and P. W. Sherman, in press. In *The Biology of Naked Mole Rats*, P. W. Sherman, J. U. M. Jarvis, and R. D. Alexander (eds.).

Table 1 After R. D. Alexander, 1974. *Annual Review of Ecology and Systematics* 5:325–383.

Table 4, Table 5 Data from H.-U. Reyer, 1984. *Animal Behaviour* 32:1163–1178.

Table 6 From S. T. Emlen, 1978. In *Behavioral Ecology: An Evolutionary Approach*, 1st ed., J. R. Krebs and N. B. Davies (eds.), Blackwell, Oxford.

Chapter 16

4 From R. L. Smith, 1984. In *Sperm Competition and the Evolution of Animal Mating Systems*, R. L. Smith (ed.), Academic Press, New York.

5 From M. Borgerhoff Mulder, 1987. In *Human Reproductive Behavior: A Darwinian Perspective,* L. L. Betzig et al. (eds.), Cambridge University Press, New York.

6 From W. Irons, 1979. In *Evolutionary Biology and Human Social Behavior: An Anthropological Perspective,* N. A. Chagnon and W. Irons (eds.), Wadsworth Publishing Co., Belmont, CA.

9 From M. Flinn, in press. *Ethology and Sociobiology.*

10 From M. Daly and M. Wilson, 1985. *Ethology and Sociobiology* 6:197–210.

11 From M. S. Smith et al., 1987. *Ethology and Sociobiology* 8:171–182.

Table 2 From J. L. Boone III, 1986. *American Anthropologist* 88:859–878. Reproduced by permission of the American Anthropological Association.

Table 3 From J. Hartung, 1982. *Current Anthropology* 23:1–12.

AUTHOR INDEX

SUBJECT INDEX

ABOUT THE BOOK

This edition of *Animal Behavior* was designed by Rodelinde Albrecht. The typeface is Century Schoolbook, with headings in Optima and illustrative material in Avant Garde Gothic Book. DEKR Corporation set the type.

Principal artists for this edition were Nancy Haver, Laszlo Meszoly, and Frederic Schoenborn. Jodi Simpson was copy-editor, and Joseph J. Vesely supervised production.

Color separations were done by Tru-Color Reproductions, Inc. New England Book Components, Inc. manufactured the cover, and the book was printed and bound by The Murray Printing Company.

1 **Startle display.** This tropical caterpillar makes itself look like a snake when touched, presumably because the transformation startles the lizard or bird that touches it. For more on startle displays, see Chapter 11. Photograph by George D. Dodge/Tom Stack & Associates.

2 **Nuptial present.** This female Australian katydid received a huge white spermatophore from the male that copulated with her recently. Why does the structure constitute a valuable gift from the male to his partner (see Chapter 13)? Photograph by Chris Codd.

3 **Female defense polygyny.** Males of most mammals attempt to practice female defense polygyny (Chapter 14), in which a male defends a group of females against rival males. Here a male kudu follows his harem across an African plain. Photograph by E. S. Ross.

4 **Maternal behavior.** The gorilla is a typical parental animal species in that females provide the great majority of the care of the young, for reasons explained in Chapter 12. Photograph by John Cancalosi/Tom Stack & Associates.